Quantum Science and Technology

AF896262

For further volumes:
http://www.springer.com/series/10039

Aims and Scope

The book series Quantum Science and Technology is dedicated to one of today's most active and rapidly expanding fields of research and development. In particular, the series will be a showcase for the growing number of experimental implementations and practical applications of quantum systems. These will include, but are not restricted to: quantum information processing, quantum computing, and quantum simulation; quantum communication and quantum cryptography; entanglement and other quantum resources; quantum interfaces and hybrid quantum systems; quantum memories and quantum repeaters; measurement-based quantum control and quantum feedback; quantum nanomechanics, quantum optomechanics and quantum transducers; quantum sensing and quantum metrology; as well as quantum effects in biology. Last but not least, the series will include books on the theoretical and mathematical questions relevant to designing and understanding these systems and devices, as well as foundational issues concerning the quantum phenomena themselves. Written and edited by leading experts, the treatments will be designed for graduate students and other researchers already working in, or intending to enter the field of quantum science and technology.

Markus Aspelmeyer · Tobias J. Kippenberg
Florian Marquardt
Editors

Cavity Optomechanics

Nano- and Micromechanical Resonators Interacting with Light

Springer

Editors
Markus Aspelmeyer
Fakultät für Physik
Universität Wien
Vienna
Austria

Florian Marquardt
Institut für Theoretische Physik II
Universität Erlangen-Nürnberg
Erlangen
Germany

Tobias J. Kippenberg
SB-PH-LPQM
École polytechnique fédérale de Lausanne
Lausanne
Switzerland

ISBN 978-3-662-52051-2 ISBN 978-3-642-55312-7 (eBook)
DOI 10.1007/978-3-642-55312-7
Springer Heidelberg New York Dordrecht London

Softcover reprint of the hardcover 1st edition 2014

Printed on acid-free paper

Springer is part of Springer Science+Business Media (www.springer.com)

Preface

This book presents the field of cavity optomechanics from the perspective of leading groups around the world. Our hope is that it will serve as a useful overview of the various approaches to this rapidly developing field at the intersection of nanophysics and quantum optics. We would like to think that especially young researchers starting in cavity optomechanics will benefit from this comprehensive presentation, as well as those more expert readers who enter the field from another area.

The idea of compiling such a volume was hatched while planning the workshop "Mechanical Systems in the Quantum Regime," which the three of us organized in 2009 and which took place as a Wilhelm-and-Else-Heraeus Seminar at the physics center of the German Physical Society in Bad Honnef, Germany, from 19 to 22 July 2009. It was one of the very first workshops that was devoted to a great extent to the then nascent field of cavity optomechanics. Even at that time, it became apparent that the number of groups working on this topic was growing quickly, and the developments have accelerated ever since then.

Admittedly, when we first sent around guidelines for writing the chapters in the late summer of 2010, we did not anticipate that it would take 3 years to finish this endeavor. In retrospect, however, it is an indicator of scientific vigor: we could have foreseen that a fast emerging field has a stronger focus on "doing the science" rather than "reviewing the science." We would like to thank all authors for their time and effort in providing such excellent overviews while they have been constantly pushing the field forward. Special thanks go to Claus Ascheron for initiating the project and to Dan Stamper-Kurn for persistently pushing us to finalize it.

We are delighted that you are now holding in your hands a view on the subject of cavity optomechanics through the eyes of some of the leading experts in the field. We are confident that their contributions, emphasizing the foundations of the field, will remain a valuable resource for beginners and experts alike, and will provide the basis for the next exciting developments in the field.

Vienna, December 2013 — Markus Aspelmeyer
Lausanne — Tobias J. Kippenberg
Erlangen — Florian Marquardt

Contents

Chapter 1
Introduction

Markus Aspelmeyer, Tobias J. Kippenberg and Florian Marquardt

Abstract We briefly guide the reader through the chapters of the book, highlighting the connections between the various approaches to cavity optomechanics.

This book about cavity optomechanics collects introductory review-style articles by most of the leading groups worldwide. During the past few years, some reviews [1–10] and brief commentary articles [11–14] on cavity optomechanics have been published, with perhaps the most comprehensive treatment offered in the recent review article written by the editors of this volume [15]. The topic has also been included in some larger reviews on nanomechanical systems [16, 17]. However, by their very nature these reviews could only briefly address the wealth of experimental systems and theoretical predictions that now exist. In the present book, beginners and experts alike will find a much more detailed discussion of many important topics that could only be covered cursorily in these reviews.

The book starts with two chapters on the theoretical description. The chapter by Clerk and Marquardt is devoted to introducing the basics of the theory of optomechanical systems. These include the Hamiltonian, the classical dynamics (both linear and nonlinear), and the elementary quantum theory for optomechanical cooling. In the chapter on "*Nonclassical States of Light and Mechanics*" (Hammerer, Genes, Vitali, Tombesi, Milburn, Simon, Bouwmeester), more advanced schemes for quantum cavity optomechanics are discussed. In particular, this chapter explains

M. Aspelmeyer (✉)
University of Vienna, Vienna, Austria
e-mail: markus.aspelmeyer@univie.ac.at

T. J. Kippenberg
EPFL Lausanne, Écublens, Lausanne, Switzerland
e-mail: tobias.kippenberg@epfl.ch

F. Marquardt
Universität Erlangen-Nürnberg, Erlangen, Germany
e-mail: Florian.Marquardt@fau.de

M. Aspelmeyer et al. (eds.), *Cavity Optomechanics*, Quantum Science and Technology,
DOI: 10.1007/978-3-642-55312-7_1, © Springer-Verlag Berlin Heidelberg 2014

the various ways of creating nonclassical quantum states of the radiation field and the mechanics, as well as light/mechanics entanglement.

The paradigmatic setup in cavity optomechanics is an optical cavity with an end-mirror that can vibrate. This kind of setup features already in the very earliest theoretical considerations and experiments, starting with Braginsky's work at the end of the 60s, and proceeding with the pioneering works in the Walther lab at the Max Planck Institute for Quantum Optics in Garching in the middle of the 80s and the experiments at Laboratoire Kastler Brossel (LKB) in the 90s and early 2000s. The modern incarnations of this setup carry micromirrors on top of flexible, vibrating nanobeams and other elements. Pierre-François Cohadon, Markus Aspelmeyer, and Roman Schnabel will present the modern perspective in the chapter "*Suspended Mirrors: from test masses to micromechanics*".

Instead of having the end-mirror vibrate, it is also possible to place a mechanical element inside the optical cavity. As the dielectric element moves back and forth, it periodically modulates the effective refractive index seen by the cavity. This approach has the advantage that it decouples the mechanical functionality from the optical functionality, strongly reducing the constraints on the size and shape of the mechanical resonator. In the chapter on "*Mechanical resonators in the middle of an optical cavity*", Eva Weig and Ivan Favero will explain how this enables cavity optomechanics with tiny nanorods, and Jack Sankey presents the "membrane-in-the-middle" setup that features a vibrating membrane of sub-wavelength thickness inside the optical cavity.

Another approach to "go small" is to produce monolithic setups, where the optical modes propagate inside some dielectric object, leading to radiation forces that produce mechanical vibrations of that object. This can produce significant coupling strengths and the possibility for integrating everything on the chip. Five chapters are devoted to experiments of this kind.

In their chapter "*Cavity optomechanics with whispering-gallery mode microresonators*", Tobias Kippenberg and Albert Schliesser recount how microtoroids feature an interaction between their optical whispering gallery modes and their mechanical breathing mode and how this can be used to perform cavity optomechanics. Ivan Favero then describes even smaller (wavelength-size) disks made of GaAs that utilize the same concept and might be exploited for embedding quantum dots in the future ("*Gallium Arsenide disks optomechanical resonators*").

In the chapter on "*Brillouin optomechanics*", Gaurav Bahl and Tal Carmon explain the novel features that result when one couples to an acoustic whispering gallery mode (instead of a breathing mode), and exploits the transitions of photons between two optical modes (instead of focussing on one). The result may be called "Brillouin optomechanics", as it derives from the physics of Brillouin scattering of photons from acoustic vibrational modes in a solid.

Instead of having 0D objects, like toroids, disks, or spherical microresonators, optomechanical interactions can also be explored for waveguides that are part of a photonic circuit. Hong Tang and Wolfram Pernice, in their chapter "*Integrated optomechanical circuits and nonlinear dynamics*", describe all the components of such systems and the various radiation forces at play. In addition, they present some first applications of optomechanics in controlling the nonlinear dynamics of these mechanical devices.

A key ingredient of photonic circuits are photonic crystals, where a bandgap for photons has been engineered by producing a periodically patterned dielectric. It turns out that such setups are a very fruitful platform for cavity optomechanics, once the photonic crystal is made free-standing and once localized defect modes for both photons and phonons are designed. These "*Optomechanical Crystal Devices*" are the focus of the chapter written by Amir Safavi-Naeini and Oskar Painter.

Our story so far has assumed that the radiation is optical. This need not be the case. In principle, any kind of radiation can be exploited. In particular, microwave cavities coupled to mechanical vibrations have been very successful, as they have benefited from routine low-temperature operation and some clever design to boost the coupling strength. Konrad Lehnert describes these developments in his chapter entitled "*Introduction to microwave cavity optomechanics*", and he frames the discussion in terms of circuit language that is appropriate in this setting.

One of the goals of cavity optomechanics is to coherently control the quantum motion of mechanical resonators. Based on the recent achievement of optomechanical ground state cooling, first steps have now been taken into this quantum domain, at the time of writing. However, the first fabricated mechanical resonator to have reached the ground state and to be manipulated in a quantum-coherent fashion was not part of an optomechanical circuit. In the chapter "*Microwave-frequency mechanical resonators operated in the quantum limit*", Andrew Cleland and Aaron O'Connell explain their 2010 experiment, where they coupled a superconducting qubit to the mechanical vibrations of a piezoelectric resonator. Recent efforts aimed at optomechanical microwave-to-optical wavelength conversion will likely lead to future setups which are based on these ideas and which merge superconducting qubits and optomechanical resonators. This would provide an important component for applications in quantum communication.

Instead of changing the type of radiation (microwave substituted for optical), it is also possible to consider completely different mechanical resonators. Going away from the solid state, Dan Stamper-Kurn introduces "*Cavity optomechanics with cold atoms*". He reminds us of the pioneering works on atomic motion in the context of cavity quantum electrodynamics, and then goes on to describe the recent experiments that deal with optomechanical effects in clouds of cold atoms. Since the total mass of such atom clouds is many orders of magnitude smaller than that of even the smallest solid state resonators, their mechanical zero-point fluctuations are large, and so is the coupling strength. At the time of writing, optomechanics with cold atomic ensembles is the only setting where the coupling between a single photon and a single phonon is larger than the dissipation rates in the problem (especially the cavity decay rate).

In the final chapter "*Hybrid Mechanical Systems*" (Treutlein, Genes, Hammerer, Poggio, and Rabl), we witness the alluring possibilities that will arise when different quantum systems are combined, with mechanical elements always being an important part of the mix. Whether it is superconducting devices, solid-state spins, atoms, ions, or molecules coupled to mechanics, there always seem to be promising applications in quantum information processing or advanced sensing.

References

1. T.J. Kippenberg, K.J. Vahala, Opt. Express **15**, 17172 (2007)
2. T.J. Kippenberg, K.J. Vahala, Science **321**, 1172 (2008)
3. F. Marquardt, S.M. Girvin, Physics **2**, 40 (2009)
4. I. Favero, K. Karrai, Nat. Photon. **3**, 201 (2009)
5. C. Genes, A. Mari, D. Vitali, P. Tombesi, Adv. At. Mol. Phys. **57**, 33 (2009)
6. M. Aspelmeyer, S. Gröblacher, K. Hammerer, N. Kiesel, J. Opt. Soc. Am. B **27**, A189 (2010)
7. A. Schliesser, T.J. Kippenberg, in Advances in atomic, molecular and optical physics (Elsevier Academic Press, Amsterdam, 2010)
8. G.D. Cole, M. Aspelmeyer, Quantum Optomechanics, in: Optical coatings and thermal noise in precision measurement, ed. by G.M. Harry, T.P. Bodiya, R.DeSalvo, (Cambridge University Press, Cambridge, 2012), pp. 259–279
9. M. Aspelmeyer, P. Meystre, K.C. Schwab, Phys. Today **65**, 29 (2012)
10. P. Meystre, Ann. Phys. (Berlin) **525**, 215233 (2013)
11. K. Karrai, Nature **444**, 41 (2006)
12. A. Cleland, Nat. Phys. **5**, 458 (2009)
13. F. Marquardt, Nature (London) **478**, 47 (2011)
14. G. Cole, M. Aspelmeyer, Nat. Nanotech. **6**, 690 (2011)
15. M. Aspelmeyer, T.J. Kippenberg, F. Marquardt (2013), arXiv:1303.0733
16. M. Poot, H.S.J. van der Zant, Phys. Rep. **511**, 273 (2012)
17. Y.S. Greenberg, Y.A. Pashkin, E. Ilichev, Phys. Uspekhi **55**, 382 (2012)

Chapter 2
Basic Theory of Cavity Optomechanics

Aashish A. Clerk and Florian Marquardt

Abstract This chapter provides a brief basic introduction to the theory used to describe cavity-optomechanical systems. This can serve as background information to understand the other chapters of the book. We first review the Hamiltonian and show how it can be approximately brought into quadratic form. Then we discuss the classical dynamics both in the linear regime (featuring optomechanical damping, optical spring, strong coupling, and optomechanically induced transparency) and in the nonlinear regime (optomechanical self-oscillations and attractor diagram). Finally, we discuss the quantum theory of optomechanical cooling, using the powerful and versatile quantum noise approach.

2.1 The Optomechanical Hamiltonian

Cavity optomechanical systems display a parametric coupling between the mechanical displacement $\hat{x}$ of a mechanical vibration mode and the energy stored inside a radiation mode. That is, the frequency of the radiation mode depends on $\hat{x}$ and can be written in the form $\omega_{\mathrm{opt}}(\hat{x})$. When this dependence is Taylor-expanded, it is usually sufficient to keep the linear term, and we obtain the basic cavity-optomechanical Hamiltonian

$$\hat{H}_0 = \hbar \left(\omega_{\mathrm{opt}}(0) - G\hat{x}\right) \hat{a}^\dagger \hat{a} + \hbar \Omega_{\mathrm{M}} \hat{b}^\dagger \hat{b} + \cdots \tag{2.1}$$

A. A. Clerk (✉)
Department of Physics, McGill University, Montreal, Canada
e-mail: clerk@physics.mcgill.ca

F. Marquardt
Institute for Theoretical Physics, Universität Erlangen-Nürnberg, Erlangen, Germany
e-mail: Florian.Marquardt@fau.de

M. Aspelmeyer et al. (eds.), *Cavity Optomechanics*, Quantum Science and Technology,
DOI: 10.1007/978-3-642-55312-7_2,

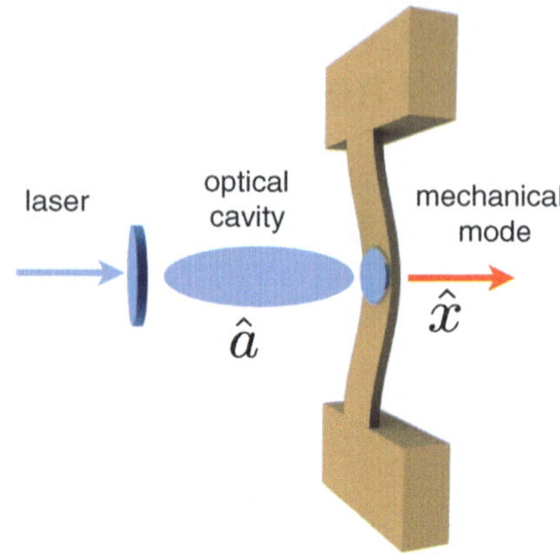

Fig. 2.1 A typical system in cavity optomechanics consists of a laser-driven optical cavity whose light field exerts a radiation pressure force on a vibrating mechanical resonator

We have used $\Omega_{\rm M}$ to denote the mechanical frequency, $\hat{a}^\dagger\hat{a}$ is the number of photons circulating inside the optical cavity mode, and $\hat{b}^\dagger\hat{b}$ is the number of phonons inside the mechanical mode of interest. Here G is the optomechanical frequency shift per displacement, sometimes also called the "frequency pull parameter", that characterizes the particular system. For a simple Fabry–Perot cavity with an oscillating end-mirror (illustrated in Fig. 2.1), one easily finds $G = \omega_{\rm opt}/L$, where L is the length of the cavity. This already indicates that smaller cavities yield larger coupling strengths. A detailed derivation of this Hamiltonian for a model of a wave field inside a cavity with a moving mirror can be found in [1]. However, the Hamiltonian is far more general than this derivation (for a particular system) might suggest: Whenever mechanical vibrations alter an optical cavity by leading to distortions of the boundary conditions or changes of the refractive index, we expect a coupling of the type shown here. The only important generalization involves the treatment of more than just a single mechanical and optical mode (see the remarks below).

A coupling of the type shown here is called 'dispersive' (in contrast to a 'dissipative' coupling, which would make κ depend on the displacement). Note that we have left out the terms responsible for the laser driving and the decay (of photons and of phonons), which will be dealt with separately in the following.

From this Hamiltonian, it follows that the radiation pressure force is

$$\hat{F}_{\rm rad} = \hbar G \hat{a}^\dagger \hat{a}\,. \tag{2.2}$$

After switching to a frame rotating at the incoming laser frequency ω_L, we introduce the detuning $\Delta = \omega_L - \omega_{\rm opt}(0)$, such that we get

$$\hat{H} = -\hbar\Delta\hat{a}^\dagger\hat{a} - \hbar G\hat{x}\hat{a}^\dagger\hat{a} + \hbar\Omega_{\rm M}\hat{b}^\dagger\hat{b} + \cdots \tag{2.3}$$

It is now possible to write the displacement $\hat{x} = x_{\rm ZPF}(\hat{b}+\hat{b}^\dagger)$ in terms of the phonon creation and annihilation operators, where $x_{\rm ZPF} = (\hbar/2m_{\rm eff}\Omega_{\rm M})^{1/2}$ is the size of the

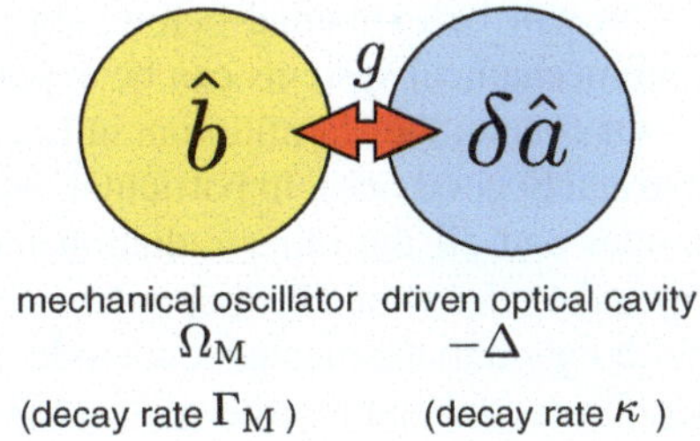

Fig. 2.2 After linearization, the standard system in cavity optomechanics represents two coupled harmonic oscillators, one of them mechanical (at a frequency Ω_{M}), the other optical (at an effective frequency given by the negative detuning $-\Delta = \omega_{\mathrm{opt}}(0) - \omega_{\mathrm{L}}$)

mechanical ground state wave function ("mechanical zero-point fluctuations"). This leads to

$$\hat{H} = -\hbar\Delta\hat{a}^{\dagger}\hat{a} - \hbar g_0 \left(\hat{b} + \hat{b}^{\dagger}\right)\hat{a}^{\dagger}\hat{a} + \hbar\Omega_{\mathrm{M}}\hat{b}^{\dagger}\hat{b} + \cdots . \tag{2.4}$$

Here $g_0 = G x_{\mathrm{ZPF}}$ represents the coupling between a single photon and a single phonon. Usually g_0 is a rather small frequency, much smaller than the cavity decay rate κ or the mechanical frequency Ω_{M}. However, the effective photon-phonon coupling can be boosted by increasing the laser drive, at the expense of introducing a coupling that is only quadratic (instead of cubic as in the original Hamiltonian). To see this, we set $\hat{a} = \alpha + \delta\hat{a}$, where α is the average light field amplitude produced by the laser drive (i.e. $\alpha = \langle\hat{a}\rangle$ in the absence of optomechanical coupling), and $\delta\hat{a}$ represents the small quantum fluctuations around that constant amplitude. If we insert this into the Hamiltonian and only keep the terms that are linear in α, we obtain

$$\hat{H}_{(\mathrm{lin})} = -\hbar\Delta\delta\hat{a}^{\dagger}\delta\hat{a} - \hbar g\left(\hat{b} + \hat{b}^{\dagger}\right)\left(\delta\hat{a} + \delta\hat{a}^{\dagger}\right) + \hbar\Omega_{\mathrm{M}}\hat{b}^{\dagger}\hat{b} + \cdots \tag{2.5}$$

This is the so-called "linearized" optomechanical Hamiltonian (where the equations of motion for $\delta\hat{a}$ and $\hat{b}$ are in fact linear). Here $g = g_0\alpha$ is the enhanced, laser-tunable optomechanical coupling strength, and for simplicity we have assumed α to be real-valued (otherwise a simple unitary transformation acting on $\delta\hat{a}$ can bring the Hamiltonian to the present form, which is always possible unless two laser-drives are involved). We have thus arrived at a rather simple system: two coupled harmonic oscillators (Fig. 2.2).

Note that we have omitted the term $-\hbar g_0 |\alpha|^2 (\hat{b} + \hat{b}^{\dagger})$, which represents a constant radiation pressure force acting on the mechanical resonator and would lead to a shift of the resonator's equilibrium position. We can imagine (as is usually done in these cases), that this shift has already been taken care of and $\hat{x}$ is measured from the new equilibrium position, or that this leads to a slightly changed "effective detuning" $\bar{\Delta}$ (which will be the notation we use further below when solving the classical equations of motion). In addition, we have neglected the term $-\hbar g_0\delta\hat{a}^{\dagger}\delta\hat{a}(\hat{b} + \hat{b}^{\dagger})$, under the assumption that this term is "small". The question when exactly this term may start to matter and lead to observable consequences is a subject of ongoing research (it seems that generally speaking $g_0/\kappa > 1$ is required).

As will be explained below, almost all of the elementary properties of cavity-optomechanical systems can be explained in terms of the linearized Hamiltonian.

Of course, the Hamiltonian in Eq. (2.1) represents an approximation (usually, an extremely good one). In particular, we have omitted all the other mechanical normal modes and all the other radiation modes. The justification for omitting the other optical modes would be that only one mode is driven (nearly) resonantly by the laser. With regard to the mechanical mode, optomechanical cooling or amplification in the resolved-sideband regime ($\kappa < \Omega_M$) usually affects only one mode, again selected by the laser frequency. Nevertheless, these simplistic arguments can fail, e.g. when κ is larger than the spacing between mechanical modes, when the distance between two optical modes matches a mechanical frequency, or when the dynamics becomes nonlinear, with large amplitudes of mechanical oscillations.

Cases where the other modes become important display an even richer dynamics than the one we are going to investigate below for the standard system (one mechanical mode, one radiation mode). Interesting experimental examples for the case of two optical modes and one mechanical mode can be found in the chapter by Bahl and Carmon (on Brillouin optomechanics), and in the contribution by Jack Sankey (on the membrane-in-the-middle setup).

In the following sections, we give a brief, self-contained overview of the most important basic features of this system, both in the classical regime and in the quantum regime. A more detailed introduction to the basics of the theory of cavity optomechanics can also be found in the recent review [2].

2.2 Classical Dynamics

The most important properties of optomechanical systems can be understood already in the classical regime. As far as current experiments are concerned, the only significant exception would be the quantum limit to cooling, which will be treated further below in the sections on the basics of quantum optomechanics.

2.2.1 Equations of Motion

In the classical regime, we assume both the mechanical oscillation amplitudes and the optical amplitudes to be large, i.e. the system contains many photons and phonons. As a matter of fact, much of what we will say is also valid in the regime of small amplitudes, when only a few photons and phonons are present. This is because in that regime the equations of motion can be linearized, and the expectation values of a quantum system evolving according to linear Heisenberg equations of motion in fact follow precisely the classical dynamics. The only aspect missing from the classical description in the linearized regime is the proper treatment of the quantum Langevin noise force, which is responsible for the quantum limit to cooling mentioned above.

We write down the classical equations for the position $x(t)$ and for the complex light field amplitude $\alpha(t)$ (normalized such that $|\alpha|^2$ would be the photon number in the semiclassical regime):

$$\ddot{x} = -\Omega_{\rm M}^2(x - x_0) - \Gamma_{\rm M}\dot{x} + (F_{\rm rad} + F_{\rm ext}(t))/m_{\rm eff} \tag{2.6}$$

$$\dot{\alpha} = [i(\Delta + Gx) - \kappa/2]\alpha + \frac{\kappa}{2}\alpha_{\rm max} \tag{2.7}$$

Here $F_{\rm rad} = \hbar G\,|\alpha|^2$ is the radiation pressure force. The laser amplitude enters the term $\alpha_{\rm max}$ in the second equation, where we have chosen a notation such that $\alpha = \alpha_{\rm max}$ on resonance ($\Delta = 0$) in the absence of the optomechanical interaction ($G = 0$). Note that the dependence on $\hbar$ in this equations vanishes once we express the photon number in terms of the total light energy $\mathscr{E}$ stored inside the cavity: $|\alpha|^2 = \mathscr{E}/\hbar\omega_{\rm L}$. This confirms that we are dealing with a completely classical problem, in which $\hbar$ will not enter any end-results if they are expressed in terms of classical quantities like cavity and laser frequency, cavity length, stored light energy (or laser input power), cavity decay rate, mechanical decay rate, and mechanical frequency. Still, we keep the present notation in order to facilitate later comparison with the quantum expressions.

2.2.2 Linear Response of an Optomechanical System

We have also added an external driving force $F_{\rm ext}(t)$ to the equation of motion for $x(t)$. This is because our goal now will be to evaluate the linear response of the mechanical system to this force. The idea is that the linear response will display a mechanical resonance that turns out to be modified due to the interaction with the light field. It will be shifted in frequency ("optical spring effect") and its width will be changed ("optomechanical damping or amplification"). These are the two most important elementary effects of the optomechanical interaction. Optomechanical effects on the damping rate and on the effective spring constant have been first analyzed and observed (in a macroscopic microwave setup) by Braginsky and co-workers already at the end of the 1960s [3].

First one has to find the static equilibrium position, by setting $\dot{x} = 0$ and $\dot{\alpha} = 0$ and solving the resulting set of coupled nonlinear algebraic equations. If the light intensity is large, there can be more than one stable solution. This 'static bistability' was already observed in the pioneering experiment on optomechanics with optical forces by the Walther group in the 1980s [4]. We now assume that such a solution has been found, and we linearize around it: $x(t) = \bar{x} + \delta x(t)$ and $\alpha(t) = \bar{\alpha} + \delta\alpha(t)$. Then the equations for δx and $\delta\alpha$ read:

$$\delta\ddot{x}(t) = -\Omega_{\rm M}^2\delta x - \Gamma_{\rm M}\delta\dot{x} + \frac{\hbar G}{m_{\rm eff}}\left[\bar{\alpha}^*\delta\alpha + \bar{\alpha}\delta\alpha^*\right] + \frac{F_{\rm ext}(t)}{m_{\rm eff}} \tag{2.8}$$

$$\delta\dot{\alpha}(t) = [i\bar{\Delta} - \kappa/2]\delta\alpha + iG\bar{\alpha}\delta x \tag{2.9}$$

Note that we have introduced the effective detuning $\bar{\Delta} = \Delta + G\bar{x}$, shifted due to the static mechanical displacement (this is often not made explicit in discussions of optomechanical systems, although it can become important for larger displacements). We are facing a linear set of equations, which in principle can be solved straightforwardly by going to Fourier space and inverting a matrix. There is only one slight difficulty involved here, which is that the equations also contain the complex conjugate $\delta\alpha^*(t)$. If we were to enter with an ansatz $\delta\alpha(t) \propto e^{-i\omega t}$, this automatically generates terms $\propto e^{+i\omega t}$ at the negative frequency as well. In some cases, this may be neglected (i.e. dropping the term $\delta\alpha^*$ from the equations), because the term $\delta\alpha^*(t)$ is not resonant (this is completely equivalent to the "rotating wave approximation" in the quantum treatment). However, here we want to display the full solution.

We now introduce the Fourier transform of any quantity $A(t)$ in the form $A[\omega] \equiv \int dt\, A(t)e^{i\omega t}$. Then, in calculating the response to a force given by $F_{\rm ext}[\omega]$, we have to consider the fact that $(\delta\alpha^*)[\omega] = (\delta\alpha[-\omega])^*$. The equation for $\delta\alpha[\omega]$ is easily solved, yielding $\delta\alpha[\omega] = \chi_c(\omega) i G\bar{\alpha}\delta x[\omega]$, with $\chi_c(\omega) = [-i\omega - i\bar{\Delta} + \kappa/2]^{-1}$ the response function of the cavity. When we insert this into the equation for $\delta x[\omega]$, we exploit $(\delta\alpha^*)[\omega] = (\delta\alpha[-\omega])^*$ as well as $(\delta x[-\omega])^* = \delta x[\omega]$, since $\delta x(t)$ is real-valued. The result for the mechanical response is of the form

$$\delta x[\omega] = \frac{F_{\rm ext}[\omega]}{m_{\rm eff}\left(\Omega_{\rm M}^2 - \omega^2 - i\omega\Gamma_{\rm M}\right) + \Sigma(\omega)} \equiv \chi_{xx}(\omega) F_{\rm ext}[\omega]. \tag{2.10}$$

Here we have combined all the terms that depend on the optomechanical interaction into the quantity $\Sigma(\omega)$ in the denominator. It is equal to

$$\Sigma(\omega) = -i\hbar G^2 |\bar{\alpha}|^2 \left[\chi_c(\omega) - \chi_c^*(-\omega)\right]. \tag{2.11}$$

Note that the prefactor can also be rewritten as $\hbar G^2 |\bar{\alpha}|^2 = 2m_{\rm eff}\Omega_{\rm M} g^2$, by inserting the expression for $x_{\rm ZPF} = (\hbar/2m_{\rm eff}\Omega_{\rm M})^{-1/2}$ and using $(Gx_{\rm ZPF}|\bar{\alpha}|)^2 = g^2$.

One may call Σ the "optomechanical self-energy" [5]. This is in analogy to the self-energy of an electron appearing in the expression for its Green's function, which summarizes the effects of the interaction with the electron's environment (photons, phonons, other electrons, ...).

If the coupling is weak, the mechanical linear response will still have a single resonance, whose properties are just modified by the presence of the optomechanical interaction. In that case, close inspection of the denominator in Eq. (2.10) reveals the meaning of both the imaginary and the real part of Σ, which we evaluate at the original resonance frequency $\omega = \Omega$. The imaginary part describes some additional optomechanical damping, induced by the light field:

$$\begin{aligned}\Gamma_{\rm opt} &= -\frac{1}{m_{\rm eff}\Omega_{\rm M}} \operatorname{Im}\Sigma(\Omega) \\ &= g^2\kappa \left\{ \frac{1}{\left(\Omega_{\rm M} + \bar{\Delta}\right)^2 + \left(\frac{\kappa}{2}\right)^2} - \frac{1}{\left(\Omega_{\rm M} - \bar{\Delta}\right)^2 + \left(\frac{\kappa}{2}\right)^2} \right\}\end{aligned} \tag{2.12}$$

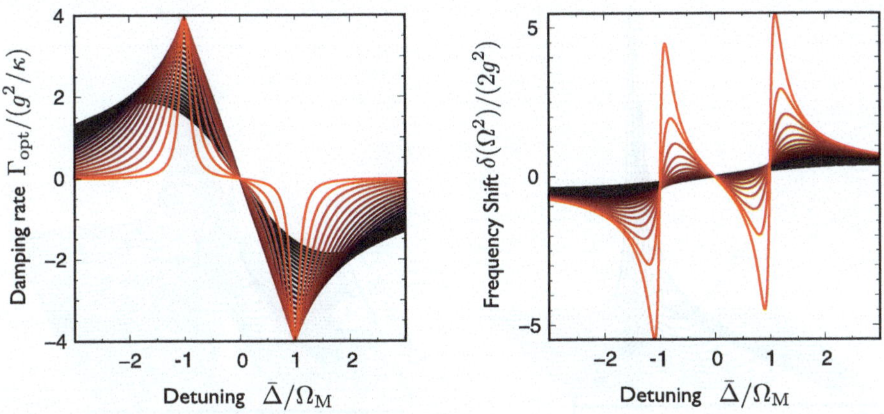

Fig. 2.3 Optomechanical damping rate (*left*) and frequency shift (*right*), as a function of the effective detuning $\bar{\Delta}$. The different curves depict the results for varying cavity decay rate, running in the interval $\kappa/\Omega_M = 0.2, 0.4, \ldots, 5$ (the largest values are shown as *black lines*). We keep the intracavity energy fixed (i.e. g is fixed). Note that the damping rate Γ_{opt} has been rescaled by g^2/κ, which represents the parametric dependence of Γ_{opt} in the resolved-sideband regime $\kappa < \Omega_M$. In addition, note that we chose to plot the frequency shift in terms of $\delta(\Omega^2) \approx 2\Omega_M \delta\Omega$ (for small $\delta\Omega \ll \Omega_M$)

The real part describes a shift of the mechanical frequency ("optical spring"):

$$\delta(\Omega^2) = \frac{1}{m_{eff}} \mathrm{Re}\,\Sigma(\Omega)$$
$$= 2\Omega_M g^2 \left\{ \frac{\Omega_M + \bar{\Delta}}{\left(\Omega_M + \bar{\Delta}\right)^2 + \left(\frac{\kappa}{2}\right)^2} - \frac{\Omega_M - \bar{\Delta}}{\left(\Omega_M - \bar{\Delta}\right)^2 + \left(\frac{\kappa}{2}\right)^2} \right\}. \quad (2.13)$$

Both of these are displayed in Fig. 2.3. They are the results of "dynamical backaction", where the (possibly retarded) response of the cavity to the mechanical motion acts back on this motion.

2.2.3 Strong Coupling Regime

When the optomechanical coupling rate g becomes comparable to the cavity damping rate κ, the system enters the strong coupling regime. The hallmark of this regime is the appearance (for red detuning) of a clearly resolved double-peak structure in the mechanical (or optical) susceptibility. This peak splitting in the strong coupling regime was first predicted in [5], then analyzed further in [6] and finally observed experimentally for the first time in [7]. This comes about because the mechanical resonance and the (driven) cavity resonance hybridize, like any two coupled harmonic oscillators, with a splitting $2g$ set by the coupling. In order to describe this correctly, we have to retain the full structure of the mechanical susceptibility, Eq. (2.10) at all

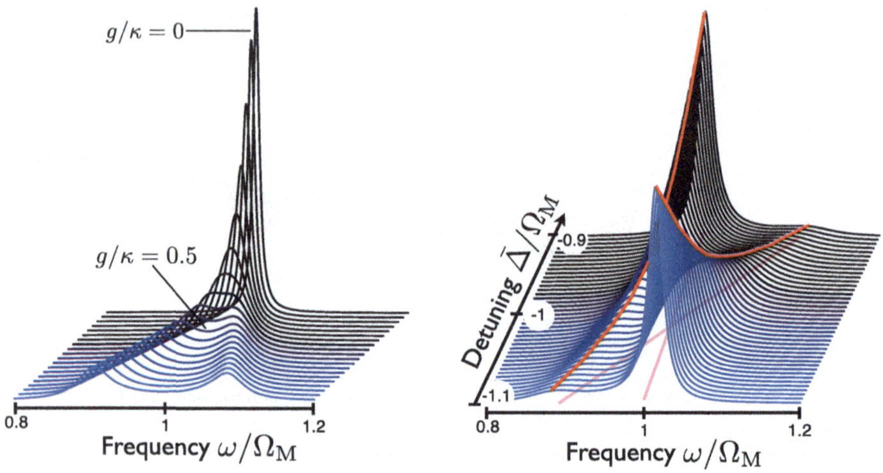

Fig. 2.4 Optomechanical strong coupling regime, illustrated in terms of the mechanical susceptibility. The figures show the imaginary part of $\chi_{xx}(\omega) = 1/(m(\Omega^2 - \omega^2 - i\omega\Gamma) + \Sigma(\omega))$. *Left* Im$\chi_{xx}(\omega)$ as a function of varying coupling strength g, set by the laser drive, for *red* detuning on resonance, $\bar{\Delta} = -\Omega$. A clear splitting develops around $g/\kappa = 0.5$. *Right* Im$\chi_{xx}(\omega)$ as a function of varying detuning $\bar{\Delta}$ between the laser drive and the cavity resonance, for fixed $g/\kappa = 0.5$

frequencies, without applying the previous approximation of evaluating $\Sigma(\omega)$ in the vicinity of the resonance (Fig. 2.4).

2.2.4 *Optomechanically Induced Transparency*

We now turn to the cavity response to a weak additional probe beam, which can be treated in analogy to the mechanical response discussed above. However, an interesting new feature develops, due to the fact that usually $\Gamma \ll \kappa$. Even for $g \ll \kappa$, the cavity response shows a spectrally sharp feature due to the optomechanical interaction, and its width is given by $\Gamma = \Gamma_M + \Gamma_{opt}$. This phenomenon is called "optomechanically induced transparency" [8, 9].

We can obtain the modified cavity response by imagining that there is no mechanical force ($F_{ext} = 0$), but instead there is an additional weak laser drive, which enters as $\cdots + \delta\alpha_L e^{-i\omega t}$ on the right-hand-side of Eq. (2.9). By solving the coupled set of equations, we arrive at a modified cavity response

$$\delta\alpha(t) = \chi_c^{eff}(\omega)\delta\alpha_L e^{-i\omega t}, \tag{2.14}$$

where we find

$$\chi_c^{eff}(\omega) = \chi_c(\omega) \cdot \left[1 + 2i m_{eff} \Omega_M g^2 \chi_c(\omega)\chi_{xx}(\omega)\right]. \tag{2.15}$$

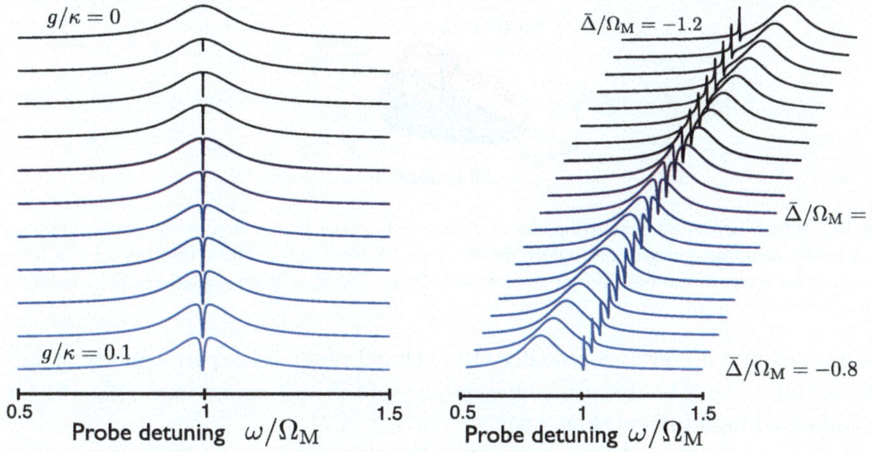

Fig. 2.5 Optomechanically induced transparency: Modification of the cavity response due to the interaction with the mechanical degree of freedom. We show $\mathrm{Re}\,\chi_c^{\mathrm{eff}}(\omega)$ as a function of the detuning ω between the weak probe beam and the strong(er) control beam, for variable coupling g of the control beam (*left*) and for variable detuning $\bar{\Delta}$ of the control beam versus the cavity resonance (*right*, at $g/\kappa = 0.1$). We have chosen $\kappa/\Omega_M = 0.2$. Note that in the *left plot*, for further increases in g, the *curves* shown here would smoothly evolve into the *double-peak* structure characteristic of the strong-coupling regime

Note that in the present section ω has the physical meaning of the detuning between the weak additional probe laser and the original (possibly strong) control beam at ω_L. That is: $\omega = \omega_{\mathrm{probe}} - \omega_L$. The result is shown in Fig. 2.5. The sharp dip goes down to zero when $g^2/(\kappa\Gamma_M) \gg 1$. Ultimately, this result is an example of a very generic phenomenon: If two oscillators are coupled and they have very different damping rates, then driving the strongly damped oscillator (here: the cavity) can indirectly drive the weakly damped oscillator (here: the mechanics), leading to a sharp spectral feature on top of a broad resonance. In the context of atomic physics with three-level atoms, this has been observed as "electromagnetically induced transparency", and thus the feature discussed here came to be called "optomechanically induced transparency".

We note that for a blue-detuned control beam, the dip turns into a peak, signalling optomechanical amplification of incoming weak radiation.

2.2.5 *Nonlinear Dynamics*

On the blue-detuned side ($\bar{\Delta} > 0$), where Γ_{opt} is negative, the overall damping rate $\Gamma_M + \Gamma_{\mathrm{opt}}$ diminishes upon increasing the laser intensity, until it finally becomes negative. Then the system becomes unstable and any small initial perturbation (e.g. thermal fluctuations) will increase exponentially at first, until the system settles into self-induced mechanical oscillations of a fixed amplitude: $x(t) = \bar{x} + A\cos(\Omega_M t)$. This is the optomechanical dynamical instability (parametric instability), which

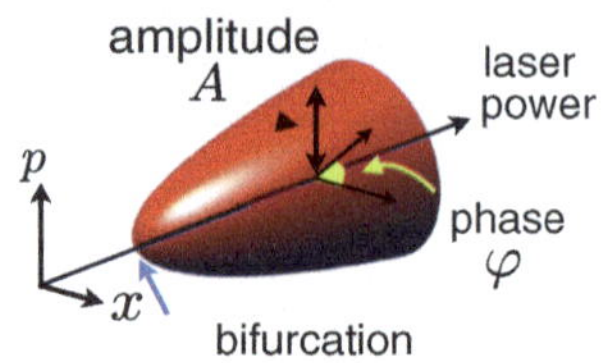

Fig. 2.6 When increasing a control parameter, such as the laser power, an optomechanical system can become unstable and settle into periodic mechanical oscillations. These correspond to a limit cycle in phase space of some amplitude A, as depicted here. The transition is called a Hopf bifurcation

has been explored both theoretically [10, 11] and observed experimentally in various settings (e.g. [12–14] for radiation-pressure driven setups and [15, 16] for photothermal light forces) (Fig. 2.6).

In order to understand the saturation of the amplitude A at a fixed finite value, we have to take into account that the mechanical ocillation changes the pattern of the light amplitude's evolution. In turn, the overall effective damping rate, as averaged over an oscillation period, changes as well. To capture this, we now introduce an amplitude-dependend optomechanical damping rate. This can be done by noting that a fixed damping rate Γ would give rise to a power loss $\langle (m_{\mathrm{eff}} \Gamma \dot{x}) \dot{x} \rangle = \Gamma m_{\mathrm{eff}} A^2/2$. Thus, we define

$$\Gamma_{\mathrm{opt}}(A) \equiv -\frac{2}{m_{\mathrm{eff}} A^2} \langle F_{\mathrm{rad}}(t) \dot{x}(t) \rangle \,. \tag{2.16}$$

This definition reproduces the damping rate Γ_{opt} calculated above in the limit $A \to 0$. The condition for the value of the amplitude on the limit cycle is then simply given by

$$\Gamma_{\mathrm{opt}}(A) + \Gamma_{\mathrm{M}} = 0. \tag{2.17}$$

The result for Eq. (2.16) can be expressed in terms of the exact analytical solution for the light field amplitude $\alpha(t)$ given the mechanical oscillations at amplitude A. This solution is a Fourier series, $|\alpha(t)| = \left| \frac{2\alpha_L}{\Omega_{\mathrm{M}}} \sum_n \tilde{\alpha}_n e^{in\Omega_{\mathrm{M}} t} \right|$, involving Bessel function coefficients. In the end, we find

$$\Gamma_{\mathrm{opt}}(A) = 4 \left(\frac{\kappa}{\Omega_{\mathrm{M}}} \right)^2 \frac{g^2}{\Omega_{\mathrm{M}}} f \left(\frac{GA}{\Omega_{\mathrm{M}}} \right), \tag{2.18}$$

where

$$f(a) = -\frac{1}{a} \sum_n \mathrm{Im} \tilde{\alpha}_{n+1}^* \tilde{\alpha}_n \tag{2.19}$$

with

$$\tilde{\alpha}_n = \frac{1}{2} \frac{J_n(-a)}{in + \kappa/(2\Omega_{\mathrm{M}}) - i\bar{\Delta}/\Omega_{\mathrm{M}}} \,. \tag{2.20}$$

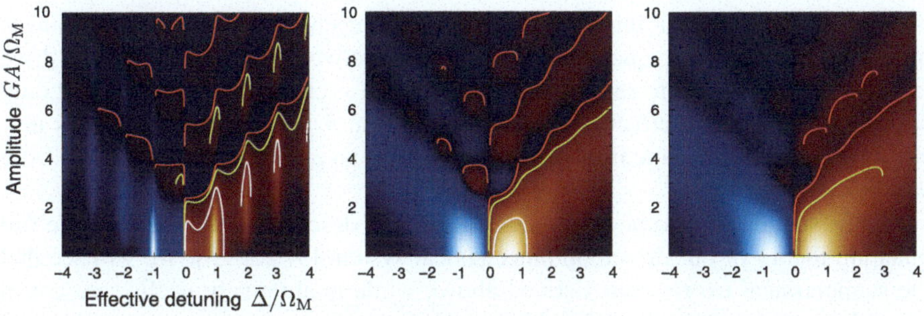

Fig. 2.7 Optomechanical Attractor Diagram: The effective amplitude-dependent optomechanical damping rate $\Gamma_{opt}(A)$, as a function of the oscillation amplitude A and the effective detuning $\bar{\Delta}$, for three different sidebands ratios $\kappa/\Omega_M = 0.2, 1, 2$, from *left* to *right* [Γ_{opt} in units of $\gamma_0 \equiv 4(\kappa/\Omega_M)^2 g^2/\Omega_M$, *blue* positive/cooling; *red* negative/amplification]. The optomechanical attractor diagram of self-induced oscillations is determined via the condition $\Gamma_{opt}(A) = -\Gamma_M$. The attractors are shown for three different values of the incoming laser power (as parametrized by the enhanced optomechanical coupling g at resonance), with $\Gamma_M/\gamma_0 = 0.1, 10^{-2}, 10^{-3}$ (*white, yellow, red*)

Note that g denotes the enhanced optomechanical coupling at resonance (i.e. for $\bar{\Delta} = 0$), i.e. it characterizes the laser amplitude. Also note that $\bar{\Delta}$ includes an amplitude-dependent shift due to a displacement of the mean oscillator position $\bar{x}$ by the radiation pressure force $\langle F_{rad} \rangle$. This has to be found self-consistently.

The resulting attractor diagram is shown in Fig. 2.7. It shows the possible limit cycle amplitude(s) as a function of effective detuning $\bar{\Delta}$, such that the self-consistent evaluation of $\bar{x}$ has been avoided.

The self-induced mechanical oscillations in an optomechanical system are analogous to the behaviour of a laser above threshold. In the optomechanical case, the energy provided by the incoming laser beam is converted, via the interaction, into coherent mechanical oscillations. While the amplitude of these oscillations is fixed, the phase depends on random initial conditions and may diffuse due to noise (e.g. thermal mechanical noise or shot noise from the laser). Interesting features may therefore arise when several such optomechanical oscillators are coupled, either mechanically or optically. In that case, they may synchronize if the coupling is strong enough. Optomechanical synchronization has been predicted theoretically [17, 18] and then observed experimentally [19, 20]. At high driving powers, we note that the dynamics is no longer a simple limit cycle but may instead become chaotic [21].

2.3 Quantum Theory

In the previous section, we have seen how a semiclassical description of the canonical optomechanical cavity gives a simple, intuitive picture of optical spring and optical damping effects. The average cavity photon number $\bar{n}_{cav}$ acts as a force on the mechanical resonator; this force depends on the mechanical position x, as changes in x change the cavity frequency and hence the effective detuning of the cavity drive laser. If $\bar{n}_{\mathrm{cav}}$ were able to respond instantaneously to changes in x, we would only have an optical spring effect; however, the fact that $\bar{n}_{\mathrm{cav}}$ responds to changes in x with a non-zero delay time implies that we also get an effective damping force from the cavity.

In this section, we go beyond the semiclassical description and develop the full quantum theory of our driven optomechanical system [5, 22, 23]. We will see that the semiclassical expressions derived above, while qualitatively useful, are not in general quantitatively correct. In addition, the quantum theory captures an important effect missed in the semiclassical description, namely the effective heating of the mechanical resonator arising from the fluctuations of the cavity photon number about its mean value. These fluctuations play a crucial role, in that they set a limit to the lowest possible temperature one can achieve via cavity cooling.

2.3.1 Basics of the Quantum Noise Approach to Cavity Backaction

We will focus here on the so-called "quantum noise" approach, where for a weak optomechanical coupling, one can understand the effects of the cavity backaction completely from the quantum noise spectral density of the radiation pressure force operator (Fig. 2.8). This spectral density is defined as:

$$S_{FF}[\omega] = \int_{-\infty}^{\infty} dt e^{i\omega t} \left\langle \hat{F}(t)\hat{F}(0) \right\rangle \tag{2.21}$$

where the average is taken over the state of the cavity at zero optomechanical coupling, and

$$\hat{F}(t) \equiv \hbar G \left(\hat{a}^\dagger \hat{a} - \left\langle \hat{a}^\dagger \hat{a} \right\rangle \right) \tag{2.22}$$

is the noise part of the cavity's backaction force operator (in the Heisenberg picture).

We start by considering the quantum origin of optomechanical damping, treating the optomechanical interaction term in the Hamiltonian of Eq. (2.3) using perturbation theory. Via the optomechanical interaction, the cavity will cause transitions between energy eigenstates of the mechanical oscillator, either upwards or downwards in energy. Working to lowest order in the optomechanical coupling G, these rates are described by Fermi's Golden rule. A straightforward calculation (see Sect. II B of Ref. [24]) shows that the Fermi's Golden rule rate $\Gamma_{n,+}$ ($\Gamma_{n,-}$) for a transition taking the oscillator from $n \to n+1$ quanta ($n \to n-1$ quanta) is given by:

$$\Gamma_{n,\pm} = \left(n + \frac{1}{2} \pm \frac{1}{2} \right) \frac{x_{\mathrm{ZPF}}^2}{\hbar^2} S_{FF}[\mp \Omega_{\mathrm{M}}] \tag{2.23}$$

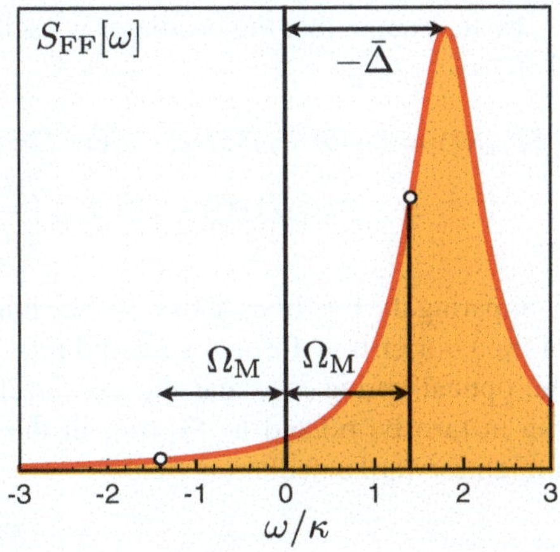

Fig. 2.8 The noise spectrum of the radiation pressure force in a driven optical cavity. This is a Lorentzian, peaked at the (*negative*) effective detuning. The transition rates are proportional to the value of the spectrum at $+\Omega_M$ (emission of energy into the cavity bath) and at $-\Omega_M$ (absorption of energy by the mechanical resonator)

The optomechanical damping rate simply corresponds to the decay rate of the average oscillator energy due to these transitions. One finds (see Appendix B of Ref. [24]):

$$\Gamma_{\text{opt}} = \Gamma_{n,-}\left(\frac{1}{n}\right) - \Gamma_{n,+}\left(\frac{1}{(n+1)}\right) \tag{2.24}$$

$$= \frac{x_{\text{ZPF}}^2}{\hbar^2}\left(S_{FF}[\Omega_M] - S_{FF}[-\Omega_M]\right) \tag{2.25}$$

Note that one obtains simple linear damping (the damping is independent of the amplitude of the oscillator's motion). Also note that our derivation has neglected the effects of the oscillator's intrinsic damping Γ_M, and thus is only valid if Γ_M is sufficiently small; we comment more on this at the end of the section.

There is a second way to derive Eq. (2.25) which is slightly more general, and which allows us to calculate the optical spring constant k_{opt}; it also more closely matches the heuristic reasoning that led to the semiclassical expressions of the previous section. We start from the basic fact that both Γ_{opt} and $\delta\Omega_{M,\text{opt}}$ arise from the dependence of the average backaction force F_{rad} on the mechanical position x. We can calculate this dependence to lowest order in G using the standard equations of quantum linear response (i.e. the Kubo formula):

$$\delta F_{\text{rad}}(t) = -\int_{-\infty}^{\infty} dt' \lambda_{FF}(t-t') \langle \hat{x}(t') \rangle \tag{2.26}$$

where the causal force-force susceptibility $\lambda_{FF}(t)$ is given by:

$$\lambda_{FF}(t) = -\frac{i}{\hbar}\theta(t)\left\langle \left[\hat{F}(t), \hat{F}(0)\right]\right\rangle \tag{2.27}$$

Next, assume that the oscillator is oscillating, and thus $\langle \hat{x}(t) \rangle = x_0 \cos \Omega_M t$. We then have:

$$\delta F_{\text{rad}}(t) = (-\text{Re}\ \lambda_{FF}[\Omega_M] \cdot x_0 \cos \Omega_M t) - (\text{Im}\ \lambda_{FF}[\Omega_M] \cdot x_0 \sin \Omega_M t) \tag{2.28}$$

$$= -\Delta k_{\text{opt}} \langle \hat{x}(t) \rangle - m_{\text{eff}} \Gamma_{\text{opt}} \left\langle \frac{d\hat{x}}{dt}(t) \right\rangle \tag{2.29}$$

Comparing the two lines above, we see immediately that the real and imaginary parts of the Fourier-transformed susceptibility $\lambda_{FF}[\Omega_M]$ are respectively proportional to the optical spring k_{opt} and the optomechanical damping Γ_{opt}. The susceptibility can in turn be related to $S_{FF}[\omega]$. In the case of the imaginary part of $\lambda_{FF}[\omega]$, a straightforward calculation yields:

$$-\text{Im}\ \lambda_{FF}[\omega] = \frac{S_{FF}[\omega] - S_{FF}[-\omega]}{2\hbar} \tag{2.30}$$

As a result, the definition of Γ_{opt} emerging from Eq. (2.29) is identical to that in Eq. (2.25). The real part of $\lambda_{FF}[\omega]$ can also be related to $S_{FF}[\omega]$ using a standard Kramers-Kronig identity. Defining $\delta\Omega_{M,\text{opt}} \equiv \frac{k_{\text{opt}}}{2m_{\text{eff}}\Omega_M}$, one finds:

$$\delta\Omega_{M,\text{opt}} = \frac{x_{\text{ZPF}}^2}{\hbar^2} \int_{-\infty}^{\infty} \frac{d\omega}{2\pi} S_{FF}[\omega] \left[\frac{1}{\Omega_M - \omega} - \frac{1}{\Omega_M + \omega} \right] \tag{2.31}$$

Thus, a knowledge of the quantum noise spectral density $S_{FF}[\omega]$ allows one to immediately extract both the optical spring coefficient, as well as the optical damping rate.

We now turn to the effects of the fluctuations in the radiation pressure force, and the effective temperature T_{rad} which characterizes them. This too can be directly related to $S_{FF}[\omega]$. Perhaps the most elegant manner to derive this is to perturbatively integrate out the dynamics of the cavity [24, 25]; this approach also has the benefit of going beyond simplest lowest-order-perturbation theory. One finds that the mechanical resonator is described by a *classical* Langevin equation of the form:

$$m\ddot{x}(t) = -(k + k_{\text{opt}})x(t) - m\Gamma_{\text{opt}}\dot{x}(t) + \xi_{\text{rad}}(t). \tag{2.32}$$

The optomechanical damping Γ_{opt} and optical spring k_{opt} are given respectively by Eqs. (2.25) and (2.31), except that one should make the replacement $\Omega_M \to \Omega'_M \equiv \Omega_M + \delta\Omega_{M,\text{opt}}$ in these equations. The last term $\xi_{\text{rad}}(t)$ above represents the fluctuating backaction force associated with photon number fluctuations in the cavity. Within our approximations of weak optomechanical coupling and weak intrinsic mechanical damping, this random force is Gaussian white noise, and is fully described by the spectral density:

$$S_{\xi_{\text{rad}}\xi_{\text{rad}}}[\omega] = m\Gamma_{\text{opt}} \coth\left(\hbar\Omega'_M / 2k_B T_{\text{rad}}\right) = m\Gamma_{\text{opt}}\left(1 + 2\bar{n}_{\text{rad}}\right). \tag{2.33}$$

Here, $T_{\rm rad}$ is the effective temperature of the cavity backaction, and $\bar{n}_{\rm rad}$ is the corresponding number of thermal oscillator quanta. These quantities are determined by $S_{FF}[\omega]$ via:

$$1 + 2\bar{n}_{\rm rad} \equiv \frac{S_{FF}\left[\Omega'_{\rm M}\right] + S_{FF}\left[-\Omega'_{\rm M}\right]}{S_{FF}\left[\Omega'_{\rm M}\right] - S_{FF}\left[-\Omega'_{\rm M}\right]} \tag{2.34}$$

Note that as the driven cavity is not in thermal equilibrium, $T_{\rm rad}$ will in general depend on the value of $\Omega_{\rm M}$; a more detailed discussion of the concept of an effective temperature is given in Ref. [24].

Turning to the stationary state of the oscillator, we note that Eq. (2.32) is identical to the Langevin equation for an oscillator coupled to a thermal equilibrium bath at temperature $T_{\rm rad}$. It thus follows that the stationary state of the oscillator will be a thermal equilibrium state at a temperature $T_{\rm rad}$, and with an average number of quanta $\bar{n}_{\rm rad}$. As far as the oscillator is concerned, $T_{\rm rad}$ is indistinguishable from a true thermodynamic bath temperature, even though the driven cavity is not itself in thermal equilibrium.

The more realistic case is of course where we include the intrinsic damping and heating of the mechanical resonator; even here, a similar picture holds. The intrinsic dissipation can be simply accounted for by adding to the RHS of Eq. (2.32) a damping term describing the intrinsic damping (rate. $\Gamma_{\rm M}$), as well as a stochastic force term corresponding to the fridge temperature T. The resulting Langevin equation still continues to have the form of an oscillator coupled to a single equilibrium bath, where the total damping rate due to the bath is $\Gamma_{\rm M} + \Gamma_{\rm opt}$, and the effective temperature $T_{\rm eff}$ of the bath is determined by:

$$n_{\rm eff} = \frac{\Gamma_{\rm M} n_0 + \Gamma_{\rm opt}\bar{n}_{\rm rad}}{\Gamma_{\rm M} + \Gamma_{\rm opt}} \tag{2.35}$$

where n_0 is the Bose-Einstein factor corresponding to the bath temperature T:

$$n_0 = \frac{1}{\exp\left(\hbar\Omega'_{\rm M}/k_{\rm B}T\right) - 1} \tag{2.36}$$

We thus see that in the limit where $\Gamma_{\rm opt} \gg \Gamma_{\rm M}$, the effective mechanical temperature tends to the backaction temperature $T_{\rm rad}$. This will be the lowest temperature possible via cavity cooling . Note that similar results may be obtained by using the Golden rule transition rates in Eq. (2.23) to formulate a master equation describing the probability $p_n(t)$ that the oscillator has n quanta at time t (see Sect. II B of Ref. [24]).

Before proceeding, it is worth emphasizing that the above results all rely on the total mechanical bandwidth $\Gamma_{\rm M} + \Gamma_{\rm opt}$ being sufficiently small that one can ignore the variance of $S_{FF}[\omega]$ across the mechanical resonance. When this condition is not satisfied, one can still describe backaction effects using the quantum noise approach, with a Langevin equation similar to Eq. (2.32). However, one now must include the variation of $S_{FF}[\omega]$ with frequency; the result is that the optomechanical damping will not be purely local in time, and the stochastic part of the backaction force will not be white.

2.3.2 Application to the Standard Cavity Optomechanical Setup

The quantum noise approach to backaction is easily applied to the standard optomechanical cavity setup, where the backaction force operator $\hat{F}$ is proportional to the cavity photon number operator. To calculate its quantum noise spectrum in the absence of any optomechanical coupling, we first write the equation of motion for the cavity annihilation operator $\hat{a}$ in the Heisenberg picture, using standard input–output theory [26, 27]:

$$\frac{d}{dt}\hat{a} = \left(-i\omega_{\mathrm{opt}} - \kappa/2\right)\hat{a} - \sqrt{\kappa}\hat{a}_{\mathrm{in}}. \tag{2.37}$$

Here, $\hat{a}_{\mathrm{in}}$ describes the amplitude of drive laser, and can be decomposed as:

$$\hat{a}_{\mathrm{in}} = e^{-i\omega_L t}\left(\bar{a}_{\mathrm{in}} + \hat{d}_{\mathrm{in}}\right), \tag{2.38}$$

where $\bar{a}_{\mathrm{in}}$ represents the classical amplitude of the drive laser (the input power is given by $P_{\mathrm{in}} = \hbar\omega_{\mathrm{opt}}|\bar{a}_{\mathrm{in}}|^2$), and $\hat{d}_{\mathrm{in}}$ describes fluctuations in the laser drive. We consider the ideal case where these are vacuum noise, i.e. there is only shot noise in the incident laser drive, and no additional thermal or phase fluctuations. One thus finds that $\hat{d}_{\mathrm{in}}$ describes operator white noise:

$$\left\langle \hat{d}_{\mathrm{in}}(t)\hat{d}^{\dagger}_{\mathrm{in}}(t')\right\rangle = \left\langle [\hat{d}_{\mathrm{in}}(t), \hat{d}^{\dagger}_{\mathrm{in}}(t')]\right\rangle = \delta(t-t') \tag{2.39}$$

It is also useful to separate the cavity field operator into an average "classical" part and a quantum part,

$$\hat{a} = e^{-i\omega_L t}e^{i\phi}\left(\sqrt{\bar{n}_{\mathrm{cav}}} + \hat{d}\right) \tag{2.40}$$

where $e^{i\phi}\sqrt{\bar{n}_{\mathrm{cav}}}$ is the classical amplitude of the cavity field, and $\hat{d}$ describes its fluctuations.

It is now straightforward to solve Eq. (2.37) for $\hat{d}$ in terms of $\hat{d}_{\mathrm{in}}$. As we will be interested in regimes where $\bar{n}_{\mathrm{cav}} \gg 1$, we can focus on the leading-order-in-$\bar{n}_{\mathrm{cav}}$ term in the backaction force operator F:

$$\hat{F} \simeq \hbar G\sqrt{\bar{n}_{\mathrm{cav}}}\left(\hat{d} + \hat{d}^{\dagger}\right) \tag{2.41}$$

Using this leading-order expression along with the solution for $\hat{d}_{\rm in}$ and Eq. (2.39), we find that the quantum noise spectral density $S_{FF}[\omega]$ (as defined in Eq. (2.21)) is given by:

$$S_{FF}[\omega] = \hbar^2 G^2 \bar{n}_{\rm cav} \frac{\kappa}{(\omega + \Delta)^2 + (\kappa/2)^2} \tag{2.42}$$

$S_{FF}[\omega]$ is a simple Lorentzian, reflecting the cavity's density of states, and is centred at $\omega = -\Delta$, precisely the energy required to bring a drive photon onto resonance. The form of $S_{FF}[\omega]$ describes the final density of states for a Raman process where an incident drive photon gains ($\omega > 0$, anti-Stokes) or loses ($\omega < 0$, Stokes) a quanta $\hbar|\omega|$ of energy before attempting to enter the cavity. From Eq. (2.25), we can immediately obtain an expression for the optomechanical damping rate; it will be large if can make the density states associated with the anti-Stokes process at frequency $\Omega'_{\rm M}$ much larger than that of the Stokes process at the same frequency. The optical spring coefficient also follows from Eq. (2.31).

We finally turn to $\bar{n}_{\rm rad}$, the effective temperature of the backaction (expressed as a number of oscillator quanta). Using Eqs. (2.34) and (2.42), we find:

$$\bar{n}_{\rm rad} = -\frac{(\Omega'_{\rm M} + \Delta)^2 + (\kappa/2)^2}{4\Omega'_{\rm M}\Delta} \tag{2.43}$$

As discussed, $\bar{n}_{\rm rad}$ represents the lowest possible temperature we can cool our mechanical resonator to. As a function of drive detuning Δ, $\bar{n}_{\rm rad}$ achieves a minimum value of

$$\bar{n}_{\rm rad}\Big|_{\rm min} = \left(\frac{\kappa}{4\Omega'_{\rm M}}\right)^2 \frac{2}{1 + \sqrt{1 + (\kappa^2/4\Omega'^2_{\rm M})}} \tag{2.44}$$

for an optimal detuning of

$$\Delta = -\sqrt{\Omega'^2_{\rm M} + \kappa^2/4}. \tag{2.45}$$

We thus see that if one is in the so-called good cavity limit $\Omega_{\rm M} \gg \kappa$, and if the detuning is optimized, one can potentially cool the mechanical resonator close to its ground state. In this limit, the anti-Stokes process is on-resonance, while the Stokes process is far off-resonance and hence greatly suppressed. The fact that the effective temperature is small but non-zero in this limit reflects the small but non-zero probability for the Stokes process, due to the Lorentzian tail of the cavity density of states. In the opposite, "bad cavity" limit where $\Omega_{\rm M} \ll \kappa$, we see that the minimum of $\bar{n}_{\rm rad}$ tends to $\kappa/\Omega_{\rm M} \gg 1$, while the optimal detuning tends to $\kappa/2$ (as anticipated in the semiclassical approach).

Note that the above results are easily extended to the case where the cavity is driven by thermal noise corresponding to a thermal number of cavity photons $n_{{\rm cav},T}$. For a drive detuning of $\Delta = -\Omega_{\rm M}$ (which is optimal in the good cavity limit), one now finds that that the $\bar{n}_{\mathrm{rad}}$ is given by [6]:

$$\bar{n}_{\mathrm{rad}} = \left(\frac{\kappa}{4\Omega_{\mathrm{M}}}\right)^2 + n_{\mathrm{cav},T}\left(1 + 2\left(\frac{\kappa}{4\Omega_{\mathrm{M}}}\right)^2\right) \tag{2.46}$$

As expected, one cannot backaction-cool a mechanical resonator to a temperature lower than that of the cavity.

2.3.3 Results for a Dissipative Optomechanical Coupling

A key advantage of the quantum noise approach is that it can be easily applied to alternate forms of optomechanical coupling. For example, it is possible have systems where the mechanical resonator modulates both the cavity frequency as well as the damping rate κ of the optical cavity [28, 29]. The position of the mechanical resonator will now couple to both the cavity photon number (as in the standard setup), as well as to the "photon tunnelling" term which describes the coupling of the cavity mode to the extra-cavity modes that damp and drive it. Because of these two couplings, the form of the effective backaction force operator $\hat{F}$ is now modified from the standard setup. Nonetheless, one can still go ahead and calculate the optomechanical backaction using the quantum noise approach. In the simple case where the cavity is overcoupled (and hence its κ is due entirely to the coupling to the port used to drive it), one finds that the cavity's backaction quantum force noise spectrum is given by [30, 31]:

$$S_{FF}[\omega] = \left(\frac{G_{\kappa}^2 \bar{n}_{\mathrm{cav}}}{4\kappa}\right) \frac{\left[\omega + 2\Delta - \frac{2G}{G_{\kappa}}\kappa\right]^2}{(\omega + \Delta)^2 + \kappa^2/4} \tag{2.47}$$

Here, $G = -d\omega_{\mathrm{opt}}/dx$ is the standard optomechanical coupling, while $G_{\kappa} = d\kappa/dx$ represents the dissipative optomechanical coupling. For $G_{\kappa} = 0$, we recover the Lorentzian spectrum of the standard optomechanical setup given in Eq. (2.42) whereas for $G_{\kappa} \neq 0$, $S_{FF}[\omega]$ has the general form of a Fano resonance. Such lineshapes arise as the result of interference between resonant and non-resonant processes; here, the resonant channel corresponds to fluctuations in the cavity amplitude, whereas the non-resonant channel corresponds to the incident shot noise fluctuations on the cavity. These fluctuations can interfere destructively, resulting in $S_{FF}[\omega] = 0$ at the special frequency $\omega = -2\Delta + 2G/G_{\kappa}$. If one tunes Δ such that this frequency coincides with $-\Omega_{\mathrm{M}}$, it follows immediately from Eq. (2.34) that the cavity backaction has an effective temperature of zero, and can be used to cool the mechanical resonator to its ground state. This special detuning causes the destructive interference to completely suppress the probability of the cavity backaction exciting the mechanical resonator, whereas the opposite process of absorption is not suppressed. This "interference cooling" does not require one to be in the good cavity limit, and thus could be potentially useful for the cooling of low-frequency (relative to κ) mechanical modes. However, the presence of internal loss in the cavity places limits on this technique, as it suppresses the perfect destructive interference between resonant and non-resonant fluctuations [30, 31].

References

1. C.K. Law, Phys. Rev. A **51**, 2537 (1995)
2. M. Aspelmeyer, T.J. Kippenberg, F. Marquardt (2013), arXiv:1303.0733
3. V.B. Braginsky, A.B. Manukin, Sov. Phys. JETP **25**, 653 (1967)
4. A. Dorsel, J.D. McCullen, P. Meystre, E. Vignes, H. Walther, Phys. Rev. Lett. **51**, 1550 (1983)
5. F. Marquardt, J.P. Chen, A.A. Clerk, S.M. Girvin, Phys. Rev. Lett. **99**, 093902 (2007)
6. J. Dobrindt, I. Wilson-Rae, T.J. Kippenberg, Phys. Rev. Lett. **101**(26), 263602 (2008)
7. S. Groblacher, K. Hammerer, M.R. Vanner, M. Aspelmeyer, Nature **460**, 724 (2009)
8. G.S. Agarwal, S. Huang, Phys. Rev. A **81**, 041803 (2010)
9. S. Weis, R. Rivière, S. Deléglise, E. Gavartin, O. Arcizet, A. Schliesser, T.J. Kippenberg, Science **330**, 1520 (2010)
10. F. Marquardt, J.G.E. Harris, S.M. Girvin, Phys. Rev. Lett. **96**, 103901 (2006)
11. M. Ludwig, B. Kubala, F. Marquardt, New J. Phys. **10**, 095013 (2008)
12. H. Rokhsari, T.J. Kippenberg, T. Carmon, K. Vahala, Opt. Express **13**, 5293 (2005)
13. T. Carmon, H. Rokhsari, L. Yang, T.J. Kippenberg, K.J. Vahala, Phys. Rev. Lett. **94**, 223902 (2005)
14. T.J. Kippenberg, H. Rokhsari, T. Carmon, A. Scherer, K.J. Vahala, Phys. Rev. Lett. **95**, 033901 (2005)
15. C. Höhberger, K. Karrai, *in Nanotechnology 2004*, Proceedings of the 4th IEEE conference on nanotechnology (2004), p. 419
16. C. Metzger, M. Ludwig, C. Neuenhahn, A. Ortlieb, I. Favero, K. Karrai, F. Marquardt, Phys. Rev. Lett. **101**, 133903 (2008)
17. G. Heinrich, M. Ludwig, J. Qian, B. Kubala, F. Marquardt, Phys. Rev. Lett. **107**, 043603 (2011)
18. C.A. Holmes, C.P. Meaney, G.J. Milburn, Phys. Rev. E **85**, 066203 (2012)
19. M. Zhang, G. Wiederhecker, S. Manipatruni, A. Barnard, P.L. McEuen, M. Lipson, Phys. Rev. Lett. **109**, 233906 (2012)
20. M. Bagheri, M. Poot, L. Fan, F. Marquardt, H.X. Tang, Phys. Rev. Lett. **111**, 213902 (2013)
21. T. Carmon, M.C. Cross, K.J. Vahala, Phys. Rev. Lett. **98**, 167203 (2007)
22. I. Wilson-Rae, N. Nooshi, W. Zwerger, T.J. Kippenberg, Phys. Rev. Lett. **99**, 093901 (2007)
23. C. Genes, D. Vitali, P. Tombesi, S. Gigan, M. Aspelmeyer, Phys. Rev. A **77**, 033804 (2008)
24. A.A. Clerk, M.H. Devoret, S.M. Girvin, F. Marquardt, R.J. Schoelkopf, Rev. Mod. Phys. **82**, 1155 (2010)
25. J. Schwinger, J. Math. Phys. **2**, 407 (1961)
26. C.W. Gardiner, M.J. Collett, Phys. Rev. A **31**(6), 3761 (1985)
27. C.W. Gardiner, P. Zoller, *Quant. Noise* (Springer, Berlin, 2000)
28. M. Li, W.H.P. Pernice, H.X. Tang, Phys. Rev. Lett. **103**(22), 223901 (2009)
29. J.C. Sankey, C. Yang, B.M. Zwickl, A.M. Jayich, J.G.E. Harris, Nat. Phys. **6**, 707 (2010)
30. F. Elste, S.M. Girvin, A.A. Clerk, Phys. Rev. Lett. **102**, 207209 (2009)
31. F. Elste, A.A. Clerk, S.M. Girvin, Phys. Rev. Lett. **103**, 149902(E) (2009)

Chapter 3
Nonclassical States of Light and Mechanics

Klemens Hammerer, Claudiu Genes, David Vitali, Paolo Tombesi, Gerard Milburn, Christoph Simon and Dirk Bouwmeester

Abstract This chapter reports on theoretical protocols for generating nonclassical states of light and mechanics. Nonclassical states are understood as squeezed states, entangled states or states with negative Wigner function, and the nonclassicality can refer either to light, to mechanics, or to both, light and mechanics. In all protocols nonclassicality arises from a strong optomechanical coupling. Some protocols rely in addition on homodyne detection or photon counting of light.

K. Hammerer (✉)
Institute for Theoretical Physics, Institute for Gravitational Physics (Albert-Einstein-Institute), Leibniz University Hannover, Hanover, Germany
e-mail: klemens.hammerer@itp.uni-hannover.de

C. Genes
University of Innsbruck, Innsbruck, Austria
e-mail: claudiu.genes@uibk.ac.at

D. Vitali · P. Tombesi
University of Camerino, Camerino, Italy
e-mail: david.vitali@unicam.it

P. Tombesi
e-mail: paolo.tombesi@unicam.it

G. Milburn
University of Queensland, Brisbane, Australia
e-mail: milburn@physics.uq.edu.au

C. Simon
University of Calgary, Calgary, Canada
e-mail: csimo@ucalgary.ca

D. Bouwmeester
Huygens-Kamerlingh Onnes Laboratory, Leiden University, Leiden, The Netherlands

Department of Physics, University of California Santa Barbara, Santa Barbara, USA
e-mail: bouwmeester@physics.ucsb.edu

M. Aspelmeyer et al. (eds.), *Cavity Optomechanics*, Quantum Science and Technology,
DOI: 10.1007/978-3-642-55312-7_3,

3.1 Introduction

An outstanding goal in the field of optomechanics is to go beyond the regime of classical physics, and to generate nonclassical states, either in light, the mechanical oscillator, or involving both systems, mechanics and light. The states in which light and mechanical oscillators are found naturally are those with Gaussian statistics with respect to measurements of position and momentum (or field quadratures in the case of light). The class of Gaussian states include for example thermal states of the mechanical mode, and on the side of light coherent states and vacuum. These are the sort of classical states in which optomechanical systems can be prepared easily. In this chapter we summarize and review means to go beyond this class of states, and to prepare *nonclassical* states of optomechanical systems.

Within the family of Gaussian states those states are usually referred to as nonclassical in which the variance of at least one of the canonical variables is reduced below the noise level of zero point fluctuations. In the case of a single mode, e.g. light or mechanics, these are *squeezed states*. If we are concerned with a system comprised of several modes, e.g. light and mechanics or two mechanical modes, the noise reduction can also pertain to a variance of a generalized canonical variable involving dynamical degrees of freedom of more than one mode. Squeezing of such a collective variable can arise in a state bearing sufficiently strong correlations among its constituent systems. For Gaussian states it is in fact true that this sort of squeezing provides a necessary and sufficient condition for the two systems to be in an inseparable, quantum mechanically *entangled state*. Nonclassicality within the domain of Gaussian states thus means to prepare squeezed or entangled states.

For states exhibiting non Gaussian statistics the notion of nonclassicality is less clear. One generally accepted criterium is based on the Wigner phase space distribution. A state is thereby classified as non classical when its Wigner function is non positive. This notion of nonclassicality in fact implies for pure quantum states that all non Gaussian states are also non classical since every pure non Gaussian quantum state has a non positive Wigner function. For mixed states the same is not true. Under realistic conditions the state of optomechanical systems will necessarily be a statistical mixture such that the preparation and verification of states with a non positive Wigner distribution poses a formidable challenge. Paradigmatic states of this kind will be states which are close to eigenenergy (Fock) states of the mechanical system.

Optomechanical systems present a promising and versatile platform for creation and verification of either sort of nonclassical states. Squeezed and entangled Gaussian states are in principle achievable with the strong, linearized form of the radiation pressure interaction, or might be conditionally prepared and verified by means of homodyne detection of light. These are all "Gaussian tools" which conserve the Gaussian character of the overall state, but are sufficient to steer the system towards Gaussian non classical states. In order to prepare non Gaussian states, possibly with negative Wigner function, the toolbox has to be enlarged in order to encompass also some non Gaussian instrument. This can be achieved either by driving the optomechanical system with a non Gaussian state of light, such as a single photon state, or by preparing states conditioned on a photon counting event. Ultimately the radiation pressure interaction itself is a nonlinear interaction (cubic in annihilation/creation operators) and therefore does in principle generate non Gausssian states for sufficiently strong coupling g_0 at the single photon level. Quite generally one can state that some sort of strong coupling condition has to be fulfilled in any protocol for achieving a nonclassical state. Fulfilling the respective strong coupling condition is thus the experimental challenge on the route towards nonclassicality in optomechanics.

In the following we will present a selection of strategies aiming at the preparation of nonclassical states. In Sect. 3.2 we review ideas of using an optomechanical cavity as a source of squeezed and entangled light. Central to this approach is the fact that the radiation pressure provides an effective Kerr nonlinearity for the cavity, which is well known to be able to generate squeezing of light. In Sect. 3.3 we discuss nonclassical states of the mechanical mode. This involves e.g. the preparation of squeezed states as well as non Gaussian states via state transfer form light, continuous measurement in a nonlinearly coupled optomechanical system, or interaction with single photons and photon counting. Section 3.4 is devoted to nonclassical states involving both systems, light and mechanics, and summarizes ideas to prepare the optomechanical system in an entangled states, either in steady state under continuous wave driving fields, or via interaction with pulsed light.

3.2 Non-classical States of Light

3.2.1 Ponderomotive Squeezing

One of the first predictions of quantum effects in cavity optomechanical system concerned ponderomotive squeezing [1, 2], i.e., the possibility to generate quadrature-squeezed light at the cavity output due to the radiation pressure interaction of the cavity mode with a vibrating resonator. The mechanical element is shifted proportionally to the intracavity intensity, and consequently the optical path inside the cavity depends upon such intensity. Therefore the optomechanical system behaves similarly to a cavity filled with a nonlinear Kerr medium. This can be seen also by inserting the formal solution of the time evolution of the mechanical displacement $\hat{x}(t)$ into the Quantum Langevin equation (QLE) for the cavity field annihilation operator $\hat{a}(t)$,

$$\dot{\hat{a}} = -\left[\frac{\kappa}{2} + i\omega_{\mathrm{opt}}(0)\right]\hat{a} + \int_{-\infty}^{t} ds\,\chi_M(t-s)\left[i\hbar G^2\hat{a}(t)\hat{a}^\dagger(s)\hat{a}(s) + iG\hat{a}(t)\hat{\xi}(s)\right] + \sqrt{\kappa}\hat{a}_{in}(t), \tag{3.1}$$

where $\hat{a}_{in}(t)$ is the driving field (including the vacuum field) and

$$\chi_M(t) = \int_{-\infty}^{\infty} \frac{d\omega}{2\pi} \frac{e^{-i\omega t}}{m_{\text{eff}}\left(\Omega_{\text{M}}^2 - \omega^2 - i\Gamma_{\text{M}}\omega\right)} = \frac{e^{-\Gamma_{\text{M}}t/2}}{m_{\text{eff}}\tilde{\Omega}_M} \sin \tilde{\Omega}_M t \tag{3.2}$$

is the mechanical susceptibility (here $\tilde{\Omega}_M = \sqrt{\Omega_{\text{M}}^2 - \Gamma_{\text{M}}^2/4}$). Equation (3.1) shows that the optomechanical coupling acts as a Kerr nonlinearity on the cavity field, but with two important differences: (1) the effective nonlinearity is delayed by a time depending upon the dynamics of the mechanical element; (2) the optomechanical interaction transmits mechanical thermal noise $\hat{\xi}(t)$ to the cavity field, causing fluctuations of its frequency. When the mechanical oscillator is fast enough, i.e., we look at low frequencies $\omega \ll \Omega_{\text{M}}$, the mechanical response is instantaneous, $\chi_M(t) \simeq \delta(t)/m_{\text{eff}}\Omega_{\text{M}}^2$, and the nonlinear term becomes indistinguishable from a Kerr term, with an effective nonlinear coefficient $\chi^{(3)} = \hbar G^2/m_{\text{eff}}\Omega_{\text{M}}^2$.

It is known that when a cavity containing a Kerr medium is driven by an intense laser, one gets appreciable squeezing in the spectrum of quadrature fluctuations at the cavity output [3]. The above analogy therefore suggests that a strongly driven optomechanical cavity will also be able to produce quadrature squeezing at its output, provided that optomechanical coupling predominates over the detrimental effect of thermal noise [1, 2].

We show this fact by starting from the Fourier-transformed linearized QLE for the fluctuations around the classical steady state

$$m_{\text{eff}}\left(\Omega_{\text{M}}^2 - \omega^2 - i\omega\Gamma_{\text{M}}\right)\hat{x}(\omega) = \hbar G\alpha_s\delta\hat{X}(\omega) + \hat{\xi}(\omega), \tag{3.3}$$

$$\left(\frac{\kappa}{2} - i\omega\right)\delta\hat{X}(\omega) = -\Delta\delta\hat{Y}(\omega) + \sqrt{\kappa}\delta\hat{X}^{in}(\omega), \tag{3.4}$$

$$\left(\frac{\kappa}{2} - i\omega\right)\delta\hat{Y}(\omega) = \Delta\delta\hat{X}(\omega) + G\alpha_s x(\omega) + \sqrt{\kappa}\delta\hat{Y}^{in}(\omega), \tag{3.5}$$

where $\Delta = \omega_L - \omega_{\text{opt}}$, and we have chosen the phase reference so that the stationary amplitude of the intracavity field α_s is real, $\delta\hat{X} = \delta\hat{a} + \delta\hat{a}^\dagger$ [$\delta\hat{Y} = -i\left(\delta\hat{a} - \delta\hat{a}^\dagger\right)$] is the amplitude (phase) quadrature of the field fluctuations, and $\delta\hat{X}^{in}$ and $\delta\hat{Y}^{in}$ are the corresponding quadratures of the vacuum input field. The output quadrature noise spectra are obtained solving Eqs. (3.3)–(3.5), and by using input-output relations [3], the vacuum input noise spectra $S_X^{in}(\omega) = S_Y^{in}(\omega) = 1$, and the fluctuation-dissipation theorem for the thermal spectrum $S_\xi(\omega) = \hbar\omega\Gamma_{\text{M}}m_{\text{eff}}\coth(\hbar\omega/2k_BT)$. The noise spectrum of a quantity X is defined through $S_X(\omega)\delta(\omega - \bar{\omega}) = \langle X(\omega)X(\bar{\omega}) + X(\bar{\omega})X(\omega)\rangle$.

The output light is squeezed at phase ϕ when the corresponding noise spectrum is below the shot-noise limit, $S_\phi^{out}(\omega) < 1$, where $S_\phi^{out}(\omega) = S_X^{out}(\omega)\cos^2\phi + S_Y^{out}(\omega)\sin^2\phi + S_{XY}^{out}(\omega)\sin 2\phi$, and the amplitude and phase noise spectra $S_X^{out}(\omega)$ and $S_y^{out}(\omega)$ satisfy the Heisenberg uncertainty theorem $S_X^{out}(\omega)S_Y^{out}(\omega) > 1 + \left[S_{XY}^{out}(\omega)\right]^2$ [4]. However, rather than looking at the noise spectrum at a fixed phase of the field, one usually performs an optimization and considers, for every frequency

ω, the field phase $\phi_{opt}(\omega)$ possessing the minimum noise spectrum, defining in this way the *optimal squeezing spectrum*,

$$S_{opt}(\omega) = \min_{\phi} S_{\phi}^{out}(\omega)$$
$$= \frac{2S_X^{out}(\omega)S_Y^{out}(\omega) - 2\left[S_{XY}^{out}(\omega)\right]^2}{S_X^{out}(\omega) + S_Y^{out}(\omega) + \sqrt{\left[S_X^{out}(\omega) - S_Y^{out}(\omega)\right]^2 + 4\left[S_{XY}^{out}(\omega)\right]^2}}. \quad (3.6)$$

The frequency-dependent optimal phase is correspondingly given by

$$\phi_{opt}(\omega) = \frac{1}{2}\arctan\left[\frac{2S_{XY}^{out}(\omega)}{S_X^{out}(\omega) - S_Y^{out}(\omega)}\right]. \quad (3.7)$$

We restrict to the resonant case $\Delta = 0$, which is always stable and where expressions are simpler. One gets

$$S_X^{out}(\omega) = 1, \qquad S_{XY}^{out}(\omega) = \frac{\kappa\hbar G^2\alpha_s^2 \mathrm{Re}\left\{\chi_M(\omega)\right\}}{\kappa^2/4 + \omega^2}, \quad (3.8)$$

$$S_Y^{out}(\omega) = 1 + S_{XY}^{out}(\omega)^2 + S_r(\omega), \quad (3.9)$$

where

$$S_r(\omega) = \left[\frac{\kappa\hbar G^2\alpha_s^2 \mathrm{Im}\left\{\chi_M(\omega)\right\}}{\kappa^2/4 + \omega^2}\right]^2 + \frac{\kappa\hbar G^2\alpha_s^2 \mathrm{Im}\left\{\chi_M(\omega)\right\}}{\kappa^2/4 + \omega^2}\coth\left(\frac{\hbar\omega}{2k_BT}\right). \quad (3.10)$$

Inserting Eqs. (3.8)–(3.9) into Eq. (3.6) one sees that the strongest squeezing is obtained when the two limits $S_r(\omega) \ll 1$ and $\left[S_{XY}^{out}(\omega)\right]^2 \gg 1$ are simultaneously satisfied. These conditions are already suggested by Eq. (3.1): $S_r(\omega) \ll 1$ means that thermal noise is negligible, which occurs at low temperatures and small mechanical damping $\mathrm{Im}\left\{\chi_M(\omega)\right\}$, i.e., large mechanical quality factor Q; $\left[S_{XY}^{out}(\omega)\right]^2 \gg 1$ means large radiation pressure, achieved at large intracavity field and small mass. Ponderomotive squeezing is therefore attained when

$$\frac{[S_{XY}^{out}(\omega)]^2}{S_r(\omega)} \sim \frac{P_{\mathrm{in}}\omega_L}{m_{\mathrm{eff}}c^2\Omega_{\mathrm{M}}^2}\frac{\mathcal{F}^2 Q}{\bar{n}_{\mathrm{th}}} \gg 1, \quad (3.11)$$

and in this ideal limit $S_{opt}(\omega) \simeq \left[S_{XY}^{out}(\omega)\right]^{-2} \to 0$, and $\phi_{opt}(\omega) \simeq -\frac{1}{2}\arctan\left[\frac{2}{S_{XY}^{out}(\omega)}\right] \to 0$. Since the field quadrature δX^{out} at $\phi = 0$ is just at the shot-noise limit (see Eq. (3.8)), one has that squeezing is achieved only within a narrow interval for the homodyne phase around $\phi_{opt}(\omega)$, of width $\sim 2\left|\phi_{opt}(\omega)\right| \sim \arctan\left|2/S_{XY}^{out}(\omega)\right|$. This extreme phase dependence is a general and well-known property of quantum squeezing, which is due to the Heisenberg principle: the width of the interval of

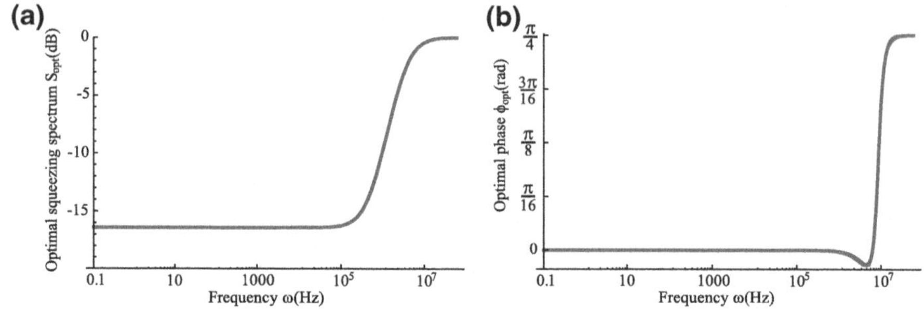

Fig. 3.1 Optimal spectrum of squeezing in dB S_{opt} (**a**), and the corresponding optimal quadrature phase ϕ_{opt} (**b**), versus frequency in the case of a cavity with bandwidth $\kappa = 1$ MHz, length $L = 1$ cm, driven by a laser at 1,064 nm and with input power $\mathscr{P}_{in} = 10$ mW. The mechanical resonator has $\Omega_M/2\pi = 1$ MHz, mass $m_{\rm eff} = 100$ ng, quality factor $Q = 10^4$, and temperature $T = 4$ K

quadrature phases with noise below the shot-noise limit is inversely proportional to the amount of achievable squeezing.

$S_{opt}(\omega)$ and the corresponding optimal phase $\phi_{opt}(\omega)$ at which best squeezing is attained for each ω, are plotted in Fig. 3.1 for a realistic set of parameter values (see figure caption). $S_{opt}(\omega)$ is below the shot-noise limit whenever $S_{XY}^{out}(\omega) \neq 0$ (see Eqs. (3.6)–(3.9)), and one gets significant squeezing at low frequencies, well below the mechanical resonance, where the optomechanical cavity becomes fully equivalent to a Kerr medium, as witnessed also by the fact that $\phi_{opt}(\omega)$ is constant in this frequency band. This equivalence is lost close to and above the mechanical resonance, where squeezing vanishes because $\mathrm{Re}\,\{\chi_M(\omega)\} \sim S_{XY}^{out}(\omega) \sim 0$, and the optimal phase shows a large variation.

The present treatment neglects technical limitations: in particular it assumes the ideal situation of a one-sided cavity, where there is no cavity loss because all photons transmitted by the input–output mirror are collected by the output mode. We have also ignored laser phase noise which is typically non-negligible at low frequencies where ponderomotive squeezing is significant. In current experimental schemes both cavity losses and laser phase noise play a relevant role and in fact. Recent experiments in cold atom optomechanics [5], photonic crystals [6] and membrane in the middle setups [7] demonstrated squeezed light along the lines outlined here. These results show that cavity optomechanical systems may become a valid alternative to traditional sources of squeezing such as parametric amplifiers and Kerr media.

3.2.2 Einstein-Podolsky-Rosen Correlated Beams of Light

Optomechanical cavities provide a source not only of squeezed light but also of entangled light, as we will explain in the following. By means of spectral filters, the continuous wave field emerging from the cavity can be split in many traveling modes thus offering the option of producing and manipulating a multipartite system [8]. In particular we focus on detecting the first two motional sidebands at frequencies $\omega_{\rm opt} \pm \Omega_{\rm M}$ and show that they posses quantum correlations of the Einstein-Podolsky-Rosen type [9].

Using the well-known input-output fields connection $\hat{a}^{out}(t) = \sqrt{\kappa}\hat{a}(t) - \hat{a}^{in}(t)$, the output mode can be split in N independent optical modes by frequency selection with a proper choice of a causal filter function:

$$\hat{a}_k^{out}(t) = \int_{-\infty}^{t} ds g_k(t-s)\hat{a}^{out}(s), \quad k = 1, \ldots N, \tag{3.12}$$

where $g_k(s)$ is the causal filter function defining the k-th output mode. The annihilation operators describe N independent optical modes when $\left[\hat{a}_j^{out}(t), \hat{a}_k^{out}(t)^\dagger\right] = \delta_{jk}$, which is fulfilled when $\int_0^\infty ds g_j(s)^* g_k(s) = \delta_{jk}$, i.e., the N filter functions $g_k(t)$ form an orthonormal set of square-integrable functions in $[0, \infty)$. As an example of a set of functions that qualify as causal filters we take

$$g_k(t) = \frac{\theta(t) - \theta(t-\tau)}{\sqrt{\tau}} e^{-i\Omega_k t}, \tag{3.13}$$

(θ denotes the Heavyside step function) provided that Ω_k and τ satisfy the condition $\Omega_j - \Omega_k = \frac{2\pi}{\tau} p$ for integer p. Such filtering is seen as a simple frequency integration around Ω_k of bandwidth $\sim 1/\tau$ (the inverse of the time integration window).

For characterization of entanglement one can compute the stationary $(2N+2) \times (2N+2)$ correlation matrix of the output modes defined as

$$V_{ij}^{out}(t) = \frac{1}{2}\left\langle u_i^{out}(t) u_j^{out}(t) + u_j^{out}(t) u_i^{out}(t)\right\rangle, \tag{3.14}$$

where

$$u^{out}(t) = \left(\hat{q}(t), \hat{p}(t), \hat{X}_1^{out}(t), \hat{Y}_1^{out}(t), \ldots, \hat{X}_N^{out}(t), \hat{Y}_N^{out}(t)\right)^T$$

is the vector formed by the mechanical position and momentum fluctuations and by the amplitude ($\hat{X}_k^{out}(t) = [\hat{a}_k^{out}(t) + \hat{a}_k^{out}(t)^\dagger]/\sqrt{2}$), and phase ($\hat{Y}_k^{out}(t) = [\hat{a}_k^{out}(t) - \hat{a}_k^{out}(t)^\dagger]/i\sqrt{2}$) quadratures of the N output modes.

We are now in position to analyze the quantum correlations between two output modes with the same bandwidth τ^{-1} and central frequencies Ω_1 and Ω_2. As a measure for entanglement we apply the logarithmic negativity $E_{\mathcal{N}}$ to the covariance matrix of the two optical modes. It is defined as $E_{\mathcal{N}} = \max[0, -\ln 2\eta^-]$, where $\eta^- \equiv 2^{-1/2}\left[\Sigma(V) - \left[\Sigma(V)^2 - 4\det V\right]^{1/2}\right]^{1/2}$, with $\Sigma(V) \equiv \det V_m + \det V_c - 2\det V_{mc}$, and we have used the 2×2 block form of the covariance matrix

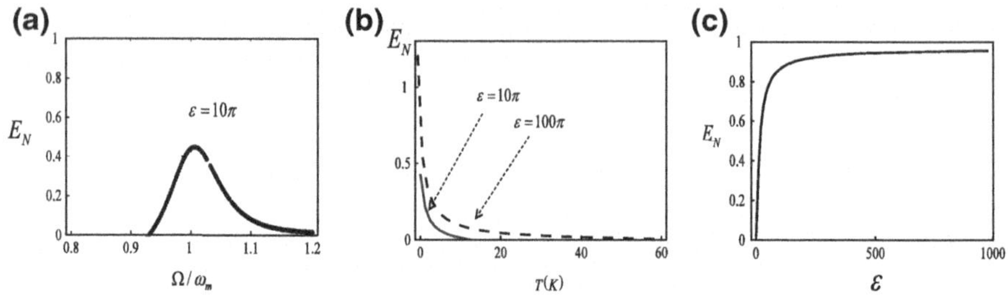

Fig. 3.2 **a** Logarithmic negativity of Stoke–Antistokes output modes when $\Omega_1 = -\Omega_M$ while Ω_2 is varied around Ω_M. The inverse bandwidth is kept constant at $\varepsilon = 10\pi$. **b** Temperature robustness of bipartite entanglement of output modes at $\pm\Omega_M$ computed for short ($\varepsilon = 10\,\pi$, *dashed line*) and long ($\varepsilon = 100\,\pi$, *solid line*) detection times. **c** The bipartite Stokes–Antistokes entanglement shows improvement and eventually saturates with increasing integration time. Parameters are $\Omega_M/2\pi = 10$ MHz, $Q = 10^5$, mass $m_{\rm eff} = 50$ ng, cavity of length $L = 1$ mm with finesse $\mathcal{F} = 2 \times 10^4$, detuning $\Delta = \Omega_M$, input power $P_{\rm in} = 30$ mW at 810 nm, and temperature $T = 0.4$ K, yielding $g_0 = 0.43$ kHz, $g = 0.41\Omega_M$, a cavity bandwidth $2\kappa = 0.75\Omega_M$, and a thermal occupation of $\bar{n}_{\rm th} \simeq 833$

$$V \equiv \begin{pmatrix} V_m & V_{mc} \\ V_{mc}^T & V_c \end{pmatrix}. \tag{3.15}$$

Therefore, a Gaussian state is entangled if and only if $\eta^- < 1/2$, which is equivalent to Simon's necessary and sufficient entanglement non-positive partial transpose criterion for Gaussian states, which can be written as $4 \det V < \Sigma - 1/4$.

The resulting quantum correlations among the upper and the lower sideband in the continuous wave output field are illustrated in Fig. 3.2. We plot the interesting and not unexpected behavior of $E_{\mathcal{N}}$ as a function of central detection frequency Ω in Fig. 3.2a, with the mirror reservoir temperature in Fig. 3.2b and with the scaled time integration window $\varepsilon = \Omega_M\tau$ in Fig. 3.2c. The conclusion of Fig. 3.2a is that indeed scattering off the mirror can produce good Stokes-Antistokes entanglement which can be optimized at the cavity output by properly adjusting the detection window. Moreover, further optimization is possible via an integration time increase as suggested by Fig. 3.2c. The temperature behavior plotted in Fig. 3.2b shows very good robustness of the mirror-scattered entangled beams that suggests this mechanism of producing Einstein-Podolsky-Rosen (EPR) entangled photons as a possible alternative to parametric oscillators.

3.3 Non-classical States of Mechanics

3.3.1 State Transfer

For a massive macroscopic mechanical resonator, just as in the case of a light field, the signature of quantum can be indicated in a first step by the ability of engineering a

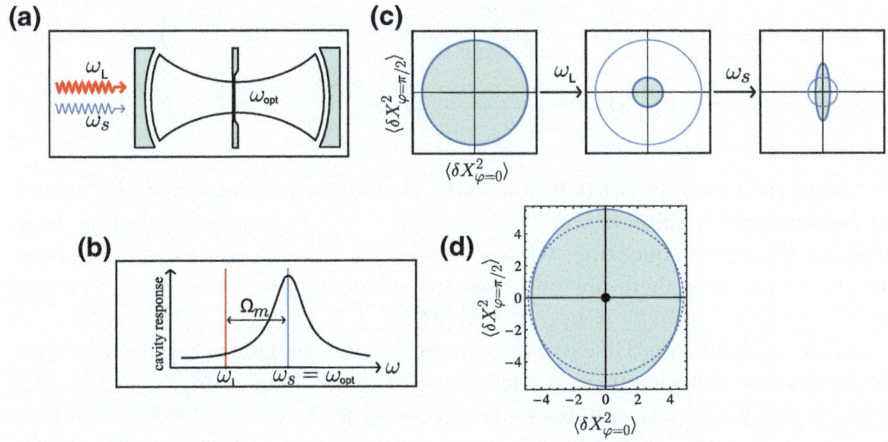

Fig. 3.3 **a** An optomechanical system is driven by a classical (coherent) field at frequency ω_L and by squeezed light of central frequency ω_S. **b** ω_s coincides with the cavity frequency, ω_L is detuned to the red by the mechanical frequency. **c** The initial thermal state of the oscillator is cooled to the ground state by passive sideband cooling. Squeezing of light will cool the oscillator to a squeezed mechanical state, cf. subfigure (**d**)

squeezed state. Such a state would also be useful in ultrahigh precision measurements or detection of gravitational waves and has been experimentally proven in only one instance for a nonlinear Duffing resonator [10] (though not in the quantum regime). Numerous proposals exist and can be categorized as (1) direct: modulated drive in optomechanical settings with or without feedback loop [11–15], and (2) indirect: mapping a squeezed state of light or atoms onto the resonator, coupling to a cavity with atomic medium within [16], coupling to a Cooper pair box [17] or a superconducting quantum interference loop [18, 19]. Mechanical squeezing can also be generated through QND measurements of the mechanical position by means of short light pulses [20], which was also experimentally explored (though not in the quantum regime) [21].

In the following we take the example of state transfer in a pure optomechanical setup where laser cooling of a mirror/membrane via a strong laser is accompanied by squeezing transfer from a squeezed vacuum second input light field [22]. While the concept is straightforward it is of interest to answer a few practical questions such as: (1) what is the resonance condition for optimal squeezing transfer and how does a frequency mismatch affect the squeezing transfer efficiency, (2) what is the optimal transfer, (3) how large should the cavity finesse be for optimal transfer etc.

To this purpose we assume an optomechanical system in cooling configuration, i.e. the driving laser is red detuned from the cavity resonance such that the Anti-stokes sideband is resonant to the cavity frequency ω_{opt}. In addition the optomechanical system is driven by squeezed light at this same frequency, that is resonant with the cavity, as shown in Fig. 3.3. The input squeezed light operators have the following correlations

$$\langle \hat{c}_{\text{in}}(t+\tau)\hat{c}_{\text{in}}(t)\rangle = \frac{M}{2}\frac{b_x b_y}{b_x^2+b_y^2}\left(b_y e^{-b_x|\tau|}+b_x e^{-b_y|\tau|}\right) \tag{3.16}$$

$$\langle \hat{c}_{\text{in}}^{\dagger}(t+\tau)\hat{c}_{\text{in}}(t)\rangle = \frac{N}{2}\frac{b_x b_y}{b_y^2-b_x^2}\left(b_y e^{-b_x|\tau|}-b_x e^{-b_y|\tau|}\right). \tag{3.17}$$

The noise operators are written in a frame rotating at ω_s and satisfy the canonical commutation relation $[\hat{c}_{\text{in}}(t), \hat{c}_{\text{in}}^{\dagger}(t')] = \delta(t-t')$. Parameters N and M determine the degree of squeezing, while b_x and b_y define the squeezing bandwidth. For pure squeezing there are only two independent parameters, as in this case $|M|^2 = N(N+1)$ and $b_y = b_x\sqrt{2(N+|M|)+1}$.

Following the standard linearized quantum Langevin equations approach for optomechanics, we first identify two conditions for optimal squeezing: (1) $\Delta = \Omega_{\text{M}}$, meaning that we require continuous laser cooling in the resolved sideband regime and (2) $\Delta_s = -\Omega_{\text{M}}$ so that the squeezing spectrum is centered around the cavity frequency. Then we look at the variances of the generalized quadrature operator

$$\delta\hat{X}_{\varphi}(t) = \frac{1}{\sqrt{2}}\left(e^{i\varphi}\hat{b}(t)+e^{-i\varphi}\hat{b}^{\dagger}(t)\right), \tag{3.18}$$

which for $\varphi = 0$ is the usual position operator $\hat{q}(t)$ and for $\varphi = -\pi/2$ is the momentum operator $\hat{p}(t)$, both taking in a rotating frame at frequency Ω_{M}. In the limit of squeezed white noise the quadrature correlations take a simple form

$$\langle \delta\hat{X}_{\varphi}(t)\delta\hat{X}_{\varphi}(t)\rangle = \left(N+\frac{1}{2}-\text{Re}\left\{Me^{2i\varphi}\right\}\right)+\frac{\Gamma_{\text{M}}}{\Gamma_{\text{eff}}}\left(\bar{n}_{\text{th}}+\frac{1}{2}\right). \tag{3.19}$$

The first term in the right hand side comes from the squeezing properties of the squeezed input vacuum while the second term is the residual occupancy after laser cooling. In view of this equation a successful squeezed mechanical state preparation automatically requires close to ground state cooling. One can follow this in Fig. 3.3c where cooling close to ground state of an initially thermal mechanical state is performed by the cooling laser, and subsequently squeezing of a quadrature is achieved via the squeezed vacuum.

To answer a practical question, when the squeezing is not white, fulfilling the resonance condition $\Delta_s = -\Omega_{\text{M}}$ is important. The deviations allowed for the frequency mismatch are smaller than the width of the cooling sideband, i.e. Γ_{eff}. A second question is the effect of the finite width of squeezing. In general there is an optimal squeezing bandwidth for which the transfer from light to membrane is maximized, but in the resolved sideband limit where $\Omega_{\text{M}} \gg \kappa$ the finite bandwidth result does not differ much from the infinite bandwidth limit result. For a large bandwidth which fully covers the motional sidebands, $b_x \gg \Omega_{\text{M}}$, the membrane sees only white squeezed input noise, whereas for smaller bandwidth, the crucial question is whether the squeezed input will touch the heating sideband or not. For a high-finesse cavity, the width is not a big issue, since the heating sideband is anyway weak, whereas for a bad cavity the squeezing transfer is much improved for an optimal, finite bandwidth where the strong heating sideband is avoided.

3.3.2 Continuous Measurements of Mechanical Oscillators

Another method for creating nonclassical states exploits the possibility to conditionally prepare states of the mechanical oscillator via measuring the output field of the optomechanical system. The coupling of a mechanical resonator to an optical cavity field enables an indirect continuous monitoring of the mechanical motion by a direct phase dependent measurement of the field leaving the cavity. The standard radiation pressure coupling is linear in the displacement of the mechanical resonator and thus enables a continuous measurement of displacement. If we wish to monitor the energy (or phonon number) of the mechanical resonator, however, we need to find an interaction Hamiltonian that is quadratic in the displacement. Such interactions can occur in a number of ways [23, 24]. We begin with the case of displacement measurements.

The standard linearised opto-mechanical coupling Hamiltonian

$$H = -\hbar\Delta\hat{a}^\dagger\hat{a} + \hbar\Omega_{\mathrm{M}}\hat{b}^\dagger\hat{b} - \hbar g(\hat{a}+\hat{a}^\dagger)(\hat{b}+\hat{b}^\dagger) \tag{3.20}$$

As the interaction part of this Hamiltonian commutes with the (dimensionless) mechanical displacement operator, $\hat{q} = \frac{1}{\sqrt{2}}(\hat{b}+\hat{b}^\dagger)$, in principle this model can be configured as a measurement of the displacement. However as $\hat{q}$ does not commute with the free mechanical Hamiltonian, this is not a strict QND measurement [25]. Nonetheless, for a rapidly damped cavity, we can effect approximate QND readout of the mechanical displacement provided the coupling constant g can be turned on and off sufficiently fast. This can be achieved by using a pulsed coherent driving field on the cavity [20].

If we include the damping of both the cavity and the mechanics, we obtain the quantum stochastic differential equations,

$$\frac{d\hat{a}}{dt} = i\Delta\hat{a} - \frac{\kappa}{2}\hat{a} + ig\hat{X} + \sqrt{\kappa}\hat{a}_{in} \tag{3.21}$$

$$\frac{d\hat{b}}{dt} = -i\Omega_{\mathrm{M}}\hat{b} - \frac{\Gamma_{\mathrm{M}}}{2}\hat{b} + ig(\hat{a}+\hat{a}^\dagger) + \sqrt{\Gamma_{\mathrm{M}}}\hat{b}_{in} \tag{3.22}$$

where we assume that the input noise to the cavity is vacuum, so that the only non zero correlation function for the cavity noise is $\langle\hat{a}_{in}(t)\hat{a}_{in}^\dagger(t')\rangle = \delta(t-t')$ but that the input noise to the mechanical resonator is thermal, $\langle\hat{b}_{in}(t)\hat{b}_{in}^\dagger(t')\rangle = (\bar{n}_{\mathrm{th}}+1)\delta(t-t')$. We expect that a good measurement will occur when the cavity field is rapidly damped so that it is slaved to the mechanical degree of freedom. We can then adiabatically eliminate the cavity degree of freedom by setting to zero the right hand side of Eq. (3.21) and formally solving for the operator $\hat{a}$,

$$\hat{a} = \frac{ige^{-i\phi}}{\sqrt{\Delta^2 + \kappa^2/4}}\hat{q} + \frac{\sqrt{\kappa}}{\kappa/2 - i\Delta}\hat{a}_{in} \tag{3.23}$$

where $\tan\phi = -2\Delta/\kappa$. The actual output field from the cavity is related to the field inside by the input/output relation, $\hat{a}_{out} = \sqrt{\kappa}\hat{a} - \hat{a}_{in}$, so that

$$\hat{a}_{out}(t) = \frac{ig\sqrt{\kappa}e^{-i\phi}}{\sqrt{\Delta^2 + \kappa^2/4}}\hat{q}(t) + e^{-2i\phi}\hat{a}_{in}(t) \tag{3.24}$$

Clearly this indicates that we need to measure a particular quadrature of the output field (for example by homodyne or heterodyne detection) and that the added noise is vacuum noise. The optimal transfer is obtained on resonance. A fast, impulsive readout of the mechanical resonator's displacement may be made by injecting a coherent pulse into the cavity and subjecting the output pulse to a homodyne measurement [20].

If we wish to measure the energy of a mechanical resonator we must find an interaction Hamiltonian that is at least quadratic in the mechanical amplitude. A number of schemes have been proposed, including trapped atoms in a standing wave [26] and a nanomechanical resonator coupled to a Cooper pair box qubit in the dispersive regime [27]. In opto-mechanics a dielectric membrane placed at the antinode of a cavity standing wave shifts the cavity frequency proportional to the square of the mechanical displacement of the membrane from equilibrium [23]. A similar interaction arises for an optically levitated particle in a standing wave [24].

The interaction Hamiltonian in this case takes the form

$$H = \hbar\omega_{opt}\hat{a}^\dagger\hat{a} + \hbar\Omega_{\rm M}\hat{b}^\dagger\hat{b} + \hbar(\varepsilon_c^*\hat{a}e^{i\omega_L t} + \varepsilon_c\hat{a}^\dagger e^{-i\omega_l t}) + \frac{\hbar}{2}G_2\hat{a}^\dagger\hat{a}(\hat{b} + \hat{b}^\dagger)^2, \tag{3.25}$$

where

$$G_2 = \frac{\hbar}{2\nu m}\left.\frac{\partial^2\omega_c(x)}{\partial x^2}\right|_{x=x_0}. \tag{3.26}$$

and where we have included a coherent driving field with amplitude ε. As usual we expand the interaction around the steady state cavity field. After the rotating wave approximation, the effective Hamiltonian in the interaction picture may then be written as

$$H_I = \frac{\hbar}{2}\chi(\bar{\hat{a}} + \bar{\hat{a}}^\dagger)\hat{b}^\dagger\hat{b}, \tag{3.27}$$

where $\chi = 2G_2\alpha_0$.

The interaction in Eq. (3.27) describes a displacement of the cavity field proportional to the number of vibrational quanta in the mechanical resonator. The average steady-state displacement is given by $\chi\bar{n}/\kappa$, where $\bar{n}$ is the mean phonon number operator for the mechanical oscillator, κ is the cavity line-width. If we continuously monitor the output field amplitude from the cavity via homodyne detection this

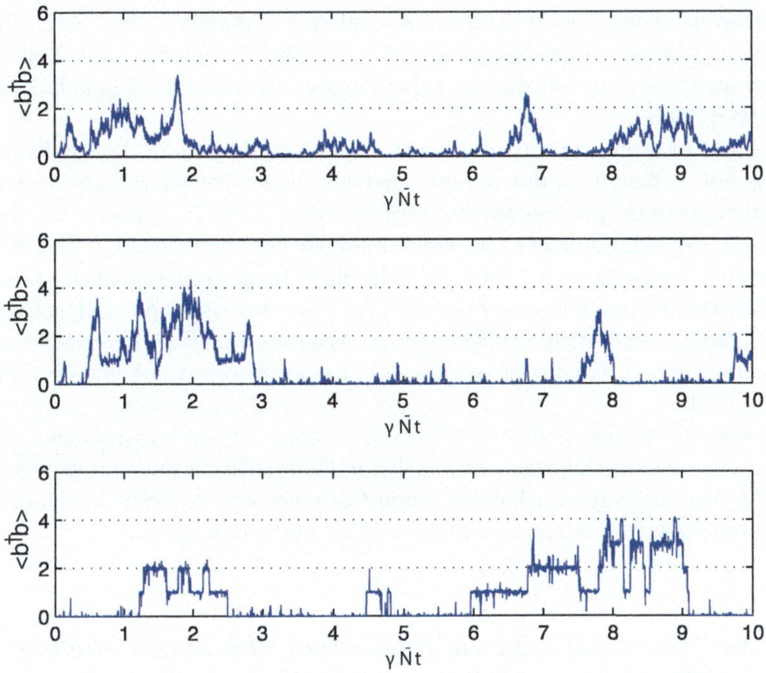

Fig. 3.4 The evolution of the conditional average phonon number with parameters $\kappa = 10^4\gamma\bar{n}_{th}$ and: **a** $\chi^2/\kappa = \gamma\bar{n}_{th}$, **b** $\chi^2/\kappa = 10\gamma\bar{N}$, **c** $\chi^2/\kappa = 10^2\gamma\bar{n}_{th}$. Jump-like behaviour occurs only when $\chi^2/\kappa \gg \gamma[\bar{n}_{th}(n+1) + (\bar{n}_{th}+1)n]$, where n is the phonon number

scheme can in principle enable a continuous monitoring of the mechanical vibrational energy, and phonon number jumps [28].

Under continuous homodyne measurement of this quadrature, the system is governed by the following stochastic master equation (SME):

$$d\rho = -\frac{i}{\hbar}[H_I, \rho]dt + \gamma(\bar{n}_{th}+1)\mathscr{D}[\hat{b}]\rho dt + \gamma\bar{n}_{th}\mathscr{D}[\hat{b}^\dagger]\rho dt + \kappa\mathscr{D}[\hat{a}]\rho dt$$
$$+ \sqrt{\kappa}dW\mathscr{H}[ae^{-i\frac{\pi}{2}}]\rho, \quad (3.28)$$

where $\mathscr{D}[c]\rho = c\rho c^\dagger - c^\dagger c\rho/2 - \rho c^\dagger c/2$ and $\mathscr{H}[c]\rho = c\rho + \rho c^\dagger - \mathrm{Tr}(c\rho + \rho c^\dagger)\rho$ is the measurement super-operator, γ and κ are the respective mechanical and cavity damping rates.

In Fig. 3.4 we show a numerical integration of the stochastic master equation with $\kappa = 10^4\gamma\bar{N}$ for three cases: $\chi^2/\kappa = \gamma\bar{N}$, $\chi^2/\kappa = 10\gamma\bar{n}_{th}$, and $\chi^2/\kappa = 10^2\gamma\bar{n}_{th}$. We start with the mechanics in the ground state, and the bath temperature is set at $\bar{n}_{th} = 0.5$. The first case, Fig. 3.4a, does not satisfy the fast-measurement condition and therefore does not resolve quantum jumps in the phonon number. The second case, Fig. 3.4b, is on the border of the fast-measurement regime for $n \sim 1$ and shows some jump-like behaviour in the phonon number. The third case, Fig. 3.4c, strongly satisfies the fast-measurement condition for low phonon numbers and shows well-resolved quantum jumps in spite of being deeply within the weak coupling regime with $\chi/\kappa = 10^{-1}$.

For jump-like behaviour to arise in the weak coupling limit, the adiabatic condition is not sufficient. In this regime, analysis shows that the rate of information acquisition about the phonon number is proportional to χ^2/κ. As in the strong coupling case, this measurement rate must dominate the thermalisation rate in order for quantum jumps to arise. Thus, in addition to being in the adiabatic limit, the weak coupling regime requires $\chi^2/\kappa \gg \gamma \bar{N}$. Note that single-phonon detection of this kind remains extremely challenging, as it requires a single-photon optomechanical coupling strength that is large compared to any absorptive photon losses in the structure [29].

The state of the mechanical system conditioned on the measurement of light will be in a non-Gaussian state. This is due to the non-linear interaction introduced in (3.25). Another way of achieving a non-Gaussian state exploits the nonlinearity provided in photon counting, as will be detailed in the next section.

3.3.3 Non-Gaussian State via Interaction with Single Photons and Photon Counting

In 1935 Schrödinger pointed out that according to quantum mechanics even macroscopic systems can be in superposition states [30]. The interference effects, characteristic of quantum mechanics, are expected to be hard to detect due to environment induced decoherence [31]. Nevertheless there have been several proposals on how to create and observe macroscopic superpositions in various physical systems. See references [32–34] for some of the first proposals. There have also been experiments on superposition states in superconducting and piezoelectric devices [35, 36] and on interference with fullerene [37] and other large molecules. One long-term motivation for this kind of experiment is the question of whether unconventional decoherence processes such as gravitationally induced decoherence or spontaneous wave-function collapse [38–42] take place.

In this section a scheme is analyzed that is close in spirit to Schrödinger's original discussion. A small quantum system (a photon) is coupled to a large system (a mirror) such that they become entangled [43]. The system consists of a Michelson interferometer in which one arm has a tiny moveable mirror. The radiation pressure of a single photon is used to displace the tiny mirror. The initial superposition of the photon being in either arm causes the system to evolve into a superposition of states corresponding to two distinct spatial locations of the mirror. A high-finesse cavity is used to enhance the interaction between the single photon and the mirror. The interference of the photon upon exiting the interferometer allows one to study the creation of coherent superposition states periodic with the motion of the mirror.

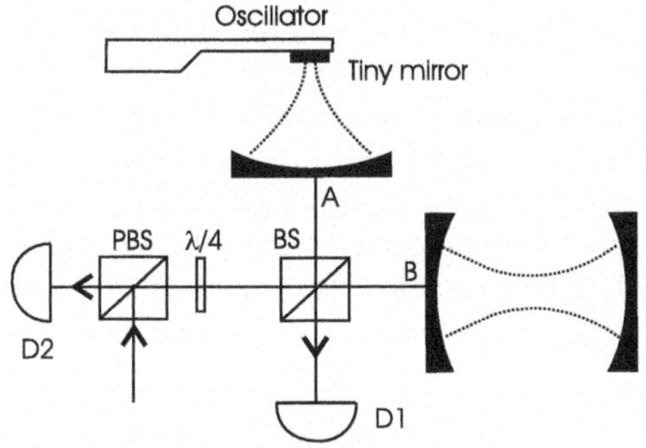

Fig. 3.5 Scheme: a Michelson interferometer for a single photon with a high-finesse cavity in each arm. The cavity in arm *A* has a very small end mirror mounted on a micro-mechanical oscillator. The single photon enters the interferometer via a polarizing beam splitter (PBS) followed by a $\lambda/4$ wave plate. This is an optical trick to allow detection of the photon leaking out of the interferometer at a later time on detector *D1* or *D2*. If the input photon is considered to be in arm *A*, the motion of the small mirror is affected by its radiation pressure. If the input photon is considered to be in arm *B*, the motion of the mirror is undisturbed. The interferometer, based on the 50 %/50 % beamsplitter (BS), leads to the entanglement between the photon being in arm *A* or in arm *B* and the state of the mirror

Consider the setup shown in Fig. 3.5, consisting of a Michelson interferometer that has a cavity in each arm. In the cavity in arm A one of the mirrors is very small and attached to a micromechanical oscillator. While the photon is inside the cavity, it exerts a radiation pressure force on the small mirror. We will be interested in the regime where the period of the mirror's motion is much longer than the roundtrip time of the photon inside the cavity, and where the amplitude of the mirror's motion is very small compared to the cavity length. Under these conditions, the system can be described by the standard optomechanical Hamiltonian [44, 45]

$$H = \omega_{\text{opt}}\hat{a}^\dagger\hat{a} + \Omega_{\text{M}}\hat{b}^\dagger\hat{b} - g_0\hat{a}^\dagger\hat{a}(\hat{b} + \hat{b}^\dagger). \tag{3.29}$$

To start with, let us suppose that initially the photon is in a superposition of being in either arm A or B, and the mirror is in a coherent state $|\beta\rangle = e^{-|\beta|^2/2}\sum_{n=0}^{\infty}\frac{\beta^n}{\sqrt{n!}}|n\rangle$, where $|n\rangle$ are the eigenstates of the harmonic oscillator. Then the initial state is

$$|\psi(0)\rangle = \frac{1}{\sqrt{2}}(|0\rangle_A|1\rangle_B + |1\rangle_A|0\rangle_B)|\beta\rangle. \tag{3.30}$$

After a time t the state of the system will be given by [46, 47]

$$\begin{aligned} |\psi(t)\rangle = &\frac{1}{\sqrt{2}}e^{-i\omega_{\text{opt}}t}\{|0\rangle_A|1\rangle_B|\beta e^{-i\Omega_{\text{M}}t}\rangle \\ &+ e^{i\eta^2(\Omega_{\text{M}}t - \sin\Omega_{\text{M}}t)}|1\rangle_A|0\rangle_B|\beta e^{-i\Omega_{\text{M}}t} + \eta(1 - e^{-i\Omega_{\text{M}}t})\rangle\}, \end{aligned} \tag{3.31}$$

where $\eta = g_0/\Omega_M$. In the second term on the right hand side the motion of the mirror is altered by the radiation pressure of the photon in cavity A. The parameter η quantifies the displacement of the mirror in units of the size of the coherent state wavepacket. In the presence of the photon the mirror oscillates around a new equilibrium position determined by the driving force.

The maximum possible interference visibility for the photon is given by twice the modulus of the off-diagonal element of the photon's reduced density matrix. By tracing over the mirror one finds from Eq. (3.31) that the off-diagonal element has the form

$$\frac{1}{2}e^{-\eta^2(1-\cos\Omega_M t)}e^{i\eta^2(\Omega_M t-\sin\Omega_M t)+i\eta \mathrm{Im}\,[\beta(1-e^{i\Omega_M t})]} \tag{3.32}$$

where Im denotes the imaginary part. The first factor is the modulus, obtaining its minimum value after half a period at $t = \pi/\Omega_M$, when the mirror is at its maximum displacement. The second factor gives the phase, which is identical to that obtained classically due to the varying length of the cavity.

For general t the phase in Eq. (3.32) depends on β, i.e. the initial condition of the mirror. However, the effect of the initial condition averages out after every full period.

In the absence of decoherence, after a full period, $t = 2\pi/\Omega_M$, the system is in the state $\frac{1}{\sqrt{2}}(|0\rangle_A|1\rangle_B + e^{i\eta^2 2\pi}|1\rangle_A|0\rangle_B)|\beta\rangle$, such that the mirror is again completely disentangled from the photon. Full interference can be observed if the photon is detected at that moment. If the environment of the mirror "remembers" that the mirror has moved, then, even after a full period, the photon will still be entangled with the mirror's environment, and thus the interference for the photon will be reduced. Therefore the setup can be used to measure the decoherence of the mirror.

In practice the mirror attached to a mechanical-resonator will be in a thermal state, which can be written as a mixture of coherent states $|\beta\rangle$ with a Gaussian probability distribution $(1/\pi\bar{n}_{th})e^{-|\beta|^2/\bar{n}_{th}}$, where $\bar{n}_{th}$ is the mean thermal number of excitations, $\bar{n}_{th} = 1/(e^{\hbar\Omega_M/kT} - 1)$. If one wants to determine the expected interference visibility of the photon at a time t for an initial mirror state which is thermal, one therefore has to average the off-diagonal element Eq. (3.32) over β with this distribution. The result is

$$\frac{1}{2}e^{-\eta^2(2\bar{n}_{th}+1)(1-\cos\Omega_M t)}e^{i\eta^2(\Omega_M t-\sin\Omega_M t)}. \tag{3.33}$$

As a consequence of the averaging of the β-dependent phase in Eq. (3.32), the off-diagonal element now decays on a timescale $1/(\eta\Omega_M\sqrt{\bar{n}_{th}})$ after $t = 0$, i.e. very fast for the realistic case of large $\bar{n}_{th}$. However, remarkably it still exhibits a revival [33] at $t = 2\pi/\Omega_M$, when photon and mirror become disentangled and the phase is independent of β, such that the phase averaging does not reduce the visibility. Figure 3.6 shows the time evolution of the visibility of the photon over one period of the mirror's motion for $\eta = 1$ and temperatures T of 1 mK, 100 μK and 10 μK.

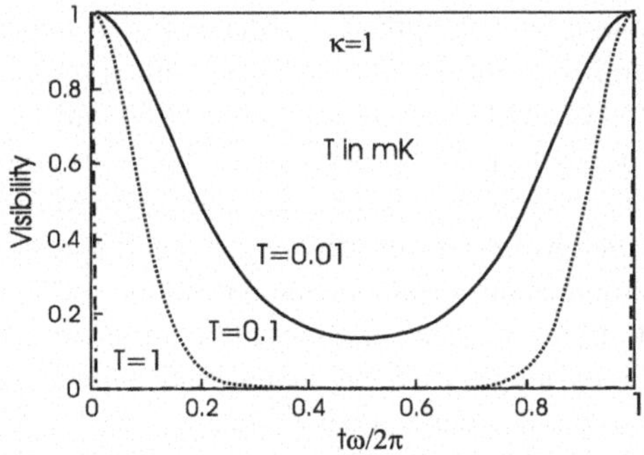

Fig. 3.6 Time evolution of the interference visibility of the photon over one period of the mirror's motion for $\eta = 1$ and temperatures T of 1 mK, 100 μK and 10 μK. The visibility decays very fast after $t = 0$, but in the absence of decoherence there is a sharp revival of the visibility after a full period (2π). The width of each peak scales like $1/\sqrt{T}$

The magnitude of the revival is reduced by any decoherence of the mirror. Furthermore the revival will also be reduced due to nonlinear terms in the mechanical oscillator. However since we will only consider extremely small displacements around the equilibrium position we assume that nonlinear effects can be ignored.

The revival demonstrates the coherence of the superposition state that exists at intermediate times. For $\eta^2 \gtrsim 1$ the state of the system is a superposition of two distinct positions of the mirror. More precisely, for a thermal mirror state, the state of the system is a mixture of such superpositions. However, this affects neither the fundamentally non-classical character of the state nor, as we have seen, the existence of the revival after a full period. We now discuss the experimental requirements for achieving such a superposition state and observing its recoherence at $t = 2\pi/\Omega_M$.

Firstly, it is required that $\eta^2 \gtrsim 1$, which physically means that the momentum kick imparted by the photon to the mirror has to be larger than the initial quantum momentum uncertainty of the mirror. Let N denote the number of roundtrips of the photon in the cavity during one period of the mirror's motion, such that $2NL/c = 2\pi/\Omega_M$. This allows us to rewrite the condition $\eta^2 \gtrsim 1$ as

$$\frac{2\hbar N^3 L}{\pi c m_{\text{eff}} \lambda^2} \gtrsim 1, \tag{3.34}$$

where λ is the wavelength of the light. The factors entering Eq. (3.34) are not all independent. The achievable N, which is determined by the quality of the mirrors, and the minimum possible mirror size depend strongly on λ. The mirror's lateral dimensions should be an order of magnitude larger than λ to limit diffraction and avoid geometrical losses. The minimum possible thickness of the mirror generally depends on the wavelength as well in order to achieve sufficiently low transmission.

Equation (3.34) allows one to compare the viability of different wavelength ranges. While the highest values for N are achievable for microwaves (up to 10^{10}), this is counteracted by their long wavelengths (of order cm). On the other hand there are no good mirrors for highly energetic photons. The optical regime seems optimal. In the following estimates we will consider a λ around 630 nm.

The cavity mode needs to have a very narrow focus on the tiny mirror, which requires the other cavity end mirror to be large due to beam divergence. The maximum cavity length is therefore limited by the difficulty of making large high quality curved mirrors. In fact from simulations it follows that indeed the surface quality of the large curved mirror is likely to be the most challenging component of the setup [48, 49]. Here we consider a cavity length of 5 cm, and a small mirror size of $10 \times 10 \times 10$ microns, leading to a mass of order 5×10^{-12} kg.

One possible path for the fabrication of such a small mirror on a good mechanical oscillator is to coat a silicon cantilever with alternating layers of SiO_2 and a metal oxide such as Ta_2O_5 by sputtering deposition. The best current mirrors are made in this way. Recently such mirrors have also been produce on Silicon Nitride cross resonators which have excellent mechanical properties [50].

For the above dimensions the condition (3.34) is satisfied for $N = 5.6 \times 10^6$. Therefore a photon loss per reflection not larger than 3×10^{-7} is needed, which is about a factor of 4 below the best reported values for such mirrors [51], and for a transmission of 10^{-7}, which is consistent with the quoted mirror thickness [52]. For these values, about 1 % of the photons are still left in the cavity after a full period of the mirror. Coupling into a high-finesse cavity with a tight focus will require carefully designed incoupling optics. For the above values of N and L one obtains a frequency $\Omega_{\mathrm{M}} = 2\pi \times 500$ Hz. This leads to a quantum uncertainty of order 10^{-13} m, which for $\eta^2 \sim 1$ corresponds to the displacement in the superposition.

Secondly, the requirement of observing the revival after a full period of the mirror's motion puts a bound on the acceptable environmental decoherence. To estimate the expected decoherence we model the mirror's environment by an (Ohmic) bath of harmonic oscillators. Applying the analysis of [53] one then finds that off-diagonal elements between different mirror positions decay with a factor

$$\exp\left[-\frac{\Gamma_{\mathrm{M}} k T m_{\mathrm{eff}} (\Delta x)^2}{\hbar^2}\left(t + \frac{\sin \Omega_{\mathrm{M}} t \cos \Omega_{\mathrm{M}} t}{\Omega_{\mathrm{M}}}\right)\right], \tag{3.35}$$

where Γ_{M} is the rate of energy dissipation for the mechanical oscillator, T is the temperature (which is constituted mainly by the internal degrees of freedom of the mirror cantilever) and Δx is the separation of two coherent states that are originally in a superposition. Note that our experiment is not in the long-time regime where decoherence is characterized simply by a rate. However, the oscillatory term in the exponent of Eq. (3.35) does not affect the order of magnitude and happens to be zero after a full period. Assuming that the experiment achieves $\eta^2 \gtrsim 1$, i.e. a separation by the size of a coherent state wavepacket, $\Delta x = \sqrt{\frac{\hbar}{m_{\mathrm{eff}} \Omega_{\mathrm{M}}}}$, the condition that the exponent in Eq. (3.35) should be at most of order 1 after a full period can be cast in the form

$$Q \gtrsim \frac{kT}{\hbar \Omega_{\mathrm{M}}} = \bar{n}_{\mathrm{th}}, \tag{3.36}$$

where $Q = \Omega_{\mathrm{M}}/\Gamma_{\mathrm{M}}$ is the quality factor of the mechanical oscillator. Bearing in mind that quality factors of the order of 10^5-10^6 have been achieved for silicon cantilevers of approximately the right dimensions and frequency, this implies that the temperature has to be approximately 3–30 mK. It will be beneficial to perform experiments at even lower temperatures to reduce the measurement time, as we will explain below.

Thirdly, the stability requirements for the experiment are very strict. The phase of the interferometer has to be stable over the whole measurement time. This means that in particular the distance between the large cavity end mirror and the equilibrium position of the small mirror has to be stable to order $\lambda/20N = 0.6 \times 10^{-14}$ m.

The required measurement time can be determined in the following way. A single run of the experiment starts by sending a weak pulse into the interferometer, such that on average 0.1 photons go into either cavity. This probabilistically prepares a single-photon state as required to a good approximation. The two-photon contribution has to be kept low because it causes noise in the interferometer. From Eq. (3.33) the width of the revival peak is $2/\eta\Omega_{\mathrm{M}}\sqrt{\bar{n}_{\mathrm{th}}}$. This implies that only a fraction $\sim 1/\pi\sqrt{\bar{n}_{\mathrm{th}}}$ of the remaining photons will leak out in the time interval of the revival. It is therefore important to work at the lowest possible temperature. Temperatures below 100 μK can be achieved with a nuclear demagnetization cryostat.

Together with the required low value of Ω_{M}, the fact that approximately 1% of the photons remain after a full period for our assumed loss, and an assumed detection efficiency of 70 %, this implies a detection rate of approximately 100 photons per hour in the revival interval. This means that a measurement time of order 30 minutes should give convincing statistics.

After every single run of the experiment the mirror has to be damped to reset it to its initial (thermal) state. This could be done by electric or magnetic fields, e.g. following Ref. [54], where a Nickel sphere was attached to the cantilever, whose Q could then be changed by 3 orders of magnitude by applying a magnetic field.

Since the width of the revival peak scales like $1/\sqrt{T}$, the required measurement time can also be decreased by decreasing the temperature below 60 μK. Passive cooling techniques may be improved. In addition, active and passive optical cooling of mirror oscillators has been proposed [55], and implemented experimentally for a large mirror [56] and for small mirrors [57–60]. Ground state cooling of the center of mass motion is achievable and reduces the required measurement time, and thus the stability requirements, by a factor of approximately 50.

Since publication of the pioneering protocol [43] presented in this paragraph a number of other theoretical works considered the generation of non-Gaussian states of mechanical oscillators via interaction with single photons, see [61–64]. Another way of achieving non-Gaussian states is to make use of the intrinsic nonlinearity of the radiation pressure force. The single-photon strong coupling regime, where $g_0 > \Omega_{\mathrm{M}}$, has been studied first in [65, 66]. When the cavity is driven with blue detuned light the *classical* nonlinear dynamics gives rise to limit cycles of the mechanical oscillator.

It was shown in [67] that the *stationary* quantum state associated with such a limit cycle can exhibit strongly sub-Poissonian phonon statistics and even a negative Wigner function. This regime was further studied in [68, 69].

3.4 Entangled States of Mechanics and Light

In this last section we turn our attention to non-classical states involving both, light and mechanics. In our discussion these will be primarily entangled states, which are generated either in steady state under a continuous drive field, or in a regime involving short pulses of light.

3.4.1 Light Mirror Entanglement in Steady State

Entanglement of a mechanical oscillator with light has been predicted in a number of theoretical studies [8, 70–78] and would be an intriguing demonstration of optomechanics in the quantum regime. These studies, as well as similar ones investigating entanglement among several mechanical oscillators [79–87], explore entanglement in the *steady-state regime*. In this regime the optomechanical system is driven by one or more continuous-wave light fields and settles into a stationary state, for which the interplay of optomechanical coupling, cavity decay, damping of the mechanical oscillator, and thermal noise forces may remarkably give rise to persistent entanglement between the intracavity field and the mechanical oscillator.

The simplest example of such a scheme involves an optomechanical system driven by one continuous wave laser field. To identify conditions for good optomechanical generation of entanglement we answer a first question that concerns the optimal detuning of the driving laser with respect to the cavity field. Given the form of the linearized radiation pressure Hamiltonian $\hbar g(\hat{a}+\hat{a}^\dagger)(\hat{b}+\hat{b}^\dagger)$, and the time evolution of operators with frequencies Δ and Ω_M, we focus on resonant processes where $\Delta=\pm\Omega_\mathrm{M}$. The first case we analyze is blue-detuning $\Delta=\Omega_\mathrm{M}$ and in which we split the interaction in two kind of interactions well known in quantum optics: (1) beam splitter interaction $\hbar g(\hat{a}^\dagger\hat{b}+h.c.)$ and (2) down-conversion interaction, $\hbar g(\hat{a}\hat{b}+h.c.)$. Since the beam splitter term is off-resonant by $2\Omega_\mathrm{M}$ and also cannot produce entanglement starting from classical states we drop it and focus on the down-conversion term, known to produce bipartite entanglement. Following a standard treatment to obtain the covariance matrix in steady state (even analytically for this particular case), it can be shown that the logarithmic negativity scales up with g as

$$E_{\mathcal{N}}\le\ln\left[\frac{1+g/\sqrt{2\kappa\Gamma_\mathrm{M}}}{1+\bar{n}_\mathrm{th}}\right]. \tag{3.37}$$

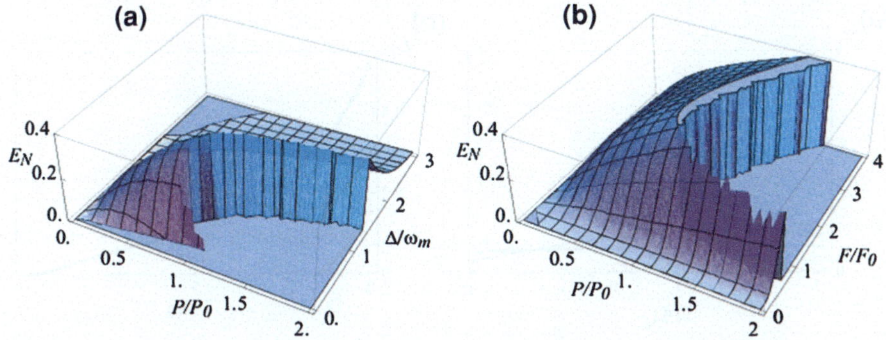

Fig. 3.7 **a** Logarithmic negativity $E_{\mathcal{N}}$ versus the normalized detuning $\Delta/\Omega_{\mathrm{M}}$ and normalized input power $P_{\mathrm{in}}/P_{\mathrm{in}}^0$, ($P_{\mathrm{in}}^0 = 50$ mW) at a fixed value of the cavity finesse $\mathscr{F} = \mathscr{F}_0 = 1.67 \times 10^4$; **b** $E_{\mathcal{N}}$ versus the normalized finesse $\mathscr{F}/\mathscr{F}_0$ and normalized input power $P_{\mathrm{in}}/P_{\mathrm{in}}^0$ at a fixed detuning $\Delta = \Omega_{\mathrm{M}}$. Parameter values are $\Omega_{\mathrm{M}}/2\pi = 10$ MHz, $\mathscr{Q} = 10^5$, mass $m_{\mathrm{eff}} = 10$ ng, a cavity of length $L = 1$ mm driven by a laser with wavelength 810 nm, yielding $g = 0.95$ kHz and a cavity bandwidth $\kappa = 0.9\Omega_{\mathrm{M}}$ when $\mathscr{F} = \mathscr{F}_0$. We have assumed a reservoir temperature for the mirror $T = 0.4$ K, corresponding to $\bar{n}_{\mathrm{th}} \simeq 833$. The sudden drop to zero of $E_{\mathcal{N}}$ corresponds to entering the instability region

However, the system is unstable in the "blue-detuned" regime owing to the fast transfer of energy from the cavity field to the mirror and an unavoidable bound is found $g < \sqrt{2\kappa\Gamma_{\mathrm{M}}}$ which in consequence limits $E_{\mathcal{N}} \leq \ln 2$. Moreover, thermal quanta $\bar{n}_{\mathrm{th}} > 0$ will eventually destroy the entanglement. One therefore concludes that the choice of the practical operation regime is dictated by the stability of the system. Thus, we move into the "red detuned" regime which allows for larger g by paying the price that, for example at $\Delta = -\Omega_{\mathrm{M}}$ the down-conversion process is $2\Omega_{\mathrm{M}}$ off-resonant. In this regime analytical results are possible but cumbersome and we settled for numerically showing the behavior of $E_{\mathcal{N}}$ in Fig. 3.7a as it scales with increasing input power and varying detunings, and in Fig. 3.7b with input power and cavity finesse.

Having concluded that intracavity optomechanical entanglement is attainable, the question of detection is to be answered next. As detailed in [72] a simple scheme can be conceived that consists of a second cavity adjacent to the main one; the second cavity output, when weakly driven, does not modify much the first cavity dynamics and its output light gives a direct measurement of the mirror dynamics. With homodyne detection of both cavities and manipulation of the two local oscillators phases one can determine all of the entries of the covariance matrix and numerically extract the logarithmic negativity.

As previously mentioned, in the "red-detuning" regime, the down-conversion process is off-resonant and its effect much washed out by the presence of the stronger beam-splitter interaction. However, proper detection around the Stokes sideband (which carries the photons entangled with the mirror) in the sense described in Sect. 3.2.2 can extract optimal light-mirror entanglement [8]. We show this by

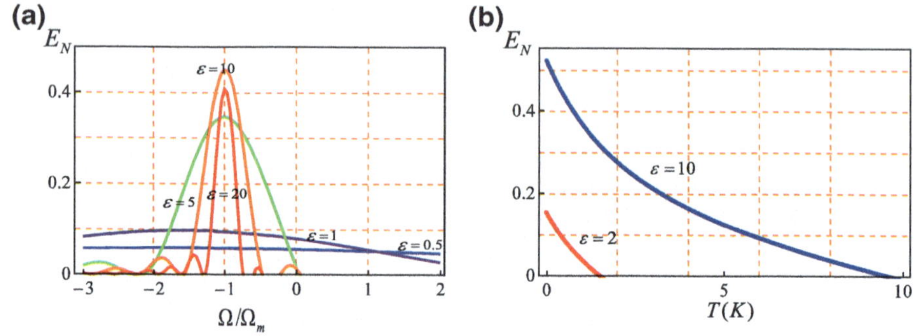

Fig. 3.8 **a** Logarithmic negativity $E_{\mathcal{N}}$ versus $\Omega/\Omega_{\mathrm{M}}$ for different values of ε. Optimal entanglement of the Stokes sideband with the mirror can be deduced. **b** $E_{\mathcal{N}}$ of output stokes mode with mirror versus T for two different values of its inverse bandwidth $\varepsilon = 2, 10$

choosing a central frequency of the detected mode Ω and its bandwidth τ^{-1} and computing the bipartite system negativity. The results are shown in Fig. 3.8a, where $E_{\mathcal{N}}$ is plotted versus $\Omega/\Omega_{\mathrm{M}}$ at five different values of $\varepsilon = \tau\Omega_{\mathrm{M}}$ and the other parameters similar to the ones used for Fig. 3.7. If $\varepsilon < 1$, i.e., the bandwidth of the detected mode is larger than Ω_{M}, the detector does not resolve the motional sidebands, and $E_{\mathcal{N}}$ has a value (roughly equal to that of the intracavity case) which does not essentially depend upon the central frequency. For smaller bandwidths (larger ε), the sidebands are resolved by the detection and the role of the central frequency becomes important. In particular $E_{\mathcal{N}}$ becomes highly peaked around the Stokes sideband $\Omega = -\Omega_{\mathrm{M}}$, showing that the optomechanical entanglement generated within the cavity is mostly carried by this lower frequency sideband. What is relevant is that the optomechanical entanglement of the output mode is significantly larger than its intracavity counterpart and achieves its maximum value at the optimal value $\varepsilon \simeq 10$, i.e., a detection bandwidth $\tau^{-1} \simeq \Omega_{\mathrm{M}}/10$. This means that in practice, by appropriately filtering the output light, one realizes an effective entanglement distillation because the selected output mode is more entangled with the mechanical resonator than the intracavity field.

It is finally important to see what the robustness of the entanglement is with increasing temperature of the thermal reservoir. This is shown by Fig. 3.8b, where the entanglement $E_{\mathcal{N}}$ of the output mode centered at the Stokes sideband is plotted versus the temperature of the reservoir at two different values of the bandwidth, the optimal one $\varepsilon = 10$, and at a larger bandwidth $\varepsilon = 2$. We see the expected decay of $E_{\mathcal{N}}$ for increasing temperature, but above all that also this output optomechanical entanglement is robust against temperature because it persists even above liquid He temperatures, at least in the case of the optimal detection bandwidth $\varepsilon = 10$.

We note that optomechanical entanglement can be enhanced by modulating the driving field as was shown in theoretical studies [15, 88, 89].

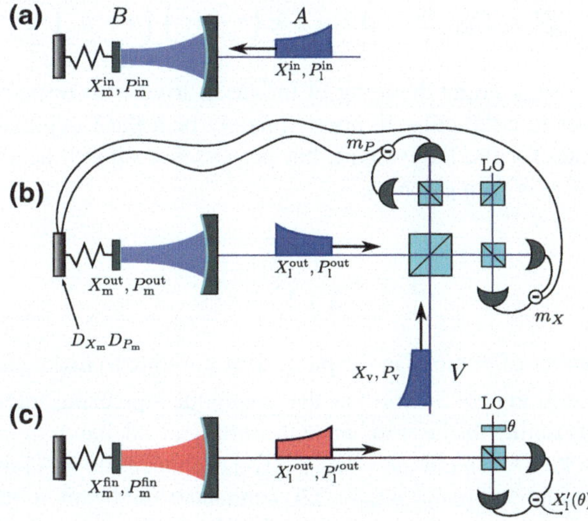

Fig. 3.9 Schematic of the system and the proposed teleportation protocol: **a** A blue detuned light pulse (A) is entangled with the mirror (B). **b** A second light pulse (V) is prepared in the input state and interferes with A on a beam-splitter. Two homodyne detectors measure $\hat{P}_{\mathrm{l}}^{\mathrm{out}} + \hat{X}_{\mathrm{v}}$ and $\hat{X}_{\mathrm{l}}^{\mathrm{out}} + \hat{P}_{\mathrm{v}}$, yielding outcomes m_X and m_P respectively. Feedback is applied by displacing the mirror state in phase space by a unitary transformation $D_{X_{\mathrm{m}}}(m_X)\, D_{P_{\mathrm{m}}}(m_P)$. **c** To verify the success of the protocol, the mirror state is coherently transferred to a red detuned laser pulse and a generalised quadrature $\hat{X}_{\mathrm{l}}(\theta) = \hat{X}_{\mathrm{l}}^{\mathrm{out}} \cos\theta + \hat{P}_{\mathrm{l}}^{\mathrm{out}} \sin\theta$ is measured. Repeating steps **(a)**–**(c)** for the same input state but for different phases θ yields a reconstruction of the mirror's quantum state

3.4.2 Entanglement with Pulsed Light

An alternative approach to achieving optomechanical entanglement works in a *pulsed regime*, as was theoretically explored in [90–93]. Here we will present a summary of the protocol of [93] which was realized experimentally in a microwave optomechanical system [94]. A pulsed scheme does not rely on the existence of a stable steady state, such that entanglement is not limited by stability requirements. In fact it is possible to operate in a parameter regime where a stationary state does not exist. This sort of optomechanical entanglement can be verified by using a pump–probe sequence of light pulses. The quantum state created in this protocol exhibits Einstein–Podolsky–Rosen (EPR) type entanglement [9] between the mechanical oscillator and the light pulse. It thus provides the canonical resource for quantum information protocols involving continuous variable (CV) systems [95]. Optomechanical EPR entanglement can therefore be used for the teleportation of the state of a propagating light pulse onto a mechanical oscillator as suggested in [90, 91].

Let us consider an optomechanical cavity in a Fabry–Perot type setup, *cf.* Fig. 3.9). A light pulse of duration τ and carrier frequency ω_{l} impinges on the cavity and interacts with the oscillatory mirror mode via radiation pressure. In a frame rotating with the laser frequency, the system is described by the (effective) Hamiltonian [1]

$$H = \Omega_M \hat{b}^\dagger \hat{b} - \Delta \hat{a}^\dagger \hat{a} + g\left(\hat{b} + \hat{b}^\dagger\right)\left(\hat{a} + \hat{a}^\dagger\right) \tag{3.38}$$

where $\Delta = \omega_l - \omega_{opt}$ is the detuning of the laser drive with respect to the cavity resonance. We assume the pulse to approximately be a flat-top pulse, which has a constant amplitude for the largest part, but possesses a smooth head and tail. The coupling constant g is then given by

$$g = g_0 \sqrt{\frac{\kappa N_{ph}/\tau}{(\Delta^2 + \kappa^2)}}, \tag{3.39}$$

with N_{ph} the number of photons in the pulse. It is possible to make either the beam-splitter like interaction $(\hat{a}\hat{b}^\dagger + \hat{b}\hat{a}^\dagger)$ or the two-mode-squeezing interaction $(\hat{a}\hat{b} + \hat{a}^\dagger\hat{b}^\dagger)$ resonant by tuning the laser to one of the motional sidebands $\omega_{opt} \pm \Omega_M$, where the blue (anti-Stokes) sideband $(\omega_l = \omega_{opt} + \Omega_M)$ enhances down-conversion, while the red (Stokes) sideband $(\omega_l = \omega_{opt} - \Omega_M)$ enhances the beam-splitter interaction [96]. In the proposed protocol we make use of both dynamics separately: Pulses tuned to the blue sideband are applied to create entanglement, while pulses on the red sideband are later used to read out the final mirror state. A similar separation of Stokes and anti-Stokes sideband was suggested in [90, 91] by selecting different angles of reflection of a light pulse scattered from a vibrating mirror in free space.

The full system dynamics, including the dissipative coupling of the mirror and the cavity decay, are described by quantum Langevin equations [97], which determine the time evolution of the corresponding operators $\hat{x}_m = (\hat{b}+\hat{b}^\dagger)/\sqrt{2}$, $\hat{p}_m = -i(\hat{b} - \hat{b}^\dagger)/\sqrt{2}$ and $\hat{a}$, $\hat{a}^\dagger$. They read

$$\dot{\hat{x}}_m = \Omega_M \hat{p}_m, \tag{3.40}$$

$$\dot{\hat{p}}_m = -\Omega_M \hat{x}_m - \Gamma_M \hat{p}_m - \sqrt{2} g \left(\hat{a} + \hat{a}^\dagger\right) - \sqrt{2\Gamma_M}\, \hat{f}, \tag{3.41}$$

$$\dot{\hat{a}} = -\left(i\Delta + \frac{\kappa}{2}\right)\hat{a} - i\sqrt{2}\, g\, x_m - \sqrt{\kappa}\, \hat{a}_{in}, \tag{3.42}$$

where we introduced the (self-adjoint) Brownian stochastic force f, and quantum noise $\hat{a}_{in}$ entering the cavity from the electromagnetic environment. Both $\hat{a}_{in}$ and—in the high-temperature limit—f are assumed to be Markovian. Their correlation functions are thus given by $\langle \hat{a}_{in}(t)\hat{a}_{in}^\dagger(t')\rangle = \delta(t-t')$ (in the optical vacuum state) and $\langle \hat{f}(t)\hat{f}(t') + \hat{f}(t')\hat{f}(t)\rangle = (2\bar{n}_{th}+1)\delta(t-t')$ (in a thermal state of the mechanics) [97].

We impose the following conditions on the system's parameters. Firstly, we drive the cavity on the blue sideband $(\Delta = \Omega_M)$ and assume to work in the resolved-sideband regime $(\kappa \ll \Omega_M)$ to enhance the down-conversion dynamics . Note that in this regime a stable steady state only exists for very weak optomechanical coupling [98], which poses a fundamental limit to the amount of entanglement that can be created in a continuous-wave scheme [8], as explained also in the previous section. In contrast, a pulsed scheme does not suffer from these instability issues. In fact, it is easy to check by integrating the full dynamics up to time τ, that working in this particular regime yields maximal entanglement, which increases with increasing sideband-resolution Ω_M/κ. Secondly we assume a weak optomechanical coupling $g \ll \kappa$, such that only first-order interactions of photons with the mechanics contribute. This minimises pulse distortion and simplifies the experimental realization of the protocol. Taken together, the conditions $g \ll \kappa \ll \Omega_M$ allow us to invoke the rotating-wave-approximation (RWA), which amounts to neglecting the beam-splitter term in 3.38. Also, we neglect mechanical decoherence effects in this section. We emphasise that this approximation is justified as long as the total duration of the protocol is short compared to the effective mechanical decoherence time $1/\Gamma_M \bar{n}_{th}$, where Γ_M is the mechanical damping rate and $\bar{n}_{th}$ the thermal occupation of the corresponding bath. Corrections to this simplified model will be addressed below.

Based on the assumptions above we can now simplify Eq. 3.40. For convenience we go into a frame rotating with Ω_M by substituting $\hat{a} \to \hat{a}e^{i\Omega_M t}$, $\hat{a}_{in} \to \hat{a}_{in}e^{i\Omega_M t}$ and $\hat{b} \to \hat{b}e^{-i\Omega_M t}$. Note that in this picture the central frequency of $\hat{a}_{in}$ is located at $\omega_l - \Omega_M = \omega_{opt}$. In the RWA the Langevin equations then simplify to

$$\dot{\hat{a}} = -\frac{\kappa}{2}\hat{a} - ig\,\hat{b}^\dagger - \sqrt{\kappa}\,\hat{a}_{in}, \qquad \dot{\hat{b}} = -ig\,\hat{a}^\dagger. \tag{3.43}$$

In the limit $g \ll \kappa$ we can use an adiabatic solution for the cavity mode and we therefore find

$$\hat{a}(t) \approx -i\frac{2g}{\kappa}\hat{b}^\dagger(t) - \frac{1}{\sqrt{\kappa}}\hat{a}_{in}(t), \tag{3.44}$$

$$\hat{b}(t) \approx e^{Gt}\hat{b}(0) + i\sqrt{2G}e^{Gt}\int_t^s ds e^{-Gs}\hat{a}_{in}^\dagger(s), \tag{3.45}$$

where we defined $G = 2g^2/\kappa$. Equation 3.45 shows that the mirror motion gets correlated to a light mode of central frequency $\omega_l - \Omega_M$ (which coincides with the cavity resonance frequency ω_{opt}) with an exponentially shaped envelope $\alpha_{in}(t) \propto e^{-Gt}$. Using the standard cavity input-output relations $\hat{a}_{out} = \hat{a}_{in} + \sqrt{\kappa}\,\hat{a}$ allows us to define a set of normalised temporal light-modes

$$\hat{A}_{in} = \sqrt{\frac{2G}{1-e^{-2G\tau}}}\int_\tau^t dt e^{-Gt}\hat{a}_{in}(t), \quad \hat{A}_{out} = \sqrt{\frac{2G}{e^{2G\tau}-1}}\int_\tau^t dt e^{Gt}\hat{a}_{out}(t), \tag{3.46}$$

which obey the canonical commutation relations $[\hat{A}_i, \hat{A}_i^\dagger] = 1$. Together with the definitions $\hat{B}_{in} = \hat{b}(0)$ and $\hat{B}_{out} = \hat{b}(\tau)$ we arrive at the following expressions, which relate the mechanical and optical mode at the end of the pulse $t = \tau$

$$\hat{A}_{\text{out}} = -e^{G\tau}\hat{A}_{\text{in}} - i\sqrt{e^{2G\tau}-1}\hat{B}^{\dagger}_{\text{in}}, \quad \hat{B}_{\text{out}} = e^{G\tau}\hat{B}_{\text{in}} + i\sqrt{e^{2G\tau}-1}\hat{A}^{\dagger}_{\text{in}}. \quad (3.47)$$

By expressing Eq. 3.47 in terms of quadratures $\hat{X}^i_{\text{m}} = (\hat{B}_i + \hat{B}^{\dagger}_i)/\sqrt{2}$ and $\hat{X}^i_{\text{l}} = (\hat{A}_i + \hat{A}^{\dagger}_i)/\sqrt{2}$, where $i \in \{\text{in}, \text{out}\}$, and their corresponding conjugate variables, we can calculate the so-called EPR-variance Δ_{EPR} of the state after the interaction. For light initially in vacuum $(\Delta\hat{X}^{\text{in}}_{\text{l}})^2 = (\Delta\hat{P}^{\text{in}}_{\text{l}})^2 = \frac{1}{2}$ and the mirror in a thermal state $(\Delta\hat{X}^{\text{in}}_{\text{m}})^2 = (\Delta\hat{P}^{\text{in}}_{\text{m}})^2 = n_0 + \frac{1}{2}$, the state is entangled iff [99]

$$\Delta_{\text{EPR}} = \left[\Delta(\hat{X}^{\text{out}}_{\text{m}} + \hat{P}^{\text{out}}_{\text{l}})\right]^2 + \left[\Delta(\hat{P}^{\text{out}}_{\text{m}} + \hat{X}^{\text{out}}_{\text{l}})\right]^2 \quad (3.48)$$

$$= 2(n_0 + 1)\left(e^r - \sqrt{e^{2r} - 1}\right)^2 < 2, \quad (3.49)$$

where $r = G\tau$ is the squeezing parameter and n_0 the initial occupation number of the mechanical oscillator. Note that in the limit of large squeezing $r \gg 1$ we find that the variance $\Delta_{\text{EPR}} \approx (n_0 + 1)e^{-2r}/2$ is suppressed exponentially, which shows that the created state asymptotically approximates an EPR-state. Therefore, this state can be readily used to conduct optomechanical teleportation. Rearranging 3.48, we find that the state is entangled as long as

$$r > r_0 = \frac{1}{2}\ln\left(\frac{(n_0 + 2)^2}{4(n_0 + 1)}\right) \sim \frac{1}{2}\ln n_0, \quad (3.50)$$

where the last step holds for $n_0 \to \infty$. This illustrates that in our scheme the requirement on the strength of the effective optomechanical interaction, as quantified by the parameter $r = \frac{g^2\tau}{\kappa}$, scales logarithmically with the initial occupation number n_0 of the mechanical oscillator. This tremendously eases the protocol's experimental realization, as neither g nor τ can be arbitrarily increased. Note that n_0 need not be equal to the mean bath occupation $\bar{n}_{\text{th}}$, but may be decreased by laser pre-cooling to improve the protocol's performance.

To verify the successful creation of entanglement a red detuned laser pulse ($\Delta = -\Omega_{\text{M}}$) is sent to the cavity where it resonantly drives the beam-splitter interaction, and hence generates a state-swap between the mechanical and the optical mode. It is straightforward to show that choosing $\Delta = -\Omega_{\text{M}}$ leads to a different set of Langevin equations which can be obtained from 3.43 by dropping the Hermitian conjugation (†) on the right-hand-side. By defining modified mode functions $\alpha'_{\text{in(out)}} = \alpha_{\text{out(in)}}$ and corresponding light modes $\hat{A}'_{\text{in(out)}}$ one obtains input/output expressions in analogy to (3.47)

$$\hat{A}'_{\text{out}} = -e^{-G\tau}\hat{A}'_{\text{in}} + i\sqrt{1 - e^{-2G\tau}}\hat{B}_{\text{in}}, \quad \hat{B}_{\text{out}} = e^{-G\tau}\hat{B}_{\text{in}} - i\sqrt{1 - e^{-2G\tau}}\hat{A}'_{\text{in}}. \quad (3.51)$$

The pulsed state-swapping operation therefore also features an exponential scaling with $G\tau$. For $G\tau \to \infty$ the expressions above reduce to $\hat{A}'_{\text{out}} = -i\hat{B}_{\text{in}}$ and

$\hat{B}_{\text{out}} = i\hat{A}'_{\text{in}}$, which shows that in this case the mechanical state—apart from a phase shift—is perfectly transferred to the optical mode. In the Schrödinger-picture this amounts to the transformation $|\varphi\rangle_{\text{m}}|\psi\rangle_{\text{l}} \rightarrow |\psi\rangle_{\text{m}}|\varphi\rangle_{\text{l}}$, where φ and ψ constitute the initial state of the mechanics and the light pulse respectively. The state-swap operation thus allows us to access mechanical quadratures by measuring quadratures of the light and therefore to reconstruct the state of the bipartite system via optical homodyne tomography. Such a quantum state transfer was experimentally realized in a microwave optomechanical system as reported in [100].

As we have shown above, pulsed operation allows us to create EPR-type entanglement, which forms the central entanglement resource of many quantum information processing protocols [95]. An immediate extension of the proposed scheme is an optomechanical continuous variables quantum teleportation protocol. The main idea of quantum state teleportation in this context is to transfer an arbitrary quantum state $|\psi_{\text{in}}\rangle$ of a travelling wave light pulse onto the mechanical resonator, without any direct interaction between the two systems, but by making use of optomechanical entanglement. The scheme works in full analogy to the CV teleportation protocol for photons [101, 102] and, due to its pulsed nature, closely resembles the scheme used in atomic ensembles [103, 104]: A light pulse (A) is sent to the optomechanical cavity and is entangled with its mechanical mode (B) via the dynamics described above. Meanwhile a second pulse (V) is prepared in the state $|\psi_{\text{in}}\rangle$, which is to be teleported. This pulse then interferes with A on a beam-splitter. In the output ports of the beam-splitter, two homodyne detectors measure two joint quadratures $\hat{P}_{\text{l}}^{\text{out}} + \hat{X}_{\text{v}}$ and $\hat{X}_{\text{l}}^{\text{out}} + \hat{P}_{\text{v}}$, yielding outcomes m_X and m_P respectively. This constitutes the analogue to the Bell-measurement in the case of qubit teleportation and effectively projects previously unrelated systems A and V onto an EPR-state [105]. Note that both the second pulse and the local oscillator for the homodyne measurements must be mode-matched to A after the interaction, i.e., they must possess the identical carrier frequency as well as the same exponential envelope. The protocol is concluded by displacing the mirror in position and momentum by m_{X} and m_{P} according to the outcome of the Bell-measurement. This can be achieved by means of short light-pulses, applying the methods described in [20, 106]. After the feedback the mirror is then described by [95]

$$\hat{X}_{\text{m}}^{\text{fin}} = \hat{X}_{\text{m}}^{\text{out}} + \hat{P}_{\text{l}}^{\text{out}} + \hat{X}_{\text{v}} = \hat{X}_{\text{v}} + \left(e^{r} - \sqrt{e^{2r} - 1}\right)(\hat{X}_{\text{m}}^{\text{in}} - \hat{P}_{\text{l}}^{\text{in}}), \quad (3.52)$$

$$\hat{P}_{\text{m}}^{\text{fin}} = \hat{P}_{\text{m}}^{\text{out}} + \hat{X}_{\text{l}}^{\text{out}} + \hat{P}_{\text{v}} = \hat{P}_{\text{v}} + \left(e^{r} - \sqrt{e^{2r} - 1}\right)(\hat{P}_{\text{m}}^{\text{in}} - \hat{X}_{\text{l}}^{\text{in}}), \quad (3.53)$$

which shows that its final state corresponds to the input state plus quantum noise contributions. It is obvious from these expressions that the total noise added to both quadratures (second term in 3.52 and 3.53 respectively) is equal to the EPR-variance. Again, for large squeezing $r \gg 1$ the noise terms are exponentially suppressed and in the limit $r \rightarrow \infty$, where the resource state approaches the EPR-state, we obtain perfect teleportation fidelity, i.e., $\hat{X}_{\text{m}}^{\text{fin}} = \hat{X}_{\text{v}}$ and $\hat{P}_{\text{m}}^{\text{fin}} = \hat{P}_{\text{v}}$. In particular this operator identity means, that *all* moments of $\hat{X}_{\text{v}}$, $\hat{P}_{\text{v}}$ with respect to the input state $|\psi_{\text{in}}\rangle$ will be transferred to the mechanical oscillator, and hence its final state will be identically given by $|\psi_{\text{in}}\rangle$. We notice that the present optomechanical entanglement in the pulsed regime may be seen as a complementary approach to that of paragraph Sect. 3.4.1, where we have discussed entanglement in a stationary regime.

We found that in the ideal scenario the amount of entanglement essentially depends only on the coupling strength (or equivalently on the input laser power) and the duration of the laser pulse and that it shows an encouraging scaling, growing exponentially with $G\tau$. This in turn means that the minimal amount of squeezing needed to generate entanglement only grows logarithmically with the initial mechanical occupation n_0. In a more realistic scenario one has to include thermal noise effects and effects of counter-rotating terms. Including the above-mentioned perturbations results in a final state which deviates from an EPR-entangled state. To minimise the extent of these deviations, the system parameters must obey the following conditions:

1. $\kappa \ll \Omega_{\text{M}}$ results in a sharply peaked cavity response and implies that the down-conversion dynamics is heavily enhanced with respect to the suppressed beam-splitter interaction.
2. $g < \kappa$ inhibits multiple interactions of a single photon with the mechanical mode before it leaves the cavity. This suppresses spurious correlations to the intracavity field. It also minimises pulse distortion and simplifies the protocol with regard to mode matching and detection.
3. $g\tau \gg 1$ is needed in order to create sufficiently strong entanglement. This is due to the fact that the squeezing parameter $r = (g/\kappa)g\tau$ should be large, while g/κ needs to be small.
4. $\bar{n}_{\text{th}}\Gamma_{\text{M}}\tau \ll 1$, where $\bar{n}_{\text{th}}$ is the thermal occupation of the mechanical bath, assures coherent dynamics over the full duration of the protocol, which is an essential requirement for observing quantum effects. As the thermal occupation of the mechanical bath may be considerably large even at cryogenic temperatures, this poses (for fixed Γ_{M} and $\bar{n}_{\text{th}}$) a very strict upper limit to the pulse duration τ.

Note however that not all of these inequalities have to be fulfilled equally strictly, but there rather exists an optimum which arises from balancing all contributions. It turns out that fulfilling (4) is critical for successful teleportation, whereas (1)–(3) only need to be weakly satisfied. Taking the above considerations into account, we find a sequence of parameter inequalities

$$\bar{n}_{\text{th}}\Gamma_{\text{M}} \ll \frac{1}{\tau} \ll g \ll \kappa \ll \Omega_{\text{M}}, \tag{3.54}$$

which defines the optimal parameter regime. Dividing this equation by Γ_{M} and taking a look at the outermost condition $\bar{n}_{\text{th}} \ll Q$, where $Q = \Omega_{\text{M}}/\Gamma_{\text{M}}$ is the mechanical quality factor, we see that the ratio $Q/\bar{n}_{\text{th}}$ defines the range which all the other parameters have to fit into. It is intuitively clear, that a high quality factor and a low bath occupation number, and consequently low effective mechanical decoherence, are favourable for the success of the protocol. Equivalently, we can rewrite the occupation number as $\bar{n}_{\text{th}} = k_{\text{B}}T_{\text{bath}}/\hbar\Omega_{\text{M}}$ and therefore find $k_{\text{B}}T_{\text{bath}}/\hbar \ll Q \cdot \Omega_{\text{M}}$, where now the $Q \cdot f$-product ($f = \Omega_M/2\pi$) has to be compared to the thermal frequency of the bath. Let us consider a numerical example: For a temperature $T_{\text{bath}} \approx 100\,\text{mK}$ the left-hand-side gives $k_B T_{\text{bath}}/\hbar \approx 2\pi \cdot 10^9\,\text{Hz}$. The $Q \cdot f$-product consequently has to be several orders of magnitude larger to successfully create entanglement. As current optomechanical systems feature a $Q \cdot f$-product of $2\pi \cdot 10^{11}$ Hz and above [107–110], this requirement seems feasible to meet.

During writing of this book chapter the protocol discussed above has been realized in the group of K. Lehnert (JILA) in a microwave optomechanical system [94]. Further theoretical studies based on stochastic master equations showed that the pulsed scheme can actually be extended to the continuous-time domain requiring feedback stabilization of the dynamics [111].

3.5 Conclusion and Outlook

The selection of protocols presented in the present chapter clearly demonstrate the feasibility—and the stringent requirements—for quantum state engineering in mesoscopic mechanical systems. During writing this book chapter some of the quantum effects discussed in this book were observed in experiments. This includes in particular cooling to the quantum ground state [112, 113], ponderomotive squeezing of light [5–7], back action noise limited position sensing [114, 115], coherent state transfer [100], and entanglement [94]. Other quantum effects such as quantum jumps, and squeezed or non-Gaussian states of mechanical oscillators have yet to be demonstrated.

From the discussion given above it should be clear that a necessary condition for observing quantum effects is a sufficiently large product of mechanical quality factor and frequency, $Q\Omega_M > k_B T/\hbar$. What exactly "large" means in this context depends crucially on the protocol to be implemented. The goal of making the quantum regime accessible for mechanical systems thus has to be approached from both sides: Experimentally, by developing optomechanical systems with a sufficiently large $Q\Omega_M$-product; and theoretically, by developing schemes which are not too demanding regarding the magnitude of this number. Once those two ends meet this will mark the birth of a new field of research, quantum optomechanics.

References

1. S. Mancini, P. Tombesi, Phys. Rev. A **49**(5), 4055 (1994)
2. C. Fabre, M. Pinard, S. Bourzeix, A. Heidmann, E. Giacobino, S. Reynaud, Phys. Rev. A **49**(2), 1337 (1994)
3. D.F. Walls, G.J. Milburn, *Quantum Optics* (Springer, Berlin, 1995)
4. V.B. Braginsky, F.Y.A. Khalili, *Quantum Measurement* (Cambridge University Press, Cambridge, 1995)

5. D.W.C. Brooks, T. Botter, S. Schreppler, T.P. Purdy, N. Brahms, D.M. Stamper-Kurn, Nature **488**(7412), 476 (2012)
6. A.H. Safavi-Naeini, S. Groeblacher, J.T. Hill, J. Chan, M. Aspelmeyer, O. Painter, arXiv:1302.6179 (2013)
7. T.P. Purdy, P.L. Yu, R.W. Peterson, N.S. Kampel, C.A. Regal, Phys. Rev. X **3**, 031012 (2013)
8. C. Genes, A. Mari, P. Tombesi, D. Vitali, Phys. Rev. A **78**(3), 32316 (2008)
9. A. Einstein, B. Podolsky, N. Rosen, Phys. Rev. **47**(10), 777 (1935)
10. R. Almog, S. Zaitsev, O. Shtempluck, E. Buks, Phys. Rev. Lett. **98**, 78103 (2007)
11. A.A. Clerk, F. Marquardt, K. Jacobs, New J. Phys. **10**, 95010 (2008)
12. D. Vitali, S. Mancini, L. Ribichini, P. Tombesi, Phys. Rev. A **65**, 63803 (2003)
13. M.J. Woolley, A.C. Doherty, G.J. Milburn, K.C. Schwab, Phys. Rev. A **78**(6), 62303 (2008)
14. L. Tian, M.S. Allman, R.W. Simmonds, New J. Phys. **10**(10), 115001 (2008)
15. A. Mari, J. Eisert, Phys. Rev. Lett. **103**, 213603 (2009)
16. H. Ian, Z.R. Gong, Y. Liu, C.P. Sun, F. Nori, Phys. Rev. A **78**(1), 13824 (2008)
17. P. Rabl, A. Shnirman, P. Zoller, Phys. Rev. B **70**(20), 205304 (2004)
18. X. Zhou, A. Mizel, Phys. Rev. Lett. **97**, 267201 (2006)
19. J. Zhang, Y. Liu, F. Nori, Phys. Rev. A **79**(5), 52102 (2009)
20. M.R. Vanner, I. Pikovski, G.D. Cole, M.S. Kim, C. Brukner, K. Hammerer, G.J. Milburn, M. Aspelmeyer, Proc. Nat. Acad. Sci. USA **108**(39), 16182 (2011)
21. M.R. Vanner, J. Hofer, G.D. Cole, M. Aspelmeyer, Nat. Commun. **4**, 2295 (2013)
22. K. Jähne, C. Genes, K. Hammerer, M. Wallquist, E.S Polzik, P. Zoller, Phys. Rev. A **79**(6), 63819 (2009)
23. J.C. Sankey, C. Yang, B.M. Zwickl, A.M. Jayich, J.G.E. Harris, Nat. Phys. **6**, 707 (2010)
24. O. Romero-Isart, A.C. Pflanzer, F. Blaser, R. Kaltenbaek, N. Kiesel, M. Aspelmeyer, J.I. Cirac, Phys. Rev. Lett. **107**(2), 020405 (2011)
25. C.M. Caves, K.S. Thorne, R.W.P. Drever, V.D. Sandberg, M. Zimmermann, Rev. Mod. Phys. **52**(2), 341 (1980)
26. P. Domokos, H. Ritsch, J. Opt. Soc. Am. B **20**(5), 1098 (2003)
27. M. Woolley, A.C. Doherty, G.J. Milburn, Phys. Rev. B **82**(9), 94511 (2010)
28. A. Gangat, T.M. Stace, G.J. Milburn, New J. Phys. **13**, 43024 (2011)
29. H. Miao, S. Danilishin, T. Corbitt, Y. Chen, Phys. Rev. Lett. **103**, 100402 (2009)
30. E. Schrödinger, Naturwissenschaften **48**, 808 (1935)
31. D. Giulini, E. Joos, C. Kiefer, J. Kupsch, I.O. Stamatescu, H.D. Zeh, *Decoherence and the Appearance of a Classical World in Quantum Theory* (Springer, Berlin, 1996)
32. M.J. Ruostekoski, R. Collett, R. Graham, D.F. Walls, Phys. Rev. A **57**(1), 511 (1998)
33. S. Bose, K. Jacobs, P.L. Knight, Phys. Rev. A **59**, 3204 (1999)
34. A.D. Armour, M.P. Blencowe, K.C. Schwab, Phys. Rev. Lett. **88**, 148301 (2002)
35. C.H. van der Wal, A.C.J. ter Haar, F.K. Wilhelm, R.N. Schouten, C. Harmans, T.P. Orlando, S. Lloyd, J.E. Mooij, Science **290**(5492), 773 (2000)
36. A.D. O'Connell, M. Hofheinz, M. Ansmann, R.C. Bialczak, M. Lenander, E. Lucero, M. Neeley, D. Sank, H. Wang, M. Weides, J. Wenner, J.M. Martinis, A.N. Cleland, Nature **464**(7289), 697 (2010)
37. M. Arndt, O. Nairz, J. Vos-Andreae, C. Keller, G. van der Zouw, A. Zeilinger, Nature **401**, 680 (1999)
38. G.C. Ghirardi, A. Rimini, T. Weber, Phys. Rev. D **34**(2), 470 (1986)
39. G.C. Ghirardi, P. Pearle, A. Rimini, Phys. Rev. A **42**(1), 78 (1990)
40. D.I. Fivel, Phys. Rev. A **56**(1), 146 (1997)
41. I.C. Percival, Proc. R. Soc. Lond. A **447**(1929), 189 (1994)
42. R. Penrose, Math. Phy, 266 (2000)
43. W. Marshall, C. Simon, R. Penrose, D. Bouwmeester, Phys. Revi. Lett. **91**(13), 130401 (2003)
44. C. Law, Phys. Rev. A **49**(1), 433 (1994)
45. C. Law, Phys. Rev. A **51**(3), 2537 (1995)
46. S. Mancini, V.I. Man'ko, P. Tombesi, Phys. Rev. A **55**(4), 3042 (1997)
47. S. Bose, K. Jacobs, P.L. Knight, Phys. Rev. A **56**, 4175 (1997)

48. D. Kleckner, W. Marshall, M. de Dood, K. Dinyari, B.J. Pors, W. Irvine, D. Bouwmeester, Phys. Rev. Lett. **96**(17), 173901 (2006)
49. D. Kleckner, W.T.M. Irvine, S.S.R. Oemrawsingh, D. Bouwmeester, Phys. Rev. A **81**(4), 043814 (2010)
50. D. Kleckner, B. Pepper, E. Jeffrey, P. Sonin, S.M. Thon, D. Bouwmeester, Opt. Expr. **19**(20), 19708 (2011)
51. G. Rempe, R.J. Thompson, H.J. Kimble, R. Lazeari, Opt. Lett. **17**(5), 363 (1992)
52. C.J. Hood, H.J. Kimble, J. Ye, Phys. Rev. A **64**(3), 033804 (2001)
53. W.T. Strunz, F. Haake, Phys. Rev. A **67**, 022102 (2003)
54. K. Wago, D. Botkin, C.S. Yannoni, D. Rugar, Appl. Phys. Lett. **72**(21), 2757 (1998)
55. S. Mancini, D. Vitali, P. Tombesi, Phys. Rev. Lett. **80**(4), 688 (1998)
56. P.F. Cohadon, A. Heidmann, M. Pinard, Phys. Rev. Lett. **83**(16), 3174 (1999)
57. D. Kleckner, D. Bouwmeester, Nature **444**(7115), 75 (2006)
58. O. Arcizet, P.F. Cohadon, T. Briant, M. Pinard, A. Heidmann, Nature **444**, 71 (2006)
59. S. Gigan, H.R. Böhm, M. Paternostro, F. Blaser, G. Langer, J.B. Hertzberg, K.C. Schwab, D. Bäuer, M. Aspelmeyer, A. Zeilinger, Nature **444**(7115), 67 (2006)
60. T.J. Kippenberg, K.J. Vahala, Science **321**(5893), 1172 (2008)
61. A. Bassi, E. Ippoliti, S.L. Adler, Phys. Rev. Lett. **94**, 030401 (2005)
62. F. Khalili, S. Danilishin, H. Miao, H. Müller-Ebhardt, H. Yang, Y. Chen, Phys. Rev. Lett. **105**, 070403 (2010)
63. J.Q. Liao, H.K. Cheung, C.K. Law, Phys. Rev. A **85**, 025803 (2012)
64. T. Hong, H. Yang, H. Miao, Y. Chen, Phys. Rev. A **88**, 023812 (2013)
65. P. Rabl, Phys. Rev. Lett. **107**, 063601 (2011)
66. A. Nunnenkamp, K. Børkje, S.M. Girvin, Phys. Rev. Lett. **107**, 063602 (2011)
67. J. Qian, A.A. Clerk, K. Hammerer, F. Marquardt, Phys. Rev. Lett. **109**, 253601 (2012)
68. N. Loerch, J. Qian, A.A. Clerk, F. Marquardt, K. Hammerer, arXiv:1310.1298 (2013)
69. P.D. Nation, Phys. Rev. A **88**, 053828 (2013)
70. M. Paternostro, D. Vitali, S. Gigan, M.S. Kim, C. Brukner, J. Eisert, M. Aspelmeyer, Phys. Rev. Lett. **99**(25), 250401 (2007)
71. H. Miao, S. Danilishin, Y. Chen, Phys. Rev. A **81**(5), 52307 (2010)
72. D. Vitali, S. Gigan, A. Ferreira, H.R. Böhm, P. Tombesi, A. Guerreiro, V. Vedral, A. Zeilinger, M. Aspelmeyer, Phys. Rev. Lett. **98**(3), 30405 (2007)
73. F. Galve, L.A. Pachón, D. Zueco, Phys. Rev. Lett. **105**(18), 180501 (2010)
74. C. Genes, D. Vitali, P. Tombesi, New J. Phys. **10**(9), 95009 (2008)
75. D. Vitali, P. Tombesi, M.J. Woolley, A.C. Doherty, G.J. Milburn, Phys. Rev. A **76**(4), 42336 (2007)
76. R. Ghobadi, A.R. Bahrampour, C. Simon, Optomechanical entanglement in the presence of laser phase noise. Phys. Rev. A **84**, 063827 (2011)
77. R. Ghobadi, A.R. Bahrampour, C. Simon, Quantum optomechanics in the bistable regime. Phys. Rev. A **84**, 033846 (2011)
78. M. Abdi, S.H. Barzanjeh, P. Tombesi, D. Vitali, Effect of phase noise on the generation of stationary entanglement in cavity optomechanics. Phys. Rev. A **84**, 032325 (2011)
79. S. Mancini, V. Giovannetti, D. Vitali, P. Tombesi, Phys. Rev. Lett. **88**(12), 120401 (2002)
80. J. Zhang, K. Peng, S.L. Braunstein, Phys. Rev. A **68**(1), 13808 (2003)
81. M. Pinard, A. Dantan, D. Vitali, O. Arcizet, T. Briant, A. Heidmann, Europhys. Lett. **72**(5), 747 (2005)
82. S. Pirandola, D. Vitali, P. Tombesi, S. Lloyd, Phys. Rev. Lett. **97**(15), 150403 (2006)
83. M.J. Hartmann, M.B. Plenio, Phys. Rev. Lett. **101**(20), 200503 (2008)
84. G. Vacanti, M. Paternostro, G. Massimo Palma, V. Vedral, New J. Phys. **10**(9), 95014 (2008)
85. S. Huang, G.S. Agarwal, New J. Phys. **11**(10), 103044 (2009)
86. D. Vitali, S. Mancini, P. Tombesi, J. Phys. A: Math. Theor. **40**(28), 8055 (2007)
87. M. Ludwig, K. Hammerer, F. Marquardt, Phys. Rev. A **82**(1), 12333 (2010)
88. A. Farace, V. Giovannetti, Phys. Rev. A **86**, 013820 (2012)
89. A. Mari, J. Eisert, New J. Phys. **14**(7), 075014 (2012)

90. S. Mancini, D. Vitali, P. Tombesi, Phys. Rev. Lett. **90**(13), 137901 (2003)
91. S. Pirandola, S. Mancini, D. Vitali, P. Tombesi, Phys. Rev. A **68**(6), 62317 (2003)
92. O. Romero-Isart, A.C. Pflanzer, M.L. Juan, R. Quidant, N. Kiesel, M. Aspelmeyer, J.I. Cirac, Phys. Rev. A **83**, 013803 (2011)
93. S.G. Hofer, W. Wieczorek, M. Aspelmeyer, K. Hammerer, Phys. Rev. A **84**, 052327 (2011)
94. T.A. Palomaki, J.D. Teufel, R.W. Simmonds, K.W. Lehnert, Science **342**(6159), 710 (2013)
95. S.L. Braunstein, P. van Loock, Rev. Mod. Phys. **77**(2), 513 (2005)
96. M. Aspelmeyer, S. Gröblacher, K. Hammerer, N. Kiesel, J. Opt. Soc. Am. B 27(A189) (2010)
97. C.W. Gardiner, P. Zoller, *Quantum Noise*, 3rd edn. (Springer, Berlin, 2004)
98. M. Ludwig, B. Kubala, F. Marquardt, New J. Phys. **10**(9), 95013 (2008)
99. L.M. Duan, G. Giedke, J.I. Cirac, P. Zoller, Phys. Rev. Lett. **84**(12), 2722 (2000)
100. T.A. Palomaki, J.W. Harlow, J.D. Teufel, R.W. Simmonds, K.W. Lehnert, Nature **495**(7440), 210 (2013)
101. L. Vaidman, Phys. Rev. A **49**(2), 1473 (1994)
102. S.L. Braunstein, H.J. Kimble, Phys. Rev. Lett. **80**(4), 869 (1998)
103. K. Hammerer, E.S. Polzik, J.I. Cirac, Phys. Rev. A **72**(5), 52313 (2005)
104. J.F. Sherson et al., Nature **443**(7111), 557 (2006)
105. D. Bouwmeester et al., Nature **390**(6660), 575 (1997)
106. J. Cerrillo et al., Pulsed laser cooling for cavity-optomechanical resonators (2011)
107. G.D. Cole, I. Wilson-Rae, K. Werbach, M.R. Vanner, M. Aspelmeyer, Nat. Commun. **2**, 231 (2011)
108. L. Ding, Others. Phys. Rev. Lett. **105**(26), 263903 (2010)
109. M. Eichenfield, J. Chan, R.M. Camacho, K.J. Vahala, O. Painter, Nature **462**(7269), 78 (2009)
110. A.H. Safavi-Naeini et al., Nature **472**(7341), 69 (2011)
111. S.G. Hofer, D.V. Vasilyev, M. Aspelmeyer, K. Hammerer, Phys. Rev. Lett. **111**, 170404 (2013)
112. J.D. Teufel, T. Donner, D. Li, J.W. Harlow, M.S. Allman, K. Cicak, A.J. Sirois, J.D. Whittaker, K.W. Lehnert, R.W. Simmonds, Nature **475**(7356), 359 (2011)
113. J. Chan, T.P.M. Alegre, A.H. Safavi-Naeini, J.T. Hill, A. Krause, S. Gröblacher, M. Aspelmeyer, O. Painter, Nature **478**(7367), 89 (2011)
114. K.W. Murch, K.L. Moore, S. Gupta, D.M. Stamper-Kurn, Nat. Phys. **4**(7), 561 (2008)
115. T.P. Purdy, R.W. Peterson, C.A. Regal, Science **339**(6121), 801 (2013)

Chapter 4
Suspended Mirrors: From Test Masses to Micromechanics

Pierre-François Cohadon, Roman Schnabel and Markus Aspelmeyer

Abstract Suspended mirrors are the most prominent model systems for optomechanical devices. During the last 10 years, microfabricated mirrors have dramatically increased the capability to exploit radiation-pressure effects on such structures. This chapter summarizes the current state-of-the-art in the performance of suspended (micro-)mirrors for cavity optomechanics experiments in terms of their optical and mechanical quality, and highlights some of the milestones experiments performed with suspended mirrors.

4.1 Introduction

Suspended mirrors are the most natural mechanical system in optomechanics. Mainly because of their simplicity, they have been at the heart of many fundamental discussions and ground-breaking experiments. To provide a few examples: the first unambiguous demonstration of radiation pressure was realized in 1901 by monitoring the motion of a torsional pendulum illuminated by a strong light source, a success achieved in two independent experiments by Nichols and Hull [1] and by Lebedew [2]. In 1909, Einstein suggested a Gedanken experiment, in which radiation pressure on a suspended mirror revealed for the first time the unavoidable wave-particle duality in the nature of light [3], and in 1931 Schrödinger, in a letter to Sommerfeld, provided

P.-F. Cohadon (✉)
Laboratoire Kastler Brossel, ENS, UPMC, CNRS, Paris, France
e-mail: cohadon@lkb.upmc.fr

R. Schnabel
Institut für Gravitationsphysik, Leibniz Universität Hannover, Hannover, Germany
e-mail: roman.schnabel@aei.mpg.de

M. Aspelmeyer
Faculty of Physics, Vienna Center for Quantum Science and Technology (VCQ), University of Vienna, Vienna, Austria
e-mail: markus.aspelmeyer@univie.ac.at

M. Aspelmeyer et al. (eds.), *Cavity Optomechanics*, Quantum Science and Technology,
DOI: 10.1007/978-3-642-55312-7_4,

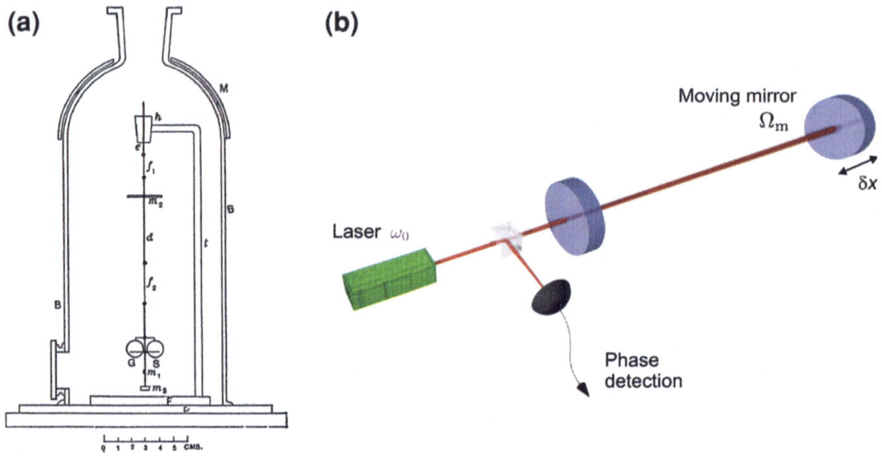

Fig. 4.1 Optomechanics then and now. **a** Drawing of one of the two early 'light-mill' experiments that demonstrated the effect of radiation pressure. The torsional mechanical oscillator consisted of two silvered and polished microscope objectives (G and S), suspended by a rod of drawn glass and each weighing around 50 mg. The mechanical displacement was measured by optical deflection [1]. **b** General scheme of a typical optomechanics experiment to date. The mechanical motion of a mirror is directly coupled to the radiation field of an optical cavity, here by making the suspended mirror the end mirror of a Fabry–Perot cavity. The mirror motion is imprinted onto the phase of the reflected beam. Quantum fluctuations of light can limit the sensitivity of the mirror position measurement, but radiation pressure can also deeply alter the mirror motion

the first hints on quantum entanglement by analyzing the reflection of a single photon off a macroscopic mirror [4]. A few decades later, large-scale Michelson laser interferometers with suspended macroscopic mirrors emerged as new designs to detect gravitational waves in an Earth-based detector [5, 6]. And it was in this latter context that the first instances of cavity optomechanics were suggested and demonstrated, primarily through the pioneering works of Braginsky [7, 8]. One direct consequence of this development was the formulation of the *Standard Quantum Limit* (SQL) in the early 80s [9].

The paradigmatic example for a cavity optomechanics scheme is a Fabry–Perot cavity with a suspended end mirror (see Fig. 4.1). Following early microwave work [8] the first cavity optomechanics experiment in this genuine configuration was performed in 1983 by Dorsel et al. [10], who used a 60-mg mirror (with a resonance frequency around 1 Hz) to demonstrate radiation-pressure induced bistability of the optical cavity. Unfortunately, laser sources were still plagued by classical noise and radiation-pressure effects quite weak so that the experiments did not proceed any further towards the observation of quantum effects.

Modern optomechanics experiments actually started at the end of the 1990s, as the combination of high-finesse optical cavities, stable laser sources and low-noise detection systems allowed to measure the Brownian motion around a given mechanical

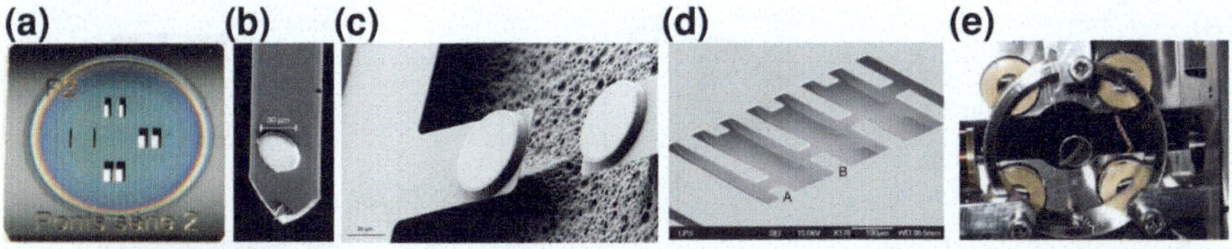

Fig. 4.2 Suspended mirrors in cavity optomechanics experiments. **a** Coated silicon micromechanical resonator [15]. **b** AFM cantilever with micromirror [16]. **c** SiN micromechanical resonator with mirror pad [17]. **d** Micromechanics from SiO_2/Ta_2O_5 Bragg mirror material [18]. **e** 1-inch mirror suspended by tapered glass fibers [19]

resonance frequency [11–13]. In the following years, a constant miniaturization of the mirror structures (see Fig. 4.2), along with the independent development of microcavities, enabled the systematic study of radiation-pressure effects in cavity optomechanics [14].

4.2 First Experiments: Mechanical Motion Sensing on the Attometer Scale

One immediate application of cavity optomechanics is the accurate monitoring of the mirror motion by monitoring the mechanically induced phase-shift of the optical cavity field. The achievable displacement sensitivities reach down to the attometer level and below. Such high sensitivities are of particular importance for instance for gravitational wave detectors.

Figure 4.3 displays two thermal noise spectra, one obtained in 1999 with a cm-size fused silica mirror as mechanical resonator, the other in 2006 with a 60-μm thick silicon micromirror. The plane-convex shape of the fused silica mirror is responsible for mode shapes that are strongly analogous to the Gaussian modes encountered in optical cavities: the strong mechanical confinement of the modes is responsible for high mechanical Q factors (around 40,000 for the first generation, up to 10^6 for the following generations), with resonance frequencies in the MHz band and effective masses at the level of hundreds of mg. Because of the thinner resonator, the effective masses of the silicon micromirror modes are notably smaller but Q are still limited to a few tens of thousands. Yet, room temperature thermal noise was easily detected, as high optical finesse values (typically 30,000 for both experiments) are possible with such mirrors, using the huge progress made in the framework of interferometric gravitational-wave detection. These experiments actually emphasize both the main advantage and the bottleneck of micromirrors: while the optical quality of macroscopic mirrors allows to achieve large cavity finesse, their relatively large size and hence effective mass restrict the available optomechanical coupling rates.

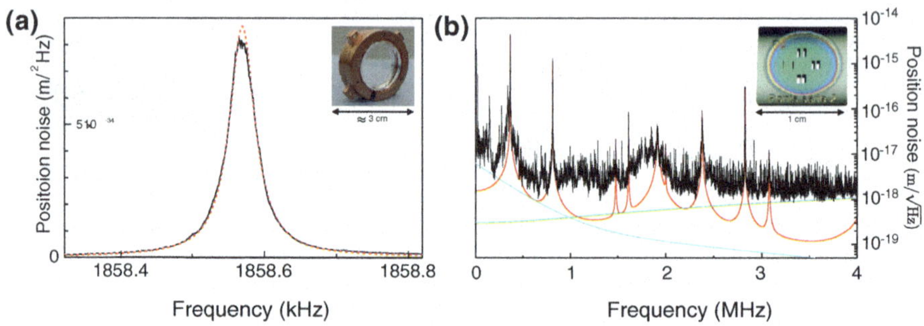

Fig. 4.3 Thermal noise spectra of a moving mirror. **a** Spectrum observed close to one of the resonances of a fused silica plane-convex mirror. The *dotted line* is the spectrum expected from the measurement of the mechanical response to an external drive. **b** Wideband spectrum (between 0 and 4 MHz) for a silicon micromirror (*black*), frequency noise (*blue*), quantum phase noise (*green*) and expected spectrum from the resonator modelling (*red*). Both insets show an optical view of the corresponding resonator. Figure adapted from Refs. [11, 20]

The detection system also gives access to raw temporal data $x(t)$ of the moving mirror. Filtered around a mechanical resonance frequency Ω_{M}, such a signal has little interest for short acquisition times as it mainly depicts an oscillation at Ω_{M}. Yet, demodulation at Ω_{M} allows to obtain the slowly-varying quadratures of motion $X(t)$ and $Y(t)$:

$$x(t) = X(t)\cos(\Omega_{\mathrm{M}}t) + Y(t)\sin(\Omega_{\mathrm{M}}t)\,. \tag{4.1}$$

These quadratures now evolve on a timescale $\tau_{\mathrm{M}} \simeq Q/\Omega_{\mathrm{M}}$, on the order of 1 s or more, and allow to see the fluctuation-dissipation theorem at work: Brownian motion now appears as the free evolution of a resonator, only slowly disturbed by its environment. Figure 4.4 presents experimental results with two different micromirrors. Figure 4.4b presents the phase-space trajectory of a silicon torsional resonator (Fig. 4.4a), with a resonance frequency $\Omega_{\mathrm{M}}/2\pi \simeq 26$ kHz. Note that the vane had to be coated 1 mm away from the torsional axis in order for the motion to change cavity length, and hence to have a non-zero effective mass (here: $m_{\mathrm{eff}} \simeq 10$ mg) [13].

As the resonance frequency is rather low and the mechanical Q is in the 10^5 range, observation over a 80-s period of time is not sufficient to fully map the phase space, and Brownian motion is quite similar to the one of a free particle. For the fused silica resonator however [21], the long-term stability of the experimental setup has allowed to monitor the trajectory over thousands of the mechanical damping time, illustrating the transition to a behaviour ruled by the ergodic theorem: the motion now appears strongly peaked at the center of phase-space, with fluctuations distributed over a 2D gaussian, with a width given by the equipartition theorem [21]:

$$\Delta X^2 = \Delta Y^2 = \frac{kT}{m_{\mathrm{eff}}\Omega_{\mathrm{M}}^2} \simeq 10^{-31}\,\mathrm{m}^2. \tag{4.2}$$

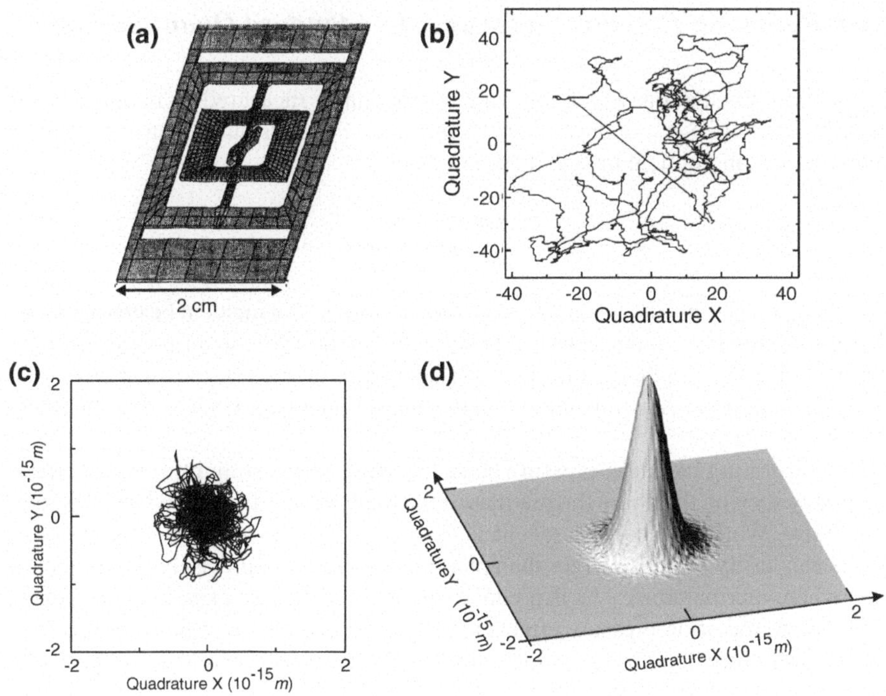

Fig. 4.4 Phase-space trajectories of the Brownian motion of a moving mirror. **a** Silicon torsion oscillator, used by Tittonen et al. [13]. **b** Trajectory over 80 s for this resonator. **c** Trajectory observed at LKB over a long timescale with a silica mirror [21]. **d** Histogram obtained with the same data as for **c**, over 10 min, corresponding to 25,000 times the characteristic time τ_M. Figure adapted from Refs. [13, 21]. Reprinted figure with permission from [13]. Copyright (1999) by the American Physical Society

4.3 Quantum Limits

Quantum fluctuations of light are responsible for quantum limits in position and phase-shift measurements. The photon statistics at the detector results in shot-noise (SN), also called *quantum measurement noise*. The photon statistics at the mirror, however, results in radiation pressure noise (*quantum back-action noise*), which was quantum mechanically analyzed in the early 80s [9, 22]. In principle (quantum) radiation pressure noise (RPN) can be observed on a simple single-ended optical cavity with a moving mirror (see Fig. 4.1). Only recently, it was demonstrated for the first time [23] in a 'membrane-in-the-middle' system (see the chapter by Sankey et al.).

4.3.1 Radiation Pressure Noise and the Standard Quantum Limit

The position fluctuations δx of a moving mirror can be measured by monitoring the phase fluctuations $\delta\varphi_{\text{out}}$ of the reflected light beam. For a resonant cavity (neglecting cavity cut-off and optical losses) these are given by

$$\delta\varphi_{\text{out}} = \delta\varphi_{\text{in}} + \frac{8\mathscr{F}}{\lambda}\left(\delta x + \delta x_{\text{cl}} + \delta x_{\text{RPN}}\right), \tag{4.3}$$

where $\delta\varphi_{\text{in}}$ represents the quantum phase fluctuations of the incident beam (responsible for shot-noise), δx_{cl} any classical position fluctuations of the mirror (for instance thermal noise δx_T) and δx_{rad} the position fluctuations created by the quantum (intracavity) radiation-pressure fluctuations of the measurement laser beam. $\mathscr{F}$ is the cavity finesse.

Demonstrating radiation-pressure noise therefore seems straightforward as it can be directly seen on the phase fluctuations of the reflected field, provided it is the dominant noise. Working close to a mechanical resonance frequency, radiation-pressure effects can easily be made larger than phase noise. Yet, it is precisely the frequency band where thermal noise δx_T is predominant and the biggest experimental issue to demonstrate radiation pressure effects usually is to make it negligible compared to radiation pressure noise δx_{rad}, as the ratio S_x^{rad}/S_x^T between the two noise spectra is:

$$\frac{S_x^{\text{RPN}}}{S_x^{\text{T}}} \simeq 2\left(\frac{\mathscr{F}}{300{,}000}\right)^2\left(\frac{800\,\text{nm}}{\lambda}\right)\left(\frac{P_{\text{in}}}{1\,\text{mW}}\right)\left(\frac{1\,\text{mg}}{m_{\text{eff}}}\right)\left(\frac{Q}{10^6}\right)\left(\frac{1\,\text{MHz}}{\Omega_{\text{M}}/2\pi}\right)\left(\frac{1\,\text{K}}{T}\right). \tag{4.4}$$

Quoted values have all been independently demonstrated on a variety of micro-optomechanical systems, but combining the excellent mechanical behaviour of micro- or nano-scale systems with the optical quality of larger mirrors is an even tougher challenge. This explains why it actually took so long to experimentally demonstrate (quantum) radiation pressure noise.

The *Standard Quantum Limit* (SQL) is defined as the minimum of the (uncorrelated) sum of δx_{SN} (due to $\delta\varphi_{\text{in}}$) and δx_{RPN}, assuming both are at the minimum level set by Heisenberg inequalities. Since the radiation pressure noise depends on the mirror's mechanical susceptibility and thus on Fourier frequency, the SQL is frequency-dependent as well. Figure 4.5 shows the quantum noise contributions as noise spectral densities normalized to displacement in $\text{m}/\sqrt{\text{Hz}}$ for a selected Fourier frequency, chosen here below the mechanical resonance of the moving mirror. It still has not yet been demonstrated. For a review we refer to [24].

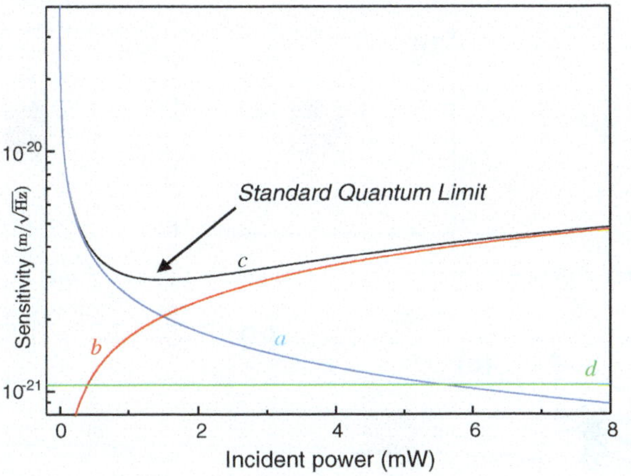

Fig. 4.5 Total displacement quantum noise spectral density (c) of an interferometric measurement versus incident optical power. The curves are computed for low frequencies ($\Omega \ll \Omega_M$) and for the following parameters: optical finesse $\mathscr{F} = 100{,}000$, temperature $T = 1$ K, and a resonator with $m_{\mathrm{eff}} = 100$ µg, $\Omega_M/2\pi = 1$ MHz and $Q = 10^6$. a quantum phase noise, b Quantum radiation-pressure noise, d thermal noise. The SQL, on the order of 10^{-21} m/$\sqrt{\mathrm{Hz}}$, is reached for an incident power close to 1.5 mW

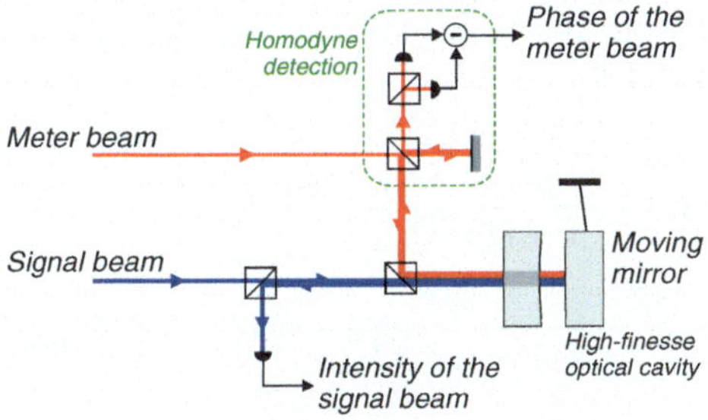

Fig. 4.6 Optomechanical correlations setup: a high-power signal beam and a weaker meter beam are sent into the moving mirror cavity. Intensity fluctuations of the signal beam are imprinted by radiation pressure onto the position fluctuations of the moving mirror, and onto the phase fluctuations of the meter beam. The two reflected beams then display intensity-phase correlations. Figure adapted from Ref. [25]

4.3.2 Optomechanical Correlations Through Radiation Pressure

In order to demonstrate RPN, an alternative approach has been developed at LKB. Two laser beams are sent into the moving mirror cavity (see Fig. 4.6): the intensity fluctuations of the high-power signal beam drive the mirror into motion by radiation pressure, whereas the phase of the weaker meter beam is used to monitor the resulting position fluctuations. Neglecting optical losses, we have the following input-output relations for the fluctuations of the various fields involved [25]:

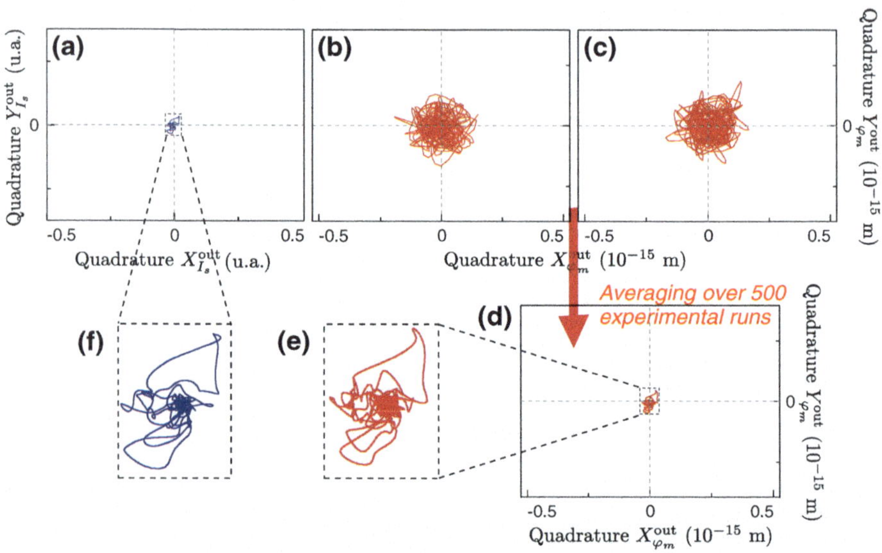

Fig. 4.7 Proof of principle of the ability of the two-beam setup to retrieve quantum optomechanical correlations in the $S_x^{\rm rad}/S_x^{\rm T} \ll 1$ limit. Phase-space trajectory of the intensity noise of the signal beam (**a**) here corresponds to a ratio $S_x^{\rm rad}/S_x^{\rm T} = 0.03$, and the resulting phase noise of the meter beam for different experimental runs (**b** and **c**) is blurred by thermal noise. The initial intensity noise pattern is retrieved by the average of 500 similar experimental runs (**d**, and close-ups **e** and **f**). Figure adapted from Ref. [26]

$$\delta I_s^{\rm out}[\Omega] = \frac{1 + i\Omega/\Omega_{\rm cav}}{1 - i\Omega/\Omega_{\rm cav}} \delta I_s^{\rm in}[\Omega], \tag{4.5}$$

$$\delta\varphi_m^{\rm out}[\Omega] = \frac{8\mathscr{F}}{\lambda(1 - i\Omega/\Omega_{\rm cav})} \left(\delta x_{\rm T}[\Omega] + \delta x_{\rm rad}[\Omega]\right), \tag{4.6}$$

with $\Omega_{\rm cav}$ the cavity bandwidth. The radiation-pressure noise $\delta x_{\rm rad}$ is related to the incident signal intensity fluctuations by:

$$\delta x_{\rm rad}[\Omega] = \frac{8\mathscr{F}}{\lambda(1 - i\Omega/\Omega_{\rm cav})} \hbar\chi[\Omega]\delta I_s^{\rm in}[\Omega], \tag{4.7}$$

where $\chi[\Omega]$ is the mechanical susceptibility of the moving mirror. The reflected signal intensity noise $\delta I_s^{\rm out}$ reproduces the incident fluctuations $\delta I_s^{\rm in}$, whereas the reflected meter phase $\delta\varphi_m^{\rm out}$ reproduces the same incident signal intensity $\delta I_s^{\rm in}$ via the mirror motion [Eqs. (4.6) and (4.7)]. The intensity-phase correlations observable between the two reflected beams therefore provide a direct measurement of the optomechanical correlations. This is true even in the regime where thermal noise prevails over QRPN. In such a case, one has to compute the correlation coefficient between $\delta I_s^{\rm out}$ and $\delta\varphi_m^{\rm out}$ (with both $\delta x_{\rm rad}$ and δx_T contributions) and average over a sufficient time for the $\delta I_s^{\rm out} \cdot \delta x_T$ term to average to zero. Details are given in Ref. [26].

Figure 4.7 presents a proof of principle of the scheme with a classical intensity noise pattern, repeated over 500 experimental runs. As $S_x^{\rm rad}/S_x^{\rm T} = 0.03$ here, the intensity noise pattern is blurred by thermal noise and cannot be observed on a single run. Yet, it can be retrieved by averaging similar experimental runs. For quantum noise, the pattern obviously wouldn't be the same for each run, meaning it is not possible to display such a proof in phase-space but the idea of demonstrating QRPN by the computation of the correlation coefficient however still holds. Experiments are still underway at LKB.

Note that in the regime $S_x^{\rm rad} \gg S_x^{\rm T}$, such a setup allows to perform a QND measurement of the signal beam intensity [27], in close analogy to the experiments performed in the 90s with nonlinear Kerr media, such as atomic beams or cold trapped atoms [28]: here, the optical nonlinearity directly stems from the radiation pressure changing the cavity length.

4.4 Towards Suspended Mirrors in the Quantum Regime

Preparing the quantum ground state (QGS) of a macroscopic mechanical resonator is a necessary precondition for manipulating and exploiting quantum states of mechanical motion [29, 30]. Experiments with suspended mirrors allowed a series of first proof-of-concept demonstrations in this direction including feedback cooling of micromechanical motion [12, 16, 31, 32], cavity cooling of micromirrors [15, 33, 34], observation of the strong optomechanical coupling regime [17], or demonstration of back-action evading interactions through pulsed quantum optomechanics [35]. Two essential conditions are required to achieve quantum ground state cooling by monitoring quantum position fluctuations. The experimental setup obviously needs a sufficient displacement sensitivity S_x:

$$S_x[\Omega_{\rm M}]\frac{\Omega_{\rm M}}{Q} \ll \frac{1}{m_{\rm eff}\Omega_{\rm M}^2}\hbar\Omega_{\rm M}, \tag{4.8}$$

where $\Omega_{\rm M}/Q$ is the mechanical bandwidth. Condensed-matter experiments typically require a sensitivity at the $10^{-16}\,{\rm m}/\sqrt{\rm Hz}$ level. As such a sensitivity is easily obtained with an optical measurement scheme, optomechanical resonators are excellent candidates, provided they can be cooled down to their QGS, which requires:

$$kT \ll \hbar\Omega_{\rm M}. \tag{4.9}$$

With resonance frequencies spanning the band from the Hz regime up to the lower part of the MHz region, suspended micromirrors require temperatures in the μK range and below (e.g. 50 μK for a 4-MHz resonator), which are not available by conventional cryogenics techniques. This section presents a number of schemes to effectively cool a resonator below its environment temperature, and some ideas to create pure quantum states of massive mirrors.

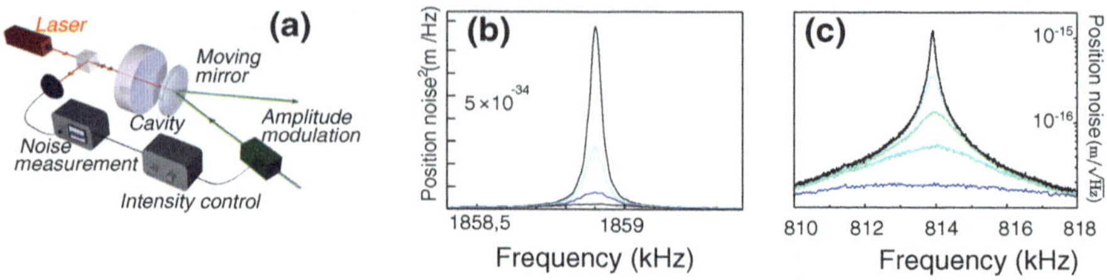

Fig. 4.8 Feedback cooling of a moving mirror. **a** Principle of the experiment. The mirror Brownian motion is detected in real-time with a high-finesse cavity. The signal is then used to apply an additional damping force to the mirror (here with radiation pressure), without any additional noise. **b** Experimental results obtained in 1999 with a silica moving mirror. The temperature reduction of the mirror (from 300 K down to about 10 K) is emphasized by the drastic decrease of the position noise spectrum area. **c** Experimental results obtained in 2006 with a silicon micromirror. The feedback force is here of electrostatic nature. The minimum effective temperature is about 5 K. Figure adapted from Refs. [12, 20]

4.4.1 Feedback Cooling

Figure 4.8a presents the principle of the feedback cooling process demonstrated at LKB in 1999. A laser beam is used to monitor in real-time the position fluctuations of the moving mirror. A second laser beam is used to cool the mirror: it is intensity-modulated by an acousto-optic modulator driven by a feedback loop using the mirror motion as error signal. Parameters are such that quantum noises such as $\delta\varphi_{\mathrm{in}}$ and δx_{rad} are negligible, so that $\delta\varphi_{\mathrm{out}}$ faithfully reproduces the position fluctuations δx.

The mirror motion in response to both the thermal Langevin force ξ and the feedback radiation pressure force is:

$$\delta x\,[\Omega] = \chi\,[\Omega]\,(\xi\,[\Omega] + F_{\mathrm{rad}}\,[\Omega])\,, \tag{4.10}$$

with the spectrum of the Langevin force related to the mechanical damping rate Γ_{M} by the fluctuation-dissipation theorem:

$$S_{\xi}\,[\Omega] = -\frac{2k_{\mathrm{B}}T}{\Omega}\,\mathrm{Im}\left(\frac{1}{\chi\,[\Omega]}\right) = 2m_{\mathrm{eff}}\Gamma_{\mathrm{M}}k_{\mathrm{B}}T. \tag{4.11}$$

The gain of the feedback loop is chosen such as to obtain a radiation-pressure force F_{rad} proportional to the actual speed $i\Omega\delta x[\Omega]$ of the mirror:

$$F_{\mathrm{rad}}\,[\Omega] = i m_{\mathrm{eff}}\Omega\Gamma_{\mathrm{M}} g \delta x\,[\Omega]\,, \tag{4.12}$$

where g is related to the electronic gain of the feedback loop. The radiation-pressure force then corresponds to an additional viscous damping for the mirror and the mirror motion now reads:

$$\delta x\,[\Omega] = \frac{1}{m_{\mathrm{eff}}\left[\Omega_{\mathrm{M}}^2 - \Omega^2 - i\Gamma_{\mathrm{M}}\,(1+g)\,\Omega\right]}\xi\,[\Omega]\,. \tag{4.13}$$

This equation is similar to the one obtained without feedback (Eq. 4.10 with $F_{\mathrm{rad}} = 0$), except that the damping is now increased by a a factor $1 + g$, without any classical fluctuations coupled to the system. This is a case of *cold damping*, obtained by coupling the mirror motion to a reservoir at zero temperature (here, the laser field), which increases its mechanical damping without altering the Langevin force, in apparent violation of the fluctuation-dissipation theorem (see Eq. 4.11).

The thermal noise spectrum $S_x\,[\Omega]$ is still Lorentzian, but with an effective mechanical bandwidth $\Gamma_{\mathrm{eff}} = \Gamma_{\mathrm{M}}(1+g)$ and a different noise power density at resonance. The corresponding motion is strictly equivalent to the one of a moving mirror of intrinsic damping $\Gamma_{\mathrm{M}}(1+g)$ at an effective equilibrium temperature T_{eff}:

$$\frac{T_{\mathrm{eff}}}{T} = \frac{\Gamma_{\mathrm{M}}}{\Gamma_{\mathrm{eff}}} = \frac{1}{1+g}. \tag{4.14}$$

Figure 4.8 presents experimental results obtained with a fused silica moving mirror (b) in 1999 [12] and with a silicon micromirror (c) in 2006 [20]. The cooling effect is evident from the widening and the overall lowering of the thermal noise resonance, clearly seen around the mechanical resonance frequency. Starting from an environment at room temperature, effective temperatures around 10 K have been obtained, mainly limited by the off-resonance thermal noise of other mechanical modes. Note that such a scheme is not limited to radiation-pressure actuation: the silicon micromirror results have indeed been obtained with an electrostatic actuation.

Similar experiments were also performed with a gram-scale flexure resonator ($\Omega_{\mathrm{M}}/2\pi \simeq 85$ Hz, $Q \simeq 45{,}000$) cooled by a modified cavity-locking feedback system [36] or an AFM cantilever [16, 31].

Even taking into account the quantum noises of the interferometric detection, such a scheme is actually likely to drive a moving mirror down to its quantum ground state [37–40].

4.4.2 Cavity Cooling

In a detuned cavity, any mirror motion is transduced into an intracavity intensity variation. This accordingly changes the radiation-pressure force upon the moving mirror (see Fig. 4.9). Such *optical spring* effects have been demonstrated for a while, for instance on the mechanical resonance frequency of a specially-designed flexure resonator [41], or on suspended mirrors [42], but in cavity-cooling experiments, the combination of a high mechanical resonance frequency with a relatively low optical bandwidth changes the radiation-pressure force, which now lags behind the mirror

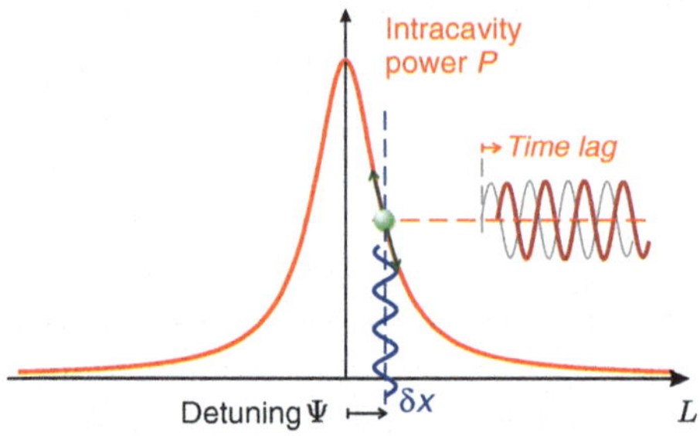

Fig. 4.9 Principle of radiation-pressure cooling. Close to a $\Psi \neq 0$ working point of the cavity, the radiation pressure force applied onto the moving mirror is proportional to its displacement δx. The finite response time of the cavity creates a lag in the force response to δx, which creates an additional damping for the mirror motion, still without any additional noise, and therefore tends to damp the mirror and decrease its effective temperature

motion:

$$F_{\text{rad}}[\Omega] = -2\frac{\varphi\varphi_{\text{NL}}}{\Delta_\Omega} m_{\text{eff}}\Omega_{\text{M}}^2\,\delta x[\Omega], \tag{4.15}$$

$$\Delta_\Omega = (1 - i\Omega/\Omega_{\text{cav}})^2 + \varphi^2, \tag{4.16}$$

where $\varphi = \Psi/\gamma$ is the cavity detuning normalized to the overall losses and φ_{NL} the dephasing corresponding to the static recoil of the mirror:

$$\varphi_{\text{NL}} = \frac{8\pi}{\lambda\gamma c}\frac{P}{m_{\text{eff}}\Omega_{\text{M}}^2}, \tag{4.17}$$

where P is the intracavity optical power.

At thermal equilibrium, the mirror motion under the effect of both the Langevin force and the radiation pressure F_{rad} is:

$$\delta x[\Omega] = \chi[\Omega]\left(\xi[\Omega] + F_{\text{rad}}[\Omega]\right). \tag{4.18}$$

In perfect analogy with Eq. (4.13), this can be seen as the motion of a mirror with an effective mechanical susceptibility $\chi_{\text{eff}}[\Omega]$:

$$\chi_{\text{eff}}[\Omega]^{-1} = \chi[\Omega]^{-1} + 2\frac{\varphi\varphi_{\text{NL}}}{\Delta_\Omega} m_{\text{eff}}\Omega_{\text{M}}^2. \tag{4.19}$$

For a large mechanical quality factor $Q \gg 1$, χ_{eff} still has a lorentzian shape, but with effective resonance frequency and damping:

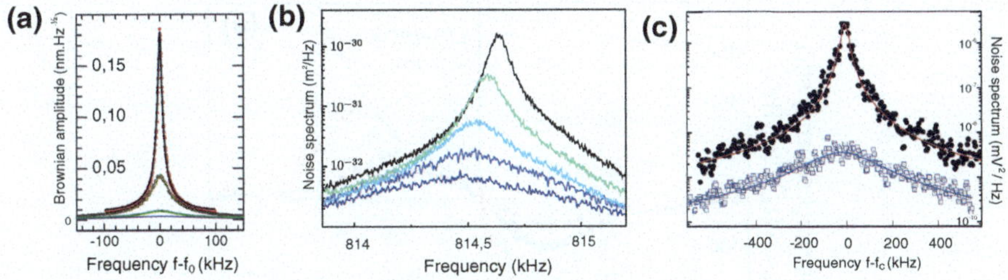

Fig. 4.10 Experimental results on intracavity cooling of a moving mirror. **a** Results obtained by the group of K. Karrai. Under the effect of the photothermal force, the mirror sees its effective temperature decreasing from 300 K (*red curve*) to 78 K (*yellow*), 36 K (*green*) and 18 K (*blue*). **b** Results obtained at Laboratoire Kastler Brossel. The *black curve* is obtained for a resonant cavity (300 K). For increasing detunings, the effective temperature decreases down to a minimum temperature of the order of 10 K (*dark blue*). **c** Results obtained by M. Aspelmeyer's group. The *black dots* are obtained at resonance. For a detuned cavity (*black dots* and *curve*), the minimum temperature is also of the order of 10 K. Figure adapted from Refs. [15, 33, 34]

$$\Omega_{\text{eff}} = \Omega_{\text{M}} \left(1 + \text{Re}\frac{\varphi\varphi_{\text{NL}}}{\Delta_{\Omega_{\text{m}}}}\right) \tag{4.20}$$

$$\Gamma_{\text{eff}} = \Gamma_{\text{M}} \left(1 - 2Q\,\text{Im}\frac{\varphi\varphi_{\text{NL}}}{\Delta_{\Omega_{\text{m}}}}\right). \tag{4.21}$$

The transition between optical spring (Ω_{eff}) and optical damping (Γ_{eff}) is due to the fact that we are no longer in the adiabatic limit $\Omega_{\text{M}} \ll \Omega_{\text{cav}}$. The Q-factor in Eq. (4.21) shows a stronger dependence of the damping to radiation pressure effects for $\Omega_{\text{m}} \simeq \Omega_{\text{cav}}$. Depending on the detuning sign, both damping or mechanical amplification (anti-damping) can be observed.

As the Langevin force is still unchanged, the effect on the resonator temperature is the same as in the feedback case:

$$\frac{T_{\text{eff}}}{T} \simeq \frac{\Gamma_{\text{M}}}{\Gamma_{\text{eff}}}. \tag{4.22}$$

Such effects were first demonstrated by Metzger and Karrai with a photothermal force (which mimics the effects of a radiation pressure force even though it is of a quite different nature) on a cantilever in a low-finesse cavity (see Fig. 4.10a) [33]. Figure 4.10b and c presents the experimental results obtained with radiation pressure cooling. As the cavity is detuned, the spectra are once again wider and lower, together with a small optical spring effect which decreases the resonance frequency. Effective temperatures were down to 10 K for these two experiments.

Experimental results at LKB can be further tested by looking at the dependence of both the damping ratio $\Gamma_{\text{eff}}/\Gamma_{\text{M}}$ and the frequency shift $(\Omega_{\text{eff}} - \Omega_{\text{M}})/2\pi$ with respect to the detuning φ. Figure 4.11 presents the LKB experimental results, for five optical powers from 0.5 mW to 3.2 mW, along with the theoretical fits. Results are in excellent agreement with Eqs. (4.20) and (4.21), the intracavity power at resonance

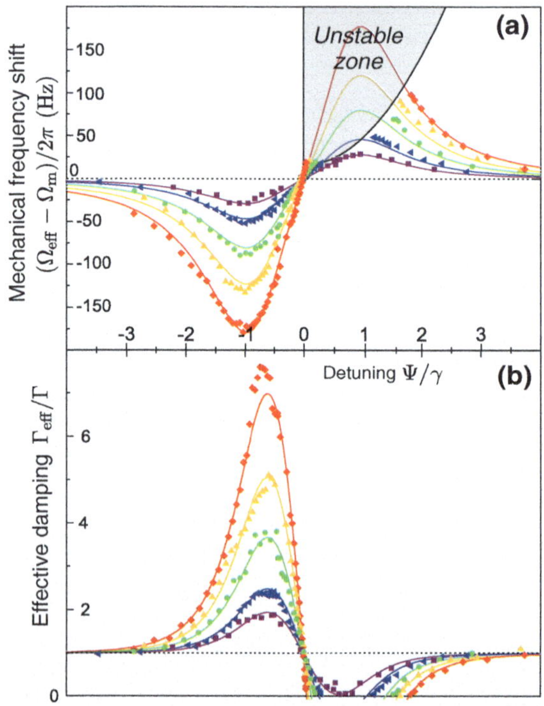

Fig. 4.11 Evolution of the cavity cooling and amplification effects with respect to the detuning φ. **a** Frequency shift $(\Omega_{eff} - \Omega_m)/2\pi$ and **b** damping ratio Γ_{eff}/Γ_M, for five values of incident power: 0.5 mW (*purple*), 0.9 mW (*blue*), 1.6 mW (*green*), 2.2 mW (*yellow*) and 3.2 mW (*red*). Points are experimental results and full lines are fits obtained from Eqs. (4.17), (4.20) and (4.21), by adjusting the intracavity power at resonance P for each *curve*. The *shaded area* in the *upper curve* shows the instability zone where Γ_{eff} vanishes

P derived from the fits exhibiting the expected linear dependence with P_{in}. This clearly shows that the observed cooling effect is solely due to radiation pressure.

The sign of the laser detuning changes the effect. The damping can be decreased as well and eventually cancel for some values of the (φ, P_{in}) parameters set: the mirror is then driven into motion at its resonance frequency Ω_{eff}, with the laser acting as an energy pump. This is the *Parametric Instability* regime (see the following subsection).

Figure 4.12a presents an alternative interpretation of the damping/anti-damping processes as a scattering of photons (at angular frequency ω_0) by the mechanical phonons of the mirror (at Ω_m), giving rise to Stokes (at $\omega_0 - \Omega_m$) and anti-Stokes (at $\omega_0 + \Omega_m$) motional sidebands.

For a resonant cavity, both sidebands have equal amplitudes: they can be used to infer the mirror motion (as a phase modulation of the reflected field) but have no net mechanical effect. As the cavity optical resonance curve acts as an effective density of output modes, detuning the cavity promotes the scattering of photons to one sideband with respect to the other. For negative detunings for instance, more photons are scattered at the $\omega_0 + \Omega_m$ frequency than at $\omega_0 - \Omega_m$ and the net effect of the

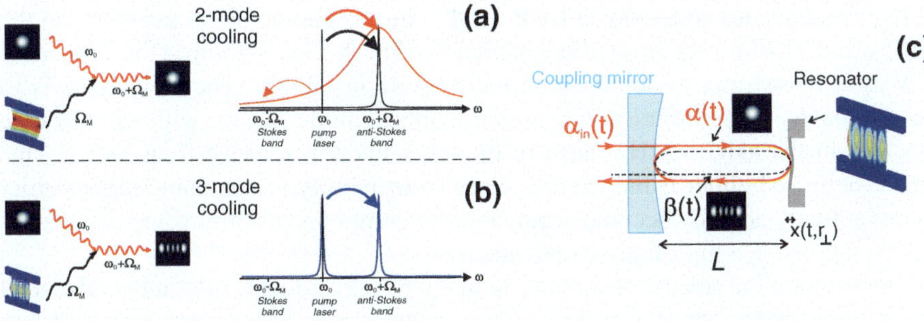

Fig. 4.12 Radiation-pressure cooling seen as a scattering process. **a** *Two-mode cooling*. The photon is scattered to the same optical mode, with a frequency offset Ω_m w.r.t. the incident photon. The cavity is detuned in order to favour the anti-Stokes process, and the pump laser is out of resonance. **b** *Three-mode cooling*. The photon is scattered to a different optical mode of the cavity. Both the pump laser and the anti-Stokes band can be simultaneously resonant with the cavity. The vibration profile of the mechanical mode must match the output mode. **c** Experimental implementation of the 3-mode cooling with a semi-confocal cavity

optomechanical coupling is a non-zero energy transfer from the mechanical motion to the optical field, i.e. damping. Anti-damping is obtained as well for positive detunings. The optimal efficiency is very simple to state in the *resolved sideband regime* ($\Omega_m \gg \Omega_c$) (Fig. 4.12a, dark curve) as the damping/anti-damping processes are resonant for a detuning matching the mechanical resonance frequency Ω_m. Figure 4.12a presents the case where the anti-Stokes process is resonant, both in the non-resolved and resolved sideband regimes. It is in this regime that the QGS was recently demonstrated (see the chapters by Lehnert and by Safavi-Naeini and Painter).

4.4.3 3-Mode Cooling and Parametric Instability

As the driving laser is strongly out of resonance, the cooling process is somehow not efficient. But the situation is different in the case with two optical modes involved, one resonant with the driving laser, the other with the (anti-)Stokes band (see Fig. 4.12b). In that case, the anti-Stokes process corresponds to an effective scattering of one photon of the pump mode ω_p to the target mode ω_t, possibly of a different optical transverse profile, by absorption of mechanical phonons ω_m. Such a process actually involves three modes (one mechanical and two optical) and naturally embodies the physics of the radiation-pressure effects. It is resonant close to an optical frequency mismatch:

$$\Delta\omega = \omega_p - \omega_t = \pm\Omega_M \tag{4.23}$$

($-$ for the Stokes process, $+$ for the anti-Stokes).

Experiments are underway at LKB (with a quasi semi-confocal cavity to set the TEM_{00} and TEM_{04} very close to degeneracy) and in D. Blair's group at the University of Western Australia to demonstrate the 3-mode instability. The main issue is to design a cavity to engineer a non-symmetric optical mode density, with typical mode spacings in the kHz to MHz range of the mechanical resonance frequencies. This can be achieved either using cavities close to instability (when some higher-order modes actually actually become degenerate) or compound cavities using a dielectric membrane inside a high-finesse optical cavity.

The 3-mode Parametric Instability is likely to play a major role in the advanced gravitational-wave interferometers. Even if they use heavy test masses (with $m_{\mathrm{eff}} \simeq 20$ kg), both Advanced Ligo and Advanced Virgo, with their substantial intracavity powers and the high optical finesses of the compound cavities, are indeed favourable playgrounds for this kind of effects, all the more since they present a high mechanical mode density in the tens of kHz band, which is the typical spacing between optical transverse modes as well. Such an instability may jeopardize the stable operation of these interferometers, as the large amplitude of the driven motion may not be possible to cope with for the stabilization schemes of the different cavities.

4.4.4 Pure Quantum States of Massive Mirrors via Conditioning

Optomechanical cooling describes the process of approaching an unconditionally pure mechanical quantum state by coupling mechanical motion to an optical field. If information about the mechanical object leaks out of the optomechanical system the efficiency of the cooling is limited. For a macroscopic or even massive, kilogram-sized mirror, which is for instance suspended as a pendulum in operating gravitational wave detectors, optical cooling even in combination with cryogenic cooling is not sufficient to bring the mirror motion into the quantum regime. The reason is the low mechanical resonance frequency of such a system. The lower the phonon energy, the lower is the effective temperature required to achieve a thermal occupation number below unity on average. Furthermore, lower frequencies make it unfeasible to operate the system in the resolved sideband regime. At least, feedback cooling together with optical trapping has led to cooling of macroscopic mirrors (10 kg individually and 2.5 kg for the differential mode) in one of the kilometer-scale LIGO gravitational wave detectors. A phonon occupation number of ~200 was achieved [43].

To reach *pure* motional quantum states of massive mirrors a conditional state preparation was proposed [44, 45]. The motional state prepared this way still has a huge *unconditional* phase space uncertainty. But when the mechanical state is conditioned on the outcome of a (continuous) measurement on the system the resulting state can be highly pure. The information that leaks out of the optomechanical system is captured and used for the state-preparation process. In this case the resolved sideband regime is not necessary. In principle the optomechanical system does not need to have optical cavity at all. The measurement of the leaked out information together with the knowledge of thermal noise and other noise coupling rates allows the definition of pure, conditional quantum states via a quantum Wiener filter process [44–47].

The purity that can be achieved via conditioning on measurement results depends on the measurement sensitivity for the displacement of the mirror. It turns out that the standard quantum limit (SQL) of the optomechanical system serves as a valuable reference. For a device that has classical noise at the SQL (but not above), it is possible to prepare a conditional state that has a von Neumann entropy identical to that of a thermal state with mean phonon occupation number of $\frac{1}{2}$. Reaching the SQL is thus a valuable milestone towards the preparation of conditional pure quantum states of mechanical motion.

Pure single-mode mechanical states such as the ground state do not offer much insight into quantum mechanics. But even in linear systems driven by Gaussian optical states also more complex pure states can be generated, for instance two-mode squeezed mechanical states. Two-mode squeezed states are also called (bi-partite) Einstein-Podolsky-Rosen (EPR) entangled states [48]. Experiments with light and atomic matter have already confirmed the prediction of quantum theory that two objects A and B can be prepared in such a state [49–52]. Measurements on EPR states are described by two distinct commutators. According to $[x_i, p_i] \neq 0$, the position and the momentum of a single object, e.g. a subsystem of the entangled state, can not have precisely defined values. For an EPR entangled state, however, the variances of the distance of two objects and the sum of their momenta do have precisely defined values, according to $[x_A - x_B, p_A + p_B] = 0$. The correlation between EPR entangled subsystems is a direct consequence of these two commutators. Optomechanical experiments with classical noise budgets below the SQL can be used to generate entanglement among mechanical objects, as described in Ref. [44]. The idea is actually very much analogous to the original gedankenexperiment of EPR, in which two mechanical objects, well isolated from the environment, have their common and differential modes measured with the classical noise budget below the SQL, and with the quantum noise budget *different from each other*, in such a way that the differential mechanical mode is squeezed in position, and common mechanical mode squeezed in momentum—therefore realizing approximately an EPR state. Using conditional states, EPR entanglement might become feasible not only with microscopic particles but even with kilogram-scale mirrors.

4.5 Current Trends in 'Suspended Mirrors'

Suspended micromirrors have outstanding optical quality but also two major drawbacks: their size—which is severely limited by diffraction losses and fabrication, masking and coating issues—and the poor mechanical quality of the dielectric coatings used for most of them. The former substantially increases the effective mass m_{eff} of the resonator with respect to their integrated or microwave competitors, while the latter often limits the Q. We present here some of the recent progress performed to tackle these issues, still with the geometry of suspended micromirrors. A summary of these new structures can be found in Fig. 4.14. Alternative approaches that are being discussed elsewhere include novel microcavity designs (see the chapter by Favero, Weig and Sankey), photonic crystal cavities (see the chapter by Safavi-Naeini and Painter) or levitated particles [53–58].

4.5.1 Monocrystalline Coatings

One approach in the past to achieve simultaneous high optical and mechanical quality has been to fabricate micrometer-scale mechanical resonators directly from optical coatings. In essence, an underlying sacrificial material is selectively removed in order to create free-standing optomechanical devices [18, 59–62]. Typical mechanical resonators achieve masses between 10 and 100 ng with eigenfrequencies up to some MHz, while optical losses due to absorption and scatter can be designed in the range of a few parts per million (ppm). Unfortunately, the best available optical coatings to date are ion-beam sputtered SiO_2/Ta_2O_5 dielectric multi-layers, which exhibit intrinsically high internal mechanical losses [63] and hence very low mechanical Q. Interestingly, the measured Q-values of these resonators have shown to be consistent with the coating loss angles observed in the LIGO studies of gravitational wave detector coatings of the same material [64–66], thereby opening a new method to determine coating loss angles through direct micro-optomechanical measurements. A recent promising development to overcome this limitation is to fabricate resonators from a single-crystal Bragg reflector, thereby minimizing the intrinsic defect density of the material and guaranteeing high mechanical Q. As a consequence, Bragg multilayer micromirrors made of the ternary alloy $Al_XGa_{1-X}As$ show significantly improved mechanical quality—approx. a factor of 10 at room temperature compared to the best ion-beam sputtered coatings—while maintaining high reflectivity comparable to other high-quality optical coatings [61, 62, 66]. Using epitaxial growth techniques such as molecular beam epitaxy (MBE) or metalorganic chemical vapor deposition (MOCVD), almost arbitrarily thick multilayers of these materials may be grown while still retaining a single-crystalline structure. Further improvements on the quality factor have been achieved by quantifying and minimizing support-induced mechanical losses [67] (see Fig. 4.14), which is another key damping mechanism in nano- and micromechanical resonators. Finally, the development of high-Q semiconductor-based micromirrors for cavity quantum optomechanics experiments has sparked the development of a new optical coating technology that allows to exploit these single-crystal multilayers as large-scale coatings [68]. The high mechanical quality leads to a significant reduction in the amplitude of the Brownian motion of the coating compared to state-of-the-art amorphous multilayer coatings. This is a long-sought improvement for high-precision interferometer applications including gravitational wave detectors and optical clocks, where the cavity noise due to thermal motion of the coatings (thermal coating noise) has become the dominant limitation for their performance [64, 69–71].

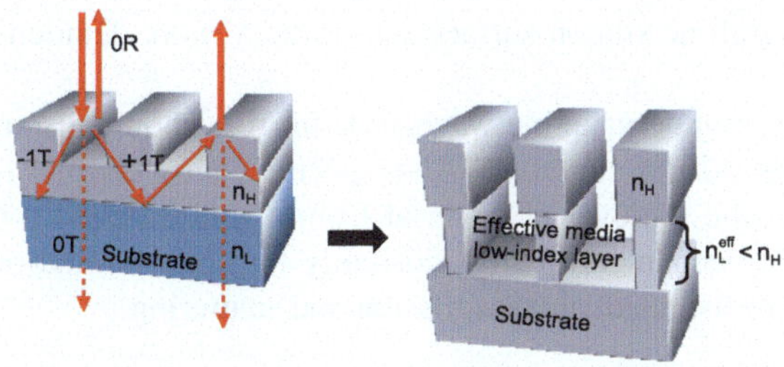

Fig. 4.13 Evolution from a standard surface wave-guide mirror (*left*) to a truly monolithic mirror from a single surface-nanostructured silicon crystal (*right*). In both cases, the grating structure at the surface produces one reflected order of refraction (0R) and three transmitted orders (−1T, 0T, +1T). In a simplified ray picture (*left*) the ±1st orders do not enter the low index layer beneath due to total internal reflection. The geometry is chosen such that for a rather broad range of wavelengths, finally, the transmitted zeroth-order beam destructively interferes, whereas the reflected zeroth-order beam constructively interferes. In theory, the concept allows for a normal incidence reflectivity of 100 %. Adapted from Ref. [72]

4.5.2 Monolithic Mirrors from Single Crystals

Another recent developement in the fabrication of mirrors with high reflectivities as well as high mechanical qualities is based on resonant surface waveguides. Reference [72] reported on the experimental realization of a high-reflectivity cavity mirror that solely consists of a single silicon crystal (Fig. 4.13, right). No material was added to the crystal. The mirror was based on resonant coupling to a guided optical mode of a surface nanostructure [73–76]. In full agreement with a rigorous model a reflectivity of (99.79 ± 0.01) % at a wavelength of 1.55 μm, and a cavity finesse of 2,784 was achieved. Since no material is added to the silicon substrate, the limiting coating Brownian thermal noise as found in Refs. [65, 77] is avoided. A coating related reduction of substrate Q-factors should also be greatly avoided, as suggested by experimental results [78].

The realized concept exploited a *broadband* guided optical mode and therefore did not increase the interaction length of the light with the mirror surface thus keeping thermorefractive noise as well as absorption low. For its fabrication, first, a standard silicon wafer with 100 mm in diameter was thermally oxidized with a 1-μm silica layer and coated with a 80-nm chromium layer. Both layers served as the mask during the silicon etching process. The grating had a period of 700 nm and a fill factor of 0.56 [79] and was defined by means of electron beam lithography for an area of (7.5 × 13) mm^2. The developed binary resist profile was then transferred into the chromium layer, the oxidized silica layer and the silicon bulk substrate by exploiting an anisotropic inductively-coupled-plasma dry etching process. Afterwards an isotropic (i.e. polydirectional) etching process enabled the undercut of the upper grating to generate the low fill factor grating beneath. Finally, the etching mask materials (silica, chromium, and resist) were removed by means of wet chemical etching baring the mono-crystalline silicon surface structure. A more detailed description can be found in [72].

The concept of monocrystalline waveguide mirrors should also be realizable for mirrors with large radii of curvature. Also specific cavity coupler reflectivities can be realized. Since the removal of material from a single (silicon) crystal does not demolish its high mechanical quality, currently the main limitation occurs on the optical side in terms of residual transmission and scattering.

4.5.3 Micropillars

Figure 4.14a presents a resonator recently designed and developed at LKB and ONERA. To reduce both clamping and coating mechanical losses, the resonator is based on a compression mechanical mode rather than a shear mode: we use the first length extension mode of a micro-pillar, clamped at its center by a thin membrane [80]. Single-crystal quartz has been chosen to benefit from its high intrinsic quality factor. The symmetry of the resonator lowers the displacement at the clamping location, decreasing clamping losses. Also, the mirror coating at the top of the pillar has a quasi-null strain, lowering coating losses. The size and geometry of both the pillar and the membrane have been carefully optimized in order to keep the pillar mechanically isolated. The geometry also respects the quartz internal trigonal symmetry.

Micro-pillars typically have a transverse size of the order of 200 μm and are 1-mm long, with a resonance frequency $\Omega_M/2\pi \simeq 3.6$ MHz. The expected mass is then around a few tens of μg. Q factors of the order of 10^6 and above have been demonstrated at room temperature. Note that such a system has a radiation-pressure over thermal-noise ratio (4.4) of a few hundreds at a cryogenic temperature of 100 mK, turning it into a promising candidate to demonstrate QRPN for instance.

4.5.4 Trampoline Resonators

A different type of resonators are *trampoline resonators* (see Fig. 4.14c), which can be used to demonstrate both very low masses and yet relatively low mechanical resonance frequencies, by using very thin resonators. The resonators are processed by etching a SiN membrane upon which layers of 2 different dielectric materials (low index/high index) have been alternatively coated to provide for the optical mirror. Resonance frequencies in the tens of kHz range and mechanical Q values of the order of a few hundreds of thousands have been demonstrated with such devices [81].

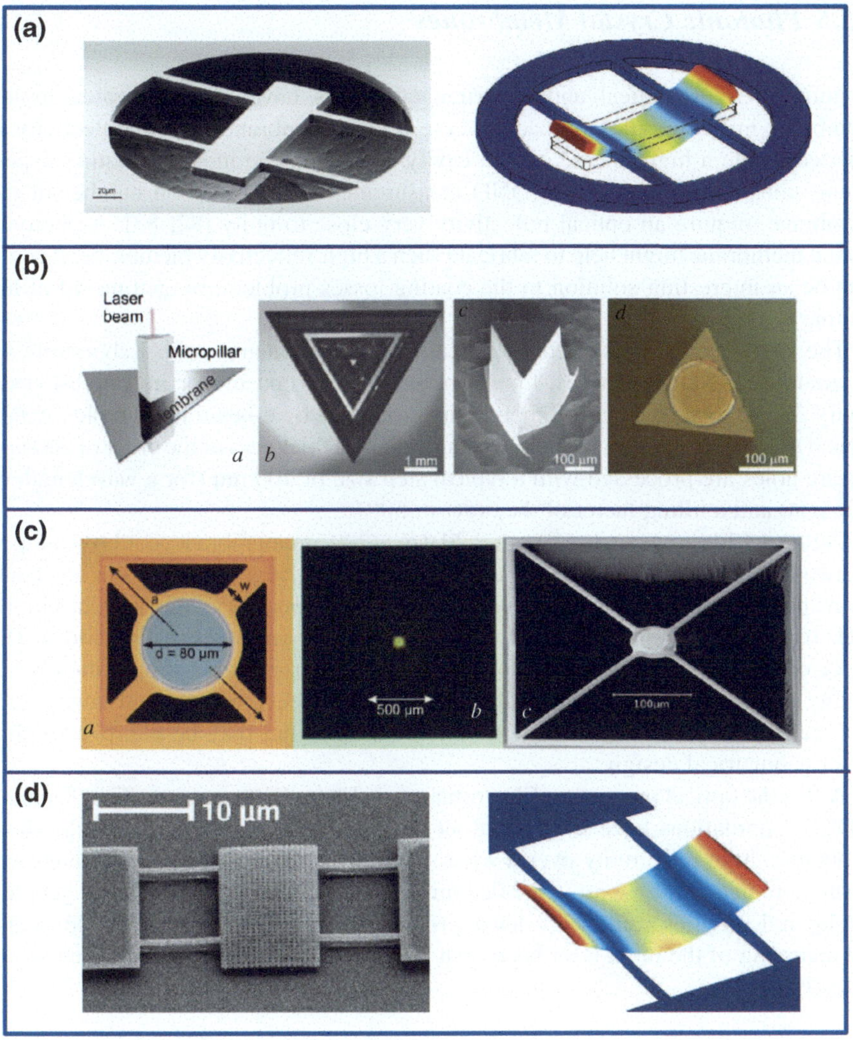

Fig. 4.14 **a** Monocrystalline optomechanical resonators fabricated from an epitaxial AlGaAs Bragg mirror. *Left* SEM image of a fabricated device. The doubly-clamped geometry is used to minimize mechanical clamping losses [67]. *Right* Numerical finite-element-modeling simulation of the mechanical motion of the fundamental mode. One can clearly see that the support structures are barely excited, hence minimizing the clamping-induced coupling to the environment. **b** *a* 3D view of the pillar concept. *b* Scanning electron microscope picture of the full structure with the micropillar at the center, its clamping membrane and dynamical frame. *c* SEM closer view of the micropillar before the optical coating. *d* *Top* view of the coating. Adapted from Ref. [80]. **c** Three micro-optomechanical trampoline resonators. *a* Optical image of a resonator with diameter $d = 80$ µm, thickness $t =$ 500 nm, resonator arms of length $a = 250$ µm and width $w = 20$ µm. *b* Optical image ($d = 80$ µm, $t = 300$ nm, $a = 2$ mm and $w = 2$ µm). *c*: Scanning electron microscope image ($d = 40$ µm, $t =$ 500 nm, $a = 500$ µm and $w = 10$ µm). Adapted from Ref. [81]. **d** SEM view of a $10 \times 20 \times 0.2\,\mu\text{m}^3$ photonic-crystal membrane, clamped by four 8-µm long and 1-µm wide auxiliary bridges. The simulated vibration profile of the fundamental mode is shown as well. Adapted from Ref. [84]

4.5.5 Photonic Crystal Membranes

A number of theoretical and experimental works have been dedicated to the membrane-in-the-middle approach, where a thin membrane of low reflectivity is inserted inside a high-finesse optical cavity. However, the most interesting experiments considered, such as the QND measurement of the phonon number of the membrane, require an optical reflectivity very close to unity [82, 83]. A photonic crystal membrane might help to fabricate such a high-reflectivity membrane. It could also be an interesting solution to the coating losses problem, by getting rid of the coating itself.

The LKB group has started investigating such resonators with Laboratoire de Photonique et Nanostructures, first to be used as an end mirror in a high-finesse cavity. The resonator is an InP membrane processed by e-beam lithography, with a typical transverse size of a few tens of microns and a thickness of the order of 200 nm. Square holes are processed with a typical step size of 700 nm (for a wavelength of 1.06 μm) and a filling factor of the order of 50 %.

Figure 4.14d shows such a 10 μm $\times$ 20 μm $\times$ 200 nm membrane, doubly-clamped by two 1-μm long strips located at vibration nodes, together with the mechanical response of its first vibration mode. Observed resonance frequencies are in the MHz band, in excellent agreement with the Finite Element Method simulations. The expected masses are of the order of 10 ng, which considerably lowers the optical quality required to demonstrate radiation-pressure effects. Mechanical quality factors are currently limited to a few tens of thousands, but may be improved with a better geometrical design.

As for the optical properties, Electromagnetic Finite Differences in Time Domain (FDTD) simulations have shown that an optimized lattice geometry should allow to obtain a high reflectivity over a wide spectral range: theoretical reflectivies can reach more than 99.99 % around 1,064 nm for instance. Processed samples yet only display reflectivities at the 95 % level, probably limited by geometry gradients and the focussing of the input laser beam, but work is still underway toward even larger reflectivities.

4.6 Conclusion and Outlook

Suspended mirrors in Fabry–Perot cavities form the most elementary system for cavity optomechanics, where they continue to contribute important proof-of-concept demonstrations for the various applications of the field. Their implementations range from test masses in gravitational wave interferometers that can weigh tens of kilogram to micrometer-sized mirrors with masses as small as some tens of nanogram. With the new developments underway also these systems will soon enter the quantum regime. The combination with the large span in accessible mass, 12 orders of magnitude, offers new opportunities: in a long term perspective, generating quantum states of motion of massive objects may bring the experimentally accessible realms of quantum physics and gravity wone step closer to each other [16, 47, 85, 86].

References

1. E.F. Nichols, G.F. Hull, Phys. Rev. (Ser. I) **13**(5), 307 (1901)
2. P. Lebedew, Annalen der Physik **311**(11), 433 (1901)
3. A. Einstein, Phys. Z. **10**(22), 817 (1909)
4. K. Meyenn (ed.), *Eine Entdeckung von ganz außerordentlicher Tragweite: Schrödingers Briefwechsel zur Wellenmechanik und zum Katzenparadoxon* (Springer, Heidelberg, 2011)
5. B.C. Barish, R. Weiss, Phys. Today **52**, 44 (1999)
6. P.R. Saulson, *Fundamentals of Interferometric Gravitational Wave Detectors* (World Scientific Publishing, Singapore, 1994)
7. V.B. Braginsky, A.B. Manukin, Sov. Phys. JETP **25**, 653 (1967)
8. V.B. Braginsky, A.B. Manukin, M.Y. Tikhonov, Sov. Phys. JETP **31**, 829 (1970)
9. C. Caves, Phys. Rev. Lett. **45**, 75 (1980)
10. A. Dorsel, J.D. McCullen, P. Meystre, E. Vignes, H. Walther, Phys. Rev. Lett. **51**, 1550 (1983)
11. Y. Hadjar, P.F. Cohadon, C.G. Aminoff, M. Pinard, A. Heidmann, Europhys. Lett. **47**, 545 (1999)
12. P.F. Cohadon, A. Heidmann, M. Pinard, Phys. Rev. Lett. **83**, 3174 (1999)
13. I. Tittonen, G. Breitenbach, T. Kalkbrenner, T. Müller, R. Conradt, S. Schiller, E. Steinsland, N. Blanc, N.F. de Rooij, Phys. Rev. A **59**(2), 1038 (1999), http://link.aps.org/abstract/PRA/v59/p1038
14. M. Aspelmeyer, T.J. Kippenberg, F. Marquardt (2013), arXiv:1303.0733
15. O. Arcizet, P.F. Cohadon, T. Briant, M. Pinard, A. Heidmann, Nature **444**, 71 (2006)
16. D. Kleckner, D. Bouwmeester, Nature **444**, 75 (2006)
17. S. Gröblacher, K. Hammerer, M.R. Vanner, M. Aspelmeyer, Nature **460**, 724 (2009)
18. S. Gröblacher, S. Gigan, H.R. Böhm, A. Zeilinger, M. Aspelmeyer, Europhys. Lett. **81**, 54003 (2008)
19. T. Corbitt, Y. Chen, E. Innerhofer, H. Müller-Ebhardt, D. Ottaway, H. Rehbein, D. Sigg, S. Whitcomb, C. Wipf, N. Mavalvala, Phys. Rev. Lett. **98**, 150802 (2007)
20. O. Arcizet, P.F. Cohadon, T. Briant, M. Pinard, A. Heidmann, J.M. Mackowski, C. Michel, L. Pinard, O. Français, L. Rousseau, Phys. Rev. Lett. **97**, 133601 (2006)
21. T. Briant, P.F. Cohadon, M. Pinard, A. Heidmann, Eur. Phys. J. D **22**, 131 (2003)
22. P. Meystre, M.O. Scully (eds.), *Quantum Optics, Experimental Gravitation and Measurement Theory* (Springer Nato Science Series B, New York, 1983)
23. T.P. Purdy, R.W. Peterson, C.A. Regal, Science (N.Y.) **339**(6121), 801 (2013)
24. A.A. Clerk, S.M. Girvin, F. Marquardt, R.J. Schoelkopf, Rev. Mod. Phys. **82**(2), 1155 (2010)
25. P. Verlot, A. Tavernarakis, T. Briant, P.F. Cohadon, A. Heidmann, Phys. Rev. Lett. **102**, 103601 (2009)
26. P. Verlot, A. Tavernarakis, C. Molinelli, A. Kuhn, T. Antoni, S. Gras, T. Briant, P.F. Cohadon, A. Heidmann, L. Pinard, C. Michel, R. Flaminio, M. Bahriz, O. Le Traon, I. Abram, A. Beveratos, R. Braive, I. Sagnes, I. Robert-Philip, C. R. Phys. **12**, 826 (2011)
27. A. Heidmann, Y. Hadjar, M. Pinard, Appl. Phys. B **64**, 173 (1997)
28. P. Grangier, J. Levenson, J. Poizat, Nature **396**, 537 (1998)
29. M.D. LaHaye, O. Buu, B. Camarota, K. Schwab, Science **304**, 74 (2004)
30. K.C. Schwab, M.L. Roukes, Phys. Today **58**, 36 (2005)
31. M. Poggio, C. Degen, H. Mamin, D. Rugar, Phys. Rev. Lett. **99**, 017201 (2007)
32. T. Corbitt, C. Wipf, T. Bodiya, D. Ottaway, D. Sigg, N. Smith, S. Whitcomb, N. Mavalvala, Phys. Rev. Lett. **99**, 160801 (2007)
33. C. Metzger, K. Karrai, Nature **432**, 1002 (2004)

34. S. Gigan, H.R. Böhm, M. Paternostro, F. Blaser, G. Langer, J.B. Hertzberg, K. Schwab, D. Bäuerle, M. Aspelmeyer, A. Zeilinger, Nature **444**, 67 (2006)
35. M.R. Vanner, J. Hofer, G.D. Cole, M. Aspelmeyer, Nat. Commun. **4**(May), 2295 (2013)
36. C. Mow-Lowry, A. Mullavey, S. Gossler, M. Gray, D. McClelland, Phys. Rev. Lett. **100**, 010801 (2008)
37. J.M. Courty, A. Heidmann, M. Pinard, Eur. Phys. J. D **17**, 399 (2001)
38. D. Vitali, S. Mancini, L. Ribichini, P. Tombesi, Phys. Rev. A **65**, 63803 (2002)
39. P.F. Cohadon, O. Arcizet, T. Briant, A. Heidmann, M. Pinard, Proc. SPIE **5846**, 124 (2005)
40. C. Genes, D. Vitali, P. Tombesi, S. Gigan, M. Aspelmeyer, Phys. Rev. A **77**, 33804 (2008)
41. B. Sheard, M. Gray, C. Mow-Lowry, D. McClelland, S. Whitcomb, Phys. Rev. A **69**, 051801 (2004)
42. A. Di Virgilio, L. Barsotti, S. Braccini, C. Bradaschia, G. Cella, C. Corda, V. Dattilo, I. Ferrante, F. Fidecaro, I. Fiori, F. Frasconi, A. Gennai, A. Giazotto, P. La Penna, G. Losurdo, E. Majorana, M. Mantovani, A. Pasqualetti, D. Passuello, F. Piergiovanni, A. Porzio, P. Puppo, P. Rapagnani, F. Ricci, S. Solimeno, G. Vajente, F. Vetrano, Phys. Rev. A **74**, 013813 (2006)
43. B.P. Abbott et al., New J. Phys. **1**, 073032 (2009)
44. H. Müller-Ebhardt, H. Rehbein, R. Schnabel, K. Danzmann, Y. Chen, Phys. Rev. Lett. **100**(1), 013601 (2008)
45. H. Müller-Ebhardt, H. Rehbein, C. Li, Y. Mino, K. Somiya, R. Schnabel, K. Danzmann, Y. Chen, Phys. Rev. A **80**(4), 043802 (2009)
46. H. Miao, S. Danilishin, H. Müller-Ebhardt, Y. Chen, New J. Phys. **12**(8), 083032 (2010)
47. Y. Chen, J. Phys. B: At. Mol. Opt. Phys. **46**(10), 104001 (2013)
48. A. Einstein, B. Podolsky, N. Rosen, Phys. Rev. **47**(10), 777 (1935)
49. Z.Y. Ou, S.F. Pereira, H.J. Kimble, K.C. Peng, Phys. Rev. Lett. **68**(25), 3663 (1992)
50. W.P. Bowen, R. Schnabel, P.K. Lam, Phys. Rev. Lett. **90**(4), 043601 (2003)
51. B. Julsgaard, A. Kozhekin, E.S. Polzik, Nature **413**, 400 (2001)
52. T. Eberle, V. Händchen, R. Schnabel, Optics Express (2013)
53. D.E. Chang, C.A. Regal, S.B. Papp, D.J. Wilson, J. Ye, O. Painter, H.J. Kimble, P. Zoller, Proc. Nat. Acad. Sci. U.S.A. **107**(3), 1005 (2010)
54. O. Romero-Isart, M.L. Juan, R. Quidant, J.I. Cirac, Quant. V **1** (2010)
55. P.F. Barker, M.N. Shneider, Phys. Rev. A **81**(2), 23826 (2010)
56. N. Kiesel, F. Blaser, U. Delić, D. Grass, R. Kaltenbaek, M. Aspelmeyer, Proc. Nat. Acad. Sci. U.S.A. **110**(35), 14180 (2013)
57. P. Asenbaum, S. Kuhn, S. Nimmrichter, U. Sezer, M. Arndt, Nat. Commun. **4**, 2743 (2013)
58. Z.Q. Yin, A.A. Geraci, T. Li, Int. J. Mod. Phys. B **27**(26), 1330018 (2013)
59. H.R. Böhm, S. Gigan, G. Langer, J.B. Hertzberg, F. Blaser, D. Bäuerle, K.C. Schwab, A. Zeilinger, M. Aspelmeyer, Appl. Phys. Lett. **89**, 223101 (2006)
60. S. Gröblacher, J.B. Hertzberg, M.R. Vanner, G. Cole, S. Gigan, K.C. Schwab, M. Aspelmeyer, Nat. Phys. **5**, 485 (2009)
61. G.D. Cole, S. Gröblacher, S. Gugler, K. Gigan, M. Aspelmeyer, Appl. Phys. Lett. **92**, 261108 (2008)
62. G.D. Cole, I. Wilson-Rae, M.R. Vanner, S. Gröblacher, J. Pohl, M. Zorn, M. Weyers, A. Peters, M. Aspelmeyer, in 2010 IEEE 23rd International Conference on Micro Electro Mechanical Systems (MEMS) (IEEE, 2010), pp. 847–850
63. S.D. Penn, P.H. Sneddon, H. Armandula, J.C. Betzwieser, G. Cagnoli, J. Camp, D.R.M. Crooks, M.M. Fejer, A.M. Gretarsson, G.M. Harry, J. Hough, S.E. Kittelberger, M.J. Mortonson, R. Route, S. Rowan, C.C. Vassiliou, Class. Quant. Gravity **20**(13), 2917 (2003)
64. G.M. Harry, A.M. Gretarsson, P.R. Saulson, S.E. Kittelberger, S.D. Penn, W.J. Startin, S. Rowan, M.M. Fejer, D.R.M. Crooks, G. Cagnoli, J. Hough, N. Nakagawa, Class. Quant. Gravity **19**, 897 (2002)
65. G.M. Harry, H. Armandula, E. Black, D.R.M. Crooks, G. Cagnoli, J. Hough, P. Murray, S. Reid, S. Rowan, P. Sneddon, M.M. Fejer, R. Route, S.D. Penn, Appl. Opt. **45**, 1569 (2006)
66. G.D. Cole, in Proceedings of SPIE 8458, Optical Trapping and Optical Micromanipulation IX, ed. by K. Dholakia, G.C. Spalding (2012), p. 845807

67. G.D. Cole, I. Wilson-Rae, K. Werbach, M.R. Vanner, M. Aspelmeyer, Nat. Commun. **2**, 231 (2011)
68. G.D. Cole, W. Zhang, M.J. Martin, J. Ye, M. Aspelmeyer, Nat. Photonics **7**(8), 644 (2013)
69. K. Numata, A. Kemery, J. Camp, Phys. Rev. Lett. **93**, 250602 (2004)
70. G. Harry, T.P. Bodiya, R. DeSalvo (eds.), *Optical Coatings and Thermal Noise in Precision Measurement* (Cambridge University Press, Cambridge, 2011)
71. T. Kessler, C. Hagemann, C. Grebing, T. Legero, U. Sterr, F. Riehle, M.J. Martin, L. Chen, J. Ye, Nat. Photonics **6**(10), 687 (2012)
72. F. Brückner, D. Friedrich, T. Clausnitzer, M. Britzger, O. Burmeister, K. Danzmann, E.B. Kley, A. Tünnermann, R. Schnabel, Phys. Rev. Lett. **104**(16), 163903 (2010)
73. G.A. Golubenko, A.S. Svakhin, V.A. Sychugov, A.V. Tishchenko, Sov. J. Quant. Electron. **15**(7), 886 (1985)
74. A. Sharon, D. Rosenblatt, A.A. Friesem, J. Opt. Soc. Am. A **14**(11), 2985 (1997)
75. C.F.R. Mateus, S. Member, M.C.Y. Huang, L. Chen, C.J. Chang-Hasnain, Y. Suzuki, IEEE Photonics Technol. Lett. **16**(7), 1676 (2004)
76. A. Bunkowski, O. Burmeister, D. Friedrich, K. Danzmann, R. Schnabel, Class. Quant. Gravity **23**(24), 7297 (2006)
77. Y. Levin, Phys. Rev. D **57**(2), 659 (1998)
78. R. Nawrodt, A. Zimmer, T. Koettig, T. Clausnitzer, A. Bunkowski, E.B. Kley, R. Schnabel, K. Danzmann, S. Nietzsche, W. Vodel, A. Tünnermann, P. Seidel, New J. Phys. **9**(7), 225 (2007)
79. F. Brückner, T. Clausnitzer, O. Burmeister, D. Friedrich, E.B. Kley, K. Danzmann, A. Tünnermann, R. Schnabel, Opt. Lett. **33**(3), 264 (2008)
80. A.G. Kuhn, M. Bahriz, O. Ducloux, C. Chartier, O. Le Traon, T. Briant, P.F. Cohadon, A. Heidmann, C. Michel, L. Pinard, R. Flaminio, Appl. Phys. Lett. **99**, 121103 (2011)
81. D. Kleckner, B. Pepper, E. Jeffrey, P. Sonin, S.M. Thon, D. Bouwmeester, Opt. Express **19**, 19708 (2011)
82. J.D. Thompson, B.M. Zwickl, A.M. Jayich, F. Marquardt, S.M. Girvin, J.G.E. Harris, Nature **452**, 72 (2008)
83. A.M. Jayich, J.C. Sankey, B.M. Zwickl, C. Yang, J.D. Thompson, S.M. Girvin, A.A. Clerk, F. Marquardt, J.G.E. Harris, New J. Phys. **10**, 95008 (2008)
84. T. Antoni, A.G. Kuhn, T. Briant, P.F. Cohadon, A. Heidmann, R. Braive, A. Beveratos, I. Abram, L. Le Gratiet, I. Sagnes, I. Robert-Philip, Opt. Lett. **36**, 3434 (2011)
85. O. Romero-Isart, Phys. Rev. A **84**(5), 1 (2011)
86. I. Pikovski, M.R. Vanner, M. Aspelmeyer, M.S. Kim, C. Brukner, Nat. Phys. **8**, 393 (2012)

Chapter 5
Mechanical Resonators in the Middle of an Optical Cavity

Ivan Favero, Jack Sankey and Eva M. Weig

Abstract The interaction of light with mechanical motion has generated a burst of interest in recent years [1–4] from fundamental questions on the quantum motion of solid objects to novel engineering concepts for sensing and optical devices. This interest was originally inspired by experimental geometries in which a mechanically compliant object acts as the back mirror of Fabry-Perot cavity. In order to maintain a stable, high-finesse cavity with this geometry, the mechanical element's transverse dimensions must be larger than the photon's wavelength and its thickness sufficient to create an appreciable reflectivity. This places a lower bound on the mass of the mechanical object, limiting the effect of individual photons. Here we explore a complementary set of geometries in which a nanomechanical element or a very thin membrane is positioned within a high-finesse, rigid optical cavity. This geometry (inspired by the success of cavity quantum electrodynamics experiments with atoms) extends Fabry-Perot-based optomechanics to smaller / sub-wavelength mechanical elements. The added complexity associated with inserting a third (movable) scatterer also affords a new set of opportunities: in addition to reproducing the physics of a two-mirror optomechanical system, several "non-standard" types of linear and non-linear optomechanical couples can be generated. Combined with the diverse set of comparatively lightweight mechanical elements that can be inserted into a cavity, this geometry offers a high degree of optomechanical versatility for potential sensing and quantum information applications.

I. Favero (✉)
CNRS, Université Paris Diderot, Paris, France
e-mail: ivan.favero@univ-paris-diderot.fr

J. Sankey
McGill University, Montreal, Canada
e-mail: jack.sankey@mcgill.ca

E. M. Weig
University of Konstanz, Konstanz, Germany
e-mail: eva.weig@uni-konstanz.de

M. Aspelmeyer et al. (eds.), *Cavity Optomechanics*, Quantum Science and Technology,
DOI: 10.1007/978-3-642-55312-7_5,

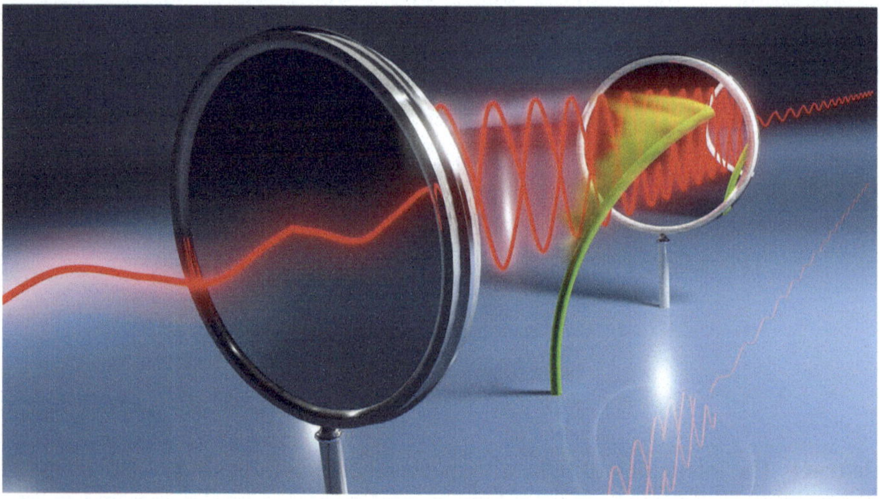

Fig. 5.1 Illustration of a hybrid cavity nano-optomechanical experiment with separate mechanical and optical resonator. The optical resonator is constituted by a Fabry-Perot cavity with rigidly mounted end mirrors. A vibrating nanomechanical resonator (*green*) is inserted into the cavity field (*red*). [Image: C. Hohmann, Nanosystems Initiative Munich]

5.1 Nanomechanical Resonators in the Middle of an Optical Cavity

Here we discuss a geometry in which a nanomechanical element is positioned within a rigid optical cavity as drawn in Fig. 5.1 and 5.2. In contrast to standard optomechanical cavities, the mechanical element and the optical resonator are two distinct bodies. The optical resonator is a Fabry-Perot cavity with rigid mirrors and the mechanical resonator is a nanomechanical object with sub-wavelength transverse dimensions. This nanomechanical resonator is positioned inside the cavity mode and acts as a scattering element for the cavity photons. This hybrid optical/nanomechanical approach offers two major benefits: First, it enables cavity optomechanical experiments with truly nanoscale mechanical resonators with dimensions far below the optical diffraction limit, such as e.g. carbon nanotubes. These molecular scale nano-objects are particularly responsive to optomechanical back-action forces. Second, a hybrid approach avoids the need for a compromise in the design of the cavity mirror and mechanical resonator, which often sets limits to the optical and/or mechanical quality for the case of a vibrating end mirror.

5.1.1 The Empty Optical Cavity

In the sub-wavelength optomechanics approach discussed here, the optical cavity can in principle be of any geometry, provided room is available for the insertion of the scattering nanomechanical element in the optical mode. As a specific example of interest, we will focus on the Fabry-Perot geometry (see experimental Sect. 5.1.4) because recent advances in fiber-based Fabry-Perot optical cavities make this platform suitable for interaction with nanomechanical systems. We remind here briefly notations that will be used throughout the chapter.

We assume a symmetric cavity composed of two lossless mirrors with amplitude reflection r and transmission it, r and t being real numbers and $i^2 = -1$. In a one-dimensional Fabry-Perot model, the standing-wave established in a resonantly pumped cavity is described by an intensity profile: $|E(x_0)|^2 = 4\mathscr{F}/\pi \cdot |E_0|^2 \sin^2(kx_0)$ where $k = 2\pi/\lambda$ is the monochromatic pump light wave number, E_0 is the pumping field amplitude and $\mathscr{F} = \pi/(1-r^2)$ the cavity finesse. In a three-dimensional approach with rotation invariance along its optical axis, the cavity supports several types of modes with different transverse geometry [5, 6]. We will use the Gaussian eigenmodes supported by cavities with spherical mirrors, and focus specifically on the TEM_{00} mode, assuming that it is the only mode to interact with the nanomechanical scatterer. Experimentally, the selection of a single mode of the cavity is obtained by spectral tuning of the pump laser and shaping the incident pump mode to overlap. In the limit of a vanishingly small divergence of the cavity mode, the three-dimensional spatial profile can be approximated by: $|E(x_0, w)|^2 = 4\mathscr{F}/\pi \cdot |E_0|^2 \sin^2(kx_0) \cdot \exp(-2w^2/w_0^2)$ with w_0 the cavity mode waist radius and w the off-axis coordinate. In this approximation, the Gaussian mode radius does not depend on the position x_0 along the optical axis and the curvature of the wavefront mode is neglected. This simplified mode profile is valid for microscale fiber-cavity experimental realizations that will be shown in Sect. 5.1.4, where the TEM_{00} mode is practically enclosed within a straight cylinder whose rotation axis is the cavity axis. More complex optical mode geometries can of course be treated with the tools introduced in this chapter, at the expense of added mathematical complexity. The wavefront curvature becomes for example important in the case of larger cavities embedding a membrane like discussed in Sect. 5.2.

5.1.2 The Nanomechanical Resonator

Here, we describe the vibrational modes of typical nanomechanical elements that can be inserted in a Fabry-Perot cavity. For simplicity, we assume that these elements are fabricated out of an isotropic elastic material with Young's Modulus Y, Poisson ratio ν and mass density ρ, and, if applicable, stress Θ. It can be diamond, carbon, silicon nitride or any other material of interest for nanomechanics. We restrict the discussion to prismatic, i.e. constant cross-section, nanoscale bars of length l, and discuss the

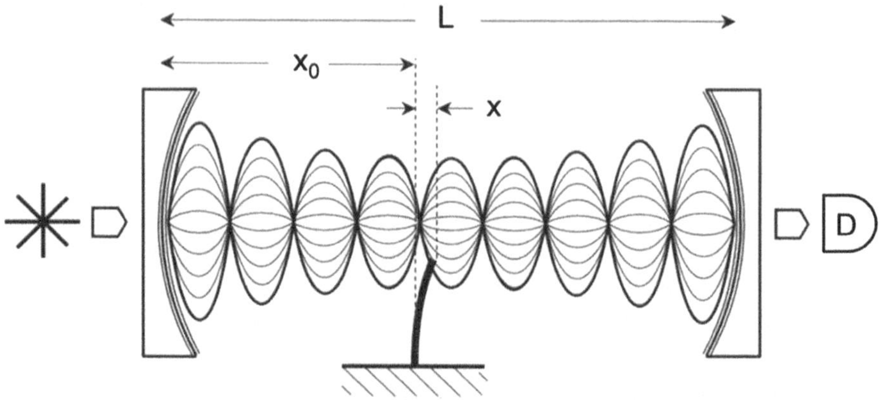

Fig. 5.2 Schematic of the amplitude envelope of the TEM_{00} mode of a Fabry-Perot cavity of length L, along with a nanomechanical resonator disturbing the mode at position x_0. In addition to the static effect of the resonator, its vibration with amplitude x dynamically modulates the cavity mode. [Figure modified from content originally published in [29]]

following two cases: Bars with rectangular cross-section (nanomechanical beams of width a and thickness b) and circular cross-section (nanomechanical plain tubes and cylinders of diameter d). In the limit where the transverse dimensions (a, b, d) of the bar are small compared to its length l, elasticity theory provides an analytical description of the flexural modes of such prismatic bars within the framework of the Euler-Bernoulli model [7, 8] in which the flexural displacement takes the general form $x(u) = C_1 \sin(\beta u) + C_2 \cos(\beta u) + C_3 \sinh(\beta u) + C_4 \cosh(\beta u)$. These modes differ depending on the end fastenings of the bar, and we will limit the discussion to three cases: singly clamped bars that are rigidly clamped at one end but free at the other end, doubly clamped bars that are rigidly held at both ends, and doubly clamped bars rigidly held at both ends that are in addition subjected to strong internal stress.

5.1.2.1 Bar Rigidly Clamped on one End

For the singly clamped boundary conditions (bar rigidly clamped at $u = 0$, see [7]), the transverse deflection x(u) at a position u along the bar takes the form

$$x(u) = C_2 [\cos(\beta u) - \cosh(\beta u)] + C_4 [\sin(\beta u) - \sinh(\beta u)] . \tag{5.1}$$

The mechanical wave vector β is a root of the implicit equation $\cos(\beta l) \cdot \cosh(\beta l) = -1$, and the condition $(d^2 x / du^2)_{u=l} = 0$ determines the relation between the two constants C_2 and C_4. The first three positive roots of this implicit equation are $\beta l = 1.875$; 4.694; 7.855. The related angular mechanical frequencies Ω_{M} are found using the relation

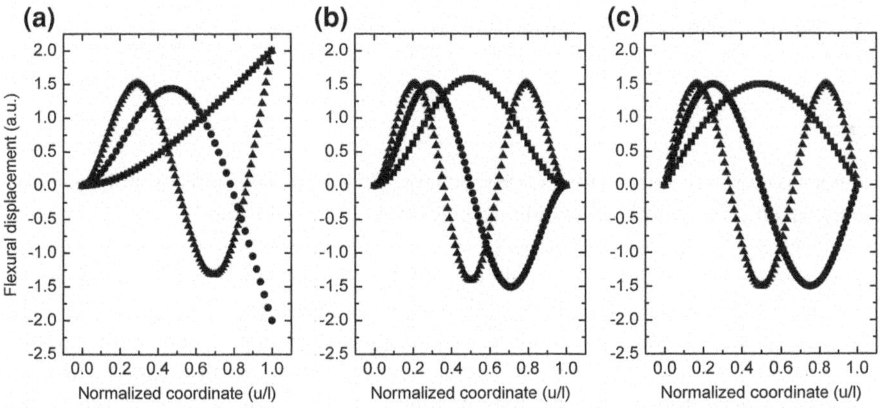

Fig. 5.3 Vibrational profile of the first three flexural modes of a **a** prismatic bar clamped rigidly on one end **b** prismatic bar rigidly clamped on both ends **c** strongly stressed prismatic bar clamped on both ends

$$\Omega_M^2 = \beta^4 \cdot \frac{YI}{\rho A} \tag{5.2}$$

with A the cross-section area and I the area moment of inertia with respect to the neutral plane of flexure ($I = \pi d^4/64$ for the plain cylinder and $I = ab^3/12$ for the rectangular beam in its out-of-plane flexural motion). The vibration profile of the first three modes for a singly clamped bar are shown in Fig. 5.3a. For the first flexural mode, of the bar (u=l). The point of maximal deflection is often used to define an effective motional mass m_{eff} through the relation

$$m_{\text{eff}} x^2(l) = \int_0^l \rho A x^2(u) du \tag{5.3}$$

Adopting this reduction procedure, the first flexural mode is represented by the motion of a point-mass located at $u = l$ with effective mass $m_{\text{eff}} = 0.23 \cdot \rho A l$. We take the opportunity here to remind that the choice of the reduction point is somewhat arbitrary, but that it directly affects the effective mass quantitatively. Here we adopt the point of maximal displacement as a reduction point, which is a rather standard choice when describing the mechanics of prismatic bars. When the bar is inserted into an optical cavity, one might want to choose another reduction point to account for the spatial selectivity of the optomechanical interaction. In the specific cases considered in this section, the reduction point will always be where both mechanical displacement and electromagnetic energy density are maximal. Note that the arbitrary character of this choice impacts not only the effective mass but also the frequency pull parameter G in the way that the product $g_0 = G x_{\text{ZPF}}$ remains a constant of the problem, independent of this choice. In case of doubt, the reader should hence re-express physical parameters in terms of the vacuum optomechanical coupling g_0. Values of g_0 will be computed for some specific experimental cases in Sect. 5.1.3.3.

Example

Let us consider here a molecular scale mechanical system: a carbon nanotube approximated by an effective plain straight cylinder (take $Y = 1\,\text{TPa}$ and $\rho = 1.925\,\text{g/cm}^3$ for a single-wall nanotube) of length $l = 5\,\mu\text{m}$ and diameter $d = 1.6\,\text{nm}$. According to the above model, the frequency of the fundamental mode is given by $f_\text{M} = \Omega_\text{M}/(2\pi) = (1.875)^2/(8\pi) \cdot \sqrt{Y/\rho} \cdot d/l^2$ and takes the value $f_\text{M} = 204\,\text{kHz}$. The corresponding effective mass is 4.45 attograms and the zero-point motion $x_\text{ZPF} = \sqrt{\hbar/(2m_\text{eff}\Omega_\text{M})} = 96\,\text{pm}$. As we see, the quantum fluctuations in position of such a bottom-up grown nanomechanical resonator are very large, much larger than those of top-down fabricated resonators that we present in the following.

5.1.2.2 Bar Rigidly Clamped on Both Ends

In the doubly clamped configuration (bar rigidly clamped at $u = 0$ and $u = l$ see [7]), the transverse deflection $x(u)$ at a position u along the bar adopts the same general form as in the singly-clamped case (see. Eq. 5.1), but with a different implicit equation $\cos(\beta l) \cdot \cosh(\beta l) = 1$. The condition $x(l) = 0$ gives the relation between the two constants C_2 and C_4. The first three positive roots of the implicit equation are $\beta l = 4.730;\ 7.853;\ 10.996$ and the relation the relation 5.2 still holds.

The vibration profiles of the first three modes for a doubly-clamped bar are shown in Fig. 5.3b. For the first flexural mode, the point of maximal deflection is now at the center of the bar ($u = l/2$) and can be employed to define an effective motional mass m_eff through the relation

$$m_\text{eff} x^2\left(\frac{l}{2}\right) = \int_0^l \rho A x^2(u) du \tag{5.4}$$

Adopting this reduction point for the flexural mode, the effective mass becomes $m_\text{eff} = 0.396 \cdot \rho A l$.

Example

Let us consider a diamond ($Y = 1.14\,\text{TPa}$ and $\rho = 3.52\,\text{g/cm}^3$) nanomechanical beam of length $l = 10\,\mu\text{m}$ and rectangular cross-section ($a = 100\,\text{nm}$ and $b = 50\,\text{nm}$). These dimensions are typical of those employed for top-down fabricated nanomechanical resonators. According to the above model, the frequency of the fundamental mode is given by $f_\text{M} = \Omega_\text{M}/(2\pi) = (4.730)^2/(4\pi\sqrt{3}) \cdot \sqrt{Y/\rho} \cdot b/l^2$ and takes the value $f_\text{M} = 9.25\,\text{MHz}$. The corresponding effective mass is 69.7 femtograms and the zero-point motion $x_\text{ZPF} = \sqrt{\hbar/(2m_\text{eff}\Omega_\text{M})} = 160\,\text{fm}$. The insertion of this diamond nanobeam in an optical cavity will also be discussed in Sect. 5.1.3.3.

5.1.2.3 Strongly Stressed Bar Rigidly Clamped on Both Ends

A third situation frequently encountered in nanomechanics is the prismatic bar clamped on both ends which, in addition, experiences strong tensile stress Θ, where stress is defined as the acting tensile force per cross-sectional area. The approximation of simply supported boundary conditions (i.e. both x(u) and its second derivative vanishing at u=0 and l) is conveniently used for long and thin stressed bars ($a,\ b \ll l$) [7]. In that case, the transverse mode profile has the form $x(u) = C_1 \sin(\beta u)$. The implicit equation determining the mechanical wave vector β is $\sin(\beta l) = 0$, such that the N-th root takes the form $\beta l = N\pi$. The vibration profile of the first three modes for the strongly stressed beam is included in Fig. 5.3c.

The corresponding angular eigenfrequencies are

$$\Omega_{M,N}^2 = \beta^4 \frac{YI}{\rho A}\left(1 + \frac{\Theta A l^2}{N^2 Y I \pi^2}\right) \tag{5.5}$$

for the N-th eigenmode. In the limit of vanishing stress, the eigenmodes of the doubly clamped simply supported bar are recovered. Interestingly, in the limit of high stress the eigenmodes of a one-dimensional stretched string $\Omega_{M,N} = N\pi/l \cdot \sqrt{\Theta/\rho}$ are obtained. This is the situation encountered later on in the experimental section for the device displayed in Fig. 5.7b. The effective motional mass of the first flexural mode of the strongly stressed bar can again be defined via the relation

$$m_{\mathrm{eff}} x^2\left(\frac{l}{2}\right) = \int_0^l \rho A x^2(u) du \tag{5.6}$$

and yields $m_{\mathrm{eff}} = 0.5 \cdot \rho A l$.

Example

Let us consider a silicon nitride nanostring ($Y = 160\,\mathrm{GPa}$, $\rho = 3\,\mathrm{g/cm^3}$ and $\Theta = 830\,\mathrm{MPa}$) of the same dimensions specified for the example of the unstressed bar in Sect. 5.1.2.2. According to the full model, the fundamental mode has an eigenfrequency of $f_{\mathrm{M}} = 26.25\,\mathrm{MHz}$ whereas the stretched string model yields $f_{\mathrm{M}} = 26.30\,\mathrm{MHz}$ irrespective of the cross sectional area A. In comparison with the example from Sect. 5.1.2.2 the tensile stress clearly increases the mechanical eigenfrequencies [9]. The effective mass is 75 femtograms and the zero-point motion yields $x_{\mathrm{ZPF}} = \sqrt{\hbar/(2m_{\mathrm{eff}}\Omega_{\mathrm{M}})} = 64\,\mathrm{fm}$.

5.1.3 Nanomechanical Resonator and Optical Cavity in Interaction

There are several ways of describing the interaction between a selected optical cavity mode and the motion of a nanomechanical object scattering photons in the cavity. A proper description should include different cases of interest listed below.

5.1.3.1 Dispersive Case

In this first case, the nanomechanical object simply phase shifts cavity photons, scattering them all selectively into the same cavity mode. This corresponds to scattering processes that are typically far from any optical resonance of the object, avoid photon absorption, and additionally do not scatter photons out of the cavity into external radiation modes.

For a non-resonant atom placed in the middle of a driven Fabry-Perot cavity, these conditions are typically met, so long as care is taken to properly match the dipole-like scattering pattern of the atom with the driven mode of the cavity. For a moderate atom-cavity interaction, the atom can be seen as inducing a tiny additional refractive index in the cavity. Depending parametrically on the atom position, the cavity mode frequency gets shifted but the cavity mode geometry and the structure of the cavity spectrum are basically left unchanged. This description has been employed to study the cavity cooling of an atom [10–12], with some classical treatment of noise induced by residual light scattering into the free-space. More recently, the same picture was employed to discuss the case of a silica nanosphere positioned in an optical cavity, for optically levitated optomechanics proposals [13–16].

A second dispersive geometry that is by now well studied is the *membrane-in-the-middle* setting [17–27] where a lossless dielectric membrane replaces the atom. Section 5.2 of this chapter is entirely devoted to this approach. If the membrane is sufficiently well-aligned with the cavity, photons are dispersively reflected back into the cavity mode without loss. Small deviations from perfect alignment generally result in a similarly-stable cavity mode [22]. But since typical membranes also have an appreciable reflectivity and lateral extension larger than the cavity waist, they can significantly modify the cavity mode. In this case, a different treatment than the atom-like picture mentioned above is required. One approach is to employ a scattering- or transfer-matrix formalism to directly calculate the cavity modes in one dimension [18, 19, 28]. This model can be extended to the three-dimensional Hermite-Gaussian modes for the highly-symmetric case of a perfectly-aligned, thin membrane positioned near the cavity waist [23]. A second approach is to calculate the full spectrum of three-dimensional cavity modes using a degenerate optical perturbation theory derived in Sect. 5.2.2. This theory well-describes the observed cavity modes for typical, modestly-reflective membranes used in experiments, positioned anywhere within the cavity and for arbitrary membrane alignment, provided a large enough subset of cavity modes is included.

5.1.3.2 Absorptive (Dissipative) Case

This second case corresponds to situations where cavity photons are lost as a consequence of optical absorption and light scattering out of the cavity by the nanomechanical element. From the cavity viewpoint, these two types of processes are equivalent. For a nanomechanical resonator placed in the middle an optical cavity, these dissipative effects become powerful optomechanical ingredients that, depending on the context, can sometimes dominate over purely dispersive effects. Absorption typically occurs near optical resonances of the nanomechanical element, and is an important parameter to consider in experiments employing sub-wavelength sized objects. Indeed nanomechanical resonators with built-in optical excitonic resonances are considered in the next generation of experiments, following the proposal made in Ref. [29]: carbon nanotubes are known to exhibit fine resonances in the near-infrared range [30], nitrogen vacancy centers in diamond nanoresonators provide optical resonances in the visible range [31] and semiconductor quantum dots mounted on nanowires have excitonic optical transition in the near-infrared [32, 33]. All these single emitters offer the possibility, by a proper choice of the optical resonance, to tune the optical properties of the nanomechanical element on which they are mounted. The relative contribution of absorptive and dispersive effects is adjusted by controlled detuning of the cavity photons to the single-emitter optical transition. Situations of purely absorptive coupling can be obtained, with dispersive coupling being turned of, but where optomechanical back-action is still present, as we will discuss below. This *dissipative optomechanics* physics has been marginally studied even though it contains several original aspects. One striking feature is the possibility of optomechanical cooling to the motional ground-state in the bad-cavity limit, in contrast to the requirement of good-cavity operation (resolved sideband regime) in the standard dispersive coupling case. This fact has been pointed out in diverse dissipative optomechanics configurations, involving either a cavity coupling to the input port modulated by mechanical motion [34–36] or optical forces of dissipative nature like the photothermal force [37, 38].

5.1.3.3 Optomechanical Model with Dispersive and Dissipative Interactions

A complete optomechanical model for the case of a nanomechanical element in the middle of an optical cavity should ideally treat dissipative and dispersive effects on the same basis, include multiple modes in the cavity, and fully account for quantum noises induced by the coupling to absorption channels and external radiation modes. Such a model does not exist at the time of writing. A first model was developed to propose the insertion of sub-wavelength sized nanomechanical resonators in the middle of an optical cavity [29] and estimate orders of magnitude of expected phenomena. Even though this first model does not fulfill all of the above requirements, it already treats absorptive and dispersive back-action effects on the same ground, allowing a discussion of many interesting features of cavity optomechanics at the sub-wavelength scale.

The problem of a deep sub-wavelength sized scattering atom or quantum dot placed in a focused Gaussian optical beam in the paraxial limit can be mapped to that of a thin plane of conductivity σ at optical frequencies placed in a plane wave at normal incidence [39–41]. Indeed, a focused laser beam travels beyond the focal plane and at a distance much greater than its optical Rayleigh length x_R like a quasi-spherical wave with a Gaussian angular distribution of the intensity. A point dipole also emits in the far field like a quasi-spherical wave with a square-of-cos θ pattern which is not much different from the Gaussian case for small angles to the optical axis. A homogeneous distribution of point dipoles illuminated by a diffraction-limited focused laser beam at normal incidence will thus also re-emit a quasi-spherical wave with a Gaussian angular distribution of the intensity at distances $\gg x_R$. Therefore, for a plane detector placed at a distance far beyond x_R and for small angles to the optical axis, the laser light coming from a single point scatterer or two-dimensional scattering plane illuminated by a diffraction-limited focused laser at normal incidence appears to originate from the same point. The phase and intensity contrast between the excited field and the scattered field can be computed effectively as if the laser illuminated a homogeneous two-dimensional plane of scatterers or, equivalently, as if the laser illuminated an effective point scatterer with a stronger oscillator strength.

Here we focus on bar-shaped nanomechanical resonators such as those introduced in Sect. 5.1.2. A bar with deep sub-wavelength diameter illuminated by a diffraction-limited focused laser beam is the exact intermediate case between the point dipole and the homogeneous scattering plane [42]. The bar is made up of a sum of point dipoles within the diffraction-limited focused spot. As in previous cases, the bar re-emits a quasi-spherical wave with Gaussian transverse profile by Rayleigh scattering. The difference is that the intensity of the re-emission depends on the polarization of the incident light with respect to the bar axis. So the scattering is effectively similar to the re-emission of an polarization anisotropic two-dimensional plane illuminated by a diffraction-limited focused laser field. After selecting one principal axis of this plane, i.e. after aligning the linear polarization of light with a principal axis of the bar, we can model the nanoscatterer as a thin plane of amplitude transmittance $1/(1+\Sigma)$ and reflectance $-\Sigma/(1+\Sigma)$, where the conductance-related term$\Sigma = \sigma/(2\varepsilon_0 c)$ consists of a real and of an imaginary part $\Sigma = \Sigma_1 + i\Sigma_2$, with Σ_1 and Σ_2 relating to absorptive (dissipative) and dispersive interaction, respectively. Once the proper light polarization is selected, the prescription to determine Σ_1 and Σ_2 is to measure the amplitude reflectance and transmittance of the nanoscatterer in a focused Gaussian beam of size matching the cavity mode. The field amplitude distribution in the cavity perturbed by the sub-wavelength scatterer is then computed using transfer matrixes for plane waves $\exp(i(kx - \omega t))$. The strength of this approach is to reduce an apparently complex scattering situation to a simple one-dimensional problem. This reduction is valid for a cavity mode well approximated by a Gaussian TEM_{00} mode with diffraction-limited waist. Note also that in this effective description, Σ_1 and Σ_2 do not only merely contain information about the nanomechanical scattering element, but also about its interaction with the considered cavity mode. This will become apparent in the following where we relate Σ_2 to the cavity mode parameters.

Experimental situations of interest correspond to small absorption and scattering $(\Sigma_1, \Sigma_2) \ll 1$ for a nanomechanical element placed at a position x_0 (see Fig. 5.2) in a high finesse cavity $\mathcal{F} \gg 1$. In this limit, the transfer matrix calculation leads to analytical expressions for the maximal cavity transmission T and cavity angular resonance shift $\delta\omega_{\text{opt}}$ as a function of Σ and x_0:

$$T(x_0) = \frac{1}{\left[1 + \frac{4\mathcal{F}}{\pi}\Sigma_1 \sin^2(kx_0)\right]^2} \tag{5.7}$$

as well as

$$\delta\omega_{\text{opt}} = \frac{2c\Sigma_2}{L}\sin^2(kx_0) \tag{5.8}$$

Here the maximal cavity transmission T must be understood as follows: for each position x_0 of the nanomechanical element, the cavity back mirror position is adjusted to maximize the transmission, with the wavelength of the pump laser $\lambda = 2\pi/k$ remaining unchanged. This transmission T is a relevant and accessible quantity in experiments where the cavity length is actively stabilized on maximum transmission. As shown by the above formula, the dependence of T on the nanomechanical position x_0 is governed by absorptive effects associated to Σ_1 while the cavity resonance shift $\delta\omega_{\text{opt}}$ is governed by dispersive effects related to Σ_2. This provides the experimentalist with a second prescription to measure Σ_1 and Σ_2, directly within the cavity. Note also that both T and $\delta\omega_{\text{opt}}$ have a $\lambda/2$ periodicity on x_0 that is a reminiscence of the standing wave periodicity in the resonantly pumped empty cavity. This periodicity of T has been observed in experiments using a nanomembrane with optical absorption [19] and on an absorptive nanomechanical rod displaced in a cavity [43]. The $\lambda/2$ periodicity of $\delta\omega_{\text{opt}}$ has been observed in Ref. [19] with a nanomembrane setting, and in some recent experiments with a nanomechanical bar in a cavity as well. For the sake of consistency, note that formula (4) of Ref. [19] is exactly the same expression as the above equation (5.8) for $\delta\omega_{\text{opt}}$ if the membrane amplitude reflectance r_d is replaced by $-i\Sigma_1$, which is indeed valid in the limit $\Sigma_2 \ll 1$. With the convention adopted here, a negative value of Σ_2 will induce an effective elongation of the cavity optical length upon insertion of the nanomechanical element in the cavity.

If the whole nanomechanical object now moves rigidly by some amount dx_0 along the optical axis, an exact expression can be given for the frequency pull parameter $G = d\delta\omega_{\text{opt}}/dx_0 = 2kc\Sigma_2/L \cdot \sin(2kx_0)$, which shows that a shorter cavity leads to stronger coupling G. It is interesting to also relate G more directly to the cavity and nanomechanical parameters. This can be done by employing a perturbative approach of Maxwell equations [44, 45] in order to express how the cavity's resonant angular frequency ω_{opt} is shifted upon introduction of the nanomechanical element, assuming the latter to be composed of an optically isotropic material of refractive index n. This shift is given by:

$$\delta\omega_{opt} = -\frac{\omega_{opt}}{2}\frac{\iiint(n^2-1)E^2d^3r}{\iiint E^2d^3r} \tag{5.9}$$

with E the unperturbed electric field, and where the upper integral is performed over the volume of the nanomechanical element, while the lower integral is performed over the whole optical mode volume.

For the sake of the discussion, let us now consider a rigid nanomechanical bar of sub-wavelength sized cross-section A, of length l, placed in the Fabry-Perot mode of volume V_{cavity} considered in Sect. 5.1.1. The bar's principal axis intersects the cavity axis orthogonally and its position along the cavity axis is x_0. This rigid bar approximation means $x = 0$ in Fig. 5.2.

As a first example, we consider a bar clamped on one endpoint with the opposing freely vibrating endpoint positioned exactly on the cavity axis. The cavity resonance shift given above is then approximated by $\delta\omega_{opt} = -(\omega_{opt}/4)\cdot(A\cdot w_0/V_{cavity})\cdot(n^2-1)\cdot\sin^2(kx_0)\cdot\sqrt{\pi/2}\cdot\text{erf}(\sqrt{2}l/w_0)$ where we remind that w_0 is the cavity mode radius, where $V_{cavity} = \pi/4\cdot L\cdot w_0^2$ for the considered idealized fundamental mode of the Fabry-Perot, and where the error function is $\text{erf}(x) = 2/\sqrt{\pi}\cdot\int_0^x\exp(-t^2)dt$. In this limit, we obtain $\Sigma_2 = -A/(\lambda w_0)\cdot(n^2-1)\cdot\sqrt{\pi/2}\cdot\text{erf}(\sqrt{2}l/w_0)$ and $G = -(4\pi c/\lambda^2)\cdot(A/Lw_0)\cdot(n^2-1)\cdot\sqrt{\pi/2}\cdot\text{erf}(\sqrt{2}l/w_0)\cdot\sin(2kx_0)$. If we now use these expressions for the carbon nanotube introduced in Sect. 5.1.2.1, for a laser wavelength of 780 nm, a cavity of length $L = 40\,\mu$m and a waist radius $w_0 = 3.5\,\mu$m, and use an approximate effective refractive index of $n = 2.42$ for the tube material, we obtain $\Sigma_2 = -4.46\times10^{-6}$. The absolute value of $|G|$ reaches 0.54 MHz/nm for positions $|\sin(2kx_0)| = 1$. The corresponding vacuum optomechanical coupling $g_0 = Gx_{ZPF}$ amounts to 51.6 kHz for the fundamental flexural mode of the tube.

As a second example, we consider a bar clamped on both ends with no built-in stress, and with its middle-point positioned exactly on the cavity axis. The cavity resonance shift is here approximated by $\delta\omega_{opt} = -\omega_{opt}\cdot A\cdot w_0\cdot(n^2-1)\cdot\sin^2(kx_0)\cdot\sqrt{\pi/2}\cdot\text{erf}(l/\sqrt{2}w_0)/(2V_{cavity})$ so that we obtain $\Sigma_2 = -2A/(\lambda w_0)\cdot(n^2-1)\cdot\sqrt{\pi/2}\cdot\text{erf}(l/\sqrt{2}w_0)$ and $G = -(8\pi c/\lambda^2)\cdot A/(Lw_0)\cdot(n^2-1)\cdot\sqrt{\pi/2}\cdot\text{erf}(l/\sqrt{2}w_0)\cdot\sin(2kx_0)$. If we now use these expressions for the diamond nanomechanical beam introduced in Sect. 5.1.2.2, for the same cavity as above, and take an optical refractive index $n = 2.419$ for diamond, we obtain $\Sigma_2 = -2.22\times10^{-2}$ and a maximal $|G|$ of 2.7 GHz/nm. When considering coupling to the fundamental flexural mode of the beam, the vacuum optomechanical coupling $g_0 = Gx_{ZPF}$ attains 434 kHz.

As a third and last example, if we now use the same expressions for the doubly-clamped silicon nitride bar with built-in stress considered in Sect. 5.1.2.3, and take a refractive index of $n = 2$ for silicon nitride, we obtain $\Sigma_2 = -1.38\times10^{-2}$ and a maximal $|G|$ of 1.54 GHz/nm. The related g_0 amounts to 98.2 kHz.

Note that the obtained g_0 are remarkably large in these three examples of solid-state optomechanical implementations. This is the result of the tiny mechanical mass and a reduced optomechanical interaction volume. At the time of writing, larger values are only obtained in nanophotonics-based devices such as silicon optomechanical crystals [46] and miniature gallium arsenide whispering gallery resonators [47, 48].

The approach of a rigid bar used in the above numerical illustrations ($x = 0$) is of course approximate, since it does not account for the exact vibrational profile of the beam when performing the above integrals leading to the G parameters. As we indicated earlier, independently of the choice of the reduction point employed to define the effective mass, the product $g_0 = Gx_{\mathrm{ZPF}}$ should be a constant of the problem. Our approximate approach does not fulfill this criterion since it combines the G computed with a rigid bar with the effective masses m_{eff} computed for the real flexural profile of the bar. Corrections to this approximation can be obtained by numerical integration for arbitrary vibrational mode profile and any cavity mode, and should be employed to precisely analyze experimental situations of interest. Still, at a first level of description, the rigid bar approximation offers very helpful analytical formulae and allows deriving correct orders of magnitude.

5.1.3.4 Fluctuation of the Transmitted Optical Power Through the Cavity

From now on, we allow for a vibration of the bar depicted in Fig. 5.2 and distinguish between the average (or rest) position of the nanomechanical resonator along the cavity axis x_0 and its vibration amplitude x such that $x_0 + x$ is its instantaneous position in the cavity mode. For each operating position x_0 the cavity back mirror position is adjusted to maximize the cavity transmission, the wavelength of the pump laser λ being fixed (as it is the case for the experiments discussed in Sect. 5.1.4 where λ is stabilized on an atomic transition reference). The nanomechanical resonator is then fluctuating upon time with a vibrational amplitude x. This operating regime is somewhat different from the standard regime of dispersive dynamical back-action where the laser is detuned from the cavity resonance. We focus however here on this *on-resonance* regime to pinpoint the original behaviors induced by dissipative optomechanical interactions. One first remark is that under these operating conditions, the dissipative interaction associated with Σ_1 is mandatory for the fluctuating mechanical motion to be imprinted onto the fluctuations of the transmitted optical intensity. This fact becomes apparent when employing transfer matrix calculations to derive the static transmission, which takes the form of

$$
\begin{aligned}
T(x_0, x) &= \frac{1}{K_1(x_0, x) + K_2(x_0, x)} \quad \text{with} \qquad (5.10)\\
K_1(x_0, x) &= \left[1 + (4\mathscr{F}/\pi)\Sigma_1 \sin^2(kx_0) + (4\mathscr{F}/\pi)\Sigma_1 \sin(2kx_0)\cdot kx\right]^2,\\
K_2(x_0, x) &= (4\mathscr{F}\Sigma_2/\pi)^2 \sin^2(kx_0)\cdot (kx)^2
\end{aligned}
$$

in the limit of $\Sigma_1 \ll 1$ [29]. If Σ_1 is set to zero in this expression, T does not depend on the motional fluctuation x at first order. Figure 5.4a–c shows the cavity transmission $T(x_0, 0)$ as a function of the nanomechanical element average position x_0 for a purely dispersive, a purely dissipative, and a hybrid dispersive/dissipative case. Figure 5.4d–f gives the corresponding cavity resonance shift $\delta\omega_{opt}$ in units of the free spectral range (FSR) for the same three cases.

5.1.3.5 Optical Force F_{rad} at Play

Dissipative optomechanical effects are also apparent in the optical force acting on the nanomechanical element. In the present model of one single cavity optical mode in the cavity, the force can also be derived using the transfer matrix formalism. The optical fields on the left and right hand side of the element are first computed and then related to an imbalance in the flow of light momentum in both directions. This leads to an analytical expression for the force acting in the direction of the incoming pump laser (see Fig. 5.2)

$$
\begin{aligned}
F_{rad} &= \frac{P_{in}}{c}\left[U_1(x_0,x)+U_2(x_0,x)\right]T(x_0) \quad \text{with} \qquad (5.11)\\
U_1(x_0,x) &= 2\Sigma_1\left[1+\frac{4\mathcal{F}}{\pi}\Sigma_1\sin^2\left(k(x_0+x)\right)\right],\\
U_2(x_0,x) &= -\frac{4\mathcal{F}}{\pi}\Sigma_2\sin\left(2k(x_0+x)\right)
\end{aligned}
$$

where P_{in} is the power impinging onto the cavity. $U_1(x_0,x)$ and $U_2(x_0,x)$ are approximate expressions valid in the limit of $(\Sigma_1,\Sigma_2)\ll 1$ and $\mathcal{F}\gg 1$ [29]. The force appears to be the sum of a "dissipative" force and of a "dispersive" force. A numerical illustration of the purely dispersive force is shown in Fig. 5.4g, while a purely dissipative force is illustrated in Fig. 5.4h. The dissipative force has a constant sign, in contrast to the dispersive force that can change its sign with the position. The sign of the dissipative force is enforced by the direction of the impinging pump laser, which breaks the left/right symmetry of the Fabry-Perot cavity. Depending on the average position x_0, the two forces can hence add up or counteract each other. Figure 5.4i shows an exemplary situation of a force consisting of both a dispersive and an absorptive contribution. Note that a similar "dissipative" force is also discussed in the chapter by Tang and Pernice (called "reactive" force). However, in their setups the motion only modifies the coupling to the incoming radiation, which is not the case in our present discussion where the total cavity decay rate is modulated by motion.

If we switch off dissipative effects for a minute letting $\Sigma_1 = U_1 = 0$, we can recover the expression of the dispersive optical force by simple arguments. In the standard dispersive optomechanical interaction scenario, the expression of the force is indeed obtained by equaling the work $F_{rad}dx$ done by the optical force on the mechanical degree of freedom upon a small displacement dx to the decrease of optical energy in the optical resonator $-\bar{n}_{cav}\hbar G dx$. Here $\bar{n}_{cav}$ is the number of photons stored in the resonator, which in the Fabry-Perot case amounts to $\bar{n}_{cav} = \mathcal{F}L\lambda P_{in}/(\hbar\pi^2c^2)$. Using the expression derived above for the frequency pull parameter G as a function

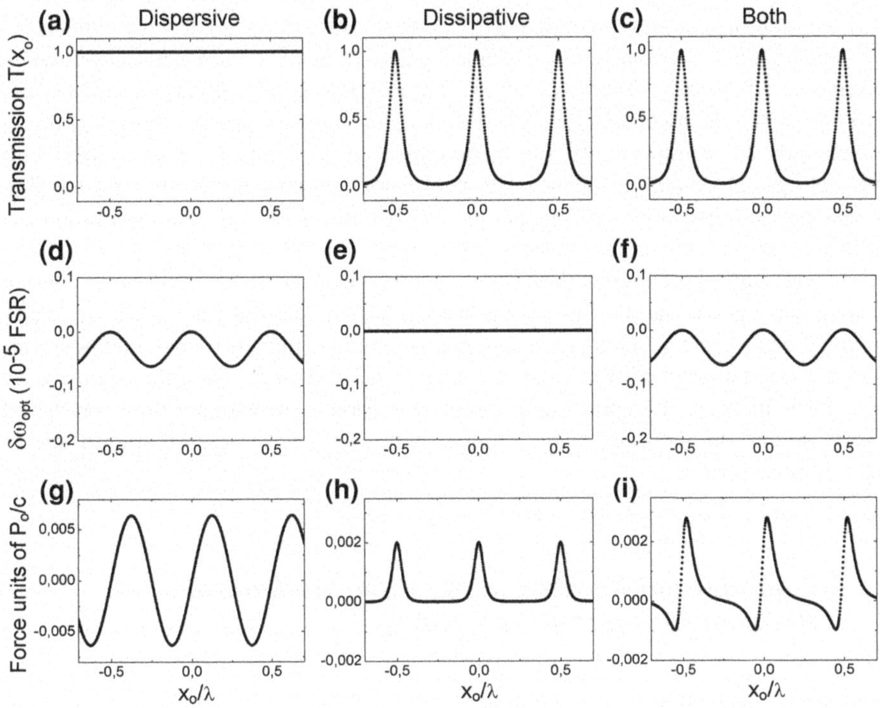

Fig. 5.4 Cavity transmission T (*upper panels*), cavity resonance shift (*middle panels*) and static optical force (*lower panels*) as a function of the nanomechanical element position x_0. The cavity finesse $\mathcal{F}$ is 5, 000 for all cases. **a**, **d** and **g** are for a purely dispersive situation where $\Sigma_1 = 0$, $\Sigma_2 = 10^{-6}$. **b**, **e** and **h** describe a dissipative situation with $\Sigma_2 = 0$, $\Sigma_1 = 10^{-3}$. **c**, **f** and **i** depict a hybrid dispersive-dissipative situation where $\Sigma_1 = 10^{-3}$ and $\Sigma_2 = 10^{-6}$

of Σ_2, the force is found to be $F_{\mathrm{rad}} = -(P_{\mathrm{in}}/c) \cdot (4\mathcal{F}/\pi) \cdot \Sigma_2 \cdot \sin 2kx_0$, which is indeed the correct expression from Eq. 5.11 for $x = 0$ and $T(x_0) = 1$.

In the above description, we have shown how both dispersive and dissipative optical effects can lead to contributions in the optical force acting on the nanomechanical element, and how this force is dependent on the fluctuating position x. This position-dependence of the force, which in case of a small motional fluctuation can be approximated as linear, leads to the typical behaviors observed in standard optomechanical settings: optical spring effect, multi-stabilities, and also dynamical back-action like cavity cooling. In case optical scattering losses and absorption could be reduced to the point where the optomechanical interaction would be purely dispersive, the *nanomechanical-resonator-in-the-middle* setting would be described formally in the same manner as the *membrane-in-the-middle* would be. The only differences would surround the reduced dimension of the mechanical element: smaller mass and larger optomechanical coupling at the expense of a reduced ability to perturb and mix the cavity modes.

The situation changes when dissipative effects are introduced, as they can lead to the appearance of non-conventional forces, and to dynamical back-action even when the standard dispersive situation would preclude it. Dissipative effects naturally come into play when the mechanical element dimensions become smaller than the photon wavelength. Most importantly, the inclusion of an artificial two-level system with discrete optical transition on the nanomechanical element allow a relative tuning of dissipative (absorptive) and dispersive contributions. A carbon nanotube with a built-in excitonic resonance is one such example. In Ref. [29] it was shown that a single-wall carbon nanotube vibrating in a Fabry-Perot cavity of small mode volume could experience a purely dissipative dynamical back-action even with the cavity operated on resonance. This prediction was obtained using a time-delayed position-dependent expansion of the force, an approach valid in the so-called bad-cavity limit [49]. In the next section, experimental efforts to investigate these predicted effects on carbon nanorods, nanotubes, as well as strongly stressed silicon nitride bars are presented.

5.1.4 Experimental Realization of a Nanomechanical Resonator in an Optical Cavity

In recent years, several experimental realizations of the cavity nano-optomechanical system schematically depicted in Fig. 5.1 have been conceived. This section will review one prominent manifestation of such an experiment, which employs a rigid Fabry-Perot cavity into which a sub-wavelength nanomechanical resonator is inserted to disturb the light field.

5.1.4.1 Fiber-Based Fabry-Perot Cavity

In order to achieve large optomechanical coupling to the nanomechanical resonator, a high finesse cavity with a small optical mode volume is desirable. A miniature Fabry-Perot cavity, which is perfectly suited to accomplish this task, has been developed for cavity quantum electrodynamics experiments with atoms [50, 51]. The cavity depicted in Fig. 5.5a is formed between the end facets of two glass fibers acting as cavity mirrors. These micro cavities are employed in several optomechanics experiments [43, 52–54]. The following description refers to the specific setup employed in Refs. [43, 52, 53]. In order to optimize both the optical cavity mode as well as the detected cavity transmission T, the input fiber consists of a single mode fiber whereas the output fiber is a multi mode fiber with a larger core. Both fiber end facets have been concavely shaped by CO_2 laser ablation followed by the deposition of dielectric Bragg mirrors consisting of 29 subsequent layers of SiO_2 and Ta_2O_5. The resulting cavity mirrors are optimized for a wavelength of 780 nm, which corresponds to the wavelength of the pump laser. Both fibers are aligned opposing each other with a

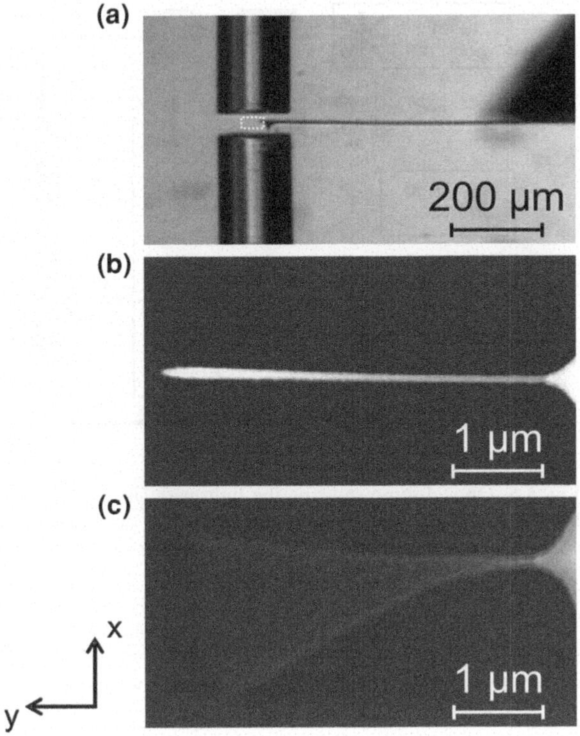

Fig. 5.5 **a** Micrograph of a fiber-based Fabry-Perot microcavity with mirror separation of 30 μm. A nanoresonator hosted by an AFM cantilever is inserted into the cavity mode. The resonator position is indicated by *the dotted box*. **b** Scanning electron micrograph showing a close-up of the nanoresonator, a carbon-based rod grown off the hosting cantilever's edge. **c** Fundamental flexural mode of the nanoresonator at $f_M = 1.9$ MHz, actuated and imaged in the scanning electron microscope. [modified from content originally published in [43]]

separation of ∼30 μm and glued to shear piezos held by a rigid mount to form a stable optical resonator. The cavity length can thus be adjusted by roughly 1.5 μm to enable a detuning of a few free spectral ranges. The resulting Fabry-Perot cavity has an estimated mode waist w_0 of 3.5 μm and a finesse $\mathscr{F}$ of up to 33,000.

The cavity is pumped by a temperature controlled external cavity laser diode stabilized on an atomic resonance of rubidium at 780 nm. While the frequency of the pump laser remains fixed the pump power can be adjusted. To detune the cavity from the pumping laser, the cavity length is altered. This allows operating the cavity both in the *on-resonance* setting discussed in detail in Sect. 5.1.3 and on the flank of its resonance. Whereas the transmitted optical intensity in the *on-resonance* experiment is sensitive to the absorptive (dissipative) effect of a nanoresonator exclusively, stabilization on the flank will yield both the absorptive (dissipative) and the dispersive contribution. After the selected detuning has been applied, a second feedback loop locks the cavity to the laser wavelength to ensure stable operation by adjusting the cavity length. Figure 5.6 shows that the reflection and transmission of the cavity are monitored by a spectrum analyzer and oscilloscope, respectively. Further details on the optical setup are described in Ref. [52].

The nanomechanical resonator is inserted into the optical cavity mode using an xyz or $xyz\theta\varphi$ piezo-positioning stage, which is also employed to scan the resonator position. For optomechanics experiments the setup has to be isolated from the envi-

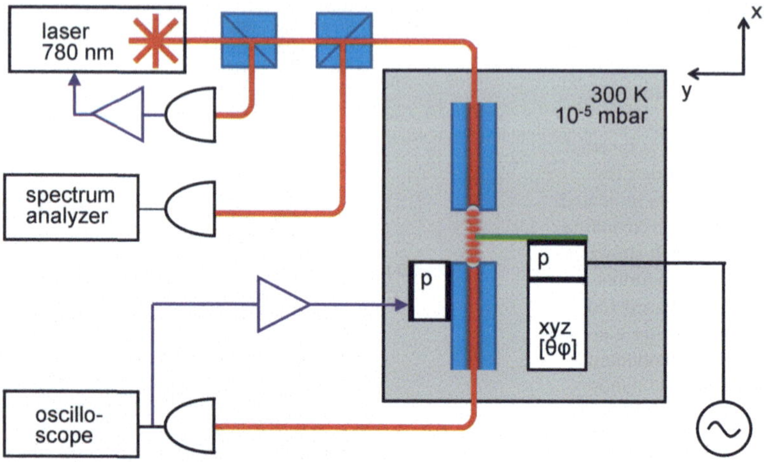

Fig. 5.6 Optical setup showing the light path (*red*) as well as major optical components (*blue*). Feedback loops are indicated by *purple lines*. The optomechanical system including cavity, nanomechanical resonator (*green*), control and actuation piezos (denoted p) and the piezo-positioning unit ($xyz[\theta\varphi]$) are placed in a vacuum cuvette (*grey*)

ronment. A vacuum cuvette keeps the pressure near $\sim 10^{-5}$ mbar to avoid mechanical damping of the resonator displacement by the surrounding gas.

5.1.4.2 Nanomechanical Resonator

One of the major challenges of the discussed experiment lies in the choice of the nanomechanical resonator. While it has become clear in Sects. 5.1.2 and 5.1.3 that in general a low effective mass nanoresonator, combined with high frequency and high mechanical quality factor Q is desirable, practical implementations are severely hindered by the cavity geometry. First of all, the notion of the nanoresonator disturbing the cavity transmission implies an isolated mechanical object. Instead of processing the device on a wafer substrate, the resonator must be fabricated to extend beyond the edge of the substrate or above a hole perforating the chip. Second, the fact that the separation between the two cavity mirrors is only of order 30 μm, most wafer-based realizations are not suitable since typical wafers employed in the semiconductor industry are roughly an order of magnitude too thick to fit the gap between the fibers.

Initial experiments have been performed using singly clamped nanoresonators (see Sect. 5.1.2.1), which can be processed as isolated tips protruding from a mount and thus seem to provide an ideal geometry. For example, carbon-based amorphous nanorods [55] such as the one displayed in Fig. 5.5b and c have been successfully employed as sub-wavelength sized mechanical elements in a fiber-based cavity [43]. These nanorods are grown along the long axis of an AFM cantilever by electron-beam deposition (ebd) [56] and can be up to several microns long with average diameters of about 100 nm. They feature eigenfrequencies f_M in the range of 1 MHz

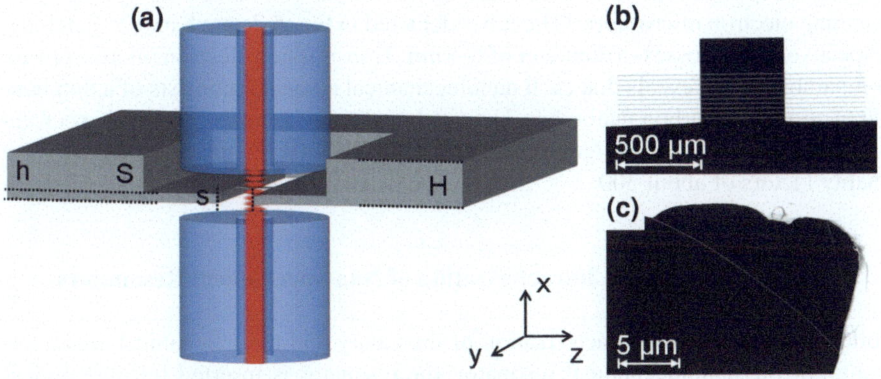

Fig. 5.7 **a** Schematic device layout for a doubly-clamped nanoresonator suitable for cavity insertion. **b** Top-down realization of (**a**) featuring a set of 10 high stress silicon nitride strings. **c** Bottom-up realization of (**a**) including two carbon nanotubes grown across a pre-processed slot. [modified from content originally published in [53]]

for the fundamental flexural mode. However, their quality factors of a few hundreds remain relatively low. Furthermore, their mechanical response eludes standard Euler-Bernoulli-based elastic theories, such that the formalism discussed in Sect. 5.1.2.1 only allows for a coarse approximation. Both is likely a consequence of the ebd growth process, leading to a varying diameter as well as incorporating an inhomogeneous distribution of a large number of unspecified defects.

A more advanced device technology going beyond singly clamped nanoresonators is schematically depicted in Fig. 5.7a. It is based on a standard silicon chip of thickness H, thinned to $h < 10\,\mu\text{m}$ in a central region S of about $200\,\mu\text{m}$ to fit the gap between the two fiber mirrors. The remaining membrane has been processed to form a slot s of at least $20\,\mu\text{m}$ width to avoid severely disturbing the cavity mode. A doubly clamped nanomechanical resonator spans across the slot.

The described general concept allows for two different strategies to realize doubly clamped nanoresonators such as the ones theoretically described in Sects. 5.1.2.2 and 5.1.2.3.

Figure 5.7b shows a set of ten high stress silicon nitride strings suitable for cavity insertion. High stress silicon nitride is known for its remarkably high mechanical quality factors at room temperature [9, 57]. Mechanical characterization of the $500\,\mu\text{m}$ long strings with 50 nm thickness and $2\,\mu\text{m}$ width in a Michelson interferometer by spectral and ringdown measurements reveal a fundamental eigenfrequency of $f_{\text{M},1} = 580\,\text{kHz}$ and linear scaling of the eigenfrequency with mode index $N = 1$. The quality factor of the fundamental mode is $Q_1 = 600{,}000$ while higher harmonics may have a larger Q, presumably as a result of the diminishing effect of clamping losses [58].

An alternative device is presented in Fig. 5.7c that displays carbon nanotubes CVD-grown across a pre-processed slotted chip, offering strongly reduced resonator dimensions. The suspended carbon nanotubes are mechanically characterized in a scanning electron microscope. The tubes depicted in Fig. 5.7c are 8 and 20 μm long, respectively, and possess a diameter of 6–8 nm. Transmission electron microscopy on comparable tubes reveals that each nanomechanical resonator consists of a thin bundle of about 5–7 carbon nanotubes. Hence, they display a rich mechanical spectrum with a set of fundamental eigenfrequencies of order 500 kHz and room temperature quality factors of about 300.

5.1.4.3 Optomechanical Characterization of Nanomechanical Resonators

Following the optical characterization of the cavity and the mechanical characterization of the nanomechanical resonator, the resonator is inserted into the optical fiber cavity. For all three devices presented in Sect. 5.1.4.2, the alignment of the nanoresonator with the cavity mode is essential. Besides the xyz piezopositioning stage illustrated in Fig. 5.6 centering the resonator (along z), setting the penetration depth into the cavity mode (along y), and adjusting the resonator position along the cavity axis x_0 (along x), a careful tilt correction ($\theta\varphi$) should be performed. From a detection point of view, this minimizes light scattering out of the cavity, whereas from a back-action perspective, it ensures that the resonator penetrates the optical mode at a well-defined position, and thus a fixed detuning Δ.

Particularly for the case of the ebd nanorod illustrated in Fig. 5.5, tilt correction is hard to accomplish. The camera system used for alignment only resolves the hosting AFM lever, but not the rod itself that typically deviates from the lever axis by several degrees. Despite these limitations, the rod can be inserted into the cavity mode with a simple xyz positioning stage. Optimum penetration is reached when the rod maximally intersects the mode without the host lever contributing to the perturbation. In this situation the nanorod position x_0 is scanned along the cavity axis. Figure 5.8a displays the optical power reflected and transmitted through the cavity mirrors, respectively, while shifting the nanorod across three modes of the cavity mode standing wave pattern. The expected $\lambda/2$ periodicity of the cavity mode profile is clearly resolved. To optomechanically detect the rod's thermal motion [29], the cavity is locked on resonance where it is sensitive to the absorptive (dissipative) signal component. Consequently the position x_0 of the rod is chosen for maximum slope dT/dx_0 of the cavity mode standing wave providing maximum signal transduction for dissipative optomechanical coupling. In Fig. 5.8a, this corresponds to a nanorod position offset to the left of the central extremum by about one tenth of a $\lambda/2$ period. As the nanorod displacement noise induced by its Brownian motion is transduced into optical frequency noise of the cavity, the noise spectrum of the cavity reveals the eigenmodes of the nanorod. Indeed several harmonics of the nanorod can be probed experimentally at room temperature. Figure 5.8b shows the Brownian vibration spectrum of the fundamental flexural mode at $f_{M,1} = 473\,\text{kHz}$ that is resolved with a displacement sensitivity of $200\,\text{fm}/\sqrt{\text{Hz}}$, limited by the noise level of the employed spectrum analyzer.

Cavity-induced back-action on the rod is not observed. Besides the described poor mechanical quality of the rod, this is mainly attributed to the geometric constraints

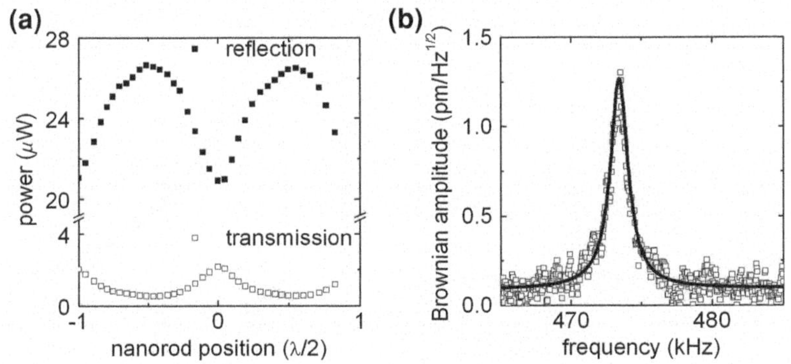

Fig. 5.8 **a** Optical power reflected off and transmitted through the cavity as a function of nanorod position x_0 along the cavity axis. **b** Brownian motion of the fundamental flexural mode of the nanorod probed via the cavity transmission noise, which has been calibrated using the equipartition theorem including a Lorentzian fit. [modified from content originally published in [43]]

and misalignments related to the growth of the rod and its relatively large diameter d of about 100 nm filling roughly a quarter period of a cavity mode.

All in all, even though the singly clamped nanorod provides an ideal geometry for cavity insertion, it is not ideally suited for optomechanics experiments: its inhomogeneous and unknown material composition leading to low quality factors and non-trivial mechanical response, as well as uncontrollable tilt angle make the obtained data hard to interpret.

A top-down fabricated nanostructure such as the SiN strings displayed in Fig. 5.7b avoids the above limitations. Furthermore, their excellent mechanical quality is expected to increase the impact of optomechanical back-action effects [49], whereas absorption has been shown to be weak even in the 780 nm wavelength regime for the case of high stress stochiometric SiN [59]. Insertion of the sample into the cavity requires careful xyz navigation as space is limited, but is relatively straightforward. Tilt compensation with a $\theta\varphi$-goniometer is simplified by an additional 30 μm wide SiN beam that is large enough to host the entire optical mode. This situation is comparable to the *membrane-in-the-middle* scheme introduced in Sect. 5.1.3 and further described in Sect. 5.2 of this chapter. Subsequently, one of the thinner strings is positioned in the cavity mode. To be sensitive to both the absorptive (dissipative) and dispersive signal component the cavity is locked on the flank of a resonance. While thin strings of width far below 1 μm can be fully inserted into the cavity mode, the slightly wider strings of Fig. 5.7b only allow for knife-edge experiments. Optimum resonator insertion is observed for their edge at about twice the optical mode waist radius of the center of the mode.

The Brownian motion of the SiN string is clearly resolved for the first 9 harmonics, with displacement sensitivities of about $1\,\text{pm}/\sqrt{\text{Hz}}$. Figure 5.9a shows the noise spectrum of the second eigenmode at $f_{M,2} = 1\,\text{MHz}$ featuring the highest quality factor of the device close to 1×10^6. The corresponding ringdown measurement is

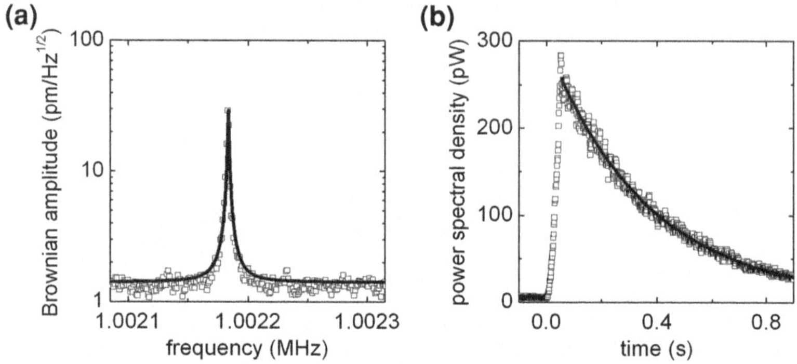

Fig. 5.9 **a** Brownian noise spectrum of a high stress SiN string with dimensions 50 nm × 2 μm × 500 μm positioned in the optical mode of the Fabry-Perot cavity. The measured cavity frequency noise signal has been calibrated using the equipartition theorem. *The solid line* displays a Lorentzian fit. **b** Ringdown measurement of the same SiN string including fit. The observed exponential decay yields a comparable mechanical quality factor

shown in Fig. 5.9b. It displays a clear exponential decay with 360 ms time constant, which corresponds to a quality factor of about 2×10^6. The observed discrepancy between the two measurements performed in the cavity is most likely a consequence of an unavoidable photon-induced back-action on the resonator. For the SiN string positioned on the flank of the cavity resonance even at the smallest possible optical input powers required to maintain the cavity lock, back-action seems to have a measurable effect on its mechanical properties.

The third device architecture goes back to bottom-up nanomechanical resonators, the carbon nanotube resonator displayed in Fig. 5.7c. Even though this implies a compromise regarding positioning and navigation in comparison with a SiN nanoresonator, the advantages prevail. As the carbon nanotube is doubly clamped, it is necessarily oriented parallel to the chip surface such that the tilt correction sequence described above can be adopted in a straightforward way. Both carbon nanotubes and SiN resonators have outstanding mechanical properties. But with a diameter of only a few nm, the carbon nanotube resonator is much better suited for cavity insertion. Indeed, its molecular size makes its interaction with the cavity mode close to the ideal picture of a perturbing atom sitting in the cavity. If aiming at a purely dispersive interaction, geometrical scattering of photons out of the cavity can be reduced as compared to SiN string resonators, which strongly modify the cavity mode once inserted. The nanotube diameter is small compared to the $\lambda/2$ period of the cavity mode, which greatly simplifies modeling of the optomechanical coupling.

The 20 μm long nanotube from Fig. 5.7c is carefully positioned in the center of the cavity mode navigating around the edge of the perforated chip and by monitoring the transmitted optical power [53]. The transmitted power on resonance of the empty cavity amounts to 0.56 μW. With the nanotube in place, the transmission drops by almost 30 %. This strong effect is predominantly attributed to residual clipping losses caused by the rim of the hole supporting the nanotube. As the nanotube length

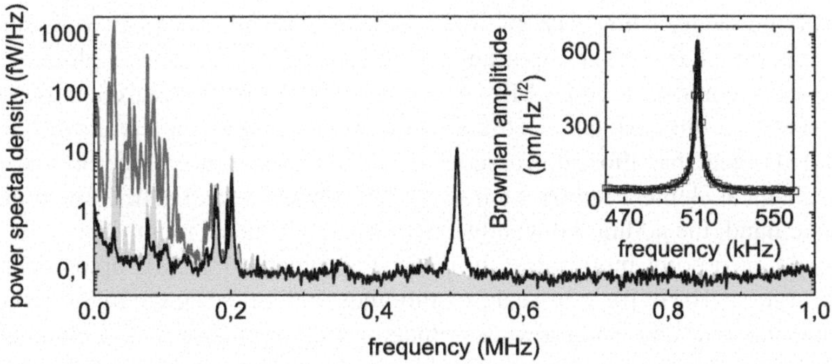

Fig. 5.10 Noise spectra of the Fabry-Perot cavity demonstrating the optomechanical effect of a carbon nanotube as a mechanical resonator. *The black line* shows the noise spectrum for the nanotube placed in the cavity mode, revealing a strong resonance at 510 kHz. *The light gray background* displays the noise spectrum of the empty cavity, whereas *the dark gray* spectrum was taken near the rim of the chip with no nanotube nearby. Clearly, technical noises and substrate resonances dominate for frequencies below 220 kHz. The inset displays the Brownian noise peak of the carbon nanotube which has been calibrated using the equipartition theorem as well as a Lorentzian fit. [modified from content originally published in [53]]

roughly coincides with the diameter of the Gaussian mode profile, even a nanotube perfectly centered in the cavity mode will experience considerable perturbation from the hosting substrate. To entirely suppress clipping losses, much larger holes would be required, leading to undesirably low nanotube eigenfrequencies. However, the response of the nanotube is strong enough to allow for displacement detection of its vibrational modes. To this end, the cavity is again locked on the flank of a resonance to be sensitive to both absorptive (dissipative) and dispersive interaction. The cavity frequency noise spectrum shown in Fig. 5.10 (black line) reveals a strong resonance at 510 kHz rising sharply above the noise floor that is attributed to a flexural mode of the nanotube. The inset of Fig. 5.10 shows a calibrated zoom along with a fit, revealing a quality factor of 300 probed with a displacement sensitivity of $50\,\mathrm{pm}/\sqrt{\mathrm{Hz}}$ [53]. The light gray background and the dark gray line show the noise level of the empty cavity and the cavity spectrum with just the substrate but no nearby carbon nanotube inserted. As discussed in Refs. [52, 53] technical noises and substrate resonances dominate for frequencies below 220 kHz, whereas clean spectra are observed for higher frequencies.

5.2 Membrane in the Middle of an Optical Cavity

Another type of sub-wavelength mechanical element that can be positioned within an optical cavity is an ultra-thin membrane. Conceptually, this so-called *membrane-in-the-middle* (MIM) geometry is not terribly different from the nanoscale mechanical systems discussed above—the drumhead motion of the membrane modulates optical modes of the cavity—however, the extension of a membrane in two dimensions (rather than one) has several important consequences we shall now explore.

A typical membrane used for MIM systems is ∼100 nm thick and extends ∼1 mm in the other two dimensions. Needless to say, this represents a *significantly* larger mechanical object than those discussed above, and so the choice of membrane over a nanomechanical element imposes a trade-off between size and added functionality.

On one hand, the additional weight associated with such a membrane reduces its mechanical zero-point motion (i.e. to a few femtometers) and its optomechanical coupling ($g_0 \sim 10\,\text{Hz}$ for mm-scale membranes and cm-scale cavities). However, in general the coupling is not always reduced for a larger mechanical element. For example, if one imagines widening a doubly-clamped nanobeam, the radiation force would scale linearly with its surface area (i.e. proportional to its mass), whereas the beam's zero-point motion would scale only as the inverse square root of its mass, meaning $g_0 \propto \sqrt{\text{mass}}$ in this case.[1] Of course, this scaling breaks down once the beam width extends beyond the cavity mode, and so the ideal mechanical element (in terms of coupling strength, anyway) would have an area comparable to the cavity mode's waist. A similar argument favors membranes with thicknesses approaching a quarter wavelength. The relatively low value of g_0 mentioned above reflects a large cavity mode volume and a large fraction of "unused" membrane mass.

With the additional material also comes a few unique opportunities. First, because membranes are planar, they scatter light in a well-defined direction. As a result, it is relatively straightforward to realize stable optical cavity modes (even when the membrane is somewhat misaligned with the cavity axis) and the optomechanical coupling can be made purely dispersive. Second, because membranes are large, they more strongly affect the modes of an optical cavity. This allows one to mix cavity modes in a predictable way, unlocking new types of optomechanical coupling that can be tuned in situ. Using such mode mixing, MIM systems have been shown to realize a relatively strong, *purely-quadratic* dispersive optomechanical coupling that could potentially be used for (among other things) quantum nondemolition (QND) readout of the membrane's phonon number state [18, 60], cooling and squeezing [61], or exploration of Landau-Zener-Stückelberg dynamics with cavity light [62].

The MIM system was first developed using low-stress SiN membranes in 2008, then achieving mechanical quality factors $\sim 10^6$ ($\sim 10^7$ at low temperature) and a cavity finesse $\sim 10^4$. The finesse in this case was limited primarily by absorption or roughness in the membrane at a wavelength of 1,064 nm [18]. By switching to high-stress stoichiometric Si_3N_4 membranes, a much lower level of optical absorption was observed at 935 nm (measured with a finesse ∼10,000 cavity) that would be compatible with cavities of finesse ∼100,000, and no apparent compromise of the mechanical Q [21]. Subsequent measurements of high-stress silicon nitride in a finesse ∼50,000 cavity at 1,064 nm found little obvious dependence of the cavity finesse on membrane position, placing a lower bound on the attainable finesse of

[1] This assumes that the beam's mechanical frequency is independent of its width, which is not out of the question because the beam's mass and spring constant should both scale ∼ linearly with its width in the simplest case.

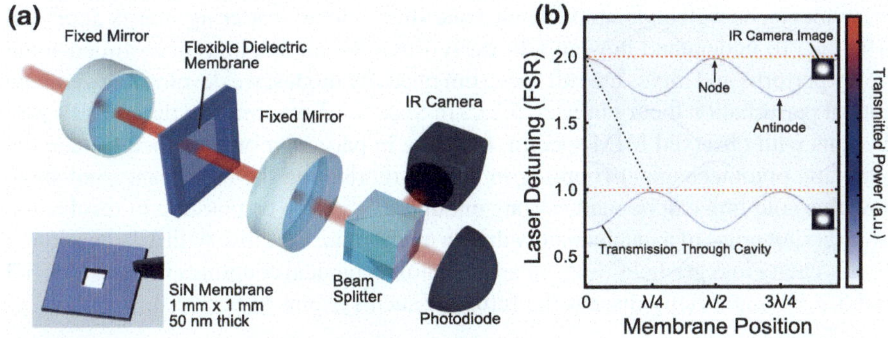

Fig. 5.11 **a** *Membrane-in-the-middle* (MIM) optomechanical geometry. A thin, flexible dielectric membrane is positioned inside a high-finesse Fabry-Perot cavity formed by two fixed mirrors. The membrane's drumhead motion is coupled to the intracavity standing wave via radiation pressure. Cavity light can be injected and collected via either fixed mirror. *Inset* shows photograph of a 1 mm × 1 mm × 50 nm silicon nitride membrane and supporting silicon frame (similar to the one used for (**b**)). **b** Transmission through the combined system as a function of laser frequency and membrane position. *Darker regions* indicate higher transmission, *red dotted lines* show the approximate locations of the cavity without the membrane, and the *black dotted line* shows the predicted frequency dependence for a perfectly reflective membrane. Here the input beam is well matched to the cavity's TEM_{00} (transverse Gaussian) mode, as verified by an infrared camera image (*insets*). The end mirrors' radius of curvature is 5 cm, the length of the cavity is 6.7 cm, the laser wavelength is 1,064 nm, and the membrane is nominally 50 nm thick, made of Si_3N_4 (index $n_{SiN} = 2$). The empty-cavity free spectral range (*FSR*) is ~4.5 GHz. Laser detuning is relative to the ~125,000-th longitudinal mode of the cavity [modified from content originally published in Refs. ([19] & [22])]

~500,000 [22]. Even with an absorptive membrane, however, a finesse of 150,000 can be achieved by positioning the membrane near a node of the intracavity standing wave [19]. Clearly, one advantage of the MIM system lies in its ability to simultaneously achieve a very high mechanical quality and very high cavity finesse. MIM systems have since been shown to be compatible with cryogenic environments [24, 25] and fiber-based optical microcavities [54], and one MIM system has behaved as a "microphone" sensitive enough to detect the "hiss" of photon shot noise from a laser [27]!

A final note about MIM systems is their relatively low cost: because free-standing silicon nitride membranes are mass-produced for other applications such as TEM sample mounts or x-ray windows, these high-quality optomechanical elements can be purchased "off the shelf" for the price of a modest meal at a restaurant.[2]

The remainder of this chapter focuses on a method to exploit the membrane's ability to perturb the modes of an optical cavity sufficiently to achieve several different types of optomechanical interactions. Section 5.2.1 introduces the basic behavior of the system for a single cavity mode and an aligned membrane; this

[2] Provided you order a minimum of ten such meals and agree to pay all shipping, duties, and brokerage fees.

behavior can be well understood using a one-dimensional scattering-matrix approach [19, 21]. To understand how an arbitrarily-tilted membrane in a three-dimensional cavity perturbs and mixes the full spectrum of cavity modes, we develop a degenerate optical perturbation theory in Sect. 5.2.2. In Sect. 5.2.3 we compare the model's predictions with observed MIM spectra, focusing in particular on a method to tune the quadratic optomechanical coupling in situ by reorienting the membrane, and show how this coupling can be made strong enough that it may be possible to resolve the *phonon* shot noise of a mechanically driven membrane. We close with a discussion of novel "cavity loss gradient" and "quartic" optomechanical couplings in Sects. 5.2.3.3 and 5.2.3.4, and briefly discuss the future of such systems in Sect. 5.3.

5.2.1 Geometry and Basic Behavior

As shown in Fig. 5.11a, the MIM geometry consists of a thin (~50 nm thick), flexible, dielectric membrane positioned between the mirrors of a high-finesse Fabry-Perot optical cavity. The membrane's "drumhead" motion is coupled to light in the cavity through radiation pressure forces. To get an idea how this works, Fig. 5.11 shows a basic characterization of the system's optical modes. When laser light landing on the left-hand mirror in Fig. 5.11a is resonant with a cavity mode, light builds up inside the cavity and some of it can exit via the right-hand mirror. Light leaving the cavity can be collected by a photodiode (to measure intensity) or an infrared camera (to measure the transverse mode profile). Figure 5.11b shows a plot of the power at the photodiode as a function of laser frequency and membrane position x_0. The profile of the incident laser is well-matched with that of the (Hermite-Gaussian) TEM_{00} cavity mode, so this is essentially the only mode to which the laser couples, as verified by the infrared camera images (inset). Two longitudinal modes of the cavity are shown in Fig. 5.11b, the upper having a *node* in the intracavity standing wave at $x_0 = 0$ and the lower having an *antinode*. The resonance frequencies of these modes vary sinusoidally with x_0 as the membrane moves between node and antinode. We will build more intuition about this behavior in the following sections, but stated briefly it arises from the membrane's dielectric response to the incident light field, which serves to decrease the mode's frequency when the membrane moves toward an antinode. The locations of the empty-cavity longitudinal modes are drawn with dashed lines for reference.

It is easy to see from Fig. 5.11b how this system can behave as a "standard" linear interferometer: if the membrane is positioned at $x_0 = \lambda/8$, the upper mode frequency ω_u *decreases* with x_0, and the system behaves as though the membrane is *elongating* the cavity as x_0 increases. Interestingly, despite the fact that the membrane is a modest reflector (only about 11 % of incident power is reflected), the optomechanical coupling $G = d\omega_u/dx_0$ is only a factor of three below the maximum possible coupling achievable with this geometry; if the membrane had a reflectivity of 100 %, the cavity mode would follow the dotted black line. For this reason it is relatively straightforward to laser cool a MIM system by a very large factor (~45,000 [18]).

Figure 5.11b also illustrates how one could tune the strength of this linear coupling continuously to $G = 0$ by displacing the membrane. When $G = 0$, the higher-order cavity response dominates and this system behaves as a nonlinear interferometer. At a node or antinode, for example, the cavity mode frequency depends only on the membrane position *squared*. As described elsewhere [18, 22, 60], this non-standard style of interferometry or "quadratic optomechanical coupling" is in principle compatible with a quantum non-demolition (QND) readout of the membrane's phonon number state, though the quadratic coupling shown in Fig. 5.11b ($d^2\omega_u/dx_0^2 \sim 2\pi \times 30$ kHz/nm^2) is far too weak to resolve an individual number state before it decoheres. As discussed below, there are several ways to increase the strength of this coupling by modifying the device geometry in situ [22, 26, 54, 64], and in a very different type of system (having its own set of advantages and challenges), the coupling can be made quite large, exceeding 1 THz/nm^2 [65].

All of the behavior discussed thus far (i.e. for a well-aligned membrane and a single transverse mode of the cavity) is well described using a one-dimensional transfer or scattering matrix formalism [18, 19, 28], similar to the framework applied in Sect. 5.1.3. In order to model the interactions between an arbitrarily-oriented membrane and the full spectrum of optical cavity modes, we require a three-dimensional model. The following section describes a perturbative approach to this problem.

5.2.2 Degenerate Perturbation Theory of Cavity Modes

The goal of this section is to develop a theory describing how a small dielectric object (e.g. a membrane) can both perturb and couple the modes of a three-dimensional optical cavity [26, 64]. This theory is similar to degenerate perturbation theory in quantum mechanics [66], with the cavity's eigenmodes playing the role of "particle wavefunctions" and the membrane's dielectric constant playing the role of "perturbing potential". This first-order model describes the observed optical modes of a MIM system quite well (see Sect. 5.2.3).

Figure 5.12 illustrates the basic idea. The end mirrors of the cavity serve as a clamped boundary condition for the electric field, thereby defining a set of standing wave normal modes $\phi_i(x, y, z)$ for the electromagnetic field inside the cavity, each satisfying the Helmholtz wave equation

$$\nabla^2\phi_i + \frac{\omega_i^2}{c^2}\phi_i = 0. \tag{5.12}$$

where c is the speed of light and ω_i is the eigenfrequency of mode i. The value of $\phi_i(x, y, z)$ is proportional to the electric field, and we normalize so that

$$\int dx \int dy \int dz \phi_i \phi_j = \delta_{ij} \tag{5.13}$$

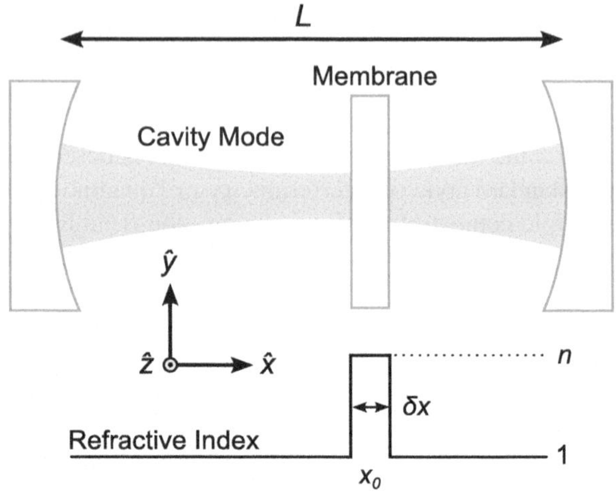

Fig. 5.12 Perturbative model. A dielectric membrane of thickness δx and refractive index n at position x_0 perturbs (and couples) the free-space optical cavity modes. The membrane's index of refraction plays the role of "perturbing potential" for the light in the cavity [modified from content originally published in Ref. [64]]

when integrated over the volume of the cavity. For the case of curved end mirrors, a convenient set of normal modes satisfying these conditions might be the Hermite-Gaussian modes [5, 26, 64].

If we now include a dielectric object such as a membrane, the non-unity index of refraction n inside the object will perturb and couple the modes of the cavity. The speed of light inside the dielectric will be given by c/n, and the perturbed Helmholtz equation takes on the form

$$\nabla^2\psi + \frac{\omega^2}{c^2}(1 + V(x, y, z))\psi = 0 \tag{5.14}$$

where $V(x, y, z) \equiv n^2(x, y, z) - 1$ is only non-zero inside the dielectric, and ψ represents a new set of eigenmodes having eigenfrequencies ω. These perturbed modes can be written as a linear combination of the original empty-cavity modes:

$$\psi = a_0\phi_0 + a_1\phi_1 + a_2\phi_2 + \cdots \tag{5.15}$$

where the a_i's are constants to be determined. As is the case for perturbation theory in quantum mechanics, only modes that are nearly degenerate with one another will contribute to a given ψ, and so the corresponding new eigenmodes will take on the form

$$\psi = \sum_i a_i\phi_i + \sum_k \varepsilon_k\phi_k \tag{5.16}$$

where i ranges over the N nearly-degenerate modes of interest, and k ranges over all other modes. The coefficients ε_k are assumed to be small because they are either far from degenerate with the N modes in the first sum or they do not couple for some other reason such as symmetry.[3] Substituting these eigenmodes ψ and eigenvalues ω into the perturbed Helmholtz equation (Eq. 5.14),

$$(\nabla^2 + (1+V)\omega^2/c^2)\left(\sum_i a_i\phi_i + \sum_k \varepsilon_k\phi_k\right) = 0. \tag{5.17}$$

If we now take an inner product of this equation with each of the N relevant modes ϕ_j we generate N new equations (indexed by j) of the form

$$(1-\omega_j^2/\omega^2)a_j + \sum_i V_{ij}a_i + \sum_k V_{kj}\varepsilon_k = 0 \tag{5.18}$$

where

$$V_{ij} \equiv \int dx \int dy \int dz V\phi_i\phi_j \tag{5.19}$$

is an integral over the volume of the dielectric (i.e. the only place where $V(x,y,z) = n^2(x,y,z) - 1$ is non-zero). For silicon nitride $n = 2$, meaning $V(x,y,z) = 3$ inside the membrane, and since the membrane generally occupies a very small fraction of a given cavity volume (it is only ~100 nm thick, compared with ~ centimeter-long cavities), the value $V_{ij} \ll 1$. Since $\varepsilon_k \ll 1$ as well, we can ignore the second-order terms of the last sum. This leaves N equations of the form

$$(V_{jj} + 1 - \omega_j^2/\omega^2)a_j + \sum_{i\neq j} V_{ij}a_i = 0 \tag{5.20}$$

At this point, it is straightforward to solve this problem (at least numerically) given any set of orthonormal cavity modes and small dielectric object of any shape; one needs only to solve N linear equations for N eigenmodes ψ having eigenvalues $1/\omega^2$.

To simplify this calculation, if we assume that all frequency perturbations and splittings within the subspace of relevant modes are small compared with their unperturbed frequencies ω_j, it is highly convenient (both numerically and symbolically) to define all frequencies relative to a fixed frequency $\omega_0 \sim \omega_j$. If we define unitless constants $\delta_j = (\omega_j - \omega_0)/\omega_0$ for each mode j and the unitless eigenvalue $\delta = (\omega - \omega_0)/\omega_0$, we have $1 - \omega_j^2/\omega^2 \sim 2\delta - 2\delta_j$ (assuming $\delta_j, \delta \ll 1$), and the N equations become

$$(V_{jj} + 2\delta - 2\delta_j)a_j + \sum_{i\neq j} V_{ij}a_i = 0. \tag{5.21}$$

If, for example (very relevant for the following section), there were $N = 4$ modes of interest, we could represent this set of equations as a matrix to be diagonalized:

[3] Note that this is generally the first assumption to reconsider if the theory doesn't match experiment.

$$\begin{pmatrix} V_{00} - 2\delta_0 + 2\delta & V_{01} & V_{02} & V_{03} \\ V_{01} & V_{11} - 2\delta_1 + 2\delta & V_{12} & V_{13} \\ V_{02} & V_{12} & V_{22} - 2\delta_2 + 2\delta & V_{23} \\ V_{03} & V_{13} & V_{23} & V_{33} - 2\delta_3 + 2\delta \end{pmatrix} \begin{pmatrix} a_0 \\ a_1 \\ a_2 \\ a_3 \end{pmatrix} = 0. \tag{5.22}$$

where we have noted a symmetry $V_{ij} = V_{ji}$. The computational bottleneck in this example is the requirement that we perform 10 integrals over the volume of the perturbing dielectric. For most applications this is quite reasonable, and for the MIM geometry, it is possible to derive closed-form approximations of the integrals V_{ij} and increase the speed of this calculation by many orders of magnitude [26, 64]. It is also a useful exercise to include only the two modes near a particular crossing, which reduces Eq. 5.22 to a 2×2 matrix, allowing one to exactly solve for the new normal modes with minimal algebra.

5.2.3 *Observed Optical Cavity Modes*

This section describes how the presence of a real membrane perturbs and couples different transverse modes of an optical cavity. We show that at some membrane positions, different transverse modes cross one another, and that some of these crossings are avoided, generating a stronger form of the quadratic coupling that can be tuned in situ. The cavity modes at avoided crossings also exhibit a gradient in cavity loss rate that can be tuned or eliminated. Finally, we discuss a configuration in which, to lowest order, the cavity mode frequency varies *quartically* with membrane displacement.

5.2.3.1 Avoided Crossings Between Different Transverse Modes

For the original measurement of Fig. 5.11b, the profile of the incident laser is shaped so that it only has a non-zero overlap with the TEM_{00} (Gaussian) modes of the optical cavity. This suppresses the higher-order (Hermite-Gaussian) transverse modes in the transmission spectrum, though they are still faintly visible. Figure 5.13a shows a similar spectrum in which the beam profile is shaped so that it more strongly overlaps with the first few higher-order transverse modes. Due to the cylindrical symmetry of the system about the cavity axis, higher-order transverse modes typically appear in the spectrum as a bundle of degenerate modes (some of these are labeled), each oscillating sinusoidally as a function of membrane position, just like the TEM_{00} mode. Owing to the Gouy phase shift [5], each bundle of transverse modes is centered at a different frequency, and different modes often cross one another. This large-scale behavior is well captured by the perturbative model (black lines) with essentially one fitting parameter: the thickness of the membrane $\delta x = 39$ nm. This seems reasonable given the nominal value of $\delta x = 50$ nm and fabrication tolerances.

Here we focus on the crossings between the TEM_{00} "singlet" mode and the threefold-degenerate $TEM_{20,11,02}$ "triplet" modes. Additional modes are examined

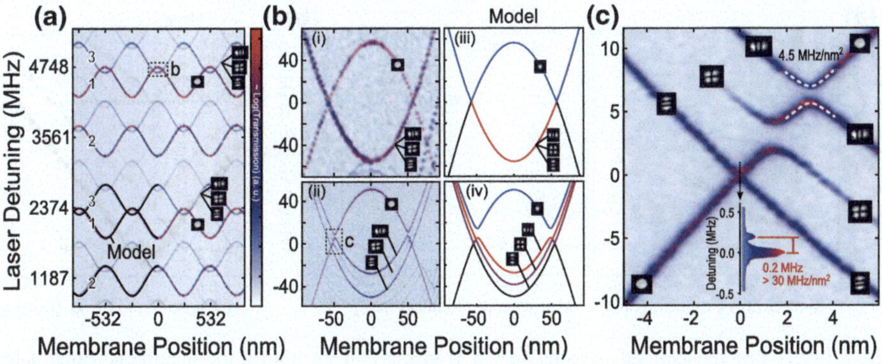

Fig. 5.13 Avoided crossings in cavity spectra. **a** Transmission spectrum with the input laser addressing several transverse modes and the membrane aligned near the center of the of the cavity. Numbers denote the degeneracy of the lowest-order transverse modes, and *dark lines* indicate eigenfrequencies calculated using the perturbative model. *Inset* infrared camera images show the intensity profile of the modes. **b** Refined data from boxed region in (**a**) for the membrane (i) aligned with the cavity axis and (ii) tilted by 0.4 mrad about the $\hat{z}$-axis. Plots (iii) and (iv) show the predicted eigenfrequencies under the same conditions, along with the intensity distributions of the empty-cavity eigenmodes. **c** Refined data from the boxed region in (**b**), showing an avoided crossing with a significantly stronger quadratic optomechanical coupling 4.5 MHz/nm^2. The lowest crossing is not resolved, and the *inset* shows the narrowest observed gap (0.2 MHz); assuming this crossing is avoided, the quadratic coupling is then at least 30 MHz/nm^2 [modified from content originally published in Ref. [22]]

in Ref. [26]. Figure 5.13b (i) shows the data in the dotted box of Fig. 5.13a in detail. For this data set, the membrane is positioned near the center of the cavity and aligned so that its surface is well-matched with the cavity modes' flat wave fronts. The cylindrical symmetry of the cavity is not broken in this case, and the triplet modes remain degenerate. Additionally, the orthogonality of the Hermite-Gaussian modes ensures that the overlap integrals $V_{i\neq j} = 0$ to lowest order and so the optical modes do not mix. Figure 5.13b (ii) shows the same spectrum after tilting the membrane about the $\hat{y}$-axis (of Fig. 5.12) by 0.4 milliradians. This breaks the cylindrical symmetry of the cavity, lifting the degeneracy of the triplet such that the mode extended along the membrane's tilt axis is least perturbed. This behavior is also well captured by the model, as shown in Fig. 5.13b (iii–iv).

When the degeneracy is lifted, we can resolve the crossings of triplet and singlet modes individually. Three of these crossings (i.e. within the dotted box of (ii)) are shown in greater detail in Fig. 5.13c. The upper two crossings are avoided, meaning $V_{i\neq j} \neq 0$; the membrane has hybridized these two modes. From this data it is possible to directly resolve the avoided gap (as labeled) and the quadratic optomechanical coupling of the hybridized mode is measured to be 4.5 MHz/nm^2. This is more than two orders of magnitude higher than can be achieved using just the TEM_{00} mode, and in principle is sufficiently strong to resolve the phonon shot noise of a mechanically driven membrane [22, 67]. Additionally, using a second laser detuned to ~-5 MHz,

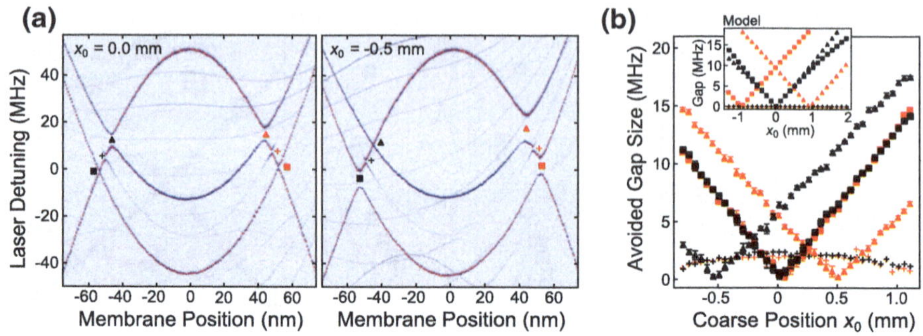

Fig. 5.14 Tuning the avoided crossing gap. **a** Transmission spectra for the membrane tilted 0.48 mrad about the $\hat{z}$ axis, (*left*) positioned near the center of the cavity ($x_0 = 0$) and (*right*) displaced by 0.5 mm to the left ($x_0 = -0.5$ mm). **b** Dependence of the avoided gaps labeled in (**a**) on the position of the membrane x_0. *Inset* shows the predictions of the perturbative model [modified from content originally published in Ref. [22]]

it would be possible to simultaneously perform linear optomechanical operations such as laser cooling, as needed [67]. Finally, while an avoided crossing has not been explicitly resolved at $x = 0$ in Fig. 5.13c, if we assume it is still avoided, the smallest measured mode splitting (inset) of 0.2 MHz places a lower bound on the curvature of 30 MHz/nm^2. This is encouraging, however we cannot improve the QND readout sensitivity indefinitely by reducing the gap size, due to the eventual onset of Landau-Zener-Stückelberg dynamics [62]; for small gaps, the optical mode will no longer adiabatically follow the curves of Fig. 5.13.

5.2.3.2 Tuning the Avoided Gaps

In addition to tilt, we also gain a good deal of control over cavity modes by varying the membrane's *position* within the cavity. Figure 5.14a (left) shows a singlet-triplet crossing structure with the membrane positioned near the cavity waist and tilted by 0.48 milliradians. An almost identical (mirrored) set of crossings are observed when the membrane moves 50 nm to the left or right, owing to the symmetry of the cavity in this configuration. If we now displace the membrane to the left by a *large* distance (0.5 mm), the avoided crossing structure becomes asymmetric: some gaps close and other gaps open. This asymmetry arises from a mismatch between the membrane's flat surface and the curvature of the cavity mode wave fronts away from the waist.

A similar set of data has been taken at other positions within the cavity, and a summary of avoided gap sizes for each of the crossings in Fig. 5.14a is shown in Fig. 5.14b. Notably, there is an essentially linear dependence of gap size on membrane position, meaning gap size is a very easy quantity to tune in situ. This linear dependence is well-captured quantitatively by the model (inset), though there is an overall offset in the avoided gap size of a few MHz. We suspect this discrepancy arises from imperfect form of the end mirrors, resulting in normal modes that are not quite Hermite-Gaussian. This would also explain why the TEM_{00}–TEM_{11} crossings are avoided when the model predicts they should not be due to the symmetries in the overlap integrals V_{ij}. In any case, by controlling the size of the avoided crossings in this manner, we also control the strength of the quadratic optomechanical coupling.

5.2.3.3 Linear Optomechanics at Avoided Crossings

When performing a QND readout of the membrane's phonon occupation, it is important to avoid an accidental position measurement, because its back-action will destroy the membrane's number states [18, 60]. As noted by the inset camera images in Fig. 5.13c, when adiabatically following an optical mode through an avoided crossing, the transverse profile of the optical mode changes. At the crossing point, it is an equal superposition the two modes, and this mixture varies linearly with membrane position. This means some information about the position of the membrane is leaving the cavity, thereby reducing the lifetime of the membrane's number states during readout [60].

An additional linear measurement may also appear in the cavity's loss rate κ. Fig. 5.15a, b shows the cavity loss of each mode (determined by individual cavity ringdown measurements) as a function of membrane position. Figure 5.15c, d shows the frequencies of these modes for reference. In general, each transverse mode has a different value of κ, which likely arises from spatial inhomogeneities in the optical performance of the mirrors and membrane. As a result, when the membrane passes through an avoided crossing, the cavity mode makes a continuous transition from one value of κ to the other, causing a gradient in κ at the center of the avoided crossing.

As described elsewhere [34], the back-action of this type of linear measurement could potentially be useful in optomechanical applications such as ground state laser cooling. In Fig. 5.15a, the dashed line shows a crossing that has not been resolved, but from the two transient data points we estimate $d\kappa/dx_0 > 600$ kHz/nm. Even under the most generous set of assumptions (i.e. no Landau-Zener transitions, and that $d\kappa/dx_0$ is entirely due to a gradient in the cavity *coupling* rate), however, this coupling would be too weak to achieve efficient laser cooling, and for the purposes of QND number state readout it represents an artifact to minimize.

In Fig. 5.15a, the membrane has been tilted about the $(\hat{x}+\hat{z})/\sqrt{2}$ axis. This defines a different axis of symmetry for the triplet modes (inset images), but otherwise the system behaves as in Fig. 5.13. Figure 5.15b shows a similar set of data for the membrane tilted about the $\hat{z}$ axis, and (perhaps not surprisingly) the transverse modes have rotated with the tilt axis. The choice of tilt *axis* therefore gives some measure of control over which portions of the mirrors are sampled by the cavity modes, allowing us to tune $d\kappa/dx_0$ as well. In Fig. 5.15b, each mode has taken on a new value of κ, and most notably, the value at the labeled avoided crossing has reversed sign, meaning it is probably possible to tune $d\kappa/dx_0 = 0$ using an intermediate tilt axis.

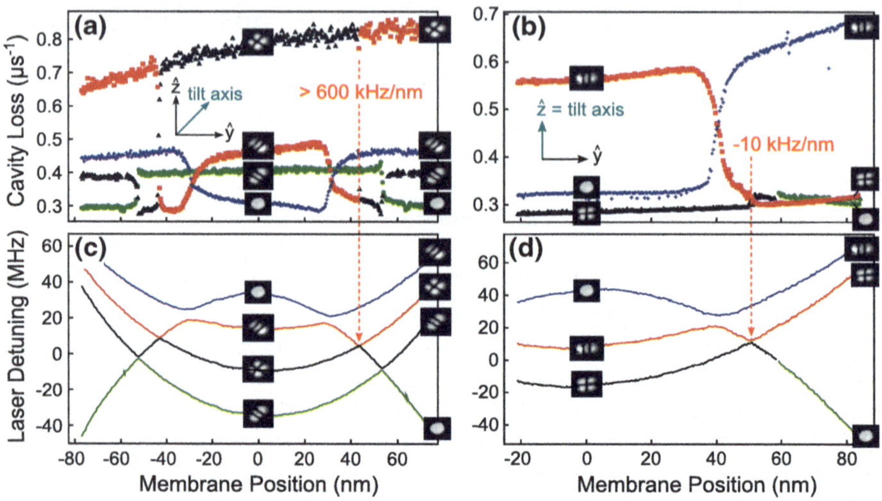

Fig. 5.15 Tunable cavity loss gradient. *Top plots* show the cavity loss rates as a function of membrane position for the four cavity modes with the membrane tilted **a** 0.66 mrad about the axis $(\hat{x}+\hat{y})/\sqrt{2}$ and **b** 0.61 mrad about the $\hat{z}$ axis. The different transverse modes (labeled with transmission images) sample different areas of the mirrors and membrane, each have their own loss rate, and the loss rates of hybrid modes (near avoided crossings) roughly tend toward a weighted average of the original modes. Plots **c** and **d** show the frequencies of the modes plotted in (**a**) and (**b**), respectively, for reference (One transverse mode is is not plotted in (**b**) and (**d**) because the input laser was not well coupled to it) [modified from content originally published in Ref. [22]]

5.2.3.4 Purely-Quartic Optomechanical Coupling

Another (rather strange) type of optomechanical coupling that can be generated with this geometry is shown in Fig. 5.16. As the membrane tilt is increased beyond ~1 milliradian, the TEM_{02} mode undergoes a smooth transition from (a) positive to (c) negative quadratic coupling. At an intermediate tilt (b), the quadratic coupling has vanished, and to lowest order the cavity mode frequency depends *quartically* on membrane position. While it is quite weak in this system (<1 Hz/nm^4), this type of nonlinearity could in principle be used to evolve coherent membrane oscillations to Schrödinger Cat states [68]. The effect is not reproduced using the four-mode degenerate perturbation theory described above, but it may be possible to explain by including more modes (e.g. another mode that repels this mode) or moving beyond a first-order theory.

5.3 Summary

The *nanomechanical resonator* and *membrane-in-the-middle* approaches described in this chapter offer several opportunities for optomechanics in addition to traditional linear dispersive coupling. These systems have been shown to generate quadratic,

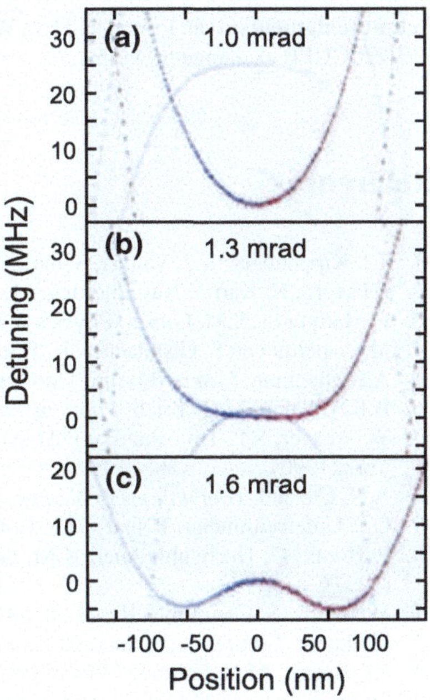

Fig. 5.16 Quartic optomechanical coupling. As the membrane tilt is increased, the position dependence of the TEM_{20} cavity mode frequency undergoes a smooth transition from **a** purely quadratic to **b** purely quartic to **c** double-well [modified from content originally published in Ref. [22]]

quartic, and absorptive (dissipative) optomechanical couplings, each of which can be readily tuned in situ by reorienting the mechanical element.

Quadratic coupling is of particular interest as a potential tool for nondemolition readout of a mechanical element's phonon number state. Due to the thermal reservoir and unintentional linear measurements, however, these number states will decay before they can be resolved, even using the strongest coupling reported with a membrane in a cavity [18, 22, 60, 67]. In order to resolve individual quantum jumps, the mechanical element must be engineered to be $\sim$1,000 times lighter and $\sim$1,000 times more flexible (in order to achieve sufficient zero-point motion and increase the resolution of the measurement), and great care must also be taken to minimize accidental linear measurements. A complementary approach would be to engineer a purely-quadratic readout of the already incredibly lightweight nanomechanical elements discussed in this chapter.

The last comment highlights a very key feature of *mechanical resonator in the middle of an optical cavity* geometry: since the mechanical element does not need to be a good mirror, essentially any mechanical element can be inserted, sub-wavelength or otherwise. All materials will have some form of absorption, scattering, or reflection, that leads to an optomechanical interaction with the cavity, so it is hard not to dream of performing optomechanics experiments on exotic or ridiculous mechanical elements – a graphene sheet, a piece of rubber, or perhaps even an optically-levitated living object. There are certainly surprises to come from future experiments that will push the concept beyond what has been imagined thus far.

Acknowledgments Ivan Favero and Eva Weig acknowledge support by DAAD/Egide Procope and BFHZ/CCUFB exchange programs.

References

1. T.J. Kippenberg, K.J. Vahala, Science **321**(5893), 1172 (2008)
2. I. Favero, K. Karrai, Nat. Photonics **3**, 201 (2009)
3. F. Marquardt, S.M. Girvin, Physics **2**, 40 (2009)
4. M. Aspelmeyer, S. Gröblacher, K. Hammerer, N. Kiesel, J. Opt. Soc. Am. B **27**, A189 (2010)
5. A.E. Siegman, *Lasers* (Oxford University Press, Oxford, 1986)
6. B.E.A. Saleh, M.C. Teich, *Fundamentals of Photonics* (Wiley, New York, 1991)
7. W. Weaver, S.P. Timoshenko, D.H. Young, *Vibration Problems in Engineering* (Wiley, New York, 1990)
8. A.N. Cleland, *Foundations of Nanomechanics* (Springer, Berlin, 2003)
9. Q.P. Unterreithmeier, T. Faust, J.P. Kotthaus, Phys. Rev. Lett. **105**, 027205 (2010)
10. P. Horak, G. Hechenblaikner, K.M. Gheri, H. Stecher, H. Ritsch, Phys. Rev. Lett. **79**, 4974 (1997)
11. V. Vuletic, S. Chu, Phys. Rev. Lett. **84**, 3787 (2000)
12. P. Maunz, T. Puppe, I. Schuster, N. Syassen, P.W.H. Pinkse, G. Rempe, Nature **428**, 50 (2004)
13. P.F. Barker, M.N. Shneider, Phys. Rev. A **81**, 023826 (2010)
14. O. Romero-Isart, M.L. Juan, R. Quidant, J.I. Cirac, New J. Phys. **12**, 033015 (2010)
15. D.E. Chang, C.A. Regal, S.B. Papp, D.J. Wilson, J. Ye, O. Painter, H.J. Kimble, P. Zoller, Proc. Natl. Acad. Sci. **107**, 1005 (2010)
16. T. Li, S. Kheifets, M.G. Raizen, Nat. Phys. **7**, 527 (2011)
17. P. Meystre, E.M. Wright, J.D. McCullen, E. Vignes, J. Opt. Soc. Am. B **2**, 1830 (1985)
18. J.D. Thompson, B.M. Zwickl, A.M. Jayich, F. Marquardt, S.M. Girvin, J.G.E. Harris, Nature **452**, 72 (2008)
19. A.M. Jayich, J.C. Sankey, B.M. Zwickl, C. Yang, J.D. Thompson, S.M. Girvin, A.A. Clerk, F. Marquardt, J.G.E. Harris, New J. Phys. **10**, 095008 (2008)
20. M. Bhattacharya, H. Uys, P. Meystre, Phys. Rev. A **77**, 033819 (2008)
21. D.J. Wilson, C.A. Regal, S.B. Papp, H.J. Kimble, Phys. Rev. Lett. **103**, 207204 (2009)
22. J.C. Sankey, C. Yang, B.M. Zwickl, A.M. Jayich, J.G.E. Harris, Nat. Phys. **6**, 707 (2010)
23. C. Biancofiore, M. Karuza, M. Galassi, R. Natali, P. Tombesi, G. Di Giuseppe, D. Vitali, Phys. Rev. A **84**, 033814 (2011)
24. A.M. Jayich, J.C. Sankey, K. Borkje, D. Lee, C. Yang, M. Underwood, L. Childress, A. Petrenko, S.M. Girvin, J.G.E. Harris, arxiv:1209.2730 (2012)
25. T.P. Purdy, R.W. Peterson, P.L. Yu, C.A. Regal, New J. Phys. **14**, 115021 (2012)
26. M. Karuza, M. Galassi, C. Biancofiore, C. Molinelli, R. Natali, P. Tombesi, G. Di Giuseppe, D. Vitali, J. Opt. **15**, 025704 (2013)
27. T.P. Purdy, R.W. Peterson, C.A. Regal, Science **339**, 801 (2013)
28. A. Xuereb, P. Domokos, New J. Phys. **14**, 095027 (2012)
29. I. Favero, K. Karrai, New J. Phys. **10**, 095006 (2008)
30. M.Y. Sfeir, F. Wang, L. Huang, C.C. Chuang, J. Hone, S.P. O'Brien, T.F. Heinz, L.E. Brus, Science **306**, 1540 (2004)
31. A. Gruber, A. Dröbenstedt, C. Tietz, L. Fleury, J. Wrachtrup, C. Borczyskowski, Science **276**, 2012 (1997)
32. M.T. Björk, B.J. Ohlsson, T. Sass, A.I. Persson, C. Thelander, M.H. Magnusson, K. Deppert, L.R. Wallenberg, L. Samuelson, Nano Lett. **2**, 87 (2002)
33. A. Högele, S. Seidl, M. Kroner, K. Karrai, R.J. Warburton, B.D. Gerardot, P.M. Petroff, Phys. Rev. Lett. **93**, 217401 (2004)
34. F. Elste, S.M. Girvin, A.A. Clerk, Phys. Rev. Lett. **102**, 207209 (2009)

35. C. Genes, H. Ritsch, D. Vitali, Phys. Rev. A **80**, 061803 (2009)
36. A. Xuereb, R. Schnabel, K. Hammerer, Phys. Rev. Lett. **107**, 213604 (2011)
37. J. Restrepo, J. Gabelli, C. Ciuti, I. Favero, C. R. Phys. **12**, 860 (2011)
38. N.L.S. De Liberato, F. Nori, Phys. Rev. A **83**, 033809 (2011)
39. S.J. van Enk, H.J. Kimble, Phys. Rev. A **63**, 023809 (2001)
40. K. Karrai, R.J. Warburton, Superlattices Microstruct. **33**, 311 (2003). Special issue dedicated to Professor Jorg Kotthaus on the occasion of his 60th Birthday, 29th May 2004
41. A. Högele, Laser spectroscopy of single charge-tunable quantum dots. Ph.D. thesis, Ludwig-Maximilians-Universität München (2006). Verlag Dr. Hut, München
42. H.C. van de Hulst, *Light Scattering by Small Particles* (Dover Publications, New York, 1981)
43. I. Favero, S. Stapfner, D. Hunger, P. Paulitschke, J. Reichel, H. Lorenz, E.M. Weig, K. Karrai, Opt. Express **17**, 12813 (2009)
44. H.B.G. Casimir, Phillips Res. Rep. **6**, 162182 (1951)
45. R.A. Waldron, Proc. IEE C Monogr. UK **107**, 272 (1960)
46. M. Eichenfeld, J. Chan, R.M. Camacho, K. Vahala, O. Painter, Nature **462**, 08524 (2009)
47. L. Ding, C. Baker, P. Senellart, A. Lemaitre, S. Ducci, G. Leo, I. Favero, Phys. Rev. Lett. **105**, 263903 (2010)
48. L. Ding, C. Baker, P. Senellart, A. Lemaitre, S. Ducci, G. Leo, I. Favero, Appl. Phys. Lett. **98**, 113108 (2011)
49. C. Metzger, I. Favero, A. Ortlieb, K. Karrai, Phys. Rev. B **78**, 035309 (2008)
50. T. Steinmetz, Y. Colombe, D. Hunger, T.W. Hänsch, A. Balocchi, R.J. Warburton, J. Reichel, Appl. Phys. Lett. **89**, 111110 (2006)
51. D. Hunger, T. Steinmetz, Y. Colombe, C. Deutsch, T.W. Hänsch, J. Reichel, New J. Phys. **12**, 065038 (2010)
52. S. Stapfner, I. Favero, D. Hunger, P. Paulitschke, J. Reichel, K. Karrai, E.M. Weig, Proc. SPIE **7727**, 772706 (2010)
53. S. Stapfner, L. Ost, D. Hunger, E.M. Weig, J. Reichel, I. Favero, Appl. Phys. Lett. **102**, 111110 (2013)
54. N.E. Flowers-Jacobs, S.W. Hoch, J.C. Sankey, A. Kashkanova, A.M. Jayich, C. Deutsch, J. Reichel, J.G.E. Harris, arXiv:1206.3558 (2012)
55. G. Jänchen, P. Hoffmann, A. Kriele, H. Lorenz, A.J. Kulik, G. Dietler, Appl. Phys. Lett. **80**, 4623 (2002)
56. NanoTools. (www.nanotools.com)
57. S.S. Verbridge, J.M. Parpia, R.B. Reichenbach, L.M. Bellan, H.G. Craighead, J. Appl. Phys. **99**, 124304 (2006)
58. I. Wilson-Rae, Phys. Rev. B **77**, 245418 (2008)
59. A. Jöckel, M.T. Rakher, M. Korppi, S. Camerer, D. Hunger, M. Mader, P. Treutlein, Appl. Phys. Lett. **99**, 143109 (2011)
60. H. Miao, S. Danilishin, T. Corbitt, Y. Chen, Phys. Rev. Lett. **103**, 100402 (2009)
61. A. Nunnenkamp, K. Børkje, J.G.E. Harris, S.M. Girvin, Phys. Rev. A **82**, 021806 (2010)
62. G. Heinrich, J.G.E. Harris, F. Marquardt, Phys. Rev. A **81**, 011801 (2010)
63. N.E.F. Jacobs, S.W. Hoch, J.C. Sankey, A. Kashkanova, A.M. Jayich, C. Deutsch, J. Reichel, J.G.E. Harris, Appl. Phys. Lett. **101**, 221109 (2012)
64. J.C. Sankey, A.M. Jayich, B.M. Zwickl, C. Yang, J.G.E. Harris, Proc. XXI Intl. Conf. Atomic Phys. (2009)
65. J. Rosenberg, Q. Lin, O. Painter, in OSA Technical Digest (CD) (Optical Society of America, 2010); doi:10.1364/cleo.2010.jmc1
66. D.J. Griffiths, *Introduction to Quantum Mechanics*, 2nd edn. (Pearson Prentice Hall, Upper Saddle River, 2005)
67. A.A. Clerk, F. Marquardt, J.G.E. Harris, Phys. Rev. Lett. **104**, 213603 (2010)
68. K. Jacobs, Phys. Rev. Lett. **99**, 117203 (2007)

Chapter 6
Cavity Optomechanics with Whispering-Gallery-Mode Microresonators

A. Schliesser and T. J. Kippenberg

Abstract Whispering gallery modes (WGM) of optical microresonators can feature long photon lifetime and small mode volume. The realisation that these devices exhibit co-localised mechanical modes coupled via radiation pressure to the WGM enabled dynamical backaction physics and a variety of other optomechanical phenomena to be observed and explored in an experimental setting. Here we provide a succinct introduction to cavity optomechanics with WGM resonators and review their use for the measurements with an imprecision below that at the standard quantum limit, cooling to the quantum regime and quantum coherent coupling. Moreover, optomechanically induced transparency (OMIT) in these resonators is described, which forms the basis for a number of quantum optomechanical protocols.

6.1 Optomechanical Coupling in Whispering-Gallery-Mode Resonators

6.1.1 Historical Perspective

Optical whispering gallery modes (WGMs) can attain quality factors in excess of 1 billion (i.e. 10^9), exhibit small optical mode volume and can be excited through their evanescent field, for example using prism coupling. These properties constitute a unique combination for experiments in quantum and nonlinear optics. Ultra high Q WGMs have first been observed in silica microspheres in the work of Braginsky

A. Schliesser (✉)
Niels Bohr Institute,Copenhagen University, Copenhagen, Denmark
e-mail: albert.schliesser@nbi.dk

T. J. Kippenberg
Ecole Polytechnique Federale de Lausanne (EPFL), Lausanne, Switzerland
e-mail: tobias.kippenberg@epfl.ch

M. Aspelmeyer et al. (eds.), *Cavity Optomechanics*, Quantum Science and Technology,
DOI: 10.1007/978-3-642-55312-7_6,

[1], and were subsequently studied in the context of cavity quantum electrodynamics (cQED) and ultralow-threshold rare earth microlasers [2]. Early work also observed their Kerr nonlinearity [3].

The extension of ultra high Q to an on-chip platform in the form of toroidal microresonators marked another enabling development [4]. These micro-fabricated resonators allowed to obtain ultra high Q ($>10^8$) with much smaller modal volumes than microspheres. Tapered optical fiber coupling was developed providing a method for efficient excitation with high ideality [5]. These developments rapidly enabled diverse applications of optical WGM microresonators, ranging from experimentally demonstrated strong coupling in cavity QED [6], atom-based single photon routing [7], low threshold on chip micro-lasers [8, 9] and biochemical dispersive sensing. Further work extended ultra high Q resonators also to other material classes such as ultrapure single-crystalline dielectrics [10]. Very low roughness was in this case obtained via polishing, instead of the laser reflow used in silica microspheres and microtoroid resonators. Moreover, the ultra high Q made a wide range of nonlinear phenomena observable at exceedingly low pump powers ($\mathcal{O}(1\,\mu\text{W})$), such as stimulated Raman lasing [11], Brillouin lasing or parametric oscillations [12]. More recently, optical frequency comb generation [13] via the Kerr nonlinearity and dissipative temporal solitons [14] have been observed in ultra high Q resonators.

Despite the wide-ranging nonlinear optical studies, effects of radiation pressure had neither been considered nor observed with optical micro-resonators.

The presence of mechanical modes in optical micro-resonators with free boundary conditions—though obvious in hindsight—had not been investigated, and the mutual coupling of optical and mechanical modes via radiation pressure in these structures was not known. Radiation pressure effects in optical microresonators thus remained unexplored until the year 2005.

Radiation pressure coupling between optical and mechanical modes were however explored theoretically in the gravity wave detection community. The pioneering work of Braginsky predicted most of its essential ramifications [15–17]. Central concepts that Braginsky developed were that of quantum and dynamical radiation pressure backaction. Quantum backaction consists in random force fluctuations that are caused by the quantum fluctuations of the laser light, giving rise to a detection limit for mechanical motion. This sets a limit which is nowadays known as the standard quantum limit (SQL) [18]. A second effect, of purely classical origin, but central to many cavity optomechanical phenomena, is referred to as "*dynamical backaction*" and occurs due to the storage lifetime of photons in the enhancement cavities in the arms of gravity wave interferometers. Granted the lifetime is on the order of the mechanical oscillator period (i.e. $\kappa^{-1} \sim \Omega_\mathrm{M}^{-1}$), dynamical backaction can lead to both amplification or damping (i.e. cooling [19]) of the cavity mirror motion. In the context of gravity wave detection, amplification can render the interferometer unstable—a phenomenon consquently known as *parametric oscillatory instability* [20]. Despite extensive theoretical activity around this phenomenon, amplification and instability due to dynamical radiation pressure backaction were not observed in gravitational wave interferometers.

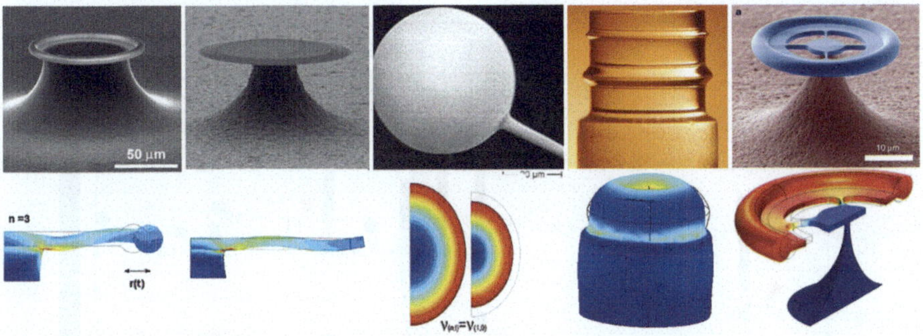

Fig. 6.1 Whispering gallery mode resonators and optomechanical coupling to co-localized mechanical modes via the effect of radiation pressure. The observed coupling in toroid resonators is a pars pro toto optomechanical coupling, found in any optical microresonator with free boundaries such that mechanical modes are supported. From *left* to *right prototypical toroid* [21, 31], *disk* [32], *sphere* [23], *crystal drum* [24], and spoke *toroid resonator* [33, 34] (*top*) and mechanical mode (*bottom*)

In 2005 the concepts developed by Braginsky in the field of gravity wave detection became an experimental reality in optical micro-resonators, specifically in toroid WGM resonators [21]. Only then was it recognized that optical micro-resonators exhibit, in addition to optical modes, also co-localized mechanical modes. It was shown that mechanical and optical modes are coupled by radiation pressure. Due to the ultra high Q and high mechanical frequencies the interaction took place in the regime where radiation pressure dynamical backaction became observable. By blue-detuned pumping, it was possible to observe amplification of the mechanical modes.

This discovery provided a general platform in which radiation pressure phenomena became accessible (Fig. 6.1). The observation of optomechanical coupling in toroidal resonators established optical microresonators in general as candidates for optomechanical coupling. It was proposed [22] that this coupling can be extended not only to other WGM resonators (such as microspheres [23], microtoroids [21] or crystalline resonators [24]), but also to other resonator geometries such as nanoscale photonic crystal cavities. Indeed, optomechanical coupling was thereafter demonstrated in 1-D and 2-D photonic crystals [25–30]; structures that were long known and had been studied for decades in their own right. These examples show that, contrary to common belief, not advances in nanofabrication lead to the rapid development of novel nano-optomechanical structures, but rather the fundamental "*pars pro toto*" insight that optical resonators exhibit intrinsic optomechanical coupling. Certainly, however, the availability of advanced micro- and nanofabrication techniques catalysed the development of novel structures with engineered optical and mechanical modes.

Optical micro-resonators possess a unique set of properties that have made them a highly suitable and successful platform for studying radiation pressure phenomena. First, the mechanical masses are extremely small (typically at the nanogram level), making these modes susceptible to weak forces. Combined with the high

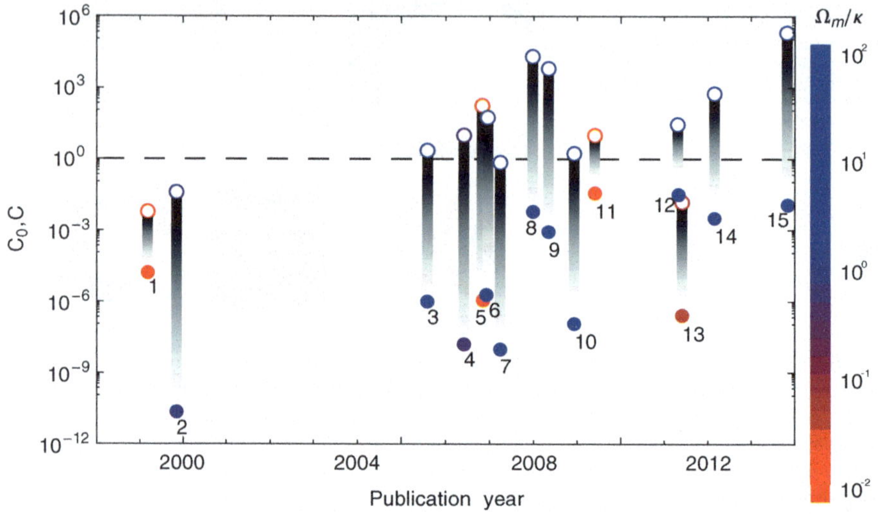

Fig. 6.2 Historical development of the single photon optomechanical cooperativity C_0 (*filled circles*) and possible range of field-enhanced cooperativity $C = C_0 \cdot \bar{n}_{\text{cav}}$ (reported maximum shown as an *empty circle*), as a function of time. The color code denotes the achieved sideband resolution factor Ω_m/κ. *Dashed line* indicates the critical level of $C = 1$, where in the resolved sideband regime, the dynamical backaction parametric oscillatory instability threshold is exceeded (as first observed in toroidal microresonators in 2006). Conversely, for red-detuning, $C \gg 1$ enables significant cooling. Shown systems are (*1*) torsional oscillator [36]—(*2*) micromirror [37]—(*3*) microtoroid [21]—(*4*) micromirror [38]—(*5*) micromirror [39]—(*6*) microtoroid [31]—(*7*) microsphere [40]—(*8*) nanomembrane [41]—(*9*) microtoroid [42]—(*10*) nanobeam coupled to microwave cavity [43]—(*11*) nanophotonic "zipper" cavity [28]—(*12*) 1D photonic crystal nanobeam [44]— (*13*) 2D photonic crystal [29]— (*14*) microtoroid [34]—(*15*) vacuum capacitor in a superconducting microwave circuit [45]

optical finesse, i.e. strong intracavity enhancement of the circulating optical power and thereby force, this gives rise to high optomechanical coupling rates $g_0/2\pi = \mathcal{O}(1\,\text{kHz})$. This, in turn, has enabled unprecedentedly large single-photon cooperativity

$$C_0 = \frac{4g_0^2}{\kappa \Gamma_{\text{M}}}, \tag{6.1}$$

a dimensionless parameter comparing the coupling strength to the dissipation rates κ, Γ_{M} of the optical, and mechanical modes, respectively. Figure 6.2 shows the historical development of this parameter. For most applications, it is also important that the mechanical frequency Ω_{M} can approach and exceed the cavity decay rate κ, a key for the observation of dynamical backaction in general, but also crucial for advanced cooling and quantum state transfer protocols (see below).

High frequencies are also beneficial to suppress thermo-optical forces [35], enabling the exclusive study of coherent radiation pressure coupling.

The compactness of optical microresonators also allowed combining optomechanics experiments with cryogenic techniques, enabling cooling of mechanical oscillators the quantum regime and providing an experimental setting in which quantum effects of mechanical oscillators could be explored. With WGM resonators this has been achieved by combining tapered fiber coupling of silica microtoroids with a helium-4 and helium-3 buffer gas cryostat [46]. The contactless evanescent coupling is in this case critical in order to ensure that the mechanical properties of the micro-resonator are not affected. Moreover, the buffer gas cooling enables efficient and bona fide thermalization of the sample, tapered fiber and coupling assembly. The successful precooling of toroidal resonators has enabled Heisenberg uncertainty limited measurements [47], cooling of mechanical oscillators to the quantum ground state with a ground state occupation of 37 % [34] and the demonstration of quantum coherent coupling between light and mechanical motion [34]. In addition, optomechanically induced transparency (OMIT) was observed [48], which has emerged as the conceptual underpinning of many current experimental protocols, such as optomechanical wavelength conversion or storage of optical fields in mechanical modes.

6.1.2 Elementary Concepts

The presence of a mechanical mode leads to a modulation of the cavity pathlength, due to the thermal fluctuations of this mode. This parametric coupling between optical and mechanical degree of freedom entails a position-dependent resonant frequency $\omega_{\mathrm{cav}}(t) = \omega_{\mathrm{cav},0} + G \cdot x(t)$. Therefore, the mutual coupling of optical and mechanical modes can simply be witnessed by recording the fluctuation spectrum of the transmitted cavity light. The "frequency pull parameter" G quantifies the dependence of the optical resonance frequency on the mechanical oscillator displacement, via $G = \partial\omega_{\mathrm{cav}}/\partial x$. For a Fabry-Perot resonator this coupling coefficient is simply given by $|G| = \omega/L$ where L is the Fabry-Perot resonator length. In the case of a WGM resonator the frequency pull parameter (with respect to a radial dilatation) is given by

$$|G| \approx \frac{\omega_{\mathrm{cav}}}{R} \tag{6.2}$$

to a good approximation.

The classical equation of motion of the optical field in the presence of the mechanical oscillator then becomes

$$\frac{d}{dt}a = i(\omega_{\mathrm{cav}} + G \cdot x(t))a + \sqrt{\kappa_{\mathrm{ex}}}s_{\mathrm{in}}e^{-i\omega_{\mathrm{L}}t}. \tag{6.3}$$

Here the field $s(a)$ is normalised to energy so that $|s_{\mathrm{in}}|^2$ ($|a|^2$) denotes the input laser power (optical energy stored in the cavity). The total energy cavity decay rate ($\kappa = \kappa_{ex} + \kappa_0$) is composed of intrinsic losses (κ_0) and useful external coupling at a rate κ_{ex}, which implies that the field transmitted by the cavity becomes [49]

$$s_{\text{out}}(t) = s_{\text{in}}(t) - \sqrt{\kappa_{\text{ex}}}a(t). \tag{6.4}$$

When exciting the optical mode, the mechanical motion $x(t)$ will cause a phase modulation of the intracavity field. This mechanism imprints the mechanical displacement fluctuations, as described by the (double-sided power) spectral density

$$S_{xx}(\Omega) \approx \frac{2\Gamma_{\text{M}} k_B T / m_{\text{eff}}}{(\Omega^2 - \Omega_{\text{M}}^2)^2 + \Gamma_{\text{M}}^2 \Omega^2} \tag{6.5}$$

in thermal equilibrium, onto the cavity field. Here m_{eff} denotes the effective mass of the mechanical mode and k_B the Boltzman factor and T the temperature of the mechanical mode. The phase (or, equivalently, frequency) modulation of the optical field emerging from the cavity can be recovered using a phase sensitive detection method, such as homodyne detection or polarization spectroscopy techniques, and analyzed with an electronic spectrum analyzer. In the simplest case, the frequency noise that the mechanical oscillator imprints into the cavity can be found as

$$\begin{aligned} S_{\omega\omega}(\Omega) &\approx G^2 S_{xx}(\Omega) \\ &= G^2 \cdot \frac{2\Gamma_{\text{M}} k_B T / m_{\text{eff}}}{(\Omega^2 - \Omega_{\text{M}}^2)^2 + \Gamma_{\text{M}}^2 \Omega^2} \\ &= g_0^2 \cdot \frac{4\Gamma_{\text{M}}}{(\Omega^2 - \Omega_{\text{M}}^2)^2 + \Gamma_{\text{M}}^2 \Omega^2} (\bar{n}\Omega_M^2), \end{aligned} \tag{6.6}$$

where the vacuum optomechanical coupling rate

$$g_0 = G \cdot x_{\text{ZPF}} = G \cdot \sqrt{\frac{\hbar}{2 m_{\text{eff}} \Omega_{\text{M}}}} \tag{6.7}$$

was used in the second parametrisation. The integrated cavity frequency noise is given by $\langle \delta\omega^2 \rangle = \int_{-\infty}^{\infty} S_{\omega\omega}(\Omega) \cdot \frac{d\Omega}{2\pi} = 2\bar{n} g_0^2$. Here, $\bar{n}$ is the mean phonon occupation of the mechanical mode, amounting to $\bar{n} = \bar{n}_{\text{th}} \approx k_{\text{B}} T / \hbar \Omega_{\text{M}}$ in thermal equilibrium (in the high temperature limit).

As a consequence, the vacuum optomechanical coupling rate can be immediately deduced from a calibrated frequency noise measurement of the cavity output, if the temperature T is precisely known. Calibration of the frequency noise can be accomplished by adding a known phase/frequency modulation to the pump laser, as detailed in Ref. [50]. This calibration technique has the unique advantage that all detection parameters (such as coupling efficiency, detector responsivity, optical losses, etc.) cancel. The method requires only accurate knowledge of phase modulation applied to the laser (and the sample temperature) in order to determine the vacuum optomechanical coupling rates for the different mechanical modes. Experimentally, for silica microtoroids, the vacuum coupling rate is highest for the radial breathing modes, with values as high as $g_0/2\pi \sim 4$kHz for optimized spoke sup-

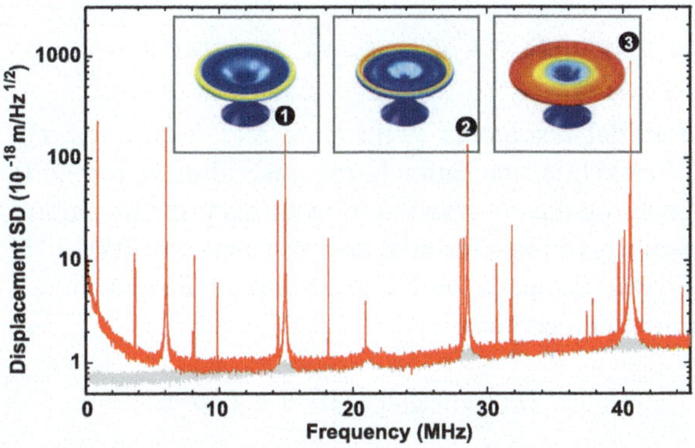

Fig. 6.3 Broadband displacement spectral density of a toroidal resonator, revealing several mechanical modes (insets). The calibration tone is attained by applying a known phase modulation to the readout laser beam (not shown). The corresponding displacement spectral density is obtained by converting the calibrated cavity frequency noise spectrum $S_{\omega\omega}(\Omega)$ into a displacement spectral density via $S_{xx}(\Omega) = S_{\omega\omega}(\Omega)/G^2$

ported micro-resonators [34]. Figure 6.3 shows an example of a broadband spectroscopy of a toroidal microresonator. Here, the recorded frequency noise spectral density has been converted back to a displacement spectral density.

Formally, the vacuum optomechanical coupling rate determines the strength of the optomechanical coupling in the full system Hamiltonian

$$H_{\mathrm{om}} = \frac{1}{2}\frac{\hat{p}^2}{m_{\mathrm{eff}}} + \frac{1}{2} m_{\mathrm{eff}} \Omega_{\mathrm{M}}^2 \hat{x}^2 + \hbar\omega_{\mathrm{cav}} \hat{a}^\dagger \hat{a} + \hbar G \hat{x} \hat{a}^\dagger \hat{a}, \tag{6.8}$$

where $\hat{a}$ is the optical annihilation operator and $\hat{x} = x_{\mathrm{ZPF}}(\hat{b} + \hat{b}^\dagger)$ the mechanical position operator. This Hamiltonian can be linearised in the intercavity field amplitude around its steady state value, i.e. $\hat{a} \rightarrow \bar{a} + \delta\hat{a}$, where $\bar{n}_{\mathrm{cav}} = |\bar{a}|^2$ denotes the average number of intracavity photons.

In the case of the radial breathing mode, the mechanical frequency greatly exceeds the cavity decay rate ($\Omega_{\mathrm{M}} \gg \kappa$), a condition which is referred to as the resolved sideband condition. In the resolved sideband regime, for the case of a blue laser detuning by $\Delta = \omega_{\mathrm{cav}} - \omega_{\mathrm{L}} = \Omega_{\mathrm{M}}$ the interaction Hamiltonian (last term in (6.8)) becomes

$$\hat{H}_{\mathrm{bsb}} = \hbar g_0 \bar{a} \cdot (\delta\hat{a}^\dagger \hat{b}^\dagger + \delta\hat{a}\hat{b}) \tag{6.9}$$

($\hat{b}$ is the mechanical annihilation operator). This Hamiltonian is thus similar to the one of parametric downconversion. Under the condition that the mechanical damping rate (Γ_{M}) is much lower than the optical damping rate (κ), this Hamiltonian leads to the parametric amplification of the mechanical oscillator (i.e. the parametric oscillatory instability). This phenomenon is analogous to a Brillioun laser, except that the roles of the mechanical oscillator and the optical modes are reversed. The threshold for the parametric oscillatory instability [20, 21] is reached when

$$C = C_0 \bar{n}_{\text{cav}} = \frac{4 g_0^2}{\kappa \Gamma_m} \bar{n}_{\text{cav}} = 1. \tag{6.10}$$

In the case of toroidal resonators (with $\kappa \sim \mathcal{O}(1\ \text{MHz})$ and $\Gamma_m \sim \mathcal{O}(1\ \text{kHz})$ and $g_0/2\pi \sim \mathcal{O}(1\ \text{kHz})$), the latter is the case already for $\mathcal{O}(10\,\mu\text{W})$ of laser power, thereby enabling the observation of radiation pressure parametric oscillatory instability as first observed in a toroidal microresonator in 2005 [21].

When pumping the resonator on the lower motional sideband ($\Delta = -\Omega_{\text{M}}$), the interaction Hamiltonian becomes

$$\hat{H}_{\text{rsb}} = \hbar g_0 \bar{a} \cdot (\delta \hat{a}^\dagger \hat{b} + \delta \hat{a} \hat{b}^\dagger) \tag{6.11}$$

in the resolved sideband regime. This Hamiltonian can lead to cooling of the mechanical oscillator, granted that the mechanical dissipation is much smaller than the optical one and granted that the laser has sufficiently low frequency (phase) noise as detailed in the next section. The optomechanical damping rate induced by dynamical backaction is given by

$$\Gamma_{\text{eff}} = \Gamma_{\text{M}}(C+1). \tag{6.12}$$

Optomechanical cooling in optical microresonators was observed in 2006 [31], concurrently with cooling in micro-mirror Fabry-Perot resonators [38, 39], and later resolved sideband cooling [42] in the regime ($\Omega_{\text{M}} \gg \kappa$) demonstrated, see below.

The broadband displacement spectra in addition reveal the extraordinary sensitivity with which the optical microresonator enables to measure the radial breathing modes due to the relatively large ratio of g_0/κ. For a quantum limited laser source, resonant laser excitation ($\Delta = 0$) and perfect overcoupling ($\kappa_{\text{ex}} \gg \kappa_0$), the shot noise limited sensitivity of frequency fluctuations is given by

$$S_{\omega\omega}^{\text{shot}}(\Omega) = \frac{\hbar\omega}{P_{in}} \frac{\kappa^2}{8} \left(1 + \frac{4\Omega^2}{\kappa^2}\right). \tag{6.13}$$

Consequently, the power to reach a sensitivity that enables recording the cavity frequency noise due to the mechanical oscillator's zero-point fluctuations with a peak power spectral density of $S_{\omega\omega}^{\text{ZPF}}(\Omega_{\text{M}}) = G^2 S_{xx}^{\text{ZPF}}(\Omega_{\text{M}}) = g_0^2/\Gamma_{\text{M}}$ are given by

$$P_{\text{SQL}} = \hbar\omega_{\text{cav}} \cdot \Gamma_{\text{M}} \left(\frac{\kappa}{2g_0}\right)^2 \cdot \left(1 + \frac{4\Omega_{\text{M}}^2}{\kappa^2}\right). \tag{6.14}$$

This power enables achieving an imprecision at the standard quantum limit (SQL). It can be compared to the power required to achieve the parametric instability threshold (P_{PI}, which corresponds to a field enhanced cooperativity of $C = 1$). In the resolved sideband regime $P_{\text{PI}} \approx 4 \times P_{\text{SQL}}$. The last term accounts for the fact that once the mechanical frequency exceeds the cavity decay rate, the buildup of the

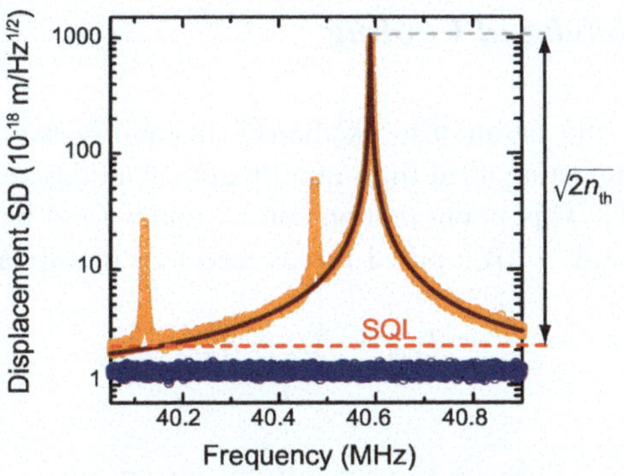

Fig. 6.4 Highly sensitive monitoring of a radial breathing mode of a toroidal microresonator at room temperature with an imprecision below that at the SQL. *Orange circles* are the measured thermal motion while the *blue circles* denote the shot noise limited imprecision. The shot-noise limited measurement imprecision of $S_{xx}^{\text{shot}}(\Omega_{\text{M}}) = (1.1\,\text{am})^2/\text{Hz}$ is below the zero point fluctuation's peak power spectral density $S_{xx}^{\text{zpm}}(\Omega_{\text{M}}) \approx (2.2\,\text{am})^2/\text{Hz}$

motional sidebands is less effective. This is not a fundamental constraint, and can be alleviated by using a multimode transducer [51], i.e. a scenario in which the mechanical oscillator is coupled to two or more optical resonances spaced by the mechanical frequency Ω_{M}, thereby yielding an effective reduction in the power to reach the SQL of $4\Omega_{\text{M}}^2/\kappa^2$ in the resolved-sideband limit.

Due to the relatively high ratio of g_0/κ very low measurement imprecision can be reached with toroidal microresonators [42, 52]. Figure 6.4 shows data for a 36 µm-diameter device with $(\kappa, \Omega_{\text{M}}, \Gamma_{\text{M}}, g_0) \approx 2\pi \cdot (10\,\text{MHz}, 40.6\,\text{MHz}, 1.2\,\text{kHz}, 1.9\,\text{kHz})$. In this measurement, the shot-noise limited imprecision of $S_{xx}^{\text{shot}}(\Omega_{\text{M}}) = (1.1\,\text{am})^2/\text{Hz}$ is below the zero-point fluctuation's peak power spectral density $S_{xx}^{\text{ZPF}}(\Omega_{\text{M}}) \approx (2.2\,\text{am})^2/\text{Hz}$. The latter is directly evidenced by the fact that the measured thermal fluctuation's peak power spectral density $S_{xx}^{\text{th}}(\Omega_{\text{M}}) = 2\bar{n}_{\text{th}} S_{xx}^{\text{ZPF}}(\Omega_{\text{M}})$ is more than a factor $2\bar{n}_{\text{th}} = 2k_{\text{B}}T/\hbar\Omega_{\text{M}}$ (or $\sqrt{2\bar{n}_{\text{th}}}$ in units of $\sqrt{S_{xx}}$) above the measurement background. For this device, the power required to reach the SQL is given by only $P_{\text{SQL}} = \mathcal{O}(1\,\mu\text{W})$. Shot-noise limited detection was achieved here with a Hänsch-Couillaud polarisation spectroscopy scheme [53]. Using near-field coupling, a sensitivity at the standard quantum limit has also been extended to objects in the near field such as nanomechanical oscillators [54].

6.2 Selected Experiments with Whispering-Gallery-Mode Resonators

The particularly high intrinsic optomechanical coupling in WGM microresonators has enabled the early observation of several novel optomechanical effects, which are discussed in the following sections.

6.2.1 Resolved-Sideband Cooling

As described above, the parametric oscillatory instability was first observed in a silica microtoroid supporting ultra-high-quality optical modes and mechanical radial breathing modes [21]. This phenomenon can be understood from the fact that for a blue-detuned laser ($\Delta > 0$), optical forces lead to a modification of mechanical damping given by

$$\Gamma_{\mathrm{opt}} = A_{-} - A_{+} \tag{6.15}$$

with

$$A_{\pm} \approx g^2 \frac{\kappa}{(\Delta \mp \Omega_{\mathrm{M}})^2 + \kappa^2/4}. \tag{6.16}$$

The optically induced damping Γ_{opt} is negative for $\Delta > 0$. Thus the total mechanical damping $\Gamma_{\mathrm{eff}} = \Gamma_{\mathrm{M}} + \Gamma_{\mathrm{opt}}$ can be eliminated, and for a sufficient coupling rate g the mechanical mode starts oscillating [21].

Analogously, for a red-detuned laser, radiation-pressure dynamical backaction gives rise to *increased* mechanical damping. As the optical force does not (or nearly not, see below) carry fluctuations, this constitutes a cold damping [37] mechanism, leading to a new effective temperature of the mechanical mode given approximately by

$$T_{\mathrm{eff}} \approx \frac{\Gamma_{\mathrm{M}}}{\Gamma_{\mathrm{M}} + \Gamma_{\mathrm{opt}}} T, \tag{6.17}$$

where T is the environment temperature. The reduced effective temperature corresponds to a suppression of fluctuations of the resonator's displacement, following the equipartition theorem $k_{\mathrm{B}} T_{\mathrm{eff}}/2 = m_{\mathrm{eff}} \Omega_{\mathrm{M}} \langle x^2 \rangle /2$.

This long-predicted effect [15, 17] has eventually been observed in 2006, by three groups, two of which used micromirror optomechanical systems [38, 39], and the authors, working with silica microtoroids [31]. While the initially demonstrated temperatures were in the Kelvin-range, the quest for ever lower temperatures—and corresponding phonon occupation $\bar{n} = k_{\mathrm{B}} T_{\mathrm{eff}} / \hbar \Omega_{\mathrm{M}}$—was opened. The ultimate goal of this endeavour is to cool a macroscopic mechanical mode to its quantum ground state ($\bar{n} = 0$) as has been achieved with single trapped ions decades before [55–57], providing an invaluable platform for quantum physics and metrology [58].

However, just as in atomic laser cooling, limits to the lowest achievable phonon occupation arise due to Stokes-scattering processes, in which a laser photon is down-converted while generating a mechanical excitation. The rate at which these processes occur is given by A_{+}, and correspondingly the minimum phonon number is given by [59, 60]

$$\bar{n}_{\mathrm{min}} = \frac{A_{+}}{\Gamma_{\mathrm{opt}}}. \tag{6.18}$$

Importantly, this limit strongly depends on the ratio of the mechanical oscillation frequency Ω_{M} and the optical cavity linewidth κ. In particular, for $\kappa \gg \Omega_{\mathrm{M}}$, the so-

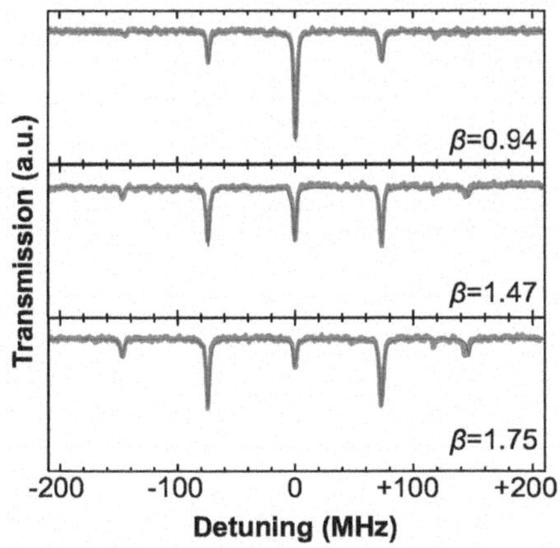

Fig. 6.5 Optical spectroscopy of a silica microtoroid whose whispering-gallery mode at 73.5 MHz is excited to oscillation of various amplitudes $\propto \beta$. The optical resonance and its mechanically-induced sidebands are only 3.2 MHz wide, much narrower than the mechanical resonance frequency. The motional sidebands are thus well resolved, essentially due to the high finesse $\mathscr{F} = 440{,}000$ of the resonator

called bad-cavity limit, the minimum occupation exceeds unity $\bar{n}_{\rm min} \approx \kappa/4\Omega_{\rm M} \gg 1$, whereas in the *resolved-sideband limit* (or, also, good cavity limit) with $\kappa \gg \Omega_{\rm M}$ one finds $\bar{n}_{\rm min} \approx \kappa^2/16\Omega_{\rm M}^2 \ll 1$. This is analogous to the case of trapped ions as first detailed in Ref. [42]. As a consequence, cooling to the quantum ground state $\bar{n} \to 0$ is fundamentally only possible if the cavity photons remain trapped in the WGM for many oscillation periods ($\kappa^{-1} \gg \Omega_{\rm M}^{-1}$). While bigger cavities realise longer photon storage times κ^{-1}, a larger radius R (or a longer linear cavity) reduces the optomechanical coupling rate, $g_0 \propto |G| = \omega_{\rm cav}/R$. It is therefore desirable to achieve as high as possible a finesse $\mathscr{F} = c/nR\kappa$ (n is refractive index here, c speed of light). Advantageously, WGM resonators are unsurpassed in terms of optical finesse, achieving values up to 10^7 in crystalline materials [24, 61]. Optimized silica microtoroids achieve slightly lower values (approaching 10^6), but combine it with large optomechanical coupling rates g_0 [50].

Figure 6.5 shows a spectroscopic signature of a silica microtoroidal resonator residing deeply in the resolved sideband regime. This $2R = 47\,\mu$m-diameter toroid supports a radial breathing mode at $\Omega_{\rm M}/2\pi = 73.5$ MHz, while the optical resonances are as narrow as $\kappa/2\pi = 3.2$ MHz. When the radial breathing mode is excited to oscillations of amplitude $\tilde{x}$ (here, with a second laser), the mean transmitted power of the probe laser can be written as [42]

$$|\bar{s}_{\rm out}|^2 = \left(1 - \eta_{\rm c}(1-\eta_{\rm c}) \sum_{n=-\infty}^{+\infty} \frac{\kappa^2 J_n(\beta)^2}{(\Delta + n\Omega_{\rm M})^2 + (\kappa/2)^2}\right) |\bar{s}_{\rm in}^2|, \tag{6.19}$$

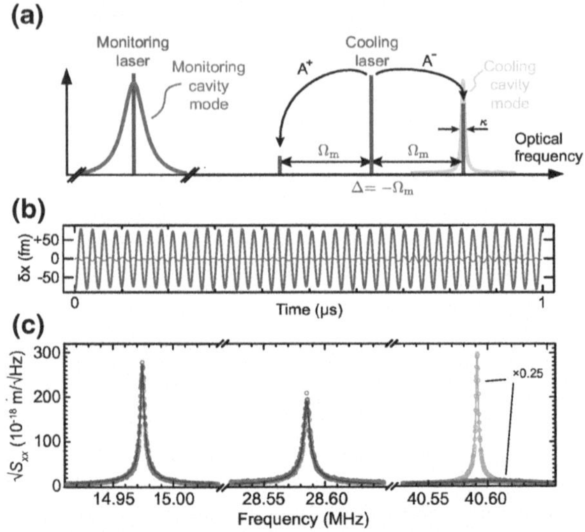

Fig. 6.6 Resolved sideband cooling. **a** The cooling laser ($\lambda = 980$ nm) is tuned to the *red sideband* of a narrow ($\kappa \ll \Omega_M$) cavity resonance, so that enhanced anti-Stokes scattering (rate $\propto A^-$) leads to cooling of the mechanical mode. The mechanical displacement fluctuations are simultaneously monitored with an independent monitoring laser ($\lambda' = 1,064$ mn) tuned to a second WGM resonance. **b** By filtering the signal from the monitoring laser in a 2-MHz-wide band around 40.6 MHz, the thermally driven displacement fluctuations of the RBM can be observed in time domain (*dark grey*; *light grey* is the noise background). **c** Spectral density of displacement fluctuations as recorded by the monitoring laser, showing the signatures of three radially symmetric mechanical modes. The fluctuations of the radial breathing mode at 40.6 MHz are strongly suppressed when the cooling laser is turned on (*dark grey*) as compared to the measurement in which the cooling laser is off (*light grey*). *Circles* are data, *lines* Lorentzian fits. From Ref. [42]

where $|\bar{s}_{\text{in}}^2|$ is the input power, J_n are the Bessel functions, $\eta_c = \kappa_{ex}/(\kappa_{ex} + \kappa_0)$ the coupling parameter and $\beta = G\tilde{x}/\Omega_M$. The transmission signal thus constitutes a series of Lorentzian dips, spaced by Ω_M (i.e. motional sidebands), with a width of κ. As evident in Fig. 6.5, the ratio Ω_M/κ is significant, ca. 23 for this device, due to its very high finesse of $\mathcal{F} = 440{,}000$.

In the absence of any external excitation, each mechanical mode still fluctuates randomly around a mean position ($\langle x^2 \rangle^{1/2} = \mathcal{O}(1\,\text{nm})$), due to its non-zero thermal occupation $\bar{n} = \mathcal{O}(10^5)$ at room temperature. Tuning a 'cooling' laser to the red motional sideband ($\Delta = -\Omega_M$), this occupation is expected to reduce according to Eq. (6.17). To evidence this effect on the mechanics, a second, 'monitoring' laser is tuned to the center of another cavity resonance. Using interferometric detection [53] the displacement fluctuations of the cavity boundary can be monitored independently, from which T_{eff} of each mode can be deduced (Fig. 6.6).

Figure 6.6b shows the displacement fluctuations of another toroid's radial breathing mode at 40.6 MHz (while $\kappa/2\pi = 5.8$ MHz), as recorded by the monitoring laser and isolated by appropriate filtering of the monitor signal. Spectral analysis of the monitor signal reveals the signatures of several modes, of which three are shown in Fig. 6.6c. The radial breathing mode features the strongest signature due its small effective mass.

When the cooling laser is turned on, as expected, the area underneath the Lorentzian peak, and therefore the mean displacement fluctuations $\langle x^2 \rangle^{1/2}$ are reduced, corresponding to cooling of this mode. Cooling rates up to $\Gamma_{\mathrm{opt}}/2\pi = 119\,\mathrm{kHz}$ are achieved in this sample, large compared to the intrinsic dissipation rate of $\Gamma_{\mathrm{M}}/2\pi = 1.3\,\mathrm{kHz}$ of this RBM. However, while a reduction to $\bar{n} = 1800$ quanta would be expected according to Eq. (6.17), a higher occupation of $\bar{n} \approx 5{,}900$ was experimentally measured [42]. A similar discrepancy was observed when cooling the toroid with the 73.5 MHz mode described above.

The origin of this discrepancy lies in the frequency fluctuations of the incoming cooling laser light. Due to the detuned operation, these fluctuations are converted into a fluctuating intracavity light power, and therefore radiation pressure force, which heats the mechanical mode. The added force fluctuations can be approximated as

$$S_{FF}^{\mathrm{fn}}(\Omega_{\mathrm{M}}) \approx \frac{S_{\omega\omega}(\Omega_{\mathrm{M}}) P^2 G^2}{\Omega_{\mathrm{M}}^4 \omega_{\mathrm{cav}}^2}, \tag{6.20}$$

in the resolved sideband regime, where $S_{\omega\omega}(\Omega_{\mathrm{M}})$ is the power spectral density of frequency fluctuations and P the incoming cooling light power. This leads to a new occupation of the mechanical mode in the presence of optical cooling of the form

$$\bar{n} = \frac{\Gamma_{\mathrm{M}} \bar{n}_{\mathrm{th}} + A_+}{\Gamma_{\mathrm{M}} + \Gamma_{\mathrm{opt}}} + n_{\mathrm{fn}}, \tag{6.21}$$

whereby $n_{\mathrm{fn}} \propto P$. When optimising over the input power P (and assuming $\Gamma_{\mathrm{M}} \bar{n}_{\mathrm{th}} \gg A_+$), a minimum occupation of [42, 62]

$$\bar{n}_{\mathrm{min}}^{\mathrm{fn}} \approx \sqrt{2 k_{\mathrm{B}} T m_{\mathrm{eff}} \Gamma_{\mathrm{M}} S_{\omega\omega}(\Omega_{\mathrm{M}})} / \hbar G \tag{6.22}$$

can be reached, consistent with our experimental observation [42].

It is therefore crucial to work with quantum-noise limited lasers ($S_{\omega\omega}(\Omega_{\mathrm{M}}) \to 0$) for ground state cooling [42, 62–65]. In particular, it follows directly from (6.22) that reaching the (somewhat arbitrary) condition $n_{\mathrm{min}}^{\mathrm{fn}} \leq 1$ requires $S_{\omega\omega}(\Omega_{\mathrm{M}}) \leq g_0^2/\Gamma_{\mathrm{M}} \bar{n}_{\mathrm{th}}$ [42, 62]. The external cavity diode lasers employed in many cooling experiments (including Ref. [42]), however, are not quantum-noise limited, actually up to high GHz Fourier frequencies due to the high relaxation oscillation frequency [66, 67]. Thus, their suitability for ground-state cooling has to be carefully assessed in each individual case [67, 68]. In addition to these noise requirements, the mode structure of WGM microresonators demand that the cooling laser be tuneable—often over several nanometers in wavelength—as the WGM resonances are far apart (free spectral range $c/2\pi n R = \mathcal{O}(10\,\mathrm{nm})$), and the precise resonance wavelengths are, in many cases, a not very well controlled result of the fabrication process. Consequently, Ti:sapphire

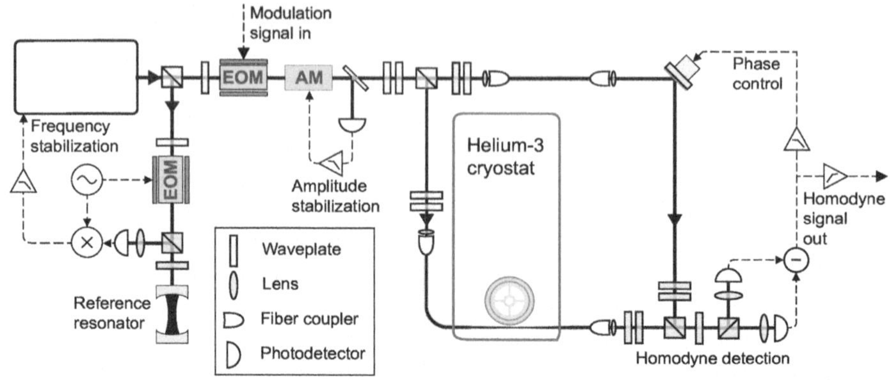

Fig. 6.7 Schematic of a cryogenic optomechanical experiment with a WGM resonator (spokes toroid). *Full lines* are optical paths, *dashed lines* electronic signals. *EOM*, electrooptic modulator; *AM*, amplitude modulator. See text for details. Adapted from Ref. [34]

lasers are frequently used in advanced cooling experiments with WGM resonators, virtually eliminating the contribution of frequency noise n_{fn} in Eq. (6.21)—due to their relatively low relaxation oscillation frequency—while maintaining wide tunability.

Furthermore, reaching the ground state ($\bar{n} \rightarrow 0$) in a room temperature environment with $\bar{n}_{\mathrm{th}} = \mathcal{O}(10^5)$ is extremely challenging. The required cooling factors $\Gamma_{\mathrm{opt}}/\Gamma_{\mathrm{M}} \gtrsim \bar{n}_{\mathrm{th}}$ are very large, often exceeding the mechanical quality factor, which is problematic as it would imply cooling rates exceeding Ω_{M}. Cryogenic precooling is therefore necessary. The relatively weak dependence of the mechanical quality factor on background gas pressure suggests to employ exchange gas cryostats, which thermalise the resonators by bringing them into contact with a cold ^{4}He [46, 47, 69, 70] or ^{3}He [34, 71] gas at a pressure on the order of 1 mbar. Coupling of light to the WGM can be achieved both using a fiber taper positioned in the WGM's near field [46], or using slightly deformed resonators [72] which couple weakly to the free-space mode of a laser beam sent into the cryostat [70]. Importantly, operation at cryogenic temperatures changes a number of material parameters of WGM resonators rather dramatically [69]. This includes thermal expansion and refractive index change—important for thermo-optical nonlinearities [1, 73], as well as mechanical dissipation [74]. Eventually, cryogenic experiments with base temperatures of a few K as provided by a ^{4}He cryostat achieved final occupations on the order of a few tens of quanta [47, 70]. Precooling in a ^{3}He cryostat to temperatures of 0.7 K enables reaching even lower occupations [34, 71].

Figure 6.7 shows a schematic of these experiments. A Ti:Sapphire laser (Sirah Matisse TX) operating at a wavelength around 780 nm is used as a laser source. It exhibits quantum limited amplitude and phase noise at relevant Fourier frequencies $\Omega/2\pi > 10\,\mathrm{MHz}$. The laser is locked to an external reference cavity such that drifts of the laser detuning Δ can be neglected during the acquisition time. Slow drifts in

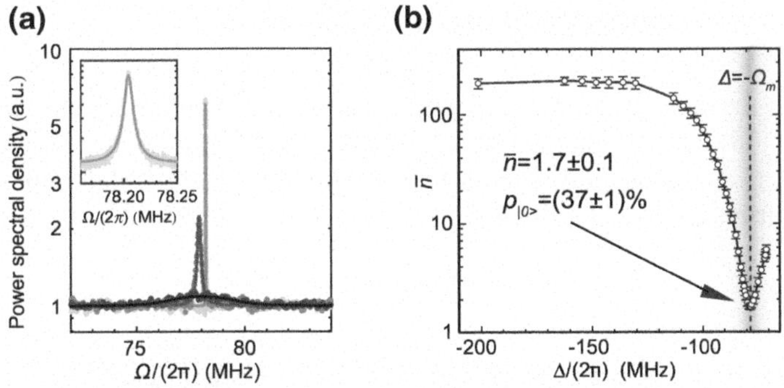

Fig. 6.8 Preparation of the quantum ground state cooling of a radial breathing mode in a silica microtoroid with 37 % probability. **a** Power spectral density of displacement fluctuations of the radial breathing mode for three different cooling laser detunings $-\Delta/\Omega_{\mathrm{M}} = \{1.89, 1.11, 1.00\}$ (*light* to *dark grey*; *points* are data and *curves* Lorentzian fits). Inset shows a zoom-in of the spectrum obtained for $-\Delta/\Omega_{\mathrm{M}} = 1.89$. **b** Occupancy of the radial breathing mode as a function of cooling laser detuning. The lowest occupation was found to be 1.7 ± 0.1 quanta on average, corresponding to a probability $p_{|0\rangle} = 37 \pm 1\,\%$ of occupying the ground state. From Ref. [34]

absolute laser power at the input of the experiment are actively cancelled to ensure operation at a constant drive amplitude. Finally, the laser's phase can be modulated with an electrooptic modulator for calibration and characterisation purposes.

The toroid resides in a (Oxford Instruments Heliox TL) helium-3 exchange gas cryostat. As it is situated directly above the surface of the liquified helium-3, the achievable temperature is directly linked to the latter's vapor pressure curve. In order to ensure sufficient thermalisation through the exchange gas, pressures $\gtrsim$0.15 mbar are favorable, corresponding to cryostat temperature setpoints $\gtrsim$650 mK. Coupling of light into the toroid is achieved via a tapered optical fiber that is approached using piezo positioners (Attocube GmbH), which are compatible with low temperature operation [46]. The fiber ends are guided through and out of the cryostat and constitute one arm of a balanced, 8-m-armlength fiber interferometer for a balanced homodyne detection scheme [75]. The detection quadrature is servo-locked using a movable mirror. This configuration allows a quantum-noise limited readout of the phase noise imprinted onto the transmitted laser field.

The closest approach to the ground state ($\bar{n} = 0$) was demonstrated with an optimised, spokes-supported [33, 34] silica microtoroid. In this work [34], a 78-MHz radial breathing mode was cooled to an occupation of $\bar{n} = 1.7 \pm 0.1$, corresponding to a probability of occupying the ground state of $p_{|0\rangle} = (1 + \bar{n})^{-1} = (37 \pm 1)\%$ (Fig. 6.8). This value is on a par with two other results reported from a nanophotonic ($p_{|0\rangle} = (54 \pm 2)\%$, Ref. [76]) and superconducting microwave system ($p_{|0\rangle} = (75 \pm 3)\%$, Ref. [77]).

6.2.2 Optomechanically Induced Transparency

Long before the most recent wave of experiments in cavity optomechanics it had been recognised [78] that the interaction described by the Hamiltonian (6.8) can assume a form that is analogous to the atom-light coupling at work in electromagnetically induced transparency (EIT), reading

$$H_{\mathrm{rsb}} = \hbar g \left(\delta \hat{a} \hat{b}^{\dagger} + \delta \hat{a}^{\dagger} \hat{b} \right) \tag{6.23}$$

in the simplest case for a red-detuned laser and a system deeply in the resolved sideband limit.

This analogy has been developed further in Ref. [79], and it was suggested to probe the dynamics of this system with a second, weak "probe" laser (amplitude δs_{p}) sent towards the optomechanical system. The Heisenberg-Langevin equations for the the mechanical displacement δx, and a small, intracavity probe field δa oscillating at a frequency Ω above a large "control" field (amplitude $\bar{a}$) can be written as [79]

$$(-i(\Omega - \Omega_{\mathrm{M}}) + \kappa/2)\, \delta a(\Omega) = -iG\bar{a}\, \delta x(\Omega) + \sqrt{\eta_{\mathrm{c}}\kappa}\, \delta s_{\mathrm{p}}(\Omega) \tag{6.24}$$

$$(-i(\Omega - \Omega_{\mathrm{M}}) + \Gamma_{\mathrm{M}}/2)\, \delta x(\Omega) = -i \frac{\hbar G \bar{a}}{2 m_{\mathrm{eff}} \Omega_{\mathrm{M}}}\, \delta a(\Omega) \tag{6.25}$$

for (i) a high-Q mechanical oscillator $\Omega_{\mathrm{M}} \gg \Gamma_{\mathrm{M}}$, (ii) the resolved-sideband regime $\Omega_{\mathrm{M}} \gg \kappa$, and (iii) a detuning $\bar{\Delta} = -\Omega_{\mathrm{M}}$ of the coupling field. The system (6.24)–(6.25) is formally equivalent to the equations describing (in the rotating wave approximation) the coherences induced in an atomic Λ-system under the conditions of EIT [79–82].

The analogy, and the terminology for the involved laser fields, becomes more clear from the illustrations in Fig. 6.9. The equivalent of the atomic levels in Λ-configuration, in the optomechanical system, is formed by three states, whose transitions $1 \leftrightarrow 2$ ($1 \leftrightarrow 3$) correspond to the excitation or removal of one mechanical (optical) excitation quantum. The starting state 1 is, in the simplest case, the ground state of the system—yet the effect can still be observed for sufficiently large laser fields if the mechanics is in a thermal state. The two laser fields are tuned close to the transitions $1 \leftrightarrow 3$ (control field, frequency ω_{l}) and $2 \leftrightarrow 3$ (probe field, frequency ω_{p}), and sent simultaneously towards the optical cavity. Their difference frequency $\Omega = \omega_{\mathrm{p}} - \omega_{\mathrm{l}}$ is close to the mechanical frequency Ω_{M}. The combined effect of the two fields, in the right circumstances, is to build up a coherence between 1 and 2, or, in other words, to excite coherent mechanical oscillation (as the mechanical mode is a harmonic oscillator, the established coherence can actually involve levels with many more mechanical excitation quanta).

This makes perfect sense from a classical point of view: The two fields' combined amplitude will oscillate at the beat frequency Ω, and so will the total radiation pressure force. If $\Omega \approx \Omega_{\mathrm{M}}$, this force is on resonance with the mechanical mode, inducing large-amplitude oscillations. This condition is referred to as the two-photon

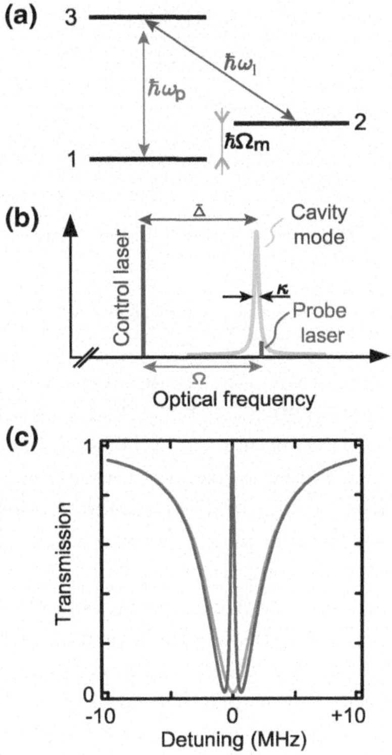

Fig. 6.9 Optomechanically induced transparency (OMIT). **a** Optomechanical Λ system. Transitions 1 ↔ 2 (1 ↔ 3) correspond to the excitation or removal of one mechanical (optical) excitation quantum. The control and probe laser fields are tuned close to the transitions 1 ↔ 2 (cavity red sideband), and 2 ↔ 3 (cavity resonance), respectively. **b** Optical frequencies of the cavity mode, control and probe lasers. **c** Transmission of the probe laser in absence (*light grey*), and presence (*dark grey*) of a control field $\bar{a}$ in the cavity. From [79]

resonance, as the energy difference between control and probe photons corresponds to a phonon's energy. As a result of an interference effect (essentially between the two driving terms on the right-hand-side of Eq. (6.24)), this oscillation prevents the buildup of an intracavity probe field δa. The probe field transmission follows from the input–output relation for the probe, $\delta s_{\mathrm{p,out}}(\Omega) = \delta s_{\mathrm{p}}(\Omega) - \sqrt{\eta_{\mathrm{c}}\kappa}\,\delta a(\Omega)$ which can be written as [83]

$$\delta s_{\mathrm{p,out}}(\Omega) = \left(1 - \eta_{\mathrm{c}}\kappa \frac{\chi_{aa}(\Omega)}{1 + g_0^2\bar{a}^2\chi_{aa}(\Omega)\chi_{xx}(\Omega)}\right)\delta s_{\mathrm{p}}(\Omega), \tag{6.26}$$

with $\chi_{xx}^{-1}(\Omega) = -i(\Omega - \Omega_{\mathrm{M}}) + \Gamma_{\mathrm{M}}/2$ and $\chi_{aa}^{-1}(\Omega) = -i(\Omega - \Omega_{\mathrm{M}}) + \kappa/2$ for $\bar{\Delta} = -\Omega_{\mathrm{M}}$. Evaluation of this expression shows that the probe field is transmitted, instead of absorbed in the cavity, when the two-photon resonance condition $\Omega = \Omega_{\mathrm{M}}$ is met. The induced transmission occurs only, however, in the presence of the field $\bar{a}$ (see Fig. 6.9), hence the name control field.

While the simplifications described above clearly expose the analogy to EIT and the underlying optomechanical interference mechanism, a more general expression for the (power) transmission of the probe field can be derived to read [79]

$$\left|t_{\mathrm{p}}\right|^2=\left|\frac{s_{\mathrm{p,out}}(\Omega)}{s_{\mathrm{p}}(\Omega)}\right|^2=\left|1-\frac{1+if(\Omega)}{-i(\Delta+\Omega)+\kappa/2+2\Delta f(\Omega)}\eta_{\mathrm{c}}\kappa\right|^2 \tag{6.27}$$

with

$$f(\Omega)=\hbar G^2\bar{a}^2\frac{\chi(\Omega)}{i(\Delta-\Omega)+\kappa/2} \tag{6.28}$$

and the full mechanical susceptibility

$$\chi(\Omega)=\left(m_{\mathrm{eff}}\left(\Omega_{\mathrm{M}}^2-\Omega^2-i\Gamma_{\mathrm{M}}\Omega\right)\right)^{-1}, \tag{6.29}$$

where it was only assumed that $\bar{a}\gg\delta a$, the system is stable, and that thermal and quantum noise can be neglected (see also Refs. [48, 84] for details of the calculation).

Experimentally, OMIT is observed by simultaneously sending a control and a probe fieldtowards an optomechanical cavity, and measuring the transmission of the probe. In the first experiment to demonstrate this effect [48], the probe field was generated by phase modulation of a Ti:sapphire laser, while the modulation carrier serves as control field. In this case, the modulation frequency sets also the frequency spacing Ω between the probe and control field. The optomechanical system was a silica microtoroid with large undercut [42] but no spokes support, and the experiment was performed at room temperature, but essentially with the same experimental setup as shown Fig. 6.7. The employed homodyne detection scheme enables recovery of the probe transmission with a high signal-to-noise ratio [48].

Figure 6.10 shows the expected and measured signatures of OMIT in a silica microtoroid with a radial breathing mode at $\Omega_{\mathrm{M}}/2\pi=51.8$ MHz. As expected, a distinct feature in the transmission, but also the detected homodyne signal, occurs at this modulation frequency. This two-photon resonance is independent of the control's detuning $\bar{\Delta}$.

Figure 6.11 shows the result of a series on systematic measurements of the OMIT feature, studying its dependence on the power of the control beam. Clearly, both the depth of the dip in the homodyne signal—corresponding to a peak in transmission—and the width of the narrow feature increase with launched control laser power. The probe transmission on resonance ($\bar{\Delta}=-\Omega_{\mathrm{M}}$, $\Delta'=0$) can be normalised to the transmission in the absence of OMIT, by defining $t_{\mathrm{p}}'=(t_{\mathrm{p}}-t_{\mathrm{r}})/(1-t_{\mathrm{r}})$ with the residual transmission $t_{\mathrm{r}}\equiv t_{\mathrm{p}}(\bar{a}=0)$. As consequence, this value is independent of the waveguide-cavity coupling η_{c} and is given simply by [48]

$$t_{\mathrm{p}}'=\frac{C}{1+C}, \tag{6.30}$$

where the cooperativity $C=4g/\Gamma_{\mathrm{M}}\kappa$ is determined by the amplitude of the control field through the coupling rate $g=\bar{a}g_0$. The data of Fig. 6.11 agree with this functional dependence. Furthermore, the width of the OMIT feature is approximately given by

$$\Gamma_{\mathrm{OMIT}}\approx(1+C)\Gamma_{\mathrm{M}}, \tag{6.31}$$

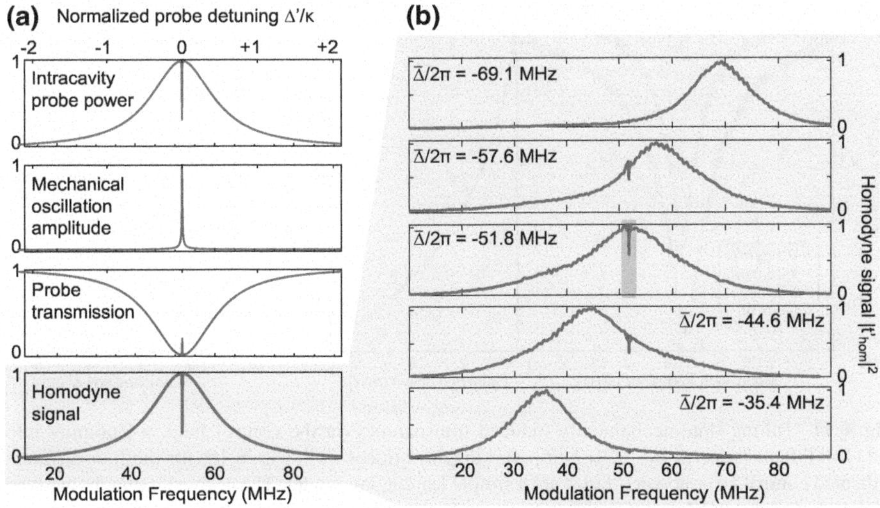

Fig. 6.10 Experimental observation of OMIT. **a** Predicted behaviour of the intracavity probe power, the mechanical oscillation amplitude, the probe transmission and the homodyne signal normalised to the interval [0, 1] versus the normalised two-photon detuning $\Delta'/\kappa = (\Omega - \Omega_\mathrm{M})/\kappa$. **b** Experimentally measured homodyne traces for different control laser detunings $\bar{\Delta}$. Independent of $\bar{\Delta}$, a sharp OMIT feature appears at zero two-photon detuning $\Delta' = 0$, when the modulation frequency equals the mechanical frequency of 51.8 MHz. From Ref. [48]

again in agreement with the experimental finding.

Two important corollaries follow immediately from the observation of these sharp, and light-tuneable dips in the probe transmission. First, the mechanically mediated interaction between the two optical fields enables switching of one light field with another, reminiscent of an optical transistor, which is of potential interest in classical applications. But remarkably, in already existing systems [76, 85] switching is induced by only $\mathcal{O}(10)$ control photons, and this mechanism should be viable all the way to the quantum regime, in which single control photons would switch the probe transmission [86]. Second, the steep variation of the probe field's phase, concomitant of the sharp transmission feature, leads to significant, and tuneable group delays $\tau_\mathrm{g} = -\partial \arg(t_\mathrm{p})/\partial\omega_\mathrm{p}$ in the vicinity of the OMIT window. Similar to atomic EIT, the generation of slow and fast light [82]—and in arrays of optomechanical systems—storage of optical pulses [87–89] can thus be expected [79].

Today, optomechanically induced transparency, and variants thereof, have been observed in a number of experimental systems, including other whispering-gallery mode resonators [90, 91], membrane-in-the-middle [92], and nanophotonic platforms [44], as well as superconducting microwave nanoelectromechanical devices [83, 85, 93]. Slow and fast light effects have been evidenced both spectroscopically [44, 85] and directly observed in time-domain pulse delay measurements [83]. Theoretically, it has been shown that the limited delay-bandwidth product $\Gamma_\mathrm{OMIT}\tau_\mathrm{g} \leq 2$ [48] can be overcome using arrays and dynamic tuning of the control field [94, 95].

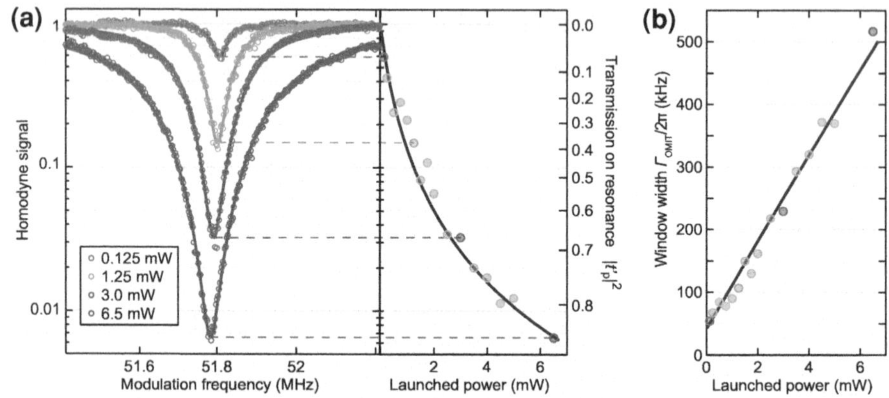

Fig. 6.11 Tuning optomechanically induced transparencyvia the control field. **a** Zooming into the OMIT feature for $\bar{\Delta} = -\Omega_M$.*Left panel* displays traces as in Fig. 6.10, for the four different indicated control laser powers. *Lines* are a simple Lorentzian model. The corresponding homodyne signal, and normalised probe (power) transmission *on resonance*, $|t'_p|^2$ (*right ordinate*), is shown in the *right panel*. The transmission on resonance is given for the for indicated powers (points connected with *dashed lines* to the *left panel*), but also for a number of other power levels, for which the spectra are not shown. The data agree well with the model (*line*). **b** Width of the OMIT transmission window extracted from the same set of probe scans, together with the theoretical model (*line*)

Looking ahead, destructive interference of multiple excitation pathways, as observed in OMIT, lies also at the heart of optomechanical dark modes [96], and is of great interest for quantum and classical wavelength conversion [91, 96, 97], in particular as it might bridge microwave and optical domains [44, 98–100], a goal towards which rapid experimental progress is currently made [101–103]. Finally, as recently pointed out [86, 104, 105], the first signature of a *generic* quantum effect (i.e. absent in the case $\hbar \to 0$) in systems approaching the strong single-photon coupling regime $g_0/\kappa \sim 1$ could be observable in an OMIT-type measurement.

6.2.3 Quantum Coherent Coupling

Profound theoretical analysis of optomechanical cooling had shown [59, 106, 107] that another limit of optomechanical cooling exists. If the mechanical mode's thermal excitation, after upconversion to an anti-Stokes photon, cannot leave the optical cavity fast enough, it may be converted back into a phonon by the presence of the cooling beam. This mechanism can be understood from the formal symmetry of the Hamiltonian (6.23): excitations of the two modes δa, b can be interconverted both ways in the presence of a red-detuned field. Only if upconverted photons are dissipated faster (rate κ) than they can be generated through conversion (rate $\Omega_c = 2g$) of mechanical excitations, cooling can be maintained. In the limit $g \gg \kappa$, in contrast, mechanical and optical excitations are cyclically interconverted at a rate Ω_c, and cooling ceases [59, 106, 107], with the final occupation given by [106]

$$\bar{n} = \left(\frac{4g^2 + \kappa^2}{4g^2 + \Gamma_M \kappa} + \frac{g^2}{\Omega_M^2} \right) \frac{\Gamma_M}{\kappa} \bar{n}_{th} + \frac{\kappa^2 + 8g^2}{16\Omega_M^2}, \tag{6.32}$$

converging to $\bar{n} \to (\Gamma_M/\kappa)\bar{n}_{th}$ for $\kappa \ll g \ll \Omega_M$.

In the frequency domain, cyclic interconversion corresponds to parametric normal-mode splitting. Therefore, in this limit, the optomechanical system enters a qualitatively new regime. In particular, the optical and mechanical modes hybridise into new eigenmodes with frequencies $\Omega_M \pm g$ (for the optical mode, in a frame rotating at the driving laser's frequency). The first signature of such hybridisation was observed in a micromirror experiment [108], in which double-peaked spectra were reported for the mechanical mode for a sufficiently large red-detuned "coupling" field leading to $\Omega_c > \kappa$. Stronger coupling (with $\Omega_c > 10\kappa$) was reported later in an electromechanical system [85].

An observation of optomechanical strong coupling in a whispering-gallery-mode resonator—a silica microtoroid [34]—is shown in Fig. 6.12. This experiment was performed in a helium-3 cryostat in a setup essentially the same as the one shown in Fig. 6.7. The detuning Δ of the coupling laser is systematically varied, which changes the frequency of the cavity mode in the frame rotating with the coupling laser frequency. Thus, by tuning Δ to $-\Omega_M$, the two modes can be effectively tuned into resonance. As expected for hybridising modes, both the mechanical displacement spectra, and the optical response of the system (recorded using a modulation–demodulation technique as in the OMIT experiments) display an avoided crossing when the optical mode is tuned through the mechanical resonance frequency $\Omega_M/2\pi = 78$ MHz.

The exchange of excitations is a fully quantum-coherent process. That implies that the quantum states of the mechanical and optical modes can be swapped by switching on the coupling field for an effective duration π/Ω_c of half a swap cycle [100, 109, 110]. It is crucial, however, that during this time, decoherence processes leave the states unaffected. Since both optical and mechanical modes are not isolated, but coupled to an environment (Fig. 6.13) this condition is far from trivial. The relevant environment of the optical mode consists of modes propagating away from the cavity, or microscopic systems (molecules, defects) in which transitions can be induced by photons getting absorbed. Similarly, the mechanical environment often consists of propagating phonon modes [111], but the details depend on the prevalent damping mechanism [33, 52, 69, 112].

However, a dramatic difference exists in the equivalent occupation of these environments: The optical environment is typically in the ground state, due to the high optical frequency, leading to the quadrature noise of the optical laser field to be quantum-limited. The mechanical environment, in contrast, has a large occupation of, e.g. $\bar{n}_{th} \approx 200$ for a 100 MHz oscillator held at 1 K temperature. As a consequence, while the optical dissipation and decoherence rates are equally given by κ, the relevant decoherence rate $\gamma \approx \bar{n}_{th}\Gamma_M = \frac{k_B T}{\hbar Q_m}$ is significantly faster than the

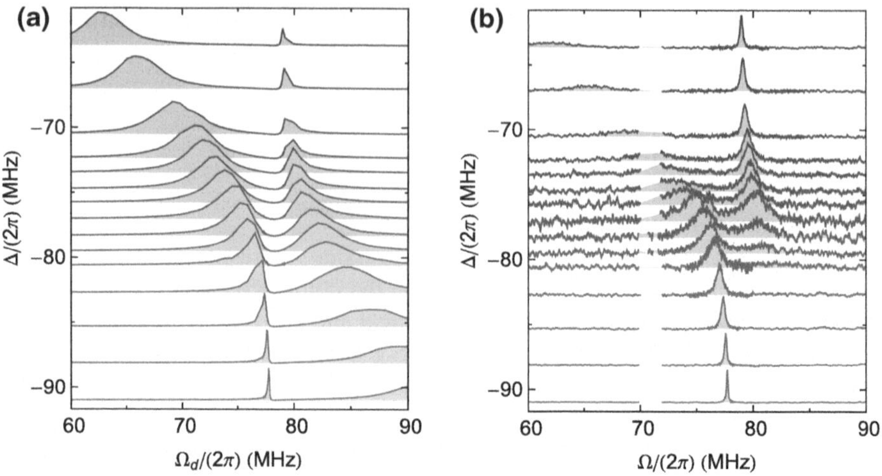

Fig. 6.12 Avoided crossing due to strong optomechanical coupling. **a** Optical response measured using a modulation-demodulation technique, with a modulation at frequency $\Omega_d/2\pi$. **b** Mechanical thermal noise spectrum as a function of Fourier frequency Ω. From Ref. [34]

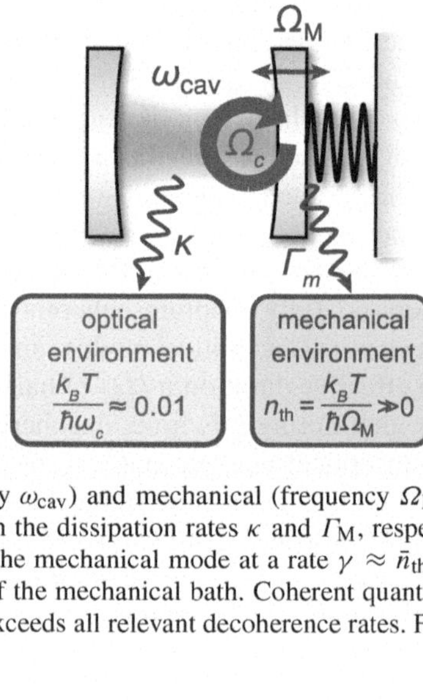

Fig. 6.13 Optical (frequency ω_{cav}) and mechanical (frequency Ω_M) modes are coupled to their respective environments with the dissipation rates κ and Γ_M, respectively. Both channels lead to decoherence, in the case of the mechanical mode at a rate $\gamma \approx \bar{n}_{th}\Gamma_M$ aggravated by the usually large mean occupation $\bar{n}_{th}$ of the mechanical bath. Coherent quantum dynamics can be observed when the coupling rate Ω_c exceeds all relevant decoherence rates. From Ref. [34]

mechanical dissipation. Thus, while $\Omega_c > (\kappa, \Gamma_M)$ is sufficient for strong coupling, coherent state swapping (as well as a number of related quantum state transfer protocols) requires $\Omega_c > \gamma$ [78, 100, 109, 110, 113], which is much more challenging to achieve experimentally.

With a silica microtoroid held at 0.65 K, we have achieved a coupling rate $\Omega_c/2\pi = (3.70 \pm 0.05)$ MHz; and extracted a significantly smaller mechanical decoherence rate $\gamma/2\pi = (2.2 \pm 0.2)$ MHz from the (carefully calibrated) ampli-

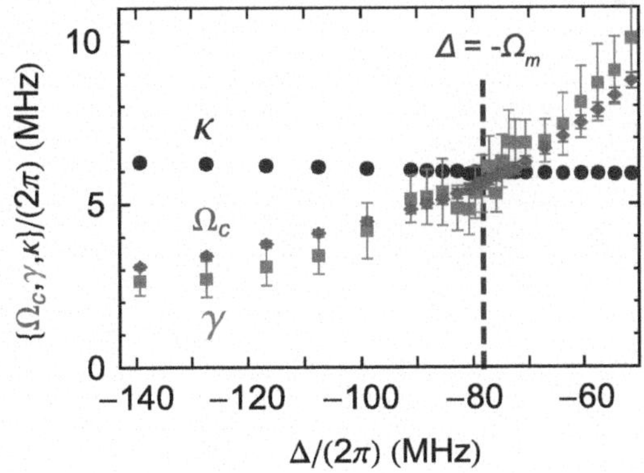

Fig. 6.14 Optomechanical coupling rate $\Omega_c = 2g$, optical (κ), and mechanical (γ) decoherence rates as a function of detuning of the coupling laser. As the laser is tuned closer to resonance, the temperature of the toroid rises due to residual absorption, leading to an increasing mechanical decoherence rate. At a detuning $\Delta = -\Omega_M$, quantum-coherent coupling can be reached. From Ref. [34]

tude of the residual thermal fluctuations of the mechanical mode [34]. If the coupling rate is further increased to also exceed κ (cf. Fig. 6.12), laser heating leads to a slightly increased temperature of the experimental system, and thus a larger mechanical decoherence rate. Yet, quantum coherent coupling, i.e. $\Omega_c \gtrsim \kappa, \gamma$ can be simultaneously achieved (Fig. 6.14), opening the door to preparation and control of non-classical states of mechanical motion with light [78, 100, 109, 110, 113].

As a proof of principle and a first classical illustration of the potential of such state-swapping experiments, the dynamical exchange of a pulsed coherent excitation between the optical and mechanical modes was demonstrated by launching a weak pulse resonant with the cavity in the presence of the coupling field (Fig. 6.15). When the coupling is weak ($\Omega_c < \kappa$), the pulse excites the mechanical mode to a finite oscillation amplitude, which decays at the optically induced damping rate. The measured homodyne signal generated by the cavity's output field is a barely modified copy of the input pulse. If the coupling laser power is increased—resulting in $\Omega_c/2\pi = 11.7\,\text{MHz} > \kappa/2\pi$—the envelope of the homodyne signal, which corresponds to the amplitude of the cavity field, features several exchange cycles with the mechanical oscillator. A numerical evaluation of our theoretical model shows excellent agreement with our measurements, and illustrates how the excitations are swapped between the optical and mechanical modes (Fig. 6.15). We have verified that this swapping protocol is working down to very low excitation amplitudes (0.9 photons per pulse) and strong coupling rates ($\Omega_c/2\pi = 16.6\,\text{MHz}$). Apart from the whispering-gallery-mode device discussed here, only optimised electromechanical systems in the microwave domain have so far demonstrated similar performance [85, 114].

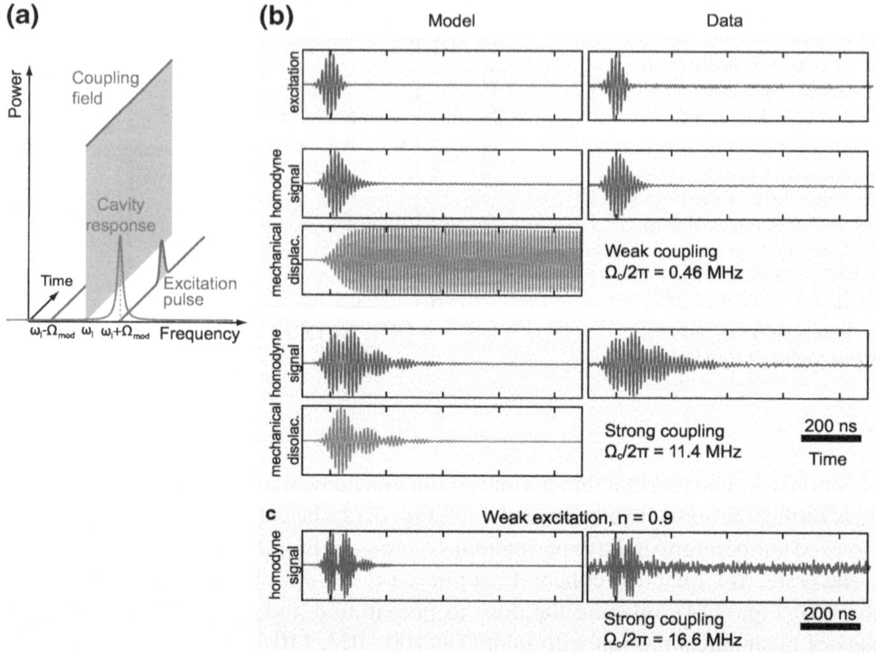

Fig. 6.15 Coherent exchange between the optical field and the micromechanical oscillator probed in time domain. **a** Schematic of the experiment. A weak, resonant, pulse is sent towards the cavity while the coupling laser beam is tuned to $\Delta = -\Omega_\mathrm{M}$. **b** The optical excitation waveform, and comparison with the optical output measured as a homodyne signal. Both results from numerical modelling (*left*) and experimental data (*right*) are shown. The mechanical displacement can not be measured independently and only the modelling results are shown. As expected, for strong coupling, the excitation cycles between optical and mechanical modes. **c** Cyclic exchange of energy persist also for very weak probe pulses, and very strong coupling. From Ref. [34]

6.3 Conclusion

The realisation that optical WGM resonators feature intrinsic optomechanical coupling has provided a novel and highly successful approach to study radiation pressure backaction phenomena, as first conceived of in the context of gravity wave detection. The co-localization of ultra high Q optical and mechanical modes in micro-scale devices has enabled the realisation of large vacuum optomechanical coupling rates and access to the regime where cavity retardation becomes important, i.e. the resolved sideband regime. As one of the first systems to simultaneously offer all these features, WGM resonators have served as a workhorse for several pioneering experiments in cavity optomechanics. This includes measurements that achieve an imprecision below that at the SQL, the preparation of a mechanical mode in its quantum ground state with ∼40 % probability, and the demonstration of strong, quantum coherent coupling. Moreover, observation of the effect of optomechanically induced transparency (OMIT) has first been reported in WGM, and the underlying destructive interference of excitation pathways is essential for many protocols for photon/phonon state conversion or storage. Without doubt, the glass WGM resonators with radial breathing modes mainly discussed here, as well as newly introduced WGM devices from other materials [24, 91, 115] and in other geometries [54, 116], will continue to contribute to the rapid development of cavity optomechanics.

Acknowledgments The authors would like to thank Hendrik Schuetz for his assistance in the preparation of Fig. 6.2, and acknowledge the invaluable contributions of their coworkers, including Georg Anetsberger, Olivier Arcizet, Samuel Deleglise, Jens Dobrindt, Emanuel Gavartin, Remi Riviere, Ewold Verhagen and Stefan Weis.

References

1. V.B. Braginsky, M.L. Gorodetsky, V.S. Ilchenko, Phys. Lett. A **137**(7–8), 393 (1989)
2. V. Sandoghdar, F. Treussart, J. Hare, V. Lefèvre-Seguin, J.M. Raimond, S. Haroche, Phys. Rev. A **54**(3), R1777 (1996)
3. F. Treussart, V.S. Ilchenko, J.F. Roch, J. Hare, V. Lefèvre-Seguin, J.M. Raimond, S. Haroche, Eur. Phys. J. D **1**, 235 (1998)
4. D.K. Armani, T.J. Kippenberg, S.M. Spillane, K.J. Vahala, Nature **421**, 925 (2003)
5. S.M. Spillane, T.J. Kippenberg, O.J. Painter, K.J. Vahala, Phys. Rev. Lett. **91**(4), 043902 (2003)
6. T. Aoki, D. Bayan, E. Wilcut, W.P. Bowen, A.S. Parkins, T.J. Kippenberg, K.J. Vahala, H.J. Kimble, Nature **443**, 671 (2006)
7. B. Dayan, A.S. Parkins, T. Aoki, E.P. Ostby, K.J. Vahala, H.J. Kimble, Science **319**(5866), 1062 (2008)
8. A. Polman, B. Min, J. Kalkman, T.J. Kippenberg, K. Vahala, Appl. Phys. Lett. **84**(7), 1037 (2004)
9. T.J. Kippenberg, J. Kalkman, A. Polman, K.J. Vahala, Phys. Rev. A **74**, 051802(R) (2006)
10. V.S. Ilchenko, A.A. Savchenkov, A.B. Matsko, L. Maleki, Phys. Rev. Lett. **92**(4), 043903 (2004)
11. S.M. Spillane, T.J. Kippenberg, K.J. Vahala, Nature **415**(6872), 621 (2002)
12. T.J. Kippenberg, S.M. Spillane, K.J. Vahala, Phys. Rev. Lett. **93**(8), 083904 (2004)
13. P. Del'Haye, A. Schliesser, O. Arcizet, T. Wilken, R. Holzwarth, T.J. Kippenberg, Nature **450**, 1214 (2007). doi:10.1038/nature06401
14. T. Herr, V. Brasch, J.D. Jost, I. Mirgorodskiy, G. Lihachev, M.L. Gorodetsky, T.J. Kippenberg, Nat. Photonics. **8**, 145–152 (2014)
15. V.B. Braginskii, A.B. Manukin, Sov. Phys. JETP Lett. **25**(4), 653 (1967)
16. V.B. Braginskii, A.B. Manukin, M.Y. Tikhonov, Sov. Phys. JETP **31**, 829 (1970)
17. V.B. Braginsky, A.B. Manukin, *Measurement of Weak Forces in Physics Experiments* (University of Chicago Press, Chicago, 1977)
18. C.M. Caves, Phys. Rev. Lett. **45**(2), 75 (1980). doi:10.1103/PhysRevLett.45.75
19. V.B. Braginsky, S.P. Vyatchanin, Phys. Lett. A **293**, 228 (2002)
20. V.B. Braginsky, S.E. Strigin, V.P. Vyatchanin, Phys. Lett. A **287**(5–6), 331 (2001)
21. T.J. Kippenberg, H. Rokhsari, T. Carmon, A. Scherer, K.J. Vahala, Phys. Rev. Lett. **95**, 033901 (2005)
22. T.J. Kippenberg, K.J. Vahala, Science **321**, 1172 (2008)
23. R. Ma, A. Schliesser, P. Del'Haye, A. Dabirian, G. Anetsberger, T.J. Kippenberg, Opt. Lett. **32**, 2200 (2007)

24. J. Hofer, A. Schliesser, T.J. Kippenberg, Phys. Rev. A **82**, 031804 (2010)
25. J.S. Foresi, P.R. Villeneuve, J. Ferrera, E.R. Thoen, G. Steinmeyer, S. Fan, J.D. Joannopoulos, L.C. Kimerling, H.I. Smith, E.P. Ippen, Nature **390**, 143 (1997)
26. M. Notomi, E. Kuramochi, H. Taniyama, Opt. Express **16**(15), 11095 (2008)
27. M. Eichenfield, J. Chan, R.M. Camacho, K.J. Vahala, O. Painter, Nature **462**, 78 (2009)
28. M. Eichenfield, R. Camacho, J. Chan, K. Vahala, O. Painter, Nature **459**, 550 (2009)
29. E. Gavartin, R. Braive, I. Sagnes, O. Arcizet, A. Beveratos, T.J. Kippenberg, I. Robert-Philip, Phys. Rev. Lett. **106**, 203902 (2011)
30. Y.G. Roh, T. Tanabe, A. Shinya, H. Taniyama, E. Kuramochi, S. Matsuo, T. Sato, M. Notomi, Phys. Rev. B **81**, 121101 (2010)
31. A. Schliesser, P. Del'Haye, N. Nooshi, K. Vahala, T.J. Kippenberg, Phys. Rev. Lett. **97**, 243905 (2006)
32. T.J. Kippenberg, S.M. Spillane, D.K. Armani, K.J. Vahala, Appl. Phys. Lett. **83**, 797 (2003)
33. G. Anetsberger, R. Rivière, A. Schliesser, O. Arcizet, T.J. Kippenberg, Nat. Photonics **2**, 627 (2008)
34. E. Verhagen, S. Deleglise, S. Weis, A. Schliesser, T.J. Kippenberg, Nature **482**, 63 (2012)
35. C. Höhberger Metzger, K. Karrai, Nature **432**, 1002 (2004)
36. I. Tittonen, G. Breitenbach, T. Kalkbrenner, T. Müller, R. Conradt, S. Schiller, E. Steinsland, N. Blanc, N.F. de Rooij, Phys. Rev. A **59**, 1038 (1999)
37. P.F. Cohadon, A. Heidmann, M. Pinard, Phys. Rev. Lett. **83**, 3174 (1999)
38. O. Arcizet, P.F. Cohadon, T. Briant, M. Pinard, A. Heidmann, Nature **444**, 71 (2006)
39. S. Gigan, H.R. Böhm, M. Paternosto, F. Blaser, G. Langer, J.B. Hertzberg, K.C. Schwab, D. Bäuerle, M. Aspelmeyer, A. Zeilinger, Nature **444**, 67 (2006)
40. T. Carmon, K.J. Vahala, Phys. Rev. Lett. **98**, 123901 (2007)
41. J.D. Thompson, B.M. Zwickl, A.M. Jayich, F. Marquardt, S.M. Girvin, J.G.E. Harris, Nature **452**, 72 (2008)
42. A. Schliesser, R. Rivière, G. Anetsberger, O. Arcizet, T.J. Kippenberg, Nat. Phys. **4**, 415 (2008)
43. J.D. Teufel, J.D. Harlow, C.A. Regal, K.W. Lehnert, Phys. Rev. Lett. **101**, 197203 (2008)
44. A.H. Safavi-Naeini, T.P. Mayer, I. Alegre, J. Chan, M. Eichenfield, M. Winger, J.Q. Lin, J.T. Hill, D.E. Chang, O. Painter, Nature **472**, 69 (2011)
45. T.A. Palomaki, J.D. Teufel, R.W. Simmonds, K.W. Lehnert, Science **342**(6159), 710 (2013)
46. R. Rivière, O. Arcizet, A. Schliesser, T.J. Kippenberg, Rev. Sci. Instr. **84**(4), 043108 (2013)
47. A. Schliesser, O. Arcizet, R. Rivière, G. Anetsberger, T.J. Kippenberg, Nat. Phys. **5**, 509 (2009)
48. S. Weis, R. Rivière, S. Deléglise, E. Gavartin, O. Arcizet, A. Schliesser, T.J. Kippenberg, Science **330**, 1520 (2010)
49. H.A. Haus, *Waves and Fields in Optoelectronics* (Prentice-Hall, New Jersey, 1984)
50. M. Gorodetsky, A. Schliesser, G. Anetsberger, S. Deleglise, T.J. Kippenberg, Opt. Express **18**, 23236 (2010)
51. J. Dobrindt, T.J. Kippenberg, Phys. Rev. Lett. **104**, 033901 (2010)
52. A. Schliesser, T.J. Kippenberg, in *Advances in Atomic, Molecular and Optical Physics*, vol. 58, ed. by E. Arimondo, P. Berman, C.C. Lin (Elsevier Academic Press, New York 2010), pp. 207–323 (Chap. 5)
53. T.W. Hänsch, B. Couillaud, Opt. Commun. **35**(3), 441 (1980)
54. G. Anetsberger, O. Arcizet, Q.P. Unterreithmeier, R. Rivière, A. Schliesser, E.M. Weig, J.P. Kotthaus, T.J. Kippenberg, Nat. Phys. **5**, 909 (2009)
55. D.J. Wineland, R.E. Drullinger, F.L. Walls, Phys. Rev. Lett. **40**(25), 1639 (1978)
56. F. Diedrich, J.C. Bergquist, W.M. Itano, D.J. Wineland, Phys. Rev. Lett. **62**(4), 403 (1989)
57. C. Monroe, D.M. Meekhof, B.E. King, S.R. Jefferts, W.M. Itano, D.J. Wineland, P. Gould, Phys. Rev. Lett. **75**(22), 4011 (1995)
58. D.J. Wineland, Ann. Phys. **525**, 739 (2013)
59. F. Marquardt, J.P. Chen, A.A. Clerk, S.M. Girvin, Phys. Rev. Lett. **99**, 093902 (2007)
60. I. Wilson-Rae, N. Nooshi, W. Zwerger, T.J. Kippenberg, Phys. Rev. Lett. **99**(9), 093901 (2007)

61. A.A. Savchenkov, A.B. Matsko, V.S. Ilchenko, L. Maleki, Opt. Express **15**, 6768 (2007)
62. P. Rabl, C. Genes, K. Hammerer, M. Aspelmeyer, Phys. Rev. A **80**, 063819 (2009)
63. L. Diósi, Phys. Rev. A **78**, 021801 (2008)
64. Z. Yin, Phys. Rev. A **80**, 033821 (2009)
65. G.A. Phelps, P. Meystre, Phys. Rev. A **83**, 063838 (2011)
66. C.E. Wieman, L. Hollberg, Rev. Sci. Instrum. **62**(1), 1 (1991)
67. T.J. Kippenberg, A. Schliesser, M.L. Gorodetsky, New J. Phys. **15**(1), 015019 (2013)
68. A.H. Safavi-Naeini, J. Chan, J.T. Hill, S. Gröblacher, H. Miao, Y. Chen, M. Aspelmeyer, O. Painter, New J. Phys. **15**(3), 035007 (2013)
69. O. Arcizet, R. Rivière, A. Schliesser, G. Anetsberger, T.J. Kippenberg, Phys. Rev. A **80**, 021803(R) (2009)
70. Y.S. Park, H. Wang, Nat. Phys. **5**, 489 (2009)
71. R. Rivière, S. Deléglise, S. Weis, E. Gavartin, O. Arciezt, A. Schliesser, T.J. Kippenberg, Phys. Rev. A **83**, 063835 (2011)
72. Y.S. Park, H. Wang, Opt. Express **15**, 16471 (2007)
73. V.S. Ilchenko, M.L. Gorodetskii, Laser Phys. **2**(2), 1004 (1992)
74. C. Enss, S. Hunklinger, *Low Temperature Physics* (Springer, Berlin, 2005)
75. A. Schliesser, G. Anetsberger, R. Rivière, O. Arcizet, T.J. Kippenberg, New J. Phys. **10**, 095015 (2008)
76. J. Chan, T.P. Mayer, J. Alegre, A.H. Safavi-Naeini, J.T.T. Hill, A. Krause, S. Gröblacher, M. Aspelmeyer, O. Painter, Nature **478**, 89 (2011)
77. J.D. Teufel, T. Donner, D. Li, J.W. Harlow, M.S. Allman, I.K. Cicak, A.J. Sirois, J.D. Whittaker, K.W. Lehnert, R.W. Simmonds, Nature **475**, 359 (2011)
78. J. Zhang, K. Peng, S.L. Braunstein, Phys. Rev. A **68**, 013808 (2003)
79. A. Schliesser, Cavity optomechanics and optical frequency comb generation with silica whispering-gallery-mode microresonators. Ph.D. thesis, Ludwig-Maximilians-Universität München (2009)
80. M.O. Scully, M.S. Zubairy, *Quantum Optics* (Cambridge University Press, Cambridge, 1997)
81. M. Fleischhauer, A. Imamoglu, J.P. Marangos, Rev. Mod. Phys. **77**, 633 (2005)
82. P.W. Milonni, *Fast Light, Slow Light and Left-handed Light* (Taylor and Francis, New York, 2005)
83. X. Zhou, F. Hocke, A. Schliesser, A. Marx, H. Huebl, R. Gross, T.J. Kippenberg, Nat. Phys. **9**, 179 (2013)
84. G.S. Agarwal, S. Huang, Phys. Rev. A **81**, 041803 (2010)
85. J.D. Teufel, D. Li, M.S. Allman, K. Cicak, J.D. Sirois, A.J. Whittaker, R.W. Simmonds, Nature **471**, 204 (2011)
86. A. Kronwald, F. Marquardt, Phys. Rev. Lett. **111**(13), 133601 (2013)
87. M. Fleischhauer, M.D. Lukin, Phys. Rev. Lett. **84**, 5094 (2000)
88. C. Liu, Z. Dutton, C. Behroozi, L. Hau, Nature **409**, 490 (2001)
89. D.F. Phillips, A. Fleischhauer, A. Mair, R.L. Walsworth, M.D. Lukin, Phys. Rev. Lett. **86**, 783 (2001)
90. C. Dong, V. Fiore, M.C. Kuzyk, H. Wang, Phys. Rev. A **87**(5), 055802 (2013)
91. Y. Liu, M. Davanco, V. Aksyuk, K. Srinivasan, Phys. Rev. Lett. **110**, 223603 (2013)
92. M. Karuza, C. Biancofiore, M. Bawaj, C. Molinelli, M. Galassi, R. Natali, P. Tombesi, G. Di Giuseppe, D. Vitali, Phys. Rev. A **88**(1), 013804 (2013)
93. F. Hocke, X. Zhou, A. Schliesser, T.J. Kippenberg, H. Huebl, R. Gross, New J. Phys. **14**, 123037 (2012)
94. M.F. Yanik, S. Fan, Phys. Rev. Lett. **92**(8), 083901 (2004)
95. D.E. Chang, A.H. Safavi-Naeini, M. Hafezi, O. Painter, New J. Phys. **13**, 023003 (2011)
96. C. Dong, V. Fiore, M.C. Kuzyk, H. Wang, Science **338**, 1609 (2012)
97. J.T. Hill, A.H. Safavi-Naeini, J. Chan, O. Painter, Nat. Commun. **3**, 1196 (2012)
98. J.M. Taylor, A.S. Sørensen, C.M. Marcus, E.S. Polzik, Phys. Rev. Lett. **107**, 273601 (2011)
99. C.A. Regal, K.W. Lehnert, J. Phys.: Conf. Ser. **264**, 012025 (2011)
100. L. Tian, H. Wang, Phys. Rev. A **82**(5), 053806 (2010). doi:10.1103/PhysRevA.82.053806

101. J. Bochmann, A. Vainsencher, D.D. Awschalom, A.N. Cleland, Nat. Phys. (2013)
102. R.W. Andrews, R.W. Peterson, T.P. Purdy, K. Cicak, R.W. Simmonds, C.A. Regal, K.W. Lehnert, Nat. Phys. **10**, 321–326 (2014)
103. T. Bagci, A. Simonsen, S. Schmid, L.G. Villanueva, E. Zeuthen, J. Appel, J.M. Taylor, A. Sørensen, K. Usami, A. Schliesser, E.S. Polzik, Nat. **507**, 81–85 (2014)
104. K. Børkje, A. Nunnenkamp, J.D. Teufel, S.M. Girvin, Phyis. Rev. Lett. **111**, 053603 (2013)
105. M.A. Lemonde, N. Didier, A.A. Clerk, Phys. Rev. Lett. **111**(5), 053602 (2013)
106. J.M. Dobrindt, I. Wilson-Rae, T.J. Kippenberg, Phys. Rev. Lett. **101**, 263602 (2008)
107. I. Wilson-Rae, N. Nooshi, J. Dobrindt, T.J. Kippenberg, W. Zwerger, New J. Phys. **10**(9), 095007 (2008)
108. S. Gröblacher, K. Hammerer, M.R. Vanner, M. Aspelmeyer, Nature **460**, 724 (2009)
109. K. Stannigel, P. Rabl, A.S. Sorensen, P. Zoller, M.D. Lukin, Phys. Rev. Lett. **105**, 220501 (2010)
110. O. Romero-Isart, A.C. Pflanzer, M.L. Juan, R. Quidant, N. Kiesel, M. Aspelmeyer, J.I. Cirac, Phys. Rev. A **83**(1), 013803 (2011)
111. I. Wilson-Rae, Phys. Rev. B **77**, 245418 (2008)
112. T. Ramos, V. Sudhir, K. Stannigel, P. Zoller, T.J. Kippenberg, Phys. Rev. Lett. **110**, 193602 (2013)
113. U. Akram, N. Kiesel, M. Aspelmeyer, G.J. Milburn, New J. Phys. **12**, 083030 (2010)
114. T.A. Palomaki, J.W. Harlow, J.D. Teufel, R.W. Simmonds, K.W. Lehnert, Nature **495**(7440), 210 (2013)
115. L. Ding, C. Baker, P. Senellart, A. Lemaitre, S. Ducci, G. Leo, I. Favero, Phys. Rev. Lett. **105**, 263903 (2010)
116. S. Rips, M. Kiffner, I. Wilson-Rae, M.J. Hartmann, New J. Phys. **14**, 023042 (2012)

Chapter 7
Gallium Arsenide Disks as Optomechanical Resonators

Ivan Favero

Abstract The interaction of light with mechanical motion—optomechanics [1–4]—is now investigated in a wide variety of experimental settings. In the last years, the field also benefited from the advances of nanophotonics. We discuss here the merits of Gallium Arsenide (GaAs) optomechanical disk resonators, which bring together high mechanical frequency, ultra-strong optomechanical coupling and low optical/mechanical dissipation. Based on a relatively simple geometry, these miniature optomechanical resonators permit a complete on-chip optical integration, a natural interfacing with optically active elements and the combination with opto-electronics architectures typical of III–V semiconductors.

7.1 Introduction

Whispering Gallery Mode (WGM) Gallium Arsenide (GaAs) optical cavities import the advantageous optical properties of GaAs into an optical mode of small volume and high quality factor Q. These cavities have thus far enabled enhanced light-matter realizations in different contexts like cavity quantum electrodynamics experiments [5–7], low-threshold semiconductor lasers [8] and more recently non-linear optics [9]. In these early works, WGMs were generally supported by micrometer-sized GaAs disks but the mechanical properties of such resonators were not a topic of interest, despite the high-quality crystalline structure of Molecular Beam Epitaxy GaAs promising low mechanical dissipation, and despite early demonstrations of performant MEMS based on GaAs [10].

As apparent from early optomechanics experiments on silica toroids (see chapter on silica toroids), WGMs resonators support not only optical but mechanical modes as well, and they are a natural geometry to obtain a strong coupling between optics

I. Favero (✉)
CNRS, Université Paris Diderot, Paris, France
e-mail: ivan.favero@univ-paris-diderot.fr

M. Aspelmeyer et al. (eds.), *Cavity Optomechanics*, Quantum Science and Technology,
DOI: 10.1007/978-3-642-55312-7_7,

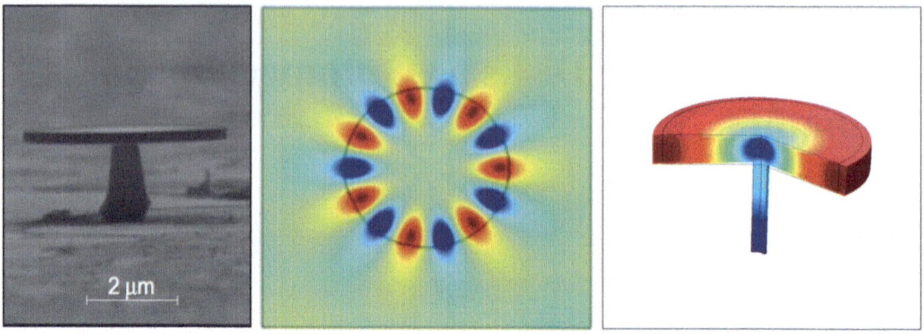

Fig. 7.1 *Left panel* A single GaAs disk sitting over an AlGaAs pedestal. *Middle panel* Top view of the electric field amplitude of a (p = 1, m = 7) whispering gallery mode supported by a GaAs disk. *Red* (*blue*) color indicates maximal positive (negative) amplitude. *Right panel* Displacement profile of the fundamental radial breathing mode of a GaAs disk. The color scale spans from maximal displacement in *red* to minimal in *blue*. The underformed disk boundary is shown for clarity

and mechanics. This principle also applies to GaAs disks but with several important differences, the most noticeable being the size. Figure 7.1 (left panel) shows a typical example of a 2.5 μm radius GaAs disk of thickness 200 nm, occupying a volume of 3.9 μm^3 that is 3–4 orders of magnitude smaller than for a typical silica toroid. This reduced volume translates into higher mechanical frequencies and enhanced optomechanical coupling, as demonstrated in a series of optomechanical experiments on GaAs disks [11–13]. If these resonators present today remarkable performances from a strict optomechanical viewpoint, they are also naturally suited to combine optomechanical functions with the well-developed assets of III–V semiconductors. Amongst these are complex optoelectronics architectures achievable through doping and electric contacts, and the rich physics of quantum wells, quantum dots and 2D electron gases.

7.2 An Optical and Mechanical Resonator

GaAs disk structures support optical WGMs of small mode-volume and high Q. Thanks to the rotation invariance of a disk around its principal axis, these modes are usually classified using a radial number p and an azimuthal number m [14, 15]. p gives the number of maxima of the electric field along the radial direction when m represents the number of nodes around the disk periphery. Figure 7.1 (middle panel) shows an example of a (p = 1, m = 7) mode. For a disk thickness in the 200–300 nm range and for the near-infrared wavelengths considered here, there is a single node of the electric along the thickness dimension, hence the third (vertical) quantization number of the electric filed will not be mentioned hereafter. The WGMs of a given disk can be computed by a semi-analytical effective index-method approach [15] or by purely numerical tools like Finite Elements Method and

Finite-Difference Frequency-Domain. These latter numerical approaches are particularly suited to compute the amount of radiative losses of a given disk resonator mode. For an optical wavelength in the telecommunications range (1.3–1.6 μm), a GaAs disk can have a radius as small as 1 μm before starting to experience important radiative losses. For example, a 1 μm-radius disk of thickness 320 nm supports TE-polarized (in plane polarization of the electric field) WGMs at a wavelength of 1.3 μm with radiative Q above 10^7.

GaAs disks support a large number of mechanical modes, having an out-of-plane (flexural type) or in-plane (radial breathing or "elliptic bulk" [16] type) nature. On top of these are the "pinch" modes where a portion of the disk is squeezed in its thickness. At a first level of description, it is possible to treat the GaAs crystal as elastically isotropic and use again the rotation invariance of the disk structure to classify mechanical modes with a radial (P) and an azimuthal (M) number. Figure 7.1 (right panel) shows the displacement profile of the fundamental radial breathing mode (M = 0). For a 1 μm-radius disk of thickness 320 nm, this mode has a frequency of about 1.4 GHz, the next radial breathing (P = 2) being at 3.5 GHz. This exemplifies that very high frequencies are easily obtained in GaAs disks thanks to miniature dimensions and high rigidity of the in-plane motion. The radial-breathing modes having a node at the disk center, they naturally minimize clamping losses associated to the pedestal's displacement. If early work attempted to estimate the clamping of a disk resonator by semi-analytical means [17, 18], a numerical FEM approach allows for direct predictive calculations [19, 20]. It shows that the fundamental radial breathing mode at 1.4 GHz reaches for example a mechanical quality factor of 10^4 for a pedestal radius of 100 nm attainable in the fabrication. This corresponds to a mechanical $Q \times f$ product in the 10^{13} range.

7.3 Optomechanical Coupling in GaAs Disk Resonators

The radial breathing of the disk couples naturally to photons stored in the gallery mode by modulating the length of the gallery. This "geometric" optomechanical coupling is at play in any whispering gallery resonator and corresponds to the picture of photons that push the walls of the gallery apart. As a consequence of the rotational symmetry of a disk structure, selection rules are present in this coupling (see [11]), leading in practice to maximal coupling for mechanical modes that have perfect rotation invariance of their displacement field (M = 0). Thanks to the reduced dimensions of GaAs disks, down to about 1 μm in radius, the vacuum optomechanical coupling g_0 associated to this purely geometric effect reaches 1 MHz, quoted for the fundamental radial breathing mode coupled to a p = 1 WGM. On top of this geometric contribution, photoelastic effects that were discussed in silica spheres [21, 22] and in silicon waveguides [23] also play an important role in GaAs disks. Indeed the GaAs crystal possesses very large photoelastic constants, which at variance with silicon have the same sign for all orientations. When filled by a finite amount of optical energy, a closed volume of GaAs will experience outward pressure (electrostriction) leading to its expansion. In a GaAs disk conversely, a radial expansion of the structure produces an increase of the refractive index experienced by photons trapped in a WGM, increasing the effective optical length. As a net result, the photoelastic optomechanical coupling in GaAs disks adds up constructively with the geometric coupling. The photoelastic effect is moderate in large disks but increases for the smallest ones where it dominates the geometric contribution. Numerical simulations predict for example a total g_0 of 2.5 MHz (geometric + photoelastic) for the fundamental radial breathing mode of a 1 μm-radius disk coupled to a $p = 1$ WGM at 1.3 μm of wavelength. The coupling can even be pushed beyond this level by shifting the operating wavelength towards the material bandgap and/or reducing further the disk radius [24]. This makes GaAs disks currently amongst the most strongly coupled optomechanical platforms in the solid state, together with silicon-based photonic crystals. Silicon disk resonators of similar size also show a very good level of coupling [25, 26] but do not benefit optimally from photoelastic effects because of the reduced symmetry of the photoelastic tensor of silicon and its smaller constants.

7.4 On-Chip Integration of GaAs Disks

GaAs disk resonators can be controlled optically by means of evanescent coupling techniques. One such technique utilizes an optical fiber taper, which can be fabricated with ultra-low optical losses employing either a flame burner rig [27] or a thermo-electric heater element for better control and reproducibility [28]. The fiber taper technique allows for in-situ tuning of the evanescent coupling strength by adjusting the disk to taper gap distance [12], but it is poorly stable mechanically, sensitive to water contamination in the air, and hence not integrable for large-scale applications. A fully integrated approach based on on-chip coupling waveguides and resonators is also manageable in a GaAs environment, even if less trivial than in a silicon-on-isulator platform. A sacrificial AlGaAs layer below a top GaAs layer can be employed and serve as optical cladding as well, but it will naturally oxidize if the Aluminum content is made large in order to guarantee refractive index mismatch with GaAs. An option is to voluntarily oxidize the AlGaAs layer to AlOx [29, 30], which has lower refractive index, but this comes at the price of less mechanical robustness and increased optical losses at the interfaces. One way out is to completely suspend the optically guiding structures and use air (or vacuum) as optical cladding. This has been realized for GaAs disks evanescently coupled to suspended GaAs tapered waveguides directly integrated on the chip [31, 32]. In [32], direct optical injection at the chip facets was implemented, allowing broadband wavelength tunability combined with fine control of the evanescent coupling and low optical losses. Such design is fully compatible with the state-of-the-art optomechanical performances mentioned above in terms of coupling, optical and mechanical dissipation. Examples of miniature GaAs disks coupled to suspended GaAs waveguides are shown in Fig. 7.2 (top left and right panel). The first structure (top left) is obtained purely by chemical etching while the second structure (top right) combines dry and wet etches. The second

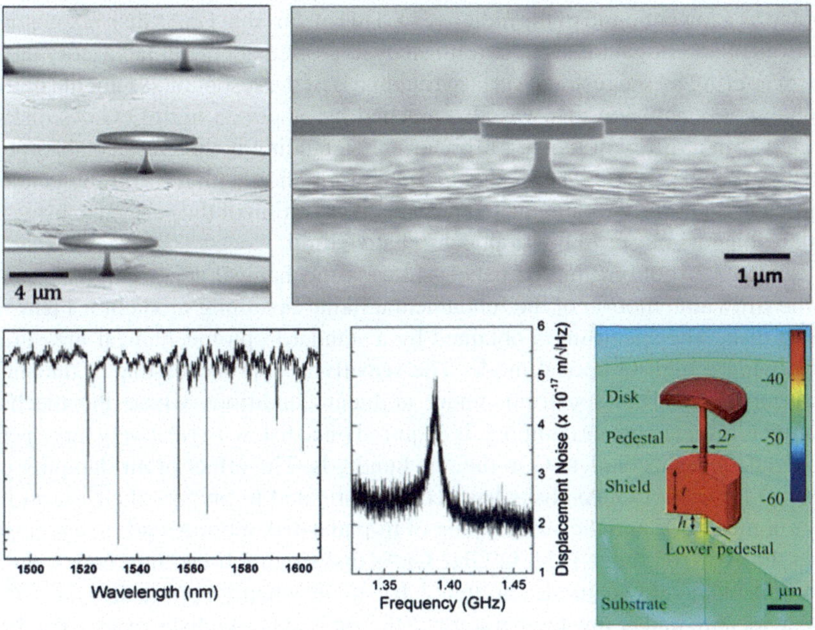

Fig. 7.2 *Top-left panel* Three GaAs disk resonators with 3.5-micron radius are integrated on the same semiconductor chip and evanescently coupled to suspended tapered waveguides. *Top-right panel* Close-up on a 1-micron radius disk with its coupling waveguide. *Bottom-left panel* Optical transmission spectrum of a waveguide/disk coupled system, showing fine dips corresponding to whispering gallery modes of the disk. *Bottom-middle panel* Brownian motion spectrum of the radial breathing mode of a 1-micron radius. *Bottom-right panel* Shielded mechanical structure for a GaAs disk with reduced clamping losses

structure involves a disk of 1 μm in radius like discussed above, critically coupled to an ultra-low loss waveguide. These structures are easily scalable for implementing arrays of resonators, optomechanical routers, switches and multiplexing schemes. They also allow easy operation with good mechanical stability in a confined environment like a cryostat chamber.

7.5 Optical and Mechanical Dissipation in GaAs Disks

Figure 7.2 (bottom-left panel) shows an optical transmission spectrum of a GaAs waveguide coupled to a disk. It displays fine WGM resonances close to critical coupling. The width of these resonances varies in experiments from hundreds of pm to a few pm, corresponding to optical Q factors evolving in the 10^3–10^5 range. The best measured intrinsic optical Q in our current experiments is 6.10^5, excluding contributions form the coupling waveguide. This is still far from the 10^7 value typically quoted as a lower bound of the radiative Q. This means that GaAs disk resonators suffer from other optical losses. A careful analysis carried out in our laboratory indicates that scattering by the residual surface roughness of the resonators is not the dominant problem, a conclusion that was also reached by others on high-Q GaAs disks fabricated with a different recipe [33]. Our understanding is that surface absorption in the disk reconstruction layer is the dominant limitation with the current technology. This calls for surface treatment and possibly passivation of the surface to bring GaAs disk resonators closer to their optical radiative limit.

Figure 7.2 (bottom-middle panel) shows a mechanical spectrum corresponding to the Brownian motion of the fundamental radial breathing mode of a 1 μm-radius GaAs disk. The spectrum is obtained by a standard optomechanical measurement employing a high-Q optical mode. The sensitivity approaches that at the standard quantum limit. This spectrum, under ambient conditions where the mechanical motion is affected by air damping, is acquired on a disk with relatively large pedestal that leads to a mechanical Q in the few hundreds. The effect of air damping on the flexural motion of GaAs disks has been investigated to prepare their use in fluidic environment [34] and the air-damping of in-plane disk motion leads to a contributed Q in the 10^3–10^4 range [35]. In GHz GaAs disks radial breathing modes, the measured mechanical Q saturates around 2.10^3 in air when it approaches the 10^4 limit under vacuum and at low temperature [20]. The best GaAs disks reach a mechanical $Q \times f$ product of 10^{13} at low temperature in current measurements, limited by clamping losses at the pedestal [20]. In order to break this limit, a mechanical shield can be integrated directly within the pedestal as shown in Fig. 7.2 (bottom-right panel). By creating destructive interferences with the disk breathing motion, the shield will block the emission of acoustic waves into the substrate, leading to reduced clamping losses. Numerical simulations indicate that this would lead to clamping-limited $Q \times f$ product between 10^{16} and 10^{18} [20]. At this level, it is expected that the non-ideality of the surface reconstruction layer will start to play a role in the mechanical dissipation.

7.6 Optomechanical Perspectives

From a standard optomechanical viewpoint, one interesting parameter to quote for GaAs disk resonators is the cooperativity in the linearized regime. Taking a GaAs disk with breathing mode at 1.4 GHz, mechanical $Q \times f$ of 10^{13} (hence Γ_M, inverse of the mechanical damping time, is 1.2 MHz), optical Q of 5×10^5 for a WGM at wavelength 1.3 μm (hence κ, inverse of photon lifetime is 2.9 GHz) and g_0 of 2.5 MHz, we obtain a single-photon cooperativity $4g_0^2/\kappa\Gamma_M$ of 8×10^{-3}. Considering that up to a few 10^4 photons are injected in the corresponding GaAs disk before facing technical problems like two-photon absorption, a linearized cooperativity of 10^2 can be reached in current experiments. At 4 K of environmental temperature, $\bar{n}_{th} = 100$, such that quantum effects in the linearized regime of optomechanics should start being observable in GaAs disk resonators. This is of course an instantaneous snapshot on experiments since several improvements are underway regarding optical and mechanical dissipation. Note also that the second order (P = 2) radial breathing mode of a 1 μm-radius GaAs disk oscillates at 3.5 GHz and shows the same amount of coupling g_0 thanks to a good photoelastic configuration, allowing in principle lower $\bar{n}_{th}$ and more pronounced quantum effects. On the non-linear optomechanics side, the g_0/κ parameter currently reaches 10^{-3}. If this makes GaAs disk resonators an appealing candidate for harnessing non-linearities at the single quantum level, other improvements will be needed to frankly enter this regime.

Apart from these established optomechnical routes, where only a mechanical and an optical mode are involved, the interest of GaAs resonators also lies in their compliancy with embedded semiconductor quantum dots or quantum wells. Indeed theoretical proposals are now merging the features of quantum optomechanics with those of solid-state cavity-QED situations. Two recent works [36, 37] investigated how the inclusion of semiconductor cavity polaritons in an optomechanical resonator influences the optical/mechanical interaction, leading to unconventional behavior. In [36] it was for example shown that the presence of a single artificial atom strongly modifies the optomechanical cooling and amplification properties, producing non-classical trajectories and strong bunching of phonons. The vast physics of semiconductor cavity polaritons based on quantum dots and quantum wells will certainly enrich the field of light-mechanics interaction, and may lead to unexpected hybrid optical/mechanical/electronic realizations in the quantum domain.

Acknowledgments I. Favero acknowledges support of the French ANR trough the Nomade and QDOM projects, of the C-Nano IdF through the Naomi project, and of the ERC through the Ganoms project.

References

1. T.J. Kippenberg, K.J. Vahala, Science **321**(5893), 1172 (2008)
2. I. Favero, K. Karrai, Nat. Photonics **3**, 201 (2009)
3. F. Marquardt, S.M. Girvin, Physics **2**, 40 (2009)
4. M. Aspelmeyer, S. Gröblacher, K. Hammerer, N. Kiesel, J. Opt. Soc. Am. B **27**(6), A189 (2010)
5. B. Gayral, J.M. Gerard, A. Lemaitre, C. Dupuis, L. Manin, J.L. Pelouard, Appl. Phys. Lett. **75**, 1908 (1999)
6. A. Kiraz, P. Michler, C. Becher, B. Gayral, A. Imamoglu, L. Zhang, E. Hu, W.V. Schoenfeld, P.M. Petroff, Appl. Phys. Lett. **78**, 3932 (2001)
7. E. Peter, P. Senellart, D. Martrou, A. Lemaître, J. Hours, J.M. Gérard, J. Bloch, Phys. Rev. Lett. **95**, 067401 (2005)
8. S.L. McCall, A.F.J. Levi, R.E. Slusher, S.J. Pearton, R.A. Logan, Appl. Phys. Lett. **60**, 289 (1992)
9. A. Andronico, I. Favero, G. Leo, Opt. Lett. **33**, 2026–2028 (2008)
10. J. Miao, H.L. Hartnagel, B.L. Weiss, R.J. Wilson, Electron. Lett. **31**(13), 1047 (1995)
11. L. Ding, P. Senellart, A. Lemaitre, S. Ducci, G. Leo, I. Favero, Phys. Rev. Lett. **105**, 263903 (2010)
12. L. Ding, P. Senellart, A. Lemaitre, S. Ducci, G. Leo, I. Favero, Proc. SPIE **7712**, 771211 (2010)
13. L. Ding, C. Baker, P. Senellart, A. Lemaitre, S. Ducci, G. Leo, I. Favero, Appl. Phys. Lett. **98**, 113108 (2011)

14. A. Andronico, X. Caillet, I. Favero, S. Ducci, V. Berger, G. Leo, J. Europ. Opt. Soc. Rap. Public. **3**, 08030 (2008)
15. A. Andronico, Etude électromagnétique d'émetteurs intégrés infrarouges et THz en AlGaAs. PhD thesis 2008
16. Z. Hao, S. Pourkamali, F. Ayazi, J. Microelectromech. Syst. **13**(6), 1043 (2004)
17. Z. Hao, F. Ayazi, Sens. Actuators A **134**, 582 (2007)
18. I. Wilson-Rae, Phys. Rev. B **77**, 245418 (2008)
19. A. Safavi-Naeini, O. Painter, Opt. Express **18**(14), 14926 (2010)
20. D.T. Nguyen, C. Baker, W. Hease, S. Sejil, P. Senellart, A. Lematre, S. Ducci, G. Leo, I. Favero, Appl. Phys. Lett. **103**, 241112 (2013)
21. G. Bahl, M. Tomes, F. Marquardt, T. Carmon, Nat. Phys. **8**, 203 (2012)
22. I. Favero, Nat. Phys. **8**, 180 (2012)
23. P.T. Rakich, P. Davids, Z. Wang, Opt. Express **18**(14), 14439 (2010)
24. C. Baker, W. Hease, T. Nguyen, A. Andronico, G. Leo, S. Ducci, I. Favero (2014) arXiv:1403.4269
25. W.C. Jiang, X. Lu, J. Zhang, Q. Lin, Opt. Express **20**(14), 15991 (2012)
26. X. Sun, X. Zhang, H.X. Tang, App. Phys. Lett. **100**, 173116 (2012)
27. G. Brambilla, V. Finazzi, D.J. Richardson, Opt. Express **12**, 2258 (2004)
28. L. Ding, C. Belacel, S. Ducci, G. Leo, I. Favero, Appl. Opt. **49**(13), 2441 (2010)
29. E. Guillotel, M. Ravaro, F. Ghiglieno, C. Langlois, C. Ricolleau, S. Ducci, I. Favero, G. Leo, Leo. Appl. Phys. Lett. **94**, 171110 (2009)
30. M. Savanier, A. Andronico, C. Manquest, A. Lemaitre, E. Galopin, I. Favero, S. Ducci, G. Leo, Opt. Lett. **36**, 2955 (2011)
31. S. Koseki, B. Zhang, K. De Greve, Y. Yamamoto, Appl. Phys. Lett. **94**, 051110 (2009)
32. C. Baker, C. Belacel, A. Andronico, P. Senellart, A. Lemaitre, E. Galopin, S. Ducci, G. Leo, I. Favero, Appli. Phys. Lett. **99**, 151117 (2011)
33. C.P. Michael, K. Srinivasan, T.J. Johnson, O. Painter, K.H. Lee, K. Hennessy, H. Kim, EHu Appl, Phys. Lett. **90**, 051108 (2007)
34. D. Parrain, C. Baker, T. Verdier, P. Senellart, A. Lemaitre, S. Ducci, G. Leo, I. Favero, Appl. Phys. Lett. **100**, 242105 (2012)
35. J. Wang, Z. Ren, IEEE Trans. Ultrason. Ferroelectr. Freq. Control **51**(12), 1607 (2004)
36. J. Restrepo, C. Ciuti, I. Favero, Phys. Rev. Lett. **112**, 013601 (2014)
37. O. Kyriienko, T.C.H. Liew, I.A. Shelykh, Phys. Rev. Lett. **112**, 076402 (2014)

Chapter 8
Brillouin Optomechanics

Gaurav Bahl and Tal Carmon

Abstract In this chapter we introduce the concept of Brillouin optomechanics, a phonon–photon interaction process mediated by the electrostrictive force exerted by light on dielectrics and the photoelastic scattering of light from an acoustic wave. We first provide a review of the phenomenon and continue with the first experiments where stimulated Brillouin optomechanical actuation was used in microdevices, and spontaneous Brillouin cooling was demonstrated.

8.1 Introduction

Stimulated Brillouin scattering (SBS) [1, 2] has been used since the 1960s as an acousto-optic gain mechanism for lasers [3]. Subsequently, it has been employed as a tool for slow light [4], for non-destructive characterization of materials [5–7], and in optical phase conjugation [8] for holography. Brillouin lasers, due to their narrow linewidth, are also employed in ring laser gyroscopes [9]. On the other hand, in fiber-based communication systems and high power fiber lasers, SBS is considered an undesirable mechanism that interferes with proper function [10]. It is important to note that SBS is an optical nonlinearity common to all dielectrics in all states of matter, and is generally considered to be the strongest optical nonlinearity [11]. In addition to optical fibers and bulk media SBS has been demonstrated in a variety of optical systems. Examples range over a variety of size scales, including nano-scale spheres [12], fluid droplets [13], photonic-crystal fibers [14], and recently in mm-scale [15, 16] and micron-scale resonators [17].

G. Bahl (✉)
University of Illinois at Urbana–Champaign, Urbana–Champaign, IL, USA
e-mail: bahl@illinois.edu

T. Carmon
Technion–Israel Institute of Technology, Haifa, Israel
e-mail: tcarmon@technion.ac.il

M. Aspelmeyer et al. (eds.), *Cavity Optomechanics*, Quantum Science and Technology, 157
DOI: 10.1007/978-3-642-55312-7_8, © Springer-Verlag Berlin Heidelberg 2014

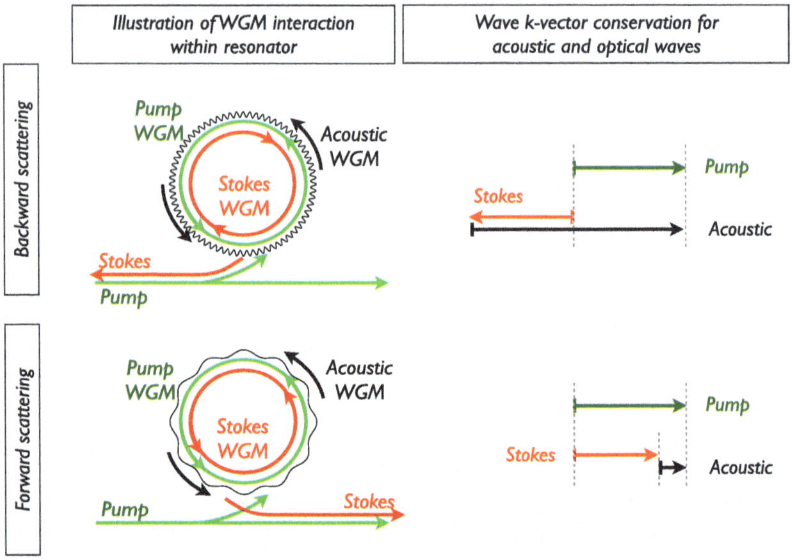

Fig. 8.1 Momentum conservation dictates the frequencies of the acoustic modes that are coupled to optical modes via stimulated Brillouin scattering. Light evanescently couples into the resonator's optical WGMs from the tapered waveguide. Photons at the appropriate Stokes frequency are scattered in either the forward- or back-scattering direction. *Top* In the case of back-scattering momentum conservation dictates that the acoustic wavevectors are long. The acoustic modes are therefore at high frequencies (>10 GHz). *Bottom* In forward scattering momentum conservation dictates that the acoustic modes are low frequencies (<1 GHz)

SBS is a process where light scatters from a sound wave in a material, resulting in a red-shifted (Stokes) laser and amplification of the original sound wave. Phase-matching considerations require that the pump optical mode and the Stokes optical mode are separated by the precise energy and momentum of the acoustic phonons populating the sound wave. Since light scattering can occur in either forward or backward direction relative to the pump, the momentum conservation constraint is a primary factor in determining the acoustic frequency. As shown in Fig. 8.1 back-scattered SBS is associated with large phonon momenta and therefore typically occurs with multi-GHz frequency acoustic waves [15–17]. For instance, in fused silica using a pump wavelength of 1,550 nm, back-scattered SBS creates an 11 GHz shifted Stokes laser relative to the pump. Forward scattered SBS, on the other hand, occurs with much lower acoustic frequencies ranging from the MHz to the low-GHz [18, 19]. Forward Brillouin scattering has been experimentally observed previously in optical fibers [20], photonic crystal fibers [14], and recently in platforms where the mechanical as well as the optical waves are resonantly enhanced [18, 19, 21].

Resonators are convenient devices with which to investigate SBS since resonant enhancement enables very low lasing thresholds. However, in nano-scale resonators [12] this process suffers from very low optical finesse, while in larger resonators

[15, 17] the mechanical finesse was seen to be very low for back-scattered SBS due to high mechanical dissipation at 10-GHz rates. Given a system where optical modes and acoustic modes are simultaneously resonant, the Brillouin interaction is significantly enhanced. It was later proposed [22] that forward-SBS would allow access to acoustic modes at low frequencies where phonon lifetimes are longer, and therefore have high mechanical quality factors. Indeed, experimental studies of forward SBS in resonators have exhibited acoustic vibrations occurring over frequencies in the 10–100 s of MHz range in amorphous silica microspheres [18], silica microcapillaries [21], and also in crystalline ($LiTaO_3$ and MgF_2) resonators [19]. These lower frequency vibrational modes were determined to have long phonon lifetimes, up to 2 orders of magnitude longer than the photon lifetimes [18, 23]. Such low phonon damping rates are needed for optomechanical cooling [24] as the rate of energy removal in the form of photons must exceed the rate of energy entering the acoustic mode from the bath. Therefore, as we will show later, these results enabled reversal of the energy flow in the Brillouin process and demonstration of cooling [23, 25].

Brillouin scattering thus joins other optomechanical actuation mechanisms in micro-devices, uniquely supporting vibration rates over a wide frequency range from 50 MHz to 12 GHz [18, 21], with the ability to cool mechanical whispering-gallery modes as well [23]. Brillouin actuation acts on azimuthally circulating density waves in these devices, compared against actuation by radiation pressure on the device walls [26]. It is interesting to note that in the very same microresonator centrifugal radiation pressure can excite the breathing mode [27, 28] while at the same time Brillouin actuation can excite its mechanical whispering gallery modes [17, 18, 21, 29]. This simultaneous actuation was recently demonstrated in microcapillaries [30].

8.2 Brillouin Optomechanics in Resonators

The Brillouin optomechanical process involves two optical modes that are separated in frequency space by the frequency of the acoustic mode of interest, and in momentum space by the momentum of the acoustic mode. These optical modes can be obtained by either designing the free spectral range (FSR) of the resonator, or by exploiting aperiodic spacing between high transverse-order optical modes in a resonator [18, 31, 32]. Non-periodic spacing between resonances is particularly helpful in suppressing scattering in the Stokes direction when only anti-Stokes scattering is desired. This selective filtering capability was exploited to achieve spontaneous Brillouin cooling of the acoustic modes [23, 25] as we will discuss in the next section.

8.2.1 Theoretical Description

We consider an optically stimulated traveling acoustic wave (i.e. sound wave) of frequency Ω_a traveling unidirectionally at the speed of sound at the equator of a whispering-gallery resonator as described in [29]. We assume this acoustic wave to be resonant, which happens when the circumference of the device is an integer multiple of the acoustic wavelength. This acoustic wave therefore forms an acoustic whispering-gallery mode (AWGM). In order to describe the momentum associated with this acoustic mode, we use the integer propagation constant M_a. This integer is equal to the number of acoustic wavelengths present along the device equator and is therefore also known as the azimuthal mode order. More formally, this integer represents the azimuthal propagation of the AWGM along the device as $e^{j(M \cdot \phi - \omega \cdot t)}$ where ϕ is the azimuthal position and ω is the wave frequency.

This traveling mechanical mode photo-elastically writes an optical grating that is capable of scattering light present in any optical whispering-gallery mode (OWGM) that has good spatial overlap with this AWGM. As a result of this scattering, pump light is scattered to red-shifted (Stokes) frequencies as shown in Fig. 8.1 for both forward and backward scattering directions. The relationship between the pump (subscript p), Stokes (subscript S), and acoustic (subscript a) wave frequencies is given by

$$\omega_S = \omega_p - \Omega_a \tag{8.1}$$

which is essentially the energy conservation relationship. Note that momentum conservation between photons and phonons must also be accounted, as we describe below. Simultaneously, the pump light and newly scattered Stokes light generate a beat note, i.e. a spatio-temporal interference pattern within the resonator (Fig. 8.2, purple "Generated pressure"). When we evaluate the electrostrictive force generated by this interference pattern, we see that the optically induced stress appears precisely at the spatial and temporal frequency of the original acoustic wave, and travels at the speed of sound. As a result of this positive feedback, this process described by photoelastic scattering and electrostriction becomes self-sustaining. The acoustic wave is amplified due to this feedback, and a Stokes laser is generated. The complete optomechanical interaction between the two OWGMs and the AWGM is illustrated for the case for forward-SBS in Fig. 8.2.

We can derive this intuitive phase-matching condition (Eq. 8.1) analytically by solving the coupled wave equations for mechanical and optical waves to reveal a synchronous solution [11]. A detailed solution of the coupled acousto-optic interaction is presented in the supplementary information of Ref. [23]. This solution reveals that the propagation constants of the pump OWGM, M_p, must equal the sum of the propagation constants of the resulting Stokes and acoustic modes, as given by

$$M_p = M_S + M_a. \tag{8.2}$$

Since one pump photon is converted into a photon and a phonon whose combined momentum must equal the pump photon's momentum, this wavevector relation is associated with momentum conservation.

This momentum relationship is graphically represented in Fig. 8.1 for both forward and backward scattering. As can be seen, the acoustic momentum vector in the case of backscattering is roughly double the length of the optical vectors. As a result

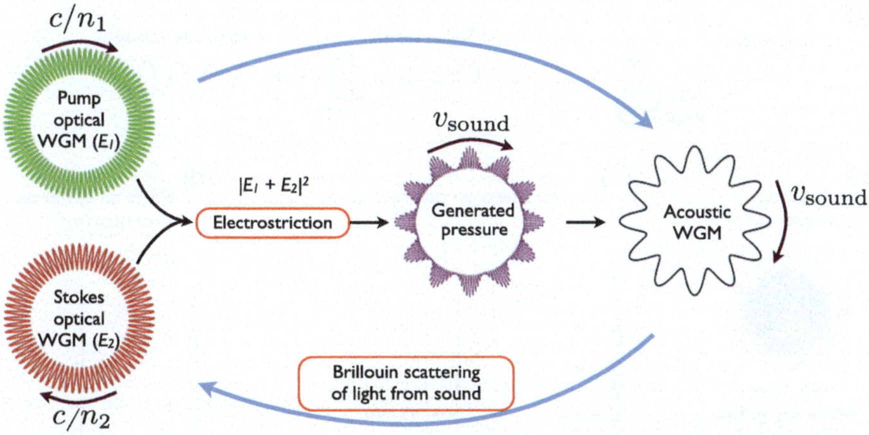

Fig. 8.2 Here we illustrate the optomechanical interaction within a resonator in the case of forward-SBS. The acoustic WGM writes a traveling photoelastic grating that scatters light from the pump OWGM to the Stokes OWGM, i.e. Brillouin scattering. Simultaneously, the electrostrictive interference pattern generated by the two OWGMs amplifies the acoustic WGM. In this manner a positive feedback is generated. Note that the two OWGMs travel at the speed of light (c/n_1 and c/n_2 where $n_{1,2}$ are the refractive indices), and the AWGM travels at the speed of sound (v_{sound})

the acoustic wavelength in the back-scattering case is half that of the optical wavelength used, and the acoustic frequencies are very high. Acoustic frequencies in the 10–17 GHz regime have been observed in various glass and crystalline resonators [15, 17]. On the other hand, since the acoustic wavevector is very short in the forward-scattering case, the acoustic frequencies are lower and the acoustic wavelengths are much longer. This prediction [22] was verified experimentally with silica microsphere resonators [18], which we describe in the next section.

In the case of back-scattering, the phonon lifetimes are low, as a result of which the propagation distance is comparable with the device circumference, implying a low mechanical finesse. This lack of acoustic finesse is a concern even for resonators in the 100 μm size scale. On the other hand, it is known that phonon dissipation from the material scales inversely as the square of frequency [11]. This reduced dissipation increases the phonon lifetimes significantly at lower acoustic frequencies. The implication is that a reduction in acoustic frequency from the 11,000 MHz (typical of backward scattering in silica) to the 50 MHz regime (typical in forward scattering SBS) may potentially provide a 50,000-fold reduction in mechanical dissipation.

The result is that phonons survive long enough to circumnavigate the WGR many times, or in other words the phonon finesse is $\gg 1$. The stimulated acoustic wave now becomes a high-finesse mechanical resonance.

As mentioned above, a key requirement of achieving SBS-based optomechanical interaction within a microresonator is the availability of two optical modes with nearly identical optical frequencies, but with propagation constants differing by M_a. However, the frequency separation between optical modes having adjacent

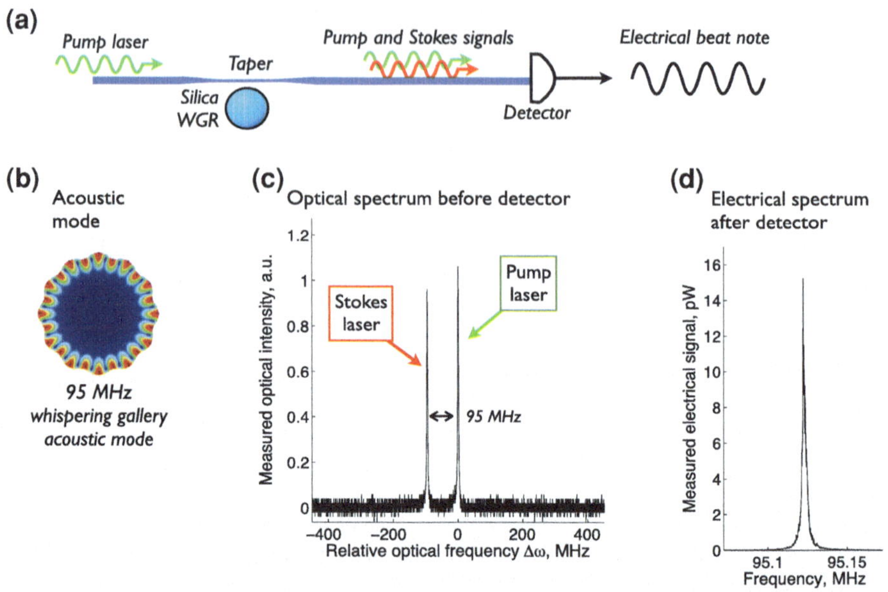

Fig. 8.3 **a** Experimental setup. **b** Finite element calculation for the acoustic whispering-gallery mode in this experiment. **c** Pump and stokes laser at output. **d** Electrical beat signal provides acoustic signature

M-numbers within one mode family is defined by the resonator FSR. Since the FSR of the optical modes of a silica resonator of 100 micron diameter is >300 GHz, the two required optical modes cannot belong to the same optical mode family, for either the forward- or back-scattering cases. This leads to a significant experimental challenge with obtaining the requisite phase match. On the other hand, optical modes with high-orders in the transverse plane exhibit different effective refractive indices. As a result, the optical wavevectors may differ significantly between two optical modes that have nearly identical frequency. The existence of such high order modes of different azimuthal wavelength but almost the same frequency was indicated by standing interference patterns experimentally generated in [31, 32]. Use of such modes in achieving phase matching for SBS was discussed in [22].

8.2.2 Experimental Implementation

We perform the experiment [18] on a silica microsphere resonator [33] having a typical optical $Q > 10^8$. A tapered optical fiber is employed to evanescently couple light into the pump OWGM [34–36], and a telecom-wavelength laser at 1,550 nm is used as the source as shown in Fig. 8.3a.

When we tune the pump laser to a resonance of the microsphere where the requisite phase match exists, pump photons are forward-scattered to the Stokes OWGM upon interaction with the thermal phonons in the AWGM. This scattered light also evanescently couples out to the tapered optical fiber and propagates to a photodetector at the opposite end of the fiber (Fig. 8.3a). The partially transmitted pump and the newly generated Stokes light (Fig. 8.3c) create a temporal beat note at the detector (Fig. 8.3d), which is at the frequency of the acoustic mode (Fig. 8.3b). This is a direct measure of the acoustic mode, and therefore the phonon profile in the resonator. We also employ optical spectrum analyzers to observe the optical signals in order to verify the SBS phenomenon (Fig. 8.3c). No modulation of the laser nor any feedback control is needed to create this stimulated vibration.

Indeed, for a given resonator, hundreds of instances of phase match can be identified experimentally. In [18] we showed that mechanical frequencies ranging from 58 MHz to 1.4 GHz (Fig. 8.4) could be generated on a single device when the laser was tuned over 1,520–1,570 nm. The lowest measured threshold power was 22.5 μW for forward Brillouin lasing at 175 MHz. Mechanical mode quality factors are estimated by observing the subthreshold linewidth as we describe later in the subsection on Brillouin cooling.

8.3 Spontaneous Brillouin Cooling

Electrostrictive Brillouin scattering is often only thought of as an amplification process [3, 11, 14, 15, 17, 18, 20, 21] since it readily appears in most optical systems when a threshold power is surpassed. On the other hand it is a known fact that spontaneous anti-Stokes Brillouin scattering also occurs [11], which based on energy conservation arguments should lead to annihilation of phonons and therefore cooling of the acoustic mode. As we see in the rest of this book, bolometric [37–39] and ponderomotive [40–45] forces are now widely used for laser cooling of the mechanical modes of microdevices. The result we present here demonstrates this cooling capability with the Brillouin optomechanical process.

8.3.1 Cooling Theory

It had been suggested that cooling would be possible in multi-resonance systems, such as those discussed in Refs. [15, 17, 18, 21, 24], provided that the phonon damping rate is lower than the damping rate of photons [24]. This is required so that the rate of thermal energy removal from the acoustic mode in the form of photons is greater than the rate at which energy re-enters the acoustic mode. However, this regime is not accessible in the SBS back-scattering experiments [3, 11, 15, 17, 20] since phonons at $>$10 GHz acoustic frequencies are generally associated with high mechanical damping [11]. On the other hand, forward Brillouin scattering [20] in

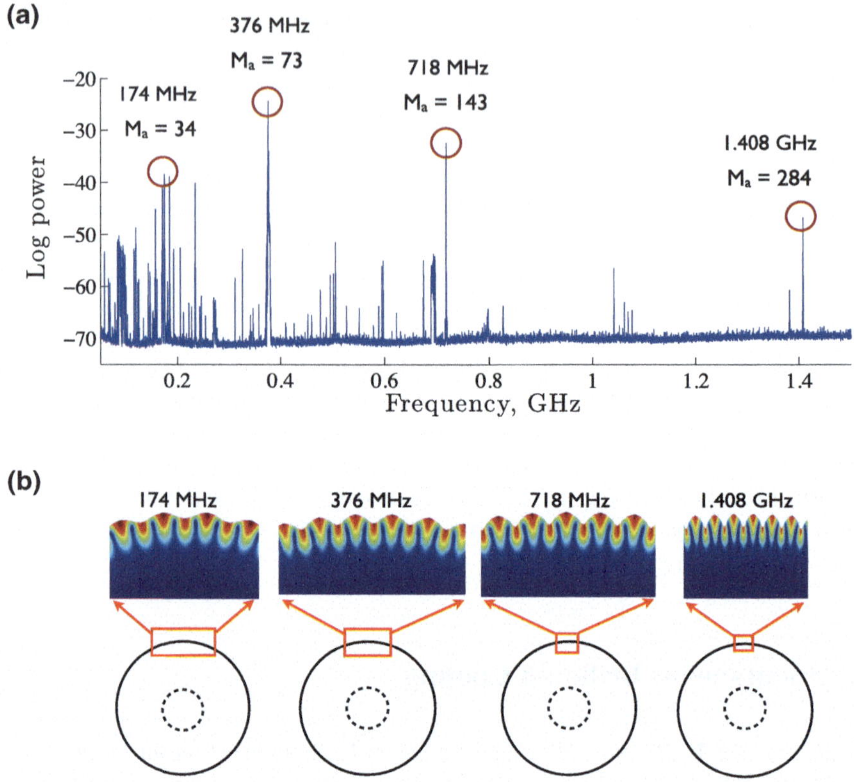

Fig. 8.4 **a** Many acoustic WGMs ranging from 50 MHz to 1.4 GHz are observed when the pump laser is scanned slowly from 1,520–1,570 nm. Some strong modes are identified here along with their propagation constant, M_a. **b** Equatorial cross-section of the finite-element calculated mode shapes for the modes identified in (**a**)

microcavities [18] provides a path to optomechanically couple 10–100 MHz range AWGMs where the mechanical dissipation is far lower, as measured through the mechanical quality factor.

A second challenge remains, which is to eliminate Stokes scattering. In such light-sound interactions incident photons are scattered to both redder (Stokes) and bluer (anti-Stokes) frequencies, resulting in heating and cooling as dictated by energy conservation. It was thought that this cooling-heating balance is always tilted towards heating (i.e. Stokes scattering) as governed by Planck's distribution [46]. This is indeed true in bulk media where all photons are almost equally transmitted. Furthermore, Stokes scattered light applies a positive feedback resulting in lasing through stimulated Brillouin scattering (SBS) that creates a large influx of phonons into the acoustic mode. We showed that this heating-cooling balance can be tilted towards cooling by exploiting the aperiodic nature of optical resonances in a resonator. In particular, rejection of Stokes scattering can be achieved by engineering the optical density of states in the system to eliminate any available phase-matched Stokes optical mode. Such filtering is challenging to do in bulk materials since relatively low (i.e. MHz regime) acoustic frequencies separate the optical modes, and sharp filtering transitions are not easily available. On the other hand, ultra-high-Q resonators readily exhibit optical modes with mode linewidths in the MHz regime, allowing for selective suppression of Stokes scattering.

We present a classical analysis in [23], showing that the cooling ratio is improved by having higher acoustic and optical quality factors and lower acoustic frequencies. Our quantum analysis in [47] provides details on the feasibility of using this system to reach ground-state [44, 45, 48].

The concept of using anti-Stokes scattering to cool a material was originally proposed by Pringsheim [49]. However, concerns regarding the thermodynamic impossibility of such a process were later raised as the second law of thermodynamics is seemingly violated. The concern was eventually resolved by Landau [50] with the explanation that the entropy of scattered light is increased through decoherence in phase, frequency, and directionality. In our experiment, the linewidth broadening of the scattered light is responsible for carrying the excess entropy as expected.

8.3.2 Cooling Experiment

We again use a silica microsphere resonator, which supports the three modes that participate in the Brillouin cooling process—two OWGMs (pump and anti-Stokes) and one AWGM. Since this is a forward-scattering interaction these three modes circulate in unison with considerable overlap. As mentioned previously, photoelastic scattering induced by density change scatters light to the anti-Stokes mode. Cooling of the acoustic mode occurs by means of the electrostrictive pressure generated by the light in the pump OWGM and the anti-Stokes OWGM, which acts to attenuate the themal mechanical motion of the phase matched AWGM.

The pump laser is positioned at the lower frequency optical mode. Light scatters in the anti-Stokes direction into the higher frequency optical mode. In the process, phonons are annihilated from the acoustic mode. Brillouin cooling is experimentally observed as the broadening of the beat-note signal between pump light and anti-Stokes lines, as a function of increasing pump power.

We employ acoustic mode linewidth as a measurement of cooling as was also shown in [38, 40]. The acoustic mode's effective temperature is inversely proportional to linewidth. In our experiment, we measured a 7.7 kHz sub-threshold linewidth for the 95 MHz acoustic mode at 294 K (Fig. 8.5b). The linewidth increased to 118 kHz when the pump power was increased to 100 μW, indicating a cooling ratio of 15 (Fig. 8.5a). As a result, the effective mode temperature at the maximum cooling point was calculated as 19 K. To ensure that the 7.7 kHz acoustic linewidth corresponds to the Brownian vibration, the Stokes (heating) experiment was performed with the pump laser positioned on the higher frequency optical resonance, and the acoustic linewidth was measured as a function of input power. The 7.7 kHz acoustic linewidth was verified by extrapolating both the heating and cooling linewidth trends

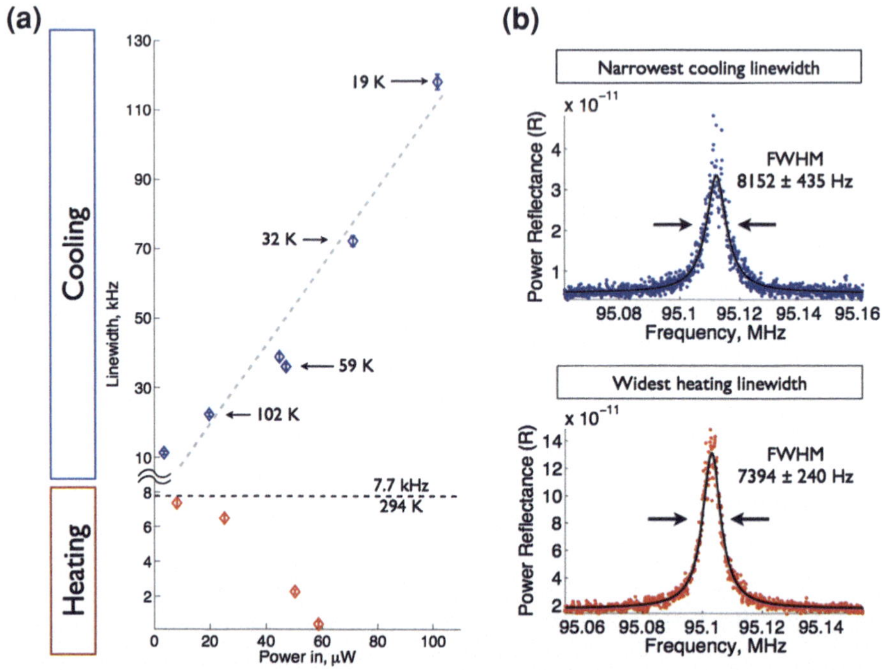

Fig. 8.5 **a** Acoustic linewidth data as a function of input optical power during heating and cooling experiments. **b** The 294 K calibration is performed by measuring the narrowest cooling linewidth and the widest heating linewidth. The trends are extrapolated back to zero input optical power in order to obtain the unperturbed intrinsic linewidth

to zero input optical power (Fig. 8.5a, b). The end result in the Brillouin cooling experiment is a net annihilation of phonons [38, 40–45, 48, 51] in a multi-resonance device [15, 17, 18, 24]. However, it is important to note that the pump does not need to be detuned with respect to the optical resonance, which greatly simplifies the experimental configuration.

Broadly speaking, Brillouin scattering belongs to a family of material-level scattering processes, including Raman scattering and Rayleigh scattering. The ability to remove phonons from a material by means of Brillouin scattering raises the question of whether similar cooling experiments would be possible with optical phonons (Raman scattering) [52] in spite of the significantly higher phonon frequencies involved. Raman cooling is attractive since it can provide a considerably higher quantum cooling efficiency.

Acknowledgments This work was supported by the Defense Advanced Research Projects Agency (DARPA) Optical Radiation Cooling and Heating in Integrated Devices (ORCHID) program through a grant from the Air Force Office of Scientific Research (AFOSR). Additional thanks go to the University of Illinois Mechanical Science and Engineering startup grant. The authors would also like to acknowledge the many contributions of Matthew Tomes, John Zehnpfennig, and Kyu Hyun Kim.

References

1. Y.R. Shen, N. Bloembergen, Phys. Rev. **137**(6A), A1787 (1965)
2. A. Yariv, IEEE J. Quantum Electron. **1**(1), 28 (1965)
3. R.Y. Chiao, C.H. Townes, B.P. Stoicheff, Phys. Rev. Lett. **12**(21), 592 (1964)
4. Y. Okawachi, M.S. Bigelow, J.E. Sharping, Z. Zhu, A. Schweinsberg, D.J. Gauthier, R.W. Boyd, A.L. Gaeta, Phys. Rev. Lett. **94**, 153902 (2005)
5. T.C. Rich, D.A. Pinnow, Appl. Phys. Lett. **20**(7), 264 (1972)
6. S.A. Lee, S.M. Lindsay, J.W. Powell, T. Weidlich, N.J. Tao, G.D. Lewen, A. Rupprecht, Biopolymers **26**(10), 1637 (1987)
7. W. Cheng, J. Wang, U. Jonas, G. Fytas, N. Stefanou, Nat. Mater. **5**(10), 830 (2006)
8. B.Y. Zel'dovich, V.I. Popocivhec, V.V. Ragul'skii, F.S. Faisullov, JETP Lett. **15**(3), 109 (1972)
9. F. Zarinetchi, S. Smith, Opt. Lett. **16**(4), 229 (1991)
10. D. Cotter, Electron. Lett. **18**(12), 495 (1982)
11. R.W. Boyd, in *Nonlinear Optics*, 3rd edn (Elsevier, 2008). Chapter 9.
12. M.H. Kuok, H.S. Lim, S.C. Ng, N.N. Liu, Z.K. Wang, Phys. Rev. Lett. **90**(25), 255502 (2003)
13. J.Z. Zhang, R.K. Chang, J. Opt. Soc. Am. B **6**(2), 151 (1989)
14. P. Dainese, P.S.J. Russell, N. Joly, J.C. Knight, G.S. Wiederhecker, H.L. Fragnito, V. Laude, A. Khelif, Nat. Phys. **2**(6), 388 (2006)
15. I.S. Grudinin, A.B. Matsko, L. Maleki, Phys. Rev. Lett. **102**(4), 043902 (2009)
16. H. Lee, T. Chen, J. Li, K.Y. Yang, S. Jeon, O. Painter, K.J. Vahala, Nat. Photonics **6**(6), 369 (2012)
17. M. Tomes, T. Carmon, Phys. Rev. Lett. **102**(11), 113601 (2009)
18. G. Bahl, J. Zehnpfennig, M. Tomes, T. Carmon, Nat. Commun. **2**, 403 (2011)
19. A.A. Savchenkov, A.B. Matsko, V.S. Ilchenko, D. Seidel, L. Maleki, Opt. Lett. **36**(17), 3338 (2011)
20. R. Shelby, M. Levenson, P. Bayer, Phys. Rev. Lett. **54**(9), 939 (1985)
21. G. Bahl, K.H. Kim, W. Lee, J. Liu, X. Fan, T. Carmon, Nat. Commun. **4**, 2994 (2013)
22. A.B. Matsko, A.A. Savchenkov, V.S. Ilchenko, D. Seidel, L. Maleki, Phys. Rev. Lett. **103**(25), 257403 (2009)
23. G. Bahl, M. Tomes, F. Marquardt, T. Carmon, Nat. Phys. **8**(3), 203 (2012)
24. I.S. Grudinin, H. Lee, O. Painter, K.J. Vahala, Phys. Rev. Lett. **104**(8) (2010)
25. M. Tomes, F. Marquardt, G. Bahl, T. Carmon, Phys. Rev. A **84**, 063806 (2011)
26. T. Carmon, H. Rokhsari, L. Yang, T.J. Kippenberg, K. Vahala, Phys. Rev. Lett. **94**(22), 223902 (2005)
27. T. Carmon, M.C. Cross, K.J. Vahala, Phys. Rev. Lett. **98**(16), 167203 (2007)
28. K.H. Kim, G. Bahl, W. Lee, J. Liu, M. Tomes, X. Fan, T. Carmon, Light: Sci. Appl. **4**, e110 (2013)
29. G. Bahl, X. Fan, T. Carmon, New J. Phys. **14**, 115026 (2012)
30. K. Han, J.H. Kim, G. Bahl, in *Frontiers in Optics 2013 / Laser Science XXIX* (2013)
31. A.A. Savchenkov, A.B. Matsko, V.S. Ilchenko, D. Strekalov, L. Maleki, Phys. Rev. A **76**, 023816 (2007)
32. T. Carmon, H.G.L. Schwefel, L. Yang, M. Oxborrow, A.D. Stone, K.J. Vahala, Phys. Rev. Lett. **100**, 103905 (2008)
33. M. Gorodetsky, V. Ilchenko, Opt. Commun. **113**(1–3), 133 (1994)
34. M. Cai, K. Vahala, Opt. Lett. **26**(12), 884 (2001)
35. S. Spillane, T.J. Kippenberg, O. Painter, K. Vahala, Phys. Rev. Lett. **91**(4), 043902 (2003)
36. J. Knight, G. Cheung, F. Jacques, T. Birks, Opt. Lett. **22**(15), 1129 (1997)
37. J. Mertz, O. Marti, J. Mlynek, Appl. Phys. Lett. **62**(19), 2344 (1993)
38. C. Metzger, K. Karrai, Nature **432**(7020), 1002 (2004)
39. C. Metzger, M. Ludwig, C. Neuenhahn, A. Ortlieb, I. Favero, K. Karrai, F. Marquardt, Phys. Rev. Lett. **101**, 133903 (2008)
40. O. Arcizet, P.F. Cohadon, T. Briant, M. Pinard, A. Heidmann, Nature **444**(7115), 71 (2006)

41. S. Gigan, H. Bohm, M. Paternostro, F. Blaser, G. Langer, J. Hertzberg, K. Schwab, D. Bauerle, M. Aspelmeyer, A. Zeilinger, Nature **444**(7115), 67 (2006)
42. D. Kleckner, D. Bouwmeester, Nature **444**(7115), 75 (2006)
43. J. Thompson, B. Zwickl, A. Jayich, F. Marquardt, S. Girvin, J. Harris, Nature **452**(7183), 72 (2008)
44. R. Riviere, S. Deleglise, S. Weis, E. Gavartin, O. Arcizet, A. Schliesser, T.J. Kippenberg, Phys. Rev. A **83**(6) (2011)
45. J. Chan, T.P.M. Alegre, A.H. Safavi-Naeini, J.T. Hill, A. Krause, S. Groeblacher, M. Aspelmeyer, O. Painter, Nature **478**, 89 (2011)
46. C. Kittel, in *Introduction to Solid State Physics*, 8th edn (Wiley, 2004)
47. M. Tomes, F. Marquardt, G. Bahl, T. Carmon, Phys. Rev. A **84**(063806) (2011)
48. J.D. Teufel, T. Donner, D. Li, J.W. Harlow, M.S. Allman, K. Cicak, A.J. Sirois, J.D. Whittaker, K.W. Lehnert, R.W. Simmonds, Nature **475**(7356), 359 (2011)
49. P. Pringsheim, Zeitschrift fur Physik **57**(11–12), 739 (1929)
50. L. Landau, On the thermodynamics of photoluminescence. J. Phys **10**(6), 503 (1946)
51. R. Epstein, M. Sheik-Bahae (eds.), in *Optical Refrigeration*, 1st edn (Wiley-VCH, 2009)
52. M.S. Kang, A. Nazarkin, A. Brenn, P.S.J. Russell, Nat. Phys. **5**(4), 276 (2009)

Chapter 9
Integrated Optomechanical Circuits and Nonlinear Dynamics

Hong Tang and Wolfram Pernice

Abstract Chip-scale optomechanical resonators have the advantage of higher Q values, smaller dimensions, and thus a greater potential for scalability and device control. This chapter discusses further integration of chip-scale optomechanic elements on a circuit level. Circuit integrated optomechanics brings a range of additional benefits for both fundamental studies and practical device applications of optomechanics. It takes advantage of many circuit components that have been developed by the nanophotonics community, such as grating couplers, splitters, combiners and low loss waveguides, which can be directly utilized to form scalable circuits for routing optical signals. High quality photonic resonators and band-gap structures, such as microring, microdisk, and photonic crystal nanocavities, can be seamlessly embedded in the circuit for direct interfacing with nanomechanical resonators without suffering from significant optical loss. Homodyne detectors, such as Mach-Zehnder interferometers, are implemented directly as part of the circuit to achieve sensitive, alignment free measurement of mechanical motion. Circuit optomechanics not only leads to further scaling of optomechanical systems but also brings new interaction dynamics in light-mechanics coupling. We will discuss the various radiation dynamics and non-radiation dynamics that are present in circuit optomechanical systems.

9.1 Integrated Optomechanics and Optomechanical Circuits

Lately we have witnessed rapid progress in cavity optomechanics with demonstration of dynamic back-action cooling and amplification of mechanical structures of vastly different length scales. In particular, chip-scale optomechanical resonators

H. Tang (✉)
Department of Electrical Engineering, Yale University, New Haven, USA
e-mail: hong.tang@yale.edu

W. Pernice
Institute of Nanotechnology (INT), Karlsruhe Institute of Technology (KIT), Karlsruhe, Germany

M. Aspelmeyer et al. (eds.), *Cavity Optomechanics*, Quantum Science and Technology,
DOI: 10.1007/978-3-642-55312-7_9,

have allowed extreme miniaturization of both mechanical resonators and photonic cavities [1–4]. With scalable top–down nanofabrication techniques, photonic crystal nanocavities with modal volume as small as 0.02 cubic wavelengths have been realized [1] whereas the mass of mechanical resonators approaches 25 femtogram [4]. The optical and mechanical quality factors of these structures on the other hand, have been consistently improved. This aggressive reduction in device size and modal volume compared to photonic crystal fiber based approaches [5–8] leads to enhanced optical forces and as a result, chip-scale optomechanical resonators have entered the strong coupling regime [9]. The attainable displacement sensitivity on chip also reached attometer levels [10–13], which previously was only achieved in free space ultra-high finesse fabry-perot cavity systems [14, 15].

The blend of nanoscale photonic and mechanical devices on a common chip represents a new trend in the development of optomechanics. Yet accessing these nanoscale devices has been a major challenge, since most advanced lasers and detectors remain stand-alone, off-chip units that often require fiber or free space coupling. A widely adopted approach is fiber-taper coupling. In this chapter, we will describe an alternative approach that employs grating couplers manufactured on chip to realize direct fiber to device coupling. These grating couplers, functioning like electrical contacts in an electronic circuit, are laid out on chip as part of the photonic circuit. With arrays of grating couplers, many devices can be simultaneously measured in parallel. Similarly, multiport measurement on a single optomechanical cavity is also possible by connecting multiple grating couplers to the same device. This allows photon transport measurement similar to electronic transport measurement in mesoscopic systems.

In a circuit platform, the optomechanical resonator is embedded in a photonic circuit like a transistor node in an electronic circuit. A range of photonic cavities have been developed during the past several years. Some of these cavities are closed loop circuit element themselves, such as microring and racetrack resonators. When a portion of the ring is released to form a nanomechanical resonator, strong optical forces are generated by the circulating light and drive the mechanical element. Miniaturized whispering-gallery-mode resonators, such as microdisk and microwheel resonators, have also been implemented with the whole volume of disk and wheel acting as vibrating mechanical resonators. Both circuit type and whispering-gallery mode resonators have relatively large mode volumes. A recent exception is a special type of sunflower photonic crystal cavity at whose center a ultrasmall whispering-gallery mode microdisk is formed [16]. Nevertheless, the smallest mode volumes are often realized in photonic crystal defect cavities, either in one dimensional crystals [2] or two dimensional crystals. One dimensional implementations of photonic crystal optomechanical cavities (such as the zipper cavity) have been discussed in preceding chapters. Here in this chapter, we discuss a new implementation of 2-d photonic crystal nanocavities in which nanoscale beams are embedded [4].

In optomechanical circuits, optical forces can originate from the evanescent gradient force [3, 17, 18] or the radiation pressure effect [19]. The two forces find their origin in Maxwell's stress tensor. Hence in ultrasmall devices where the wave is confined at wavelength scale in all dimensions, it is difficult to distinguish these two types of forces. Fortunately, the backaction effect can be universally described by a set of coupled equation of motions. In this chapter we will focus on the discussion of the gradient force which applies to propagating waves and circuit resonators. We also show that in certain circuit configurations, optomechanical interaction not only dispersively modifies the optical resonance center frequency but also changes its damping rate. We will present the cooperation of these two coupling mechanism (dispersive and dissipative coupling) with a microdisk coupled waveguide resonator.

It is known that in optomechanics, both optical resonators and mechanical resonators can exhibit highly nonlinear behavior [20, 21]. This is in addition to the nonlinear nature of cavity backaction. Circuit integrated optomechanics allows separate design of mechanical resonators and optical resonators. Utilizing the extreme mechanical nonlinearity in doubly clamped beams and the amplification effect of the cavity, we investigate the dynamic coupling of light and mechanical resonator in the ultrahigh amplitude regime, which is a highly coherent phenomenon not accessible in conventional nanomechanical systems.

It is foreseeable that future optomechanical systems will also include light sources and detectors. Indeed, both have been actively pursued in the photonics community. For example, high quantum efficiency Germanium detectors have already been implemented on nanophotonic waveguides [22]. Laser diodes on the other hand, have been integrated on a silicon photonics platform by relying on bonding techniques. Foundry services have already started offering these components on shared wafer runs [23]. In this sense, manufacturable optomechanical circuits are already on the horizon.

9.2 Components of Optomechanical Circuits

The design of optomechanical circuits follows a methodology derived from the microelectronics industry. Complex optical systems are composed of simpler building blocks, contained within a library of photonic basic elements. Therefore the overall layout resembles a photonic computer aided design (CAD) approach. In the following we will review some of the most important components required for the assembly of full circuits.

9.2.1 Input/output Coupling

When working with on-chip devices one of the most important engineering tasks is coupling light into the chip. A fundamental difficulty arises due to the small cross-section of a nanophotonic waveguides, which is significantly smaller than an optical fiber or a free-space laser beam. The large size-mismatch implies, that direct coupling will lead to high loss because only a small portion of the input light will be eventually guided by the waveguide. Therefore a number of techniques have been developed to improve the coupling efficiency. Here we will focus on coupling from optical fibers into on-chip circuits, which is the dominant method employed in integrated optics. In principal, two different coupling possibilities exist: (1) surface coupling, also called longitudinal coupling, and (2) end coupling, also called end-fire coupling or transverse coupling. In the first scheme the optical wave is coupled out along the longitudinal surface of a waveguide. In the second approach, the optical wave is coupled directly through an exposed cross-section of the end of the waveguide.

9.2.1.1 Grating Coupling

Grating coupler devices allow for out-of-plane coupling to optical fibers by exploiting Bragg scattering. From the point of coupled-mode theory, the coupling relies on the interaction between guided modes in the waveguide and radiation modes. Any non-guided optical mode (such as the input mode coming from an optical fiber) can be expanded in terms of the radiation modes of the waveguide. In order to achieve coupling to the guided mode in the waveguide phase-matching has to be fulfilled. This is however not possible without perturbing the waveguide because the propagation constant of any radiation mode is always smaller than the propagation constant of the guided mode. Here the propagation constant is the wavenumber of the mode passing along the waveguide. A grating coupler overcomes this difficulty by providing additional wavevector components resulting from a periodic modulation of the effective index of the optical waveguide mode. This is illustrated schematically in Fig. 9.1a, where we show a cross-sectional view of a waveguide as well as grating input and output couplers. An example of a fabricated structure is shown in Fig. 9.1b. In this structure the interaction between the input mode and the guided modes is governed by the Bragg condition, which describes the relation between the wave-vectors of the incident and diffracted waves. For a grating with uniaxial periodicity along the z-axis with period Λ between the materials with refractive indices n_1, n_2 and n_3 the relevant equation states

$$k_{in} \sin \theta_{in} + \frac{2\pi}{\Lambda} = \beta \tag{9.1}$$

where $\beta = \frac{2\pi}{\lambda_0} n_{eff}$ is the propagation constant of the guided mode, λ_0 the desired wavelength and $k_{in} = \frac{2\pi}{\lambda_0} n_3$ is the wave vector of the incident wave. In the above equation we assume scattering on first order, as illustrated for the phase-matching diagram in Fig. 9.1c. The first order diffraction is near-vertically coupled out of the waveguiding plane and thus well suited to couple into an optical fiber. The above equation therefore states, that the horizontal component of the incoming wave vector k_z combined with the grating wave component K must equal the wave vector of the propagating mode.

It is important to note that the above equation is only strictly exact for infinite periodic structures. For finite grating structures there is not only one discrete k-vector for which diffraction occurs, but a range of k-vectors around the predicted value

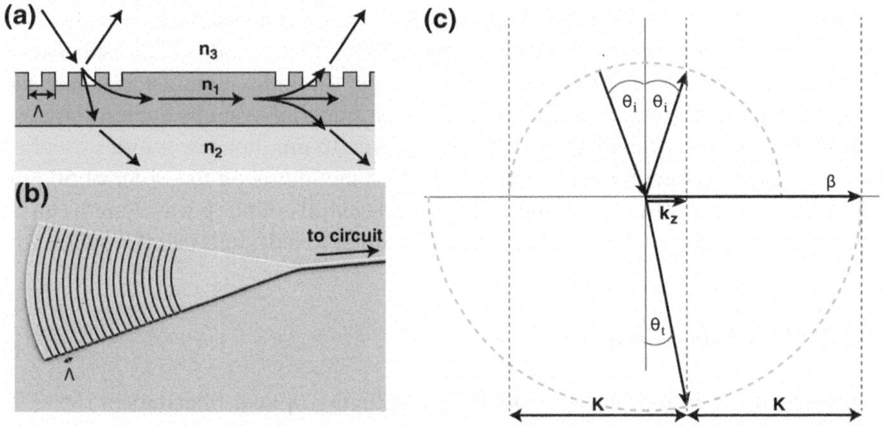

Fig. 9.1 **a** Cross-sectional view of a waveguide with input and output grating coupler ports. The optical wave is guided in the high refractive index layer (n_1) and diffracted in the regions where the grating structure with period Λ is inscribed. **b** SEM image of a fabricated focussing grating coupler device. **c** Schematic illustration of the phase-matching requirements for the structure in (**a**)

Table 9.1 Overview over grating coupler implementations

Type of grating	Research group	Efficiency (%)	Reference
Bottom mirror	IMEC	69 (measured)	[24, 25]
Top mirror	University of California	95 (theoretical)	[26]
Metal grating	IMEC	60 (theoretical), 34 (measured)	[27]
Slanted grating	Osaka University, IMEC	59 (theoretical), 16 (measured)	[28, 29]
Chirped grating	University of Hong Kong	34 (measured)	[30, 31]
Rear reflector	IMEC	70 (theoretical), 33 (measured)	[32]
1D focussing grating	Luxtera	70 (measured)	[33]
2D focussing grating	IMEC	27 (measured)	[34]

Reported efficiencies are given in %, with a best value of 70 % reported for the focussing grating coupler by Luxtera

resulting from the Bragg condition. The range of these k-vectors or in other words the bandwidth of the grating coupler will depend on the coupling strength of the chosen grating design.

Several designs of grating couplers have been investigated in the last decade to develop a photonic platform that is directly compatible with integrated electronic circuits. Depending on the complexity of the chosen design and employed materials the coupling efficiency can be very high. A summary of representative results is collected in Table 9.1.

Out of the reported coupler implementations the lowest insertion loss for a single coupling port amounts to −1.5 dB (a focussing grating coupler design as also shown in Fig. 9.1b). When simpler designs are employed with a single lithography plus subsequent etching steps are involved, typical insertion loss is on the order of −7 dB.

As mentioned above, the grating couplers provide a certain usable bandwidth. The width of the coupling region depends hereby on the refractive index contrast between the coupling structure and the optical fiber: when the index contrast is small, the coupling bandwidth increases. For high-contrast material systems such as silicon-on-insulator (SOI), typical bandwidth values cover 30 nm. For lower index contrast materials, such as silicon nitride, the coupling bandwidth is on the order of 50 nm. While the optical bandwidth is limited, the actual central coupling wavelength can be adjusted by varying the period of the grating coupler to a desired target wavelength.

9.2.1.2 Fiber Taper Coupling

Bandwidth restrictions can be avoided by employing optical fiber tapers [35–37]. Such a device is realized by heating and stretching a section of optical fiber to form a narrow thread, or waist, which is joined to the untreated ends of the fiber by a gradual taper transition. In the waist region the fiber core is no longer significant, and the light travels in the fundamental mode along the waveguide formed by the silica waist surrounded by air. The taper waist can be less than a micrometer in diameter. Silica nanotapers with diameters in the range of several tens to several hundreds of nanometres have previously been fabricated using a variety of methods, however most of these have exhibited an irregular profile along their length and this has limited their usefulness for optical applications [38–40]. On reaching the other end of the waist, the light remaining in the waist is returned to the guided mode in the fiber core. If the waist is small, the fundamental mode will have an evanescent tail extending significantly out into the free space surrounding the taper, and the propagation constant of the mode will be a function of the waist radius.

The evanescent tail of the optical mode within the tapered section can be efficiently used to couple evanescently to nanophotonic structures. When the effective index of the fiber mode matches one of the modes in the waveguide, resonant coupling between the optical fiber and the waveguide will occur, leading to intensity transfer from the fiber into the waveguide. Because optical fibers are broadband transparent, a wide wavelength range of guided modes can be excited in this fashion. Condition, however, is that the evanescent tail of the waveguide mode shows sufficient overlap with the fiber modes. Therefore typical coupling distances between waveguide and fiber taper are in the sub-micron to nanometer range.

Because close proximity is required between the fiber taper and the optical waveguide difficulties arise during optical alignment when the waveguide forms part of an optical chip. Traditional fiber taper couples do not allow for local or point-contact coupling, which is often desirable to investigate optical cavities. In order to overcome these difficulties curved fiber taper probes have been designed to reduce parasitic coupling loss into the substrate in unwanted chip regions [41–43]. This also includes dimpled taper optical probes used for microresonator coupling [44].

9.2.1.3 Butt Coupling

One of the most widely used traditional optical coupling methods is end-fire or butt coupling. End-fire coupling from free-space or optical fibers can be made highly efficient, even to high index contrast semiconductor waveguides, through the use of tapered waveguide sections [45] or other non-adiabatic mode converters [46, 47]. Such devices are limited to coupling at the periphery of the chip where a cleaved facet can be formed. The coupling efficiency will depend on the spatial overlap between the transverse fields of the optical modes in the fiber and in the channel waveguide. A complicating factor is the geometric mismatch between the core of the diffused waveguides, which tend to have rectangular aspect ratios, and the core of the fiber, which is typically circular. Coupling to and from these devices usually involves high losses resulting from mode-size and effective-index mismatch between the optical fiber and the waveguide structure, which induces coupling to radiation modes and back reflection. To date, most of the on-chip structures suggested to alleviate this coupling problem [48–51] have suffered from at least one of the following drawbacks: they are very long (hundreds of micrometers), are difficult to fabricate or have strong backreflection. Therefore tapered devices that transition from waveguide dimensions to the fiber dimensions have been suggested. However, to avoid excessive coupling to radiation modes in the taper, the required typical taper length must be of the order of millimeters. Almeida et al. [52] recently proposed a micrometre-long nano-taper coupler that converts both the mode size and the effective index of the waveguide to that of the optical fibre. The nano-taper consists of a highly confining waveguide tapered to a nanometre-sized tip at the facet, in contact with an optical fibre. Due to the subwavelength dimensions of the tip, the field profile expands, inducing a very large mode similar in effective index and profile to that of the fibre. By pointing the tip to the center of the fibre, the light from the fibre is completely drawn into the nano-sized waveguide. The nano-taper coupler is the shortest mode converter with high coupling efficiency on silicon. The results demonstrate the principle of mode delocalization using high confinement waveguides tapered to nanometre sizes for bridging between optical structures across size scales. A fundamental problem with butt-coupling, however, lies in the sophisticated alignment techniques that need to be used to orient each single fiber towards the coupling region. Therefore the approach is mostly suited for optical circuits with a limited number of input ports.

9.2.2 Optomechanical Resonators

Having light coupled into the chip the layout can be adjusted to form complex photonic circuits for optomechanical applications. In a chip-scale framework the fundamental building block for light-mechanical interactions are free-standing waveguides. Nanophotonic waveguides are a natural choice for nanoscale mechanical resonators, because the typical cross-sections for single-mode waveguideing are on the same order as NEMS resonators. In order to enable mechanical vibrations, resonant devices are formed by selectively removing the substrate underneath the waveguiding layer. The remaining free-standing waveguide is then free to move under the influence of the optical gradient force.

In order to enhance optomechanical interactions it is advantageous to embed the nanoscale resonator inside optical cavities. Cavity feedback can then be employed for optical cooling or heating, while at the same time the optical readout-sensitivity of the photonic circuit is increased. Furthermore, due to optical power recycling inside the cavity and subsequent intensity buildup, significant gradient optical forces can be reached at reduced optical input powers. In the following a number of on-chip implementations of coupled NEMS-cavity systems will be presented.

9.2.2.1 Beam in a Ring Resonator

Within integrated nanophotonics one of the most frequently used cavity implementation is the microring resonator. Such a device is formed by cyclic connection of a nanophotonic waveguide, thus forcing light coupled into the waveguide to circulate until optical losses dissipate the guided wave. Microring resonators are able to maintain high optical quality factors in excess of several million [53–55], with record Q-values approaching 30 Million at 1,310 nm wavelength [56]. From such devices on-chip cavity opto-mechanical systems can be conveniently prepared by releasing part of the resonator [57]. In this case the NEMS beam is coupled evanescently to the underlying substrate, providing the basis for gradient force actuation out-of-plane. In Fig. 9.2a we show a typical fabricated device. The mechanical resonator is part of a silicon resonator, released by removing the underlying oxide layer.

Stronger gradient optical forces can however be obtained by increasing the electric field gradient across the waveguide boundary. In a cavity framework this can be conveniently achieved by employing slot waveguides [58], which consist of closely coupled dielectric waveguides separated by a nanogap. Electric fields are tightly confined inside the gap, leading to gradient force enhancement on the waveguide inner surface. In contrast to substrate coupled devices, nanoscale mechanical resonators fabricated from slot waveguides experience in-plane movement. In this case the magnitude of the actuation force can be precisely controlled through the design of the slot waveguide geometry. Because of the resulting strong optical forces, this design is suitable for actuation of stiff mechanical resonators and therefore short devices that resonate at high mechanical frequencies [59]. A typical device is shown in Fig. 9.2d, illustrating the dual mechanical resonators freely suspended above the substrate. By reducing the length of the release window ultrashort devices with low mass and GHz frequency can be fabricated.

In addition to waveguides separated by a vertical slot, a horizontal slot waveguide can be employed to realize out-of-plane movement. In this case the waveguides are realized in a multi-layer material system, consisting of two high refractive index layers separated by a low index layer [60]. After fabricating waveguides in this design mechanical resonators can be formed by selectively removing the separating

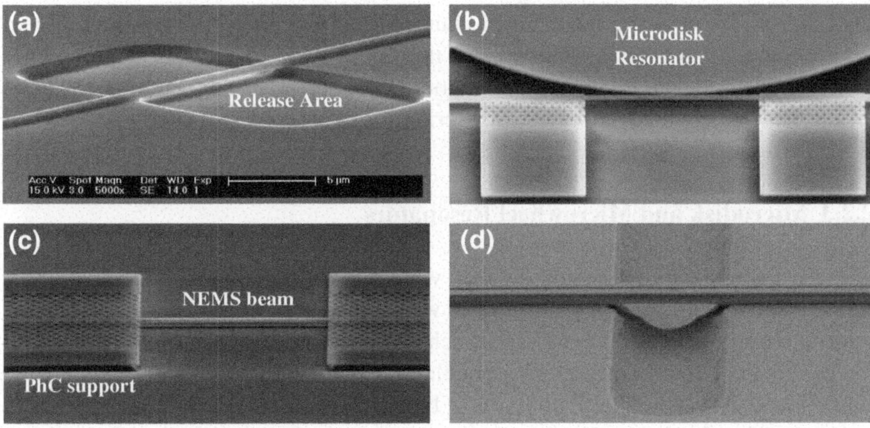

Fig. 9.2 **a** SEM image of a released section of a nanophotonic waveguide. By selectively removing the underlying oxide layer a free-standing mechanical beam resonator is realized. **b** SEM image of a nanomechanical resonator coupled to a microdisk resonator. The beam waveguide is clamped on the edges using single-sided photonic crystal mirrors. **c** A beam resonator clamped between two photonic crystal waveguide couplers. Due to low insertion loss the coupling structures do not significantly degrade the optical quality factor of the resonator. **d** An ultra-short double-beam resonator formed by underetching a slot-ring resonator

interlayer [61]. This approach leads to low scattering loss between the released and unreleased waveguide sections thus maintaining the optical quality factor of the unperturbed cavity.

9.2.2.2 Beam in a Photonic Crystal Cavity

In the ring resonator structure the length of the mechanical resonator is defined through the size of the etching window for the subsequent release. During the actual substrate removal process underetching of the resist layer may occur, leading to enlarged resonator length or poor mechanical clamping and the substrate-waveguide interface. In order to improve the geometric control of the mechanical resonator it is therefore desirable to provide lithographically defined clamping points. Suitable structures are multi-mode interference (MMI) devices [3], which however do not provide ultra-low loss required for embedment in ring resonator cavities. Alternatively, photonic crystal (PhC) designs can be employed, as shown in Fig. 9.2c. Here a defect waveguide is enclosed between photonic crystal mirrors, allowing for negligible scattering loss at the waveguide-PhC interface. By using two such structures a mechanical resonator of desired length can be fabricated with controlled clamping points [62]. The approach allows for simultaneous high optical and mechanical quality factors and precise geometric control of the resonance frequency.

Alternatively, photonic crystal structures can be employed in single-mirror configuration for side-coupling to on-chip photonic cavities. As shown in Fig. 9.2b such devices allow for defining mechanical clamping points, while consuming less real estate on the chip. This is for example of importance when side-coupling nanomechanical waveguides to microdisk resonators, where the clamping point needs to be kept far from the disk edge to avoid scattering loss [63].

9.2.2.3 Microdisk and Microwheel Resonators

Alternatively to beam-resonator devices which provide flexural mode vibrations circular mechanical resonators can be used with different mode structures. By relying on non-flexural resonant designs high mechanical frequencies can be achieved with good mechanical quality factors. In particular mechanical resonators operating in the radial contour and wine-glass modes have been reported to have high mechanical Q factors resulting from their in-plane modal vibration and immunity to air damping. The angular mechanical frequency is in this case given as $\Omega_m = (k/m_{eff})^{1/2}$, where k denotes the modal spring constant and m_{eff} is the effective modal mass. Therefore in order to obtain high frequency optomechanical resonators one needs to minimize m_{eff}. This strategy has recently been implemented in Gallium Arsenide disks [64], where the resulting effective mass is reduced to a few picograms or even less. The other route to high mechanical frequency consists in exploring vibrating modes that possess very high stiffness. This concept is exploited in optomechanical resonators that consist of a microring resonator with supporting-spoke structure. Such wheel-shaped optomechanical resonators operate at GHz frequency with high mechanical Q factor in ambient air [13].

While wheel resonators are promising candidates for optomechanical applications, the relatively large mass limits the resolution for mass sensing applications. To reduce the resonator mass microdisks with a much smaller size are an attractive alternative [65]. Another advantage of smaller devices is the larger optomechanical coupling that can be achieved with wrap-around waveguides. By reducing the clamping loss from the disk pedestal through optimized fabrication, silicon disk resonators are able to vibrate at GHz frequencies with high mechanical Q factors. The disk's motion is optically transduced using specially designed coupling waveguides, which not only facilitate coupling to the disk's optical whispering gallery modes but also help to excite and sense the disk's in-plane mechanical modes.

9.3 Optical Force Induced Dynamic Backaction in Optomechanical Circuits

In an optical circuit, the interaction between mechanical resonator and light can be mediated by the radiation force or the gradient force, depending on the specific device configuration. The former often describes the coupling in whispering-gallery-mode resonators whereas the latter is an evanescent dipole force generated by guided waves. Sections 9.3.1 and 9.3.2 present the generation principle of these two types of optical forces. The forces are responsible for the dispersive coupling between mechanical vibration and optical field. In certain geometries, both forces are manifested. In particular, in an open optomechanical system, evanescent coupling could bring additional damping to the optomechanical resonator. This dissipative coupling [66], which we term as "reactive coupling" will be described in Sect. 9.3.3. Finally Sect. 9.3.4 discusses a new regime of impulse optical force backaction where the mechanical vibration reaches extremely high amplitudes and the cavity dynamics become highly nonlinear.

9.3.1 Radiation Pressure Force in Micro and Nano-optical Resonators

The radiation pressure force has been extensively studied in free space optomechanical systems. In an optomechanical circuit, radiation pressure force takes place mostly in whispering-gallery-mode resonators where light circulates along the circumference of the resonator and as a result of momentum transfer from photons to mechanical structure, an optical force applies along the radial direction of the resonator. The radiation pressure force is a surface normal force which applies perpendicular to the surface. It can be precisely quantified by the photon number within the cavity. The force per photon for the fundamental breathing mode is often described by the $\hbar g$, where $g = \omega_0/R$ and $R \gg \lambda$ is radius of the whispering-gallery-resonator. We note that this diverges as $R \rightarrow 0$ and should be modified when R approaches the optical wavelength. For the other modes, an effective optomechanical coupling length l_{om} can be used to describe the optomechanical coupling strength $g = \omega_0/l_{om}$. In high Q optical resonators, the radiation force is significantly enhanced since the number of photons residing in the optical cavity can reach very high values. In terms of input power, the total radiation pressure force is described by, $F = P_{in}\kappa/(\omega_0^2 + \Delta^2)\hbar g$.

9.3.2 Gradient Optical Force

Unlike the radiation pressure force, the gradient optical force originates from the lateral gradient of a propagating light field. This force finds important use in optomechanical circuits since the force can be generated along the transverse direction of a waveguide without the need for making a bend or suffering loss. In submicron scale photonic waveguides, the gradient of an optical field is enhanced by orders of magnitude due to the strong field confinement, as theoretically predicted by Povinelli et al. [67, 68]. It is also not necessary to employ a cavity to obtain strong force. Hence the bandwidth of force generation can be very wide. The experimental verification of the gradient optical force in a waveguide geometry was achieved by us at Yale [3]. In this experiment, we exploited the high force sensitivity of a nanomechanical resonator to sense and quantify the optical force. To do this, a Mach-Zehnder interferometer (Fig. 9.3a) is fabricated on a Silicon-on-Insulator substrate, which is a commonly used substrate for silicon photonics. A portion of the upper arm of the Mach-Zehnder interferometer (MZI) is suspended to form a free standing nanomechanical resonator (Fig. 9.3b). The gradient optical force, arising from the evanescent coupling between the waveguide and the substrate, will move the mechanical resonator and introduce a tiny change of the effective refractive index (δn_{eff}) of the released portion of the waveguide. Hence a phase change of $\delta n_{eff} L$ is introduced to the propagating light in the upper arm. The lower arm of the MZI serves as a local oscillator (reference arm) to detect this phase modulation. To enable a homodyne detection circuit, the optical path length difference between the two arms needs to be adjusted to operate at the quadrature point. In free space interferometers, this is achieved by using a feedback circuit to adjust the reference mirror position. For on-chip implementations, instead, the physical path length difference is fixed at 100 μm, whereas the wavelength of the input laser for the interferometer is scanned to set the operating point.

The transmission of the MZI is displayed in Fig. 9.3c. The high extinction ratio indicates good balance of the interferometer. During the measurement two separate light beams are applied to the circuit: one modulated pump beam with a wavelength set at a dark fringe to generate a time-varying force; and a probe light with a wavelength set at the quadrature to detect the mechanical motion. Figure 9.3d shows the frequency response of the device to modulated pump light at various amplitudes. Clearly the optical force is significant—the device can be driven to a very high amplitude that exceeds the Duffing nonlinearity threshold. The force on the beam can be calculated by $F = k\Delta z(\omega_0)/Q$, where $\Delta z(\omega_0)$ is the beam's resonance amplitude calibrated by a separate thermomechanical noise measurement. For the beam shown in Fig. 9.3b with a substrate separation of 360 nm, the optical force normalized to the beam length and the optical power is $0.5 \pm 0.1\,\mathrm{pN\,\mu m^{-1}\,mW^{-1}}$. This value is in line with our modelling [69] and theoretical values predicted by [67].

The magnitude of the gradient optical force depends strongly on the evanescent coupling gap. The optical force can be calculated numerically by integrating the Maxwell stress tensor of the surface of the waveguide, which is obtained from finite element simulations of the mode profile with varying gap size. The force on the waveguide is then given as $F_{opt} = \int_S T \cdot d\vec{n}$ where S is the surface of the waveguide, $d\vec{n}$ is the surface normal and T denotes the Maxwell stress tensor given as $T_{ij} = D_i E_j + H_i B_j \frac{1}{2}(\vec{D} \cdot \vec{E} + \vec{H} \cdot \vec{B})\delta_{ij}$. Alternatively, the force can be calculated semi-analytically using an effective index method as described in Refs. [62, 67]. In this approach the force is given in terms of the effective index and the optical energy U, $F_{opt} = \frac{1}{n_{eff}} \frac{\partial n_{eff}}{\partial g} U = \frac{n_g l}{n_{eff} c} \frac{\partial n_{eff}}{\partial g} P$, where the energy $U = n_g P l/c$ is expressed in terms of optical power and the group index n_g. Hence the force normalized by the beam length and laser input power is given by $F_{opt}/lP = \frac{1}{n_{eff}} \frac{\partial n_{eff}}{\partial g}$. The corresponding force per photon is expressed as $F_0 = \hbar\omega_0(\frac{1}{n_{eff}} \frac{\partial n_{eff}}{\partial g})$. In waveguide devices, $\frac{dn_{eff}}{dx}$ could reach 10^{-3}/nm [62]. Hence the force per photon could be as high as 130 fN.

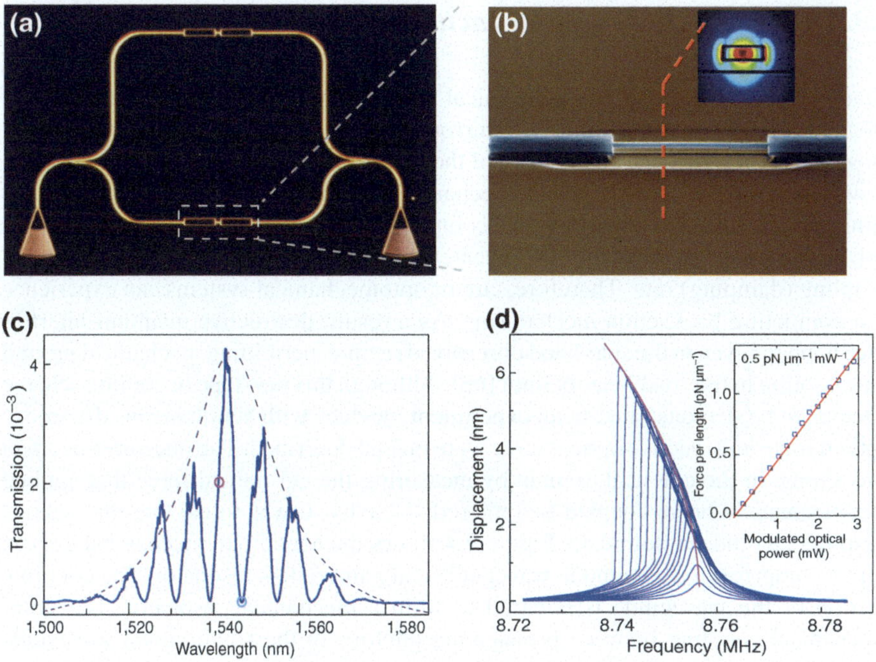

Fig. 9.3 Photonic interferometric circuit for quantifying the gradient optical force. **a** Optical micrograph of a Mach-Zehnder interferometer (*MZI*) device. **b** SEM image of the suspended nanomechanical resonator. The *inset* shows the optical modal profile. **c** Transmission spectrum of the MZI, showing interference fringes with high extinction ratio. The probe wavelength is set at the quadrature point (*red circle*) whereas the actuation laser wavelength is set at a dark fringe (*blue circle*). **d** Resonance response curves of a 10-μm-long waveguide beam at varying modulation levels of the actuation light. Softening nonlinearity appears as a result of compressive stress in the SOI structure

Theory also predicted that if two waveguides are evanescently coupled, the gradient optical force could be either attractive or repulsive depending on the relative phase in the propagating wave in each waveguide. This bipolar optical force was experimentally demonstrated in carefully designed optomechanical circuits in which the relative phase of two waveguide modes can be continuously varied. When a symmetric mode (in phase) is launched into the coupled structure, a normal attractive force was recorded whereas the out-of-phase asymmetric mode yielded a repulsive force.

The gradient optical force can be engineered by structural design and therefore is highly flexible for use in a photonic circuit. The attractive and repulsive forces discussed above are not restricted to a certain, specific geometry. They have been extended to other device geometries such as ring resonators [57], slot waveguides [59], and microdisk geometries [70].

9.3.3 Complex Optical Force Including Reactive Coupling

Gradient optical force in optomechanical circuits also brings about a new cooling mechanism—"reactive" cooling. In mirror cavity systems, as the mirror moves, it mostly modifies the center frequency of the cavity resonance. In planar devices such as we described in Sect. 9.2 and optomechanical zipper cavities, due to the strong coupling between mechanical and optical components, the mechanical displacement not only dispersively modifies the cavity center frequency, but also modulates the cavity coupling (damping) rate. Therefore, circuit optomechanical systems can experience two competing backaction mechanisms. As a result, destructive quantum interference happens even in the side-band unresolved regime, permitting mechanical ground state cooling in the "bad"-cavity limit [66]. Although this new type of cooling scheme remains to be demonstrated in an experiment, devices with simultaneous dispersive optical force and reactive optical force were indeed described. The interference effect was shown in the classical domain by measuring the net optical force that applied to a mechanical beam coupled to a microdisk cavity. It was found that the reactive coupling can indeed dominate. Figure 9.4 shows the actual optomechanical device, which comprises a single mode waveguide and a microdisk resonator. The coupling portion of the waveguide is released to form a mechanical resonator. The optomechanical coupling between propagating photons in the single-mode waveguide and a cavity mode in the micro-disk can be described by a Hamiltonian, [63, 66] $H/\hbar = \omega_R(x)a^+a + \omega_M b^+b + (\omega L/c)n_g(x)a_{in}^+a_{in} + i\sqrt{2\kappa(x)}(a^+a_{in} - a_{in}^+a)$, where the first three terms represent the free Hamiltonian of the intra-cavity field (a), the mechanical resonator (b), and the propagating waveguide mode (a_{in}). The fourth term describes the transport of photons from the waveguide to the cavity, or the "driving term". At small displacement x, the Hamiltonian can be linearized to the first order of x and yield an optomechanical coupling Hamiltonian, $H_{int}/\hbar = \hat{x}(g_{om}a^+a + k_{om}a_{in}^+a_{in} + i\kappa_{om}\sqrt{1/2/\kappa(x)}(a^+a_{in} - a_{in}^+a))$. Here g_{om} is the dispersive coupling coefficient, and k_{om} is perturbation to the waveguide mode arising from gradient optical force. The new coupling term, $\kappa_{om} = d\kappa(x)/dx$ describes the reactive optomechanical coupling due to the modulation to the cavity decay rate.

The total optical force is then calculated from the static solution of the the linearized Langevin equation $F_s = -\hbar g_{om}|a_s|^2 - i\hbar\gamma_{om}\sqrt{1/2/\gamma}a_{in}(a_s^* - a_s) - \hbar k_{om}|c_s|^2$, where a_s is the stationary solution of of cavity field and a_in is the input field amplitude. Hence the total force applied on the waveguide has three components. The first term is associated with the backaction resulting from the intracavity photon energy change, $F_{cav} = -\hbar g_{om}|a_s|^2 = -2\kappa P_{in}g_{om}/\omega(\Delta^2 + \gamma^2)$, where γ is the cavity decay rate and Δ is the cavity detuning. This force is attractive since the resonance frequency will decrease as the waveguide moves towards the disk ($g_{om} = \partial g/\partial x > 0$). The second force term stems from the reactive coupling between the waveguide and the cavity, $F_{reactive} = -2\gamma P_{in}\kappa_{om}\Delta/(\Delta^2 + \gamma^2)$. When the waveguide moves towards the disk, the waveguide coupling rate κ will increase so $\kappa_{om} = \partial\kappa/\partial x < 0$. Therefore this reactive force is attractive when the cavity is red detuned ($\Delta < 0$) and repulsive when the cavity is blue detuned ($\Delta > 0$).

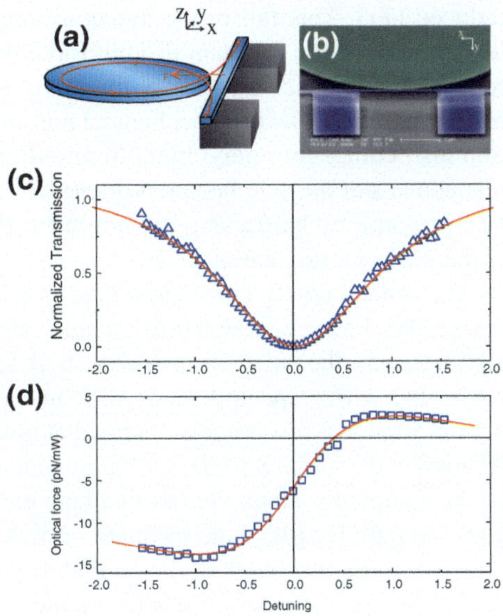

Fig. 9.4 **a** Schematic diagram of the microdisk-waveguide optomechanical system. **b** SEM image of the device. **c** The measured transmission of optical resonance. **d** The measured total optical force on the waveguide beam versus detuning ($\phi = \Delta/\gamma$)

The third term $F_{ev} = -k_{om}P_{in}/\omega$ constitutes a constant gradient optical force resulting from the perturbed waveguide mode due to the presence of the microdisk. Thus the total force applied at the waveguide normalized to the input optical power is given by, $f_{total} = f_{ev} - (g_{om} + 2\gamma_{om}\phi)/\omega\gamma(1+\phi^2)$, where critical coupling is assumed and $\phi = \Delta/\gamma$ is normalized detuning. All these different force terms were clearly manifested in the measurement data shown in Fig. 9.4d. By fitting the transmission curve and the measured force as a function of detuning, the parameters g_{om} and κ_{om} are found to be $g_{om}/2\pi = 2.0 \pm 0.4$ MHz/nm and $\kappa_{om}/2\pi = -26.6 \pm 0.5$ MHz/nm, indicating a dominant reactive force in this specific system. On the other hand, the offset in Fig. 9.4 gives a gradient optical force of -5.2 pN/mW.

9.3.4 Optomechanics at High Amplitude Regime

Circuit integration of optomechanical systems provides an important platform to manipulate mechanical resonators with optical forces. We already know that the interplay between the mechanical resonator and the cavity leads to damping and amplification of the motion of the resonator. In particular, the use of optical circuits to amplify mechanical motion is highly attractive for room temperature device applications. A fundamental question to answer is to what limit the optomechanical resonator motion can be amplified by the backaction force. In a traditional nanoscale mechanical system, it is difficult to reach the high-amplitude regime because the dynamic range of the system decreases dramatically as the dimensions of the resonator are reduced [71]. This fall in the dynamic range may be explained by the fact that as the strength of the resonant driving force increases, Duffing nonlinearities shift the resonance frequency away from the drive frequency, so the amplitude of the motion hardly increases. Besides mechanical nonlinearity, the cavity nonlinear transfer function also comes into play: high amplitude motion could drive the cavity to be off-resonance and the gain becomes quenched. Here, we show that these limitations can be overcome by harnessing the nonlinear dynamics of nanomechanical resonators embedded in a ring cavity.

The optical cavity considered here is a silicon waveguide in the form of a racetrack. The device is coupled to a nanoscale flexural resonator that is fabricated by etching away the substrate below a short length of the waveguide (Fig. 9.5a). The cavity has a free spectral range of 2 nm and a typical linewidth of 10 GHz. The nanomechanical resonator is clamped at both ends and has a fundamental resonance frequency of $\omega_m/2\pi\,8$ MHz, and a mechanical damping rate of $\Gamma_0/2\pi = 2.1$ kHz. In this coupled system, the out-of-plane motion of the mechanical resonator modulates the path length of the resonant optical field inside the cavity by modifying the effective refractive index of the optical waveguide as it moves towards or away from the substrate. The separation between the resonator and the substrate is 250 nm; this distance is small enough for the optical gradient forces between the resonator and the substrate to be quite strong ($g_{om} = d\omega_0/dx = 2\pi\,1\,\mathrm{GHz\,nm}^{-1}$), but it is also large enough to allow large-amplitude oscillations of the resonator. Owing to the residual compressive stress introduced by the SOI wafer bonding process, the resonators are slightly buckled and have two stable configurations at rest: the buckled up and buckled down states. This means that the out-of-plane motion of the resonator can be described by a double-well potential, with the up and down states corresponding to the two minima in the potential (Fig. 9.5e, f).

The two mechanical states can be discriminated in optical transmission measurements, since the effective refractive index is larger in the down state because the interaction between the optical field and the substrate is stronger (due to the reduced separation) than in the up state. The optical transmission spectrum is measured when the input optical power is well above threshold (600 µW) for self-sustained oscillations. These oscillations start when the laser becomes blue detuned with respect to the cavity when the resonator is in the up state, and the cavity resonant frequency ω_c oscillates back and forth with an amplitude $A_{pp}g$, where A_{pp} is the mechanical resonator oscillation amplitude. The self-sustained oscillations are observed for all laser frequencies between the resonant frequency of the cavity when the resonator is in the up state and the resonant frequency when the resonator is in the down state, which indicates that the energy of the resonator exceeds the energy barrier between the two states. The high-amplitude optomechanical system that we study has very distinctive nonlinear dynamic properties in both the mechanical and optical domains. In the conventional framework of cavity optomechanics, a blue-detuned pump laser resonantly enhances the sideband at $\omega_p - \Omega_m$ (the Stokes sideband), while the off-resonant sideband ($\omega_p + \Omega_m$, (Fig. 9.5a)) is suppressed. The imbalance between the Stokes and anti-Stokes photon generation rates results in net phonon emission, thereby amplifying the resonator motion. This dynamical backaction of the cavity on the resonator can be represented by a negative damping rate Γ_{BA}, which reduces the total mechanical damping rate $\Gamma = \Gamma_0 + \Gamma_{BA}$. When the pump power is large enough, Γ_{BA} can become so negative that the total damping rate vanishes, leading to regenerative oscillations sustained by the static pump power.

In cavity optomechanics, the resolved sideband regime (Fig. 9.5b) is regarded as the most efficient way to amplify the mechanical resonator. However, it is only possible for each pump photon to generate a single phonon in this regime, [19] and the unresolved sideband regime (Fig. 9.5c) is preferable for amplifying the motion of the resonator to reach the high-amplitude regime because each photon can generate many phonons [72, 73] (Fig. 9.2c). In a system operated in the unresolved sideband regime, the oscillations can actually shift the cavity resonance significantly beyond the cavity linewidth (that is, $A_{pp} \gg g/\gamma$; Fig. 9.5c. As the oscillations move the cavity out of resonance with the pump laser, the cavity is empty most of the time and the resonator undergoes damped harmonic motion. However, when the resonator motion brings the cavity into resonance with the pump laser, the cavity fills quickly with photons that generate (or absorb) large numbers of phonons in an avalanche process, exciting self-sustained oscillations at large amplitudes. In this work, we were able to experimentally excite the nanomechanical resonator to high amplitudes by using a deeply unresolved sideband cavity system ($g/2\pi = 10$ GHz vs. $\Omega_m/2\pi \sim 8$ MHz) in combination with a large cavity dynamic range ($\gamma/g = 10$ nm), so that very large oscillation amplitudes (>300 nm, equivalent to more than 10^{12} phonons) were obtained.

In this experiment, the mechanical resonator is able to gain so much energy from the optical field that it can overcome the double-well potential barrier and oscillate between the up and down states. When the pump laser is removed, the resonator can randomly relax into either the down state or up state. Deterministic selection of the final state is possible by applying a cooling laser to the red-side of the optical resonance of selected state. In this way, one can switch the mechanical device back and forth simply by setting the wavelength of the cooling laser, despite of a high energy barrier equivalent to 350,000,000 K which separates the two states.

9.4 Non-radiation Dynamics Processes in Silicon Optomechanical Circuits

While optomechanical interactions lead to a complex dynamic interplay between optical and mechanical modes, in silicon further material specific properties influence the vibration dynamics. In particular photothermal effects are significantly more pronounced compared to silica or silicon nitride, which results in a much larger thermo-optical coefficient. In addition strong two-photon absorption leads to the generation of charge carriers inside the material, influencing both the thermal and optical properties of nanophotonic components. In the following we will focus on free-carrier effects which start to manifest even at moderate optical intensities.

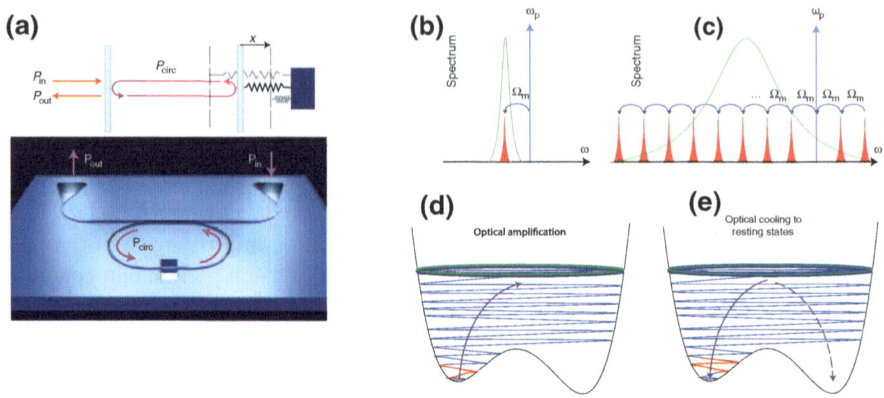

Fig. 9.5 **a** Circuit cavity optomechanical system with a nanomechanical resonator embedded in a racetrack-shaped waveguide that serves as the optical cavity. **b** Optomechanical amplification in the resolved sideband regime where only one phonon is exchanged between the optical field and the resonator. **c** In the unresolved sideband regime, when the line width of the cavity spectrum is much larger than Ω_m, many sidebands can lie inside the cavity linewidth, so more than one of them can be amplified. **e**, **f** Optomechanical amplification and cooling in the high amplitude regime where the mechanical device experiences a double-well potential

9.4.1 Free Carrier Effect and Thermal Instability

In silicon three absorption mechanisms are predominantly encountered: (1) bandgap or intrinsic absorption where the photon energy must be higher than the bandgap, (2) impurity level-to band absorption and (3) free carrier absorption, in which the photon energy is absorbed by free carriers in either conduction or valence band. Free carrier absorption usually does not result in any useful photoresponse and is thus a performance degrading mechanism. While free carrier absorption can generally be neglected in the intrinsic regime, it can become significant at longer wavelengths and high carrier concentrations.

In silicon waveguides the two major nonlinear optical loss mechanisms at communications wavelengths are two-photon absorption (TPA) and free-carrier absorption (FCA) [74, 75]. Two-photon absorption is a result of the proximity of the band edge to communication wavelengths and is unavoidable when operating in this wavelength range. Free-carriers are generated via TPA, and these carriers can efficiently absorb subsequent photons through linear absorption. For most demonstrations, the severity of the FCA mechanism is determined by the lifetime of the photo-excited free-carriers present in the silicon. The free-carrier lifetime is on the order of microseconds in lightly doped bulk silicon [76]. A number of techniques have been applied to reduce the free-carrier lifetime in silicon waveguides, including ion implantation [77, 78], geometric modification of the waveguide surface [79, 80] and carrier removal through the implementation of p–i–n diodes [80–83]. Using these approaches the carrier lifetime was reduced into the picosecond range, with negligible increase in propagation loss.

The generation of free carriers and subsequent optical absorption lead to a strong photothermal response in silicon nanophotonic waveguides [84]. These effects significantly impact the transport dynamics of optical microcavities. In particular, under pulsed excitation build-up of free carriers and heat in the waveguides leads to a beating oscillation of the cavity resonance frequency. Because the underlying nonlinear phenomena are governed by different characteristic time constants, they can be distinguished in a time-domain framework [85]. Nonlinearities due to carrier effects occur on a fast time-scale on the order of nano-seconds or even picoseconds. Thermally induced nonlinearities on the other hand occur on a microsecond scale in bulk optical materials. The interplay between both leads to a carrier-induced optical bistability effect, with nanosecond transition times in silicon ring resonators [86]. Free carriers generated by TPA induce a decrease of the refractive index of silicon and cause a blueshift of the resonance spectrum of the ring resonator. If the wavelength of the optical pulses is shorter than the resonance when no free carriers are present, this blueshift increases the optical coupling into the ring, which increases the carrier generation in the ring. The higher carrier generation in turn causes more blueshift of the resonance. This positive feedback process occurs until the blueshift induced by the input power is large enough that the wavelength of the input pulses becomes longer than the resonant wavelength of the ring. At this point, the previous positive feedback effect becomes a negative feedback effect, and the blueshifting process can be stabilized.

Because TPA is a strong effect in silicon, the optical bistability can be observed at relatively low input powers on a mW scale. The power requirement is essentially dependent on the optical quality factor of the resonator employed and can thus be further reduced for high Q cavities.

9.4.2 Relaxation Oscillation at Low Temperatures

While geometric modification and carrier removal can be employed to reduce the carrier lifetime, the temperature dependence of the carrier dynamics can also be exploited. When operating silicon photonic devices at cryogenic temperatures two effects lead to improved optical stability of the resonator: the carrier lifetime is significantly reduced to picosecond levels, while the thermo-optical response is much slower [87]. Hence the resonance condition of optical ring resonators is stable for much longer times at low temperatures and thus the resonator is photo-thermally stable on a 10s of nanoseconds timescale, during which the resonance wavelength does not shift significantly with respect to the optical linewidth.

9.5 Optomechanical Circuits in Other Material Systems

Silicon is a prime choice for the realization of integrated optomechanical circuits due to mature processing technology and the superior crystalline structure of commercially available substrates. However, the presence of strong two-photon absorption and a relatively small electronic bandgap restrict the use of silicon for high-intensity optical drive as well as applications in the visible wavelength range. Therefore several alternative material systems are commonly employed. Because in integrated optical circuits waveguiding requires buffer layers of lower refractive index above and below the actual photonic layer, materials with relatively high refractive index are used. In the following we will consider two nitride based compounds.

9.5.1 Silicon Nitride

Silicon nitride offers optical transparency above 350 nm into the infrared wavelength regime. To date silicon nitride is deposited onto oxidized silicon wafers using chemical vapour deposition (CVD). Because the underlying thermal oxide is amorphous, the resulting nitride layer does not provide a crystallographic preference orientation and therefore the material is isotropic. The most commonly used CVD methods rely on low-pressure deposition at higher temperatures (LPCVD) or on plasma-enhanced CVD (PECVD). Depending on the deposition method used, the refractive index varies between 2.0 and 2.2. Furthermore, by controlling the growth conditions the internal stress in the nitride film can be adjusted, from both compressive stress to highly tensile stressed films. Tensile silicon nitride thin films are of particular importance for integrated optomechanics, because they allow the preparation of the mechanical resonators with ultra-high quality factors [88]. Due to the high internal stress, however, the achievable thickness of the nitride layer is limited.

Photonic circuitry in silicon nitride is realized following the same design methodology as described above. Photonic waveguides and cavities are defined by electron beam lithography and subsequent dry etching. When mechanical structures are needed, the underlying oxide is removed in hydrofluric acid, leading to free-standing structures that can be actuated via the gradient optical force. Initial optomechanical resonators consisted of coupled microring resonators, anchored on a central pedestal through thin spokes [70]. The resulting wheel structure was also previously used to define optomechanical resonators in silica glass with improved mechanical quality factors [89]. In the design employed by M. Lipson's group at Cornell University, two wheel resonators are stacked vertically, separated by a nanogap. Evanescent coupling of the whispering-gallery modes in the rings leads to strong optical force actuation on resonance. Depending on the symmetry of the optical modes the resulting force can be either attractive or repulsive [18], thus decreasing or increasing the gap between the rings. This results in large optical tuning of the resonance wavelength at moderate optical input power. By adjusting the detuning from the resonance wavelength at fixed optical input power the gap between the wheel resonators was decreased by 20 nm, corresponding to a wavelength tuning of 2 nm.

While wheel resonators driven in their flexural mechanical modes allow for maintaining high optical quality factors, the mechanical properties of optomechanical resonators can be much improved by relying on clamped resonators. When processing nanoscale beams from highly stressed, stoichiometric silicon nitride films, ultra-low mechanical dissipation can be readily achieved, leading to mechanical quality factors in excess of 1 million. In order to employ such resonators in integrated optomechanical circuits, the beam resonators can either be directly used as optical waveguides or coupled evanescently to suitable optical cavities. A configuration where evanescent coupling is employed is shown in Fig. 9.6a. Here a long, free-standing beam resonator is coupled evanescently to an on-chip Mach-Zehnder interferometer. The beam is separated from the upper MZI arm by a nanogap. Due to the evanescent coupling the effective refractive index of the mode propagating inside the MZI is dependent on the distance between the beam and the waveguide. A modulation of the refractive index due to mechanical movement of the beam therefore corresponds to a modulation of the phase delay of the MZI arm. This phase modulation can be read out with high sensitivity in the optical transmission signal.

As mentioned above, the mechanical beams show very high mechanical quality factors in vacuum at room temperature. As shown in Fig. 9.6b, the dissipation can be further significantly reduced by cooling the device down to cryogenic temperatures [90]. For the device shown in Fig. 9.6a an increase in mechanical Q by a factor of ten is observed. Even higher Q factors are feasible by increasing the length of the resonator.

In addition to sensitive motion readout the high internal stress of silicon nitride enables the fabrication of ultralong nanomechanical resonators. For optical purposes such devices offer the possibility of realizing tunable photonic components. A fundamental optical building block is here a tunable directional coupler [91]. Here the coupling ratio is controlled by the gap between evanescently coupled nanophotonic waveguides. By controlling the separation between the waveguides via gradient optical forces an all-optically tunable device can therefore be fabricated. By increasing the optical power inside the slot waveguide and thus increasing the optical attraction between the waveguides the coupling ratio can be continuously adjusted from 0 to 1.

9.5.2 Aluminum Nitride

While silicon nitride offers optical transparency throughout the visible wavelength spectrum, even smaller wavelengths can be employed in nitride compounds with a larger bandgap. Among the available candidates aluminum nitride (AlN) offers the widest bandgap of all semiconductors (6.12 eV [92]), corresponding to a transparency window spanning from 220 nm to 13.6 μm. In addition, AlN offers attractive optical properties, such as low material absorption, second order nonlinearity and the piezoelectric effect.

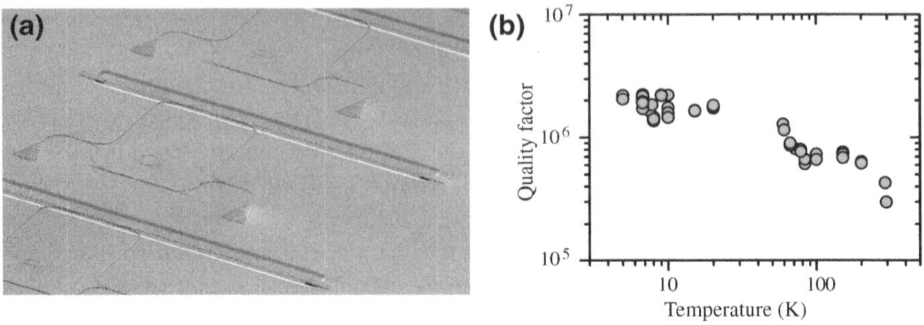

Fig. 9.6 **a** SEM image of a Silicon Nitride optomechanical circuit, comprising an on-chip Mach-Zehnder interferometer and a long free-standing nanobeam coupled evanescently to the top MZI arm. Grating coupler input ports in analogy to silicon focussing grating couplers are used to couple light into the circuit. **b** Due to the high internal tensile stress, ultra-high quality factors can be obtained. By operating the circuit at low temperatures the mechanical Q increases roughly by factor of ten when proceeding to liquid Helium temperatures

The refractive index of AlN in the telecoms window is similar to silicon nitride and amounts to roughly 2.1. Therefore the design procedures developed for silicon nitride can be transferred to an AlN photonic platform with little adaptation. Furthermore, AlN can be sputter-deposited onto suitable substrates on a wafer-scale. This procedure holds promise for low-cost, larger area fabrication of wideband optical integrated circuits.

Using electron beam lithography and subsequent dry etching, high quality photonic circuits can be fabricated. Because the AlN thin film is sputter-deposited, the waveguiding layer does not contain a crystallographic preference orientation. Therefore structural anisotropy due to preferential etching along different crystal planes is not observed and hence smooth waveguide sidewalls can be obtained. For optical microring cavities optical quality factors up to 600,000 have been measured [93], which corresponds to propagation loss of 0.6 dB/cm. Similarly, in one-dimensional photonic crystal cavities high optical Q factors in excess of 10^5 were measured [94], illustrating that also in small modal volumes low material loss can be sustained. A fabricated device is shown in in the SEM image in Fig. 9.7a. The image shows the one-dimensional photonic crystal nanobeam coupled to a feeding waveguide. The feeding waveguide is laid out in a ring pattern in order to provide a point contact of coupling. Mechanical clamping of the beam waveguide is achieved by using a two-step etching procedure, leaving a thin ridge everywhere except for the release window. After wet-chemical undercut the beam is freestanding, but still mechanically clamped at the ridge support.

The ability to fabricate high quality optical cavities in AlN provides the basis for the realization of optomechanical resonators. Relying on the wheel structure introduced above high frequency mechanical resonators can be implemented [12]. In order to access high frequency resonances, radial breathing modes can be employed when the ring width is sufficiently wide. As a result the wet-chemical undercut needs to

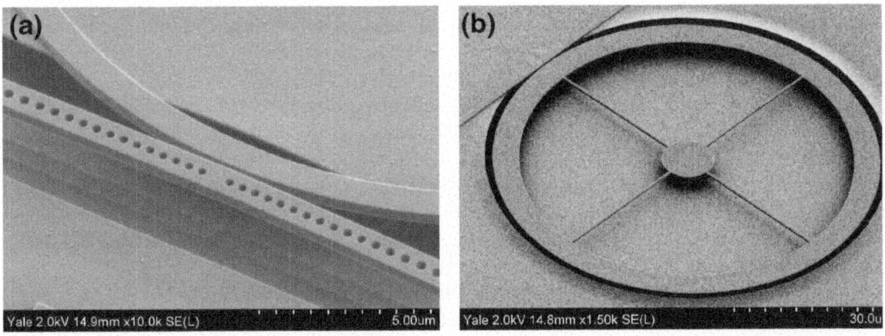

Fig. 9.7 **a** SEM image of an Aluminum Nitride optomechanical device, consisting of a one-dimensional photonic crystal nanobeam cavity coupled to a feeding waveguide. The beam resonator is mechanically clamped through a thin AlN ridge. **b** A SEM image of an AlN optomechanical wheel resonator coupled to a feeding waveguide. In order to realize defined undercut the two-step etching procedure used for the structure in (**a**) is also employed, leading to a wide ring resonator supported by spokes and a central pedestal

be performed for sufficiently long times. In order to obtain clearly defined photonic structures the two-step etching procedure is also employed in this case. A released device is shown in Fig. 9.7b, illustrating the free-standing wheel structure and the feeding waveguide in the top of the image. The thin ridge support is clearly visible as an inherent etch mask for the wet release. For the fabricated devices optical quality factors on the order of 125,000 are routinely obtained for critically coupled devices. When measuring the mechanical spectrum of the resonator, high frequency modes in excess of 1 GHz are obtained, corresponding to the wine-glass mode of the ring structure. At the same time, good mechanical quality factors above 2,000 are measured. For the high frequency mode displacement sensitivity of 6.2×10^{-18} m/Hz$^{1/2}$ is obtained. Such optomechanical resonators are suitable for applications in precision oscillators and high speed ultra-sensitive systems. Furthermore, the optomechanical coupling can be further enhanced by employing rings with smaller radii, because the coupling strength $g_{om} = d\omega/dR$ is inversely proportional to the radius R. Alternatively, even higher mechanical Q can be obtained by mechanical optimization of the radiation loss resulting from the supporting spokes [13]. This will allow for operation of optomechanical high frequency resonators under ambient conditions.

References

1. M. Eichenfield, R. Camacho, J. Chan, K. Vahala, O. Painter, Nature **459**(7246), 550 (2009)
2. M. Eichenfield, J. Chan, R. Camacho, K. Vahala, O. Painter, Nature **462**(7269), 78 (2009)
3. M. Li, W. Pernice, C. Xiong, T. Baehr-Jones, M. Hochberg, H. Tang, Nature **456**(7221), 480 (2008)
4. X. Sun, J. Zheng, M. Poot, C.W. Wong, H.X. Tang, Nano Lett. **12**(5), 2299 (2012)

5. P. Dainese, P.S.J. Russell, N. Joly, J. Knight, G. Wiederhecker, H.L. Fragnito, V. Laude, A. Khelif, Nat. Phys. **2**(6), 388 (2006)
6. M. Kang, A. Butsch, P.S.J. Russell, Nat. Photonics **5**(9), 549 (2011)
7. A. Butsch, C. Conti, F. Biancalana, P.S.J. Russell, Phys. Rev. Lett. **108**(9), 093903 (2012)
8. A. Butsch, M. Kang, T. Euser, J. Koehler, S. Rammler, R. Keding, P.S.J. Russell, Phys. Rev. Lett. **109**(18), 183904 (2012)
9. E. Verhagen, S. Deleglise, S. Weis, A. Schliesser, T.J. Kippenberg, Nature **482**(7383), 63 (2012)
10. O. Arcizet, P. Cohadon, T. Briant, M. Pinard, A. Heidmann, Nature **444**(7115), 71 (2006)
11. A. Schliesser, G. Anetsberger, R. Riviere, O. Arcizet, T. Kippenberg, New J. Phys. **10** (2008)
12. C. Xiong, X. Sun, K.Y. Fong, H.X. Tang, Appl. Phys. Lett. **100**(17) (2012)
13. X. Sun, K. Fong, C. Xiong, W. Pernice, H. Tang, Opt. Express **19**(22), 22316 (2011)
14. O. Arcizet, P.F. Cohadon, T. Briant, M. Pinard, A. Heidmann, J.M. Mackowski, C. Michel, L. Pinard, O. Français, L. Rousseau, Phys. Rev. Lett. **97**, 133601 (2006)
15. L.S. Collaboration, Nat. Phys. **7**(12), 962 (2011)
16. X. Zhang, X. Sun, H.X. Tang, Opt. Lett. (2012)
17. M. Eichenfield, C.P. Michael, R. Perahia, O. Painter, Nat. Photonics **1**(7), 416 (2007)
18. M. Li, W. Pernice, H. Tang, Nat. Photonics **3**(8), 464 (2009)
19. T.J. Kippenberg, H. Rokhsari, T. Carmon, A. Scherer, K.J. Vahala, Phys. Rev. Lett. **95**(3) (2005)
20. R. Lifshitz, M. Cross, *Reviews of Nonlinear Dynamics and Complexity* (Wiley, New York, 2008), pp. 1–52
21. A. Dorsel, J.D. McCullen, P. Meystre, E. Vignes, H. Walther, Phys. Rev. Lett. **51**(17), 1550 (1983)
22. S. Assefa, F. Xia, Y.A. Vlasov, Nature **464**(7285), 80 (2010)
23. http://depts.washington.edu/uwopsis/
24. F. Van Laere, G. Roelkens, M. Ayre, J. Schrauwen, D. Taillaert, D. Van Thourhout, T. Krauss, R. Baets, J. Lightwave Technol. **25**(1), 151 (2007)
25. F. Van Laere, G. Roelkens, J. Schrauwen, D. Taillaert, P. Dumon, W. Bogaerts, D. Van Thourhout, R. Baets, in *Optical Fiber Communication Conference* (Optical Society of America, 2006)
26. A. Narasimha, E. Yablonovitch, in *CLEO* (Optical Society of America, 2003)
27. S. Scheerlinck, J. Schrauwen, D. Taillaert, D. Van Thourhout, R. Baets, in *CLEO* (Optical Society of America, 2007)
28. M. Matsumoto, IEEE J. Quantum Electron. **28**(10), 2016 (1992)
29. F. Van Laere, M. Kotlyar, D. Taillaert, D. Van Thourhout, T. Krauss, R. Baets, IEEE Photonics Technol. Lett. **19**(6), 396 (2007)
30. X. Chen, C. Li, H. Tsang, in *IEEE/LEOS International Conference on Optical MEMs and Nanophotonics* (IEEE, 2008), pp. 56–57
31. X. Chen, C. Li, H. Tsang, IEEE Photonics Technol. Lett. **20**(23), 1914 (2008)
32. D. Taillaert, F. Van Laere, M. Ayre, W. Bogaerts, D. Van Thourhout, P. Bienstman, R. Baets, Jpn. J. Appl. Phys. **45**(8A), 6071 (2006)
33. C. Gunn, in *IEEE Symposium on VLSI Technology* (IEEE, 2007), pp. 6–9
34. F. Van Laere, T. Stomeo, D. Taillaert, G. Roelkens, D. Van Thourhout, T. Krauss, R. Baets, IEEE Photonics Technol. Lett. **20**(4), 318 (2008)
35. L. Bobb, P. Shankar, Microwave J. **35**, 218 (1992)
36. J. Knight, G. Cheung, F. Jacques, T. Birks, Opt. Lett. **22**(15), 1129 (1997)
37. P. Barclay, K. Srinivasan, M. Borselli, O. Painter, Opt. Lett. **29**(7), 697 (2004)
38. J. Bures, R. Ghosh, JOSA A **16**(8), 1992 (1999)
39. Z. Wang, R. Gao, J. Gole, J. Stout, Adv. Mater. **12**(24), 1938 (2000)
40. L. Tong, R. Gattass, J. Ashcom, S. He, J. Lou, M. Shen, I. Maxwell, E. Mazur et al., Nature **426**(6968), 816 (2003)
41. C. Grillet, C. Smith, D. Freeman, S. Madden, B. Luther-Davies, E. Magi, D. Moss, B. Eggleton, Opt. Express **14**(3), 1070 (2006)
42. I. Hwang, S. Kim, J. Yang, S. Kim, S. Lee, Y. Lee, Appl. Phys. Lett. **87**(13), 131107 (2005)
43. I. Hwang, G. Kim, Y. Lee, IEEE J. Quantum Electron. **42**(2), 131 (2006)

44. C. Michael, M. Borselli, T. Johnson, C. Chrystal, O. Painter, Opt. Express **15**(8), 4745 (2007)
45. G. Masanovic, G. Reed, W. Headley, B. Timotijevic, V. Passaro, R. Atta, G. Ensell, A. Evans, Opt. Express **13**(19), 7374 (2005)
46. J. Leuthold, J. Eckner, E. Gamper, P. Besse, H. Melchior, J. Lightwave Technol. **16**(7), 1228 (1998)
47. B. Offrein, G. Bona, R. Germann, I. Massarek, D. Erni et al., J. Lightwave Technol. **16**(9), 1680 (1998)
48. K. Kasaya, O. Mitomi, M. Naganuma, Y. Kondo, Y. Noguchi, IEEE Photonics Technol. Lett. **5**(3), 345 (1993)
49. G. Vawter, C. Sullivan, J. Wendt, R. Smith, H. Hou, J. Klem, IEEE J. Sel. Top. Quantum Electron. **3**(6), 1361 (1998)
50. R. Hauffe, U. Siebel, K. Petermann, R. Moosburger, J. Kropp, F. Arndt, in *Proceedings of 50th Electronic Components and Technology Conference* (IEEE, 2000), pp. 238–243
51. T. Alder, A. Stohr, R. Heinzelmann, D. Jager, IEEE Photonics Technol. Lett. **12**(8), 1016 (2000)
52. V. Almeida, R. Panepucci, M. Lipson, Opt. Lett. **28**(15), 1302 (2003)
53. M. Soltani, S. Yegnanarayanan, A. Adibi, Opt. Express **15**(8), 4694 (2007)
54. A. Gondarenko, J. Levy, M. Lipson, Opt. Express **17**(14), 11366 (2009)
55. E. Shah Hosseini, S. Yegnanarayanan, A. Atabaki, M. Soltani, A. Adibi, Opt. Express **17**(17), 14543 (2009)
56. M. Tien, J. Bauters, M. Heck, D. Spencer, D. Blumenthal, J. Bowers, Opt. Express **19**(14), 13551 (2011)
57. M. Bagheri, M. Poot, M. Li, W.P.H. Pernice, H.X. Tang, Nat. Nanotechnol. **6**(11), 726 (2011)
58. V. Almeida, Q. Xu, C. Barrios, M. Lipson, Opt. Lett. **29**(11), 1209 (2004)
59. M. Li, W.H.P. Pernice, H.X. Tang, Appl. Phys. Lett. **97**(18), 183110 (2010)
60. C. Xiong, W. Pernice, M. Li, H. Tang, Opt. Express **18**(20), 20690 (2010)
61. C. Xiong, W. Pernice, M. Li, M. Rooks, H.X. Tang, Appl. Phys. Lett. **96**(26), 263101 (2010)
62. W. Pernice, M. Li, H. Tang, Opt. Express **17**(15), 12424 (2009)
63. M. Li, W. Pernice, H. Tang, Phys. Rev. Lett. **103**(22), 223901 (2009)
64. L. Ding, C. Baker, P. Senellart, A. Lemaitre, S. Ducci, G. Leo, I. Favero, Appl. Phys. Lett. **98**(11), 113108 (2011)
65. X. Sun, X. Zhang, H. Tang, Appl. Phys. Lett. **100**(17), 173116 (2012)
66. F. Elste, S. Girvin, A. Clerk, Phys. Rev. Lett. **102**(20) (2009)
67. M.L. Povinelli, M. Loncar, M. Ibanescu, E.J. Smythe, S.G. Johnson, F. Capasso, J.D. Joannopoulos, Opt. Lett. **30**(22), 3042 (2005)
68. A. Mizrahi, L. Schachter, Opt. Express **13**(24), 9804 (2005)
69. W. Pernice, M. Li, H. Tang, Opt. Express **17**(3), 1806 (2009)
70. G.S. Wiederhecker, L. Chen, A. Gondarenko, M. Lipson, Nature **462**(7273), 633 (2009)
71. H.W.C. Postma, I. Kozinsky, A. Husain, M.L. Roukes, Appl. Phys. Lett. **86**(22) (2005)
72. F. Marquardt, J. Harris, S. Girvin, Phys. Rev. Lett. **96**(10) (2006)
73. C. Metzger, I. Favero, A. Ortlieb, K. Karrai, Phys. Rev. B **78**(3) (2008)
74. Q. Lin, J. Zhang, P. Fauchet, G. Agrawal, Opt. Express **14**(11), 4786 (2006)
75. R. Jones, H. Rong, A. Liu, A. Fang, M. Paniccia, D. Hak, O. Cohen, Opt. Express **13**(2), 519 (2005)
76. D. Dimitropoulos, R. Jhaveri, R. Claps, J.C.S. Woo, B. Jalali, Appl. Phys. Lett. **86**(7), 071115 (2005)
77. Y. Liu, H. Tsang, Opt. Lett. **31**(11), 1714 (2006)
78. N. Wright, D. Thomson, K. Litvinenko, W. Headley, A. Smith, A. Knights, F. Gardes, G. Mashanovich, R. Gwilliam, G. Reed, in *5th IEEE International Conference on Group IV Photonics* (IEEE, 2008), pp. 122–124
79. T. Johnson, O. Painter, in *CLEO* (Optical Society of America, 2009)
80. S. Preble, Q. Xu, B. Schmidt, M. Lipson, Opt. Lett. **30**(21), 2891 (2005)
81. H. Rong, A. Liu, R. Jones, O. Cohen, D. Hak, R. Nicolaescu, A. Fang, M. Paniccia, Nature **433**(7023), 292 (2005)

82. H. Rong, R. Jones, A. Liu, O. Cohen, D. Hak, A. Fang, M. Paniccia, Nature **433**(7027), 725 (2005)
83. A. Turner-Foster, M. Foster, J. Levy, C. Poitras, R. Salem, A. Gaeta, M. Lipson, Opt. Express **18**(4), 3582 (2010)
84. W. Pernice, M. Li, H. Tang, Opt. Express **18**(17), 18438 (2010)
85. M. Foster, A. Turner, J. Sharping, B. Schmidt, M. Lipson, A. Gaeta, Nature **441**(7096), 960 (2006)
86. Q. Xu, M. Lipson, Opt. Lett. **31**(3), 341 (2006)
87. W. Pernice, C. Schuck, M. Li, H. Tang, Opt. Express **19**(4), 3290 (2011)
88. S.S. Verbridge, H.G. Craighead, J.M. Parpia, Appl. Phys. Lett. **92**(1), 013112 (2008)
89. G. Anetsberger, R. Rivière, A. Schliesser, O. Arcizet, T.J. Kippenberg, Nat. Photonics **2**(10), 627 (2008)
90. K. Fong, W. Pernice, H. Tang, Phys. Rev. B **85**(16), 161410(R) (2012)
91. K. Fong, W. Pernice, M. Li, H. Tang, Opt. Express **19**(16), 15098 (2011)
92. J. Li, K.B. Nam, M.L. Nakarmi, J.Y. Lin, H.X. Jiang, P. Carrier, S.H. Wei, Appl. Phys. Lett. **83**(25), 5163 (2003)
93. C. Xiong, W.H.P. Pernice, H.X. Tang, Nano Letters **0**(0), null (0). doi:10.1021/nl3011885
94. W.H.P. Pernice, C. Xiong, C. Schuck, H.X. Tang, Appl. Phys. Lett. **100**(9), 091105 (2012)

Chapter 10
Optomechanical Crystal Devices

Amir H. Safavi-Naeini and Oskar Painter

Abstract We present the basic ideas and techniques utilized in recent work on optomechanical crystals. Optomechanical crystals are nanofabricated cavity optomechanical systems where the confinement of light and motion is obtained by nanopatterning periodic structures in thin-films. In this chapter we start from a basic review of the properties of optical and elastic waves in nanostructures, before introducing the properties and design of periodic structures. After reviewing fabrication and characterization methods, experimental results in 1D and 2D systems are presented.

10.1 Introduction

Nanofabricated devices and systems comprise a major recent strand of research in the field of cavity optomechanics. The first motivation pushing these developments is the promise of lower mass resonators with significantly larger optical and mechanical field localization. These structures would in principle offer larger optomechanical coupling strengths than would be typically possible in a Fabry-Pérot geometry. A second motivation arises from opportunities in development of more complex optomechanical systems with many elements. In addition to basic linear information conversion [1] and storage [2], these systems would allow observation

A. H. Safavi-Naeini (✉) · O. Painter
California Institute of Technology, Pasadena, USA
e-mail: safavi@stanford.edu

A. H. Safavi-Naeini
ETH Zürich, Zürich, Switzerland

A. H. Safavi-Naeini
Stanford University, Stanford, USA

O. Painter
The Max Planck Institute for the Science of Light (MPL), Erlangen, Germany
e-mail: opainter@caltech.edu

M. Aspelmeyer et al. (eds.), *Cavity Optomechanics*, Quantum Science and Technology,
DOI: 10.1007/978-3-642-55312-7_10,

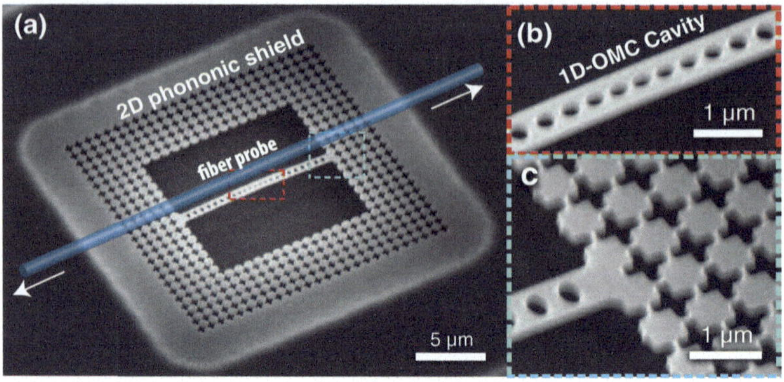

Fig. 10.1 **a** A scanning electron microscope (SEM) image of an optomechanical crystal cavity formed from the *top* Si layer of a silicon-on-insulator chip. The device consists of **b** a 1D-OMC cavity embedded in **c** a phononic crystal shield. Coupling to the 1D-OMC cavity is achieved by bringing a fiber taper probe (illustrated in **a**) into the near field of the optical cavity

of nontrivial collective dynamics [3, 4], as well as realization of novel many-body interactions involving spatially separated phonons and photons [5]. To control and engineer phonons, photons, and their interactions in a scalable manner, development of easily realizable structures providing control over localized excitations on the surface of a chip is required.

In this chapter one approach to the development of such structures—called "optomechanical crystals" (OMCs) is outlined. A typical optomechanical crystal system used in several recent experiments [6–10] is shown in Fig. 10.1. A pattern has been etched into a thin-film of silicon (Si) with the spacing and shape of the holes designed to give rise to localized optical and mechanical resonances in the center of the 1D-OMC beam. We will show in this chapter that the interaction between these resonances is described well by the optomechanical Hamiltonian. To understand the operation of these devices, we introduce the theoretical and experimental framework required to design, fabricate, and characterize OMC systems. This chapter is intended to give the reader an understanding of the function and design of every element of the OMC device shown in Fig. 10.1. Our main goal is to outline the device physics, and thus details of the experiments and potential applications of optomechanical crystals are not presented. For a thorough investigation of these details and the effects of quantum and classical noise in OMC experiments, we refer the reader to Ref. [11].

We start in this section by reviewing the quantized theory of photons and phonons in structures. In Sect. 10.2.1–10.2.2 the role of periodicity and symmetry in engineering the spectrum and interactions of light and sound in a solid-state setting are studied. In Sect. 10.2.3 we see that by engineering local modifications in such substrates, optical and mechanical excitations can be localized in 0- and 1-dimensional defects, i.e. optomechanical crystal cavities and waveguides. Recipes and formulas for calculating the couplings between phonons and photons are presented in Sect. 10.2.4. The experimental demonstrations of these ideas are reviewed in Sect. 10.3, where after a brief review of fabrication and optical probing techniques, we turn our focus onto "nanobeam" (1D-OMC) and "snowflake" (2D-OMC) structures in Sect. 10.3.3. Finally, in Sect. 10.4, we conclude this chapter by presenting future developments that are expected to allow these engineered structures to realize their full potential.

First we begin with a review of the classical field theory of light and motion in a solid. The artificially patterned nanostructures studied in this chapter do not permit exact closed-form solutions for the electromagnetic and displacement fields and their resonant frequencies. Instead, computational software packages such as COMSOL [12] and MPB [13], developed to understand the *classical* electromagnetics and elastodynamics of solids, are used. The quantization of these fields, which is treated below, allows us to make a link between the computational methods developed over decades for classical engineering applications, and the quantum optomechanical Hamiltonian.

10.1.1 Maxwell's Equations for a Photonic Resonator and Second Quantization

Optically, we describe the nanofabricated structure with an inhomogeneous dielectric tensor $\overline{\overline{\varepsilon}}(\boldsymbol{r})$. In most of this work, $\overline{\overline{\varepsilon}}(\boldsymbol{r}) = \varepsilon_{\mathrm{Si}}$ in regions containing Si, and $\overline{\overline{\varepsilon}}(\boldsymbol{r}) = \varepsilon_0$ otherwise. The time-harmonic Maxwell's equations for such a system are given by:

$$\text{curl}\, \boldsymbol{E}(\boldsymbol{r}) = i\omega\mu_0 \boldsymbol{H}(\boldsymbol{r}), \qquad \text{curl}\, \boldsymbol{H}(\boldsymbol{r}) = -i\omega\overline{\overline{\varepsilon}}(\boldsymbol{r}) \cdot \boldsymbol{E}(\boldsymbol{r}). \tag{10.1}$$

In the absence of a source, these equations lead to self-sustaining fields, or modes, represented by transverse (or solenoidal) and longitudinal (irrotational) eigenvectors, with the eigenvalue problem for the transverse magnetic fields given by

$$\mathsf{L}\boldsymbol{h}_j = \omega_j^2 \boldsymbol{h}_j, \qquad \mathsf{L}(\cdot) = c^2 \text{curl}\left[\frac{\varepsilon_0}{\overline{\overline{\varepsilon}}(\boldsymbol{r})} \text{curl}\,(\cdot)\right]. \tag{10.2}$$

Additionally, for completeness, the irrotational fields given by $\boldsymbol{g}_j = \text{grad}\, \psi_j$, satisfying appropriate boundary conditions, where ψ_m is an eigenvector of div grad $\psi_j = -\nu_j \psi_j$ must also be considered [14], though these solutions don't play a role in the optical response of the system. This normal mode prescription is always valid for a volume enclosed inside a perfectly reflective boundary. In the limit of an open system, such as the dielectric cavities presented here, complications arise since the resonances are typically not normal modes, but so-called 'quasi-normal' resonances. We use a finite-element-method (FEM) software package, COMSOL [12] to solve Eq. (10.2) with open or 'scattering' boundary conditions enclosing the simulation space. The very small loss rates for the obtained resonances, informs us of the validity of the normal-mode approximation which is used throughout this work. As such, we take the numerically calculated high-Q solutions with frequency ω_j and field profile $\boldsymbol{h}_j(\boldsymbol{r})$ to be normal modes of the structures, and then add in losses 'by hand' (by assuming linear coupling to a bath of linear oscillators) to take into account the effects of a finite Q.

Quantization of the transverse electromagnetic field is accomplished in the standard way (see for example Ref. [15]) by associating bosonic creation and annihilation operators, $\hat{a}_j^\dagger$ and $\hat{a}_j$ respectively, with each modal solution $\{\boldsymbol{h}_j(\boldsymbol{r}), \boldsymbol{e}_j(\boldsymbol{r})\}$ of Maxwell's equation. The field operators are expressed in the Heisenberg picture as

$$\hat{\boldsymbol{H}}(\boldsymbol{r}) = \sum_j \boldsymbol{h}_j(\boldsymbol{r})\hat{a}_j e^{-i\omega_j t} + \text{h.c.}, \tag{10.3}$$

$$\hat{\boldsymbol{E}}(\boldsymbol{r}) = \sum_j \boldsymbol{e}_j(\boldsymbol{r})\hat{a}_j e^{-i\omega_j t} + \text{h.c.}, \tag{10.4}$$

and we obtain from Eq. (10.1) that $\boldsymbol{e}_j(\boldsymbol{r}) = i\omega^{-1}\overline{\overline{\varepsilon^{-1}}}(\boldsymbol{r}) \cdot \text{curl}\, \boldsymbol{h}_j(\boldsymbol{r})$. To calculate the proper normalization of the field profiles, we assume a single photon state $|\psi\rangle = |1\rangle_j \prod_{\forall k \neq j} |0\rangle_k$ and find the expected value of additional field energy above vacuum ($|\text{vac}\rangle = \prod_k |0\rangle_k$):

$$\begin{aligned} U_{\text{em}} &= \langle\psi| \int \text{d}\boldsymbol{r}\ \hat{\boldsymbol{E}}(\boldsymbol{r})\overline{\overline{\varepsilon}}(\boldsymbol{r})\hat{\boldsymbol{E}}(\boldsymbol{r})|\psi\rangle - \langle\text{vac}| \int \text{d}\boldsymbol{r}\ \hat{\boldsymbol{E}}(\boldsymbol{r})\overline{\overline{\varepsilon}}(\boldsymbol{r})\hat{\boldsymbol{E}}(\boldsymbol{r})|\text{vac}\rangle \\ &= 2\int \text{d}\boldsymbol{r}\ \boldsymbol{e}_j^*(\boldsymbol{r})\overline{\overline{\varepsilon}}(\boldsymbol{r})\boldsymbol{e}_j(\boldsymbol{r}) \\ &= 2V_{\text{eff}}\max[\boldsymbol{e}_j^*(\boldsymbol{r})\overline{\overline{\varepsilon}}(\boldsymbol{r})\boldsymbol{e}_j(\boldsymbol{r})]. \end{aligned} \tag{10.5}$$

Assuming a maximum field amplitude inside the isotropic dielectric with index $\varepsilon_{\text{diel}}$, and $U_{\text{em}} = \hbar\omega_j$, we obtain the maximum single-photon field

$$\max[|\boldsymbol{e}_j(\boldsymbol{r})|] = \sqrt{\frac{\hbar\omega_j}{2V_{\text{eff},j}\varepsilon_{\text{diel}}}}, \tag{10.6}$$

where we have defined the effective mode volume for mode j to be

$$V_{\text{eff},j} = \frac{\int \text{d}\boldsymbol{r}\ \boldsymbol{e}_j^*(\boldsymbol{r})\overline{\overline{\varepsilon}}(\boldsymbol{r})\boldsymbol{e}_j(\boldsymbol{r})}{\max[\boldsymbol{e}_j^*(\boldsymbol{r})\overline{\overline{\varepsilon}}(\boldsymbol{r})\boldsymbol{e}_j(\boldsymbol{r})]}. \tag{10.7}$$

In many of the structures demonstrated in this work, the maximum field amplitude for a single photon can be surprisingly large, reaching values of 10^5 V/m *per photon*.

10.1.2 Mechanical Waves and Their Quantization

Analyzing mechanical vibrations, and their quantum description in terms of phonons in a crystal lattice can be approached in two ways. The first begins with the atomic structure of the crystal, the forces sensed by each ion, and studies the motion of these ions. The interested reader is referred to any of a wide array of excellent textbooks on condensed matter physics for such a treatment [16]. A second approach, which is less general,[1] as it only applies to phonons with wavelengths much longer than the atomic spacing, involves starting with an effective continuum mechanics or elastodynamics description. This is the approach we take, as the motion of single ions is not of interest in this work. This is conceptually no different than the approach taken in Sect. 10.1.1 where the macroscopic electromagnetic fields are quantized and the motion of the charges in the structure are only considered in terms of how they contribute to the electric susceptibility of the material at optical frequencies. At this continuum limit, the material is characterized by an elasticity tensor $\overline{\overline{c}}$ (a fourth-rank tensor with components c_{ijkl} which depend on the Young's modulu E, and Poisson's ratio ν), its density $\rho(\boldsymbol{r})$, and its dynamical state represented by a time-dependent displacement vector field $\boldsymbol{Q}(\boldsymbol{r}, t)$ often also denoted $\boldsymbol{u}(\boldsymbol{r}, t)$ in the literature.

Fundamentally, it is the strain, a measure of the local deformation in a structure which is of interest. Local deformations arise from the spatial variation of $\boldsymbol{Q}(\boldsymbol{r}, t)$, and are found by taking the derivative of this vector field. This total derivative is symmetrized to do away with rotations, and is represented by a unitless 3×3 matrix with components

$$S_{ij} = \frac{1}{2}\left(\partial_i Q_j + \partial_j Q_i\right). \tag{10.8}$$

Stress in a structure is also a 3×3 matrix T, which gives for every infinitesimal volume element in the structure the local forces which act on its surfaces. For a surface with normal vector $\hat{n}$, this force is $T \cdot \hat{n}$. Hooke's law, i.e., the linear relationship between forces on a structure and the resulting deformation, extended to this formalism can be expressed as a linear relation between strain and stress, which is compactly stated as[2]

$$T_{ij} = c_{ijkl} S_{kl}. \tag{10.9}$$

At this point Newton's law can be used to express the acceleration of volume element due to the stress in the structure,

$$\rho \partial_t^2 Q_i = \partial_j T_{ji} = \frac{1}{2} \partial_j c_{jikl} \left(\partial_k Q_l + \partial_l Q_k\right), \tag{10.10}$$

leading to,

$$\rho \partial_t^2 \boldsymbol{Q} = (\lambda + \mu) \nabla (\nabla \cdot \boldsymbol{Q}) + \mu \nabla^2 \boldsymbol{Q}. \tag{10.11}$$

---

[1] In a way, this approach is more general, since it applies to non-crystalline materials as well, so long as the wavelengths are larger than the interatomic spacing.

[2] Summations are implied over repeated indices.

This equation is a full vectorial wave equation for acoustic waves in an isotropic material, where the components c_{ijkl} are reduced to μ and λ, the two Lamé constants. Equation can be written as an eigenvector equation much like Eq. (10.2):

$$\omega_j^2 \boldsymbol{Q}_j(\boldsymbol{r}) = \mathsf{L}\boldsymbol{Q}_j(\boldsymbol{r}), \quad \mathsf{L}(\cdot) = -\frac{\lambda+\mu}{\rho}\nabla(\nabla\cdot(\cdot)) - \frac{\mu}{\rho}\nabla^2(\cdot). \tag{10.12}$$

A solution at frequency ω_j has a mechanical mode profile $\boldsymbol{Q}_j(\boldsymbol{r})$.

Quantizing the motion follows an approach similar to that used for the electromagnetic field. We define phonon creation and annihilation operators, $\hat{b}_j^\dagger$ and $\hat{b}_j$ respectively, for each modal solution $\boldsymbol{Q}_j(\boldsymbol{r})$ of the equations of elasticity in the structure. The field operator is then expressed in the Heisenberg picture as

$$\hat{\boldsymbol{Q}}(\boldsymbol{r}) = \sum_j \boldsymbol{Q}_j(\boldsymbol{r})\hat{b}_j e^{-i\omega_j t} + \text{h.c.} \tag{10.13}$$

To calculate the proper normalization of the field profiles, we assume a single phonon state $|\psi\rangle = |1\rangle_j \prod_{\forall k\neq j} |0\rangle_k$ and find the expected value of additional field energy above vacuum ($|\text{vac}\rangle = \prod_k |0\rangle_k$):

$$\begin{aligned} U_{\text{mech}} &= \langle\psi| \int \mathrm{d}\boldsymbol{r}\; \dot{\hat{\boldsymbol{Q}}}(\boldsymbol{r})\rho(\boldsymbol{r})\dot{\hat{\boldsymbol{Q}}}(\boldsymbol{r})|\psi\rangle - \langle\text{vac}| \int \mathrm{d}\boldsymbol{r}\; \dot{\hat{\boldsymbol{Q}}}(\boldsymbol{r})\rho(\boldsymbol{r})\dot{\hat{\boldsymbol{Q}}}(\boldsymbol{r})|\text{vac}\rangle \\ &= 2\omega_j^2 \int \mathrm{d}\boldsymbol{r}\; \boldsymbol{Q}_j^*(\boldsymbol{r})\rho(\boldsymbol{r})\boldsymbol{Q}_j(\boldsymbol{r}) \\ &= 2m_{\text{eff}}\omega_j^2 \max[|\boldsymbol{Q}_j(\boldsymbol{r})|^2]. \end{aligned} \tag{10.14}$$

Here we have taken the energy to be twice the kinetic energy. It is easy to show that for a mechanical mode, half of the energy will be kinetic, and the other half potential. Assuming the energy of a phonon to be $U_{\text{mech}} = \hbar\omega_j$, we obtain the maximum single-phonon displacement

$$x_{\text{zpf},j} \equiv \max[|\boldsymbol{Q}_j(\boldsymbol{r})|] = \sqrt{\frac{\hbar}{2m_{\text{eff},j}\omega_j}}, \tag{10.15}$$

where we've defined the effective mass for mode j to be

$$m_{\text{eff},j} = \frac{\int \mathrm{d}\boldsymbol{r}\; \boldsymbol{Q}_j^*(\boldsymbol{r})\rho(\boldsymbol{r})\boldsymbol{Q}_j(\boldsymbol{r})}{\max[|\boldsymbol{Q}_j(\boldsymbol{r})|^2]}. \tag{10.16}$$

Intriguingly, all the structures demonstrated in this work have a zero-point fluctuation amplitude on the order of a femtometer (10^{-15} m) regardless of their effective mass (ranging from 10^{-18} to 10^{-15} kg).

10.2 Photonic and Phononic Crystals

In the late 1980s and early 1990s, analogies between engineered structures and condensed matter systems led to the prediction of novel optical [17, 18] and mechanical [19, 20] phenomena in periodic structures patterned at the nanoscale. Initially, the focus was obtaining Purcell enhanced (or inhibited) spontaneous emission from quantum emitters via localization in such optical structures by engineering the local density of states. It was soon realized that the photonic and phononic bandgaps arising in periodic structures provide a powerful design paradigm where an engineer can simply design circuits in a periodic metamaterial by careful generation of defects [21]. Due to the presence of the bandgap, defect waveguides were predicted to exhibit excellent performance (e.g., loss-less sharp bends and mitigation of cross-talk), while defect cavities with wavelength-scale localization of light were predicted and soon demonstrated.

Despite significant progress, three-dimensional crystals remain difficult to fabricate in a reliable fashion, and many of the experiments have focused on quasi-1D and 2D systems made from patterns defined in thin-films of dielectric material. These thin-films are often suspended via an under-etching of the substrate beneath. In this setting, the out-of-plane confinement of light is due to total internal reflection at the boundary between a high-index dielectric and vacuum (lack of phase matching to radiation modes), while the out-of-plane confinement of phonons follows naturally from lack of propagating phonon modes in vacuum. From here on, we focus on these thin-film structures and the quasi-1D and 2D crystals that are fabricated on them. We begin first with a review of the basic properties of guided electromagnetic and acoustic waves in simple geometries with continuous translation symmetry. The focus in Sect. 10.2.1 is understanding these guided modes before the introduction of periodicity. Periodicity gives rise to a mode structure best described by band diagrams and Bloch waves as reviewed in Sect. 10.2.2. In Sect. 10.2.3, we consider how defects engineered into a periodic structure can be used to localize light and motion effectively in the same volume. Finally, we calculate the origin and strength of interactions between such co-localized excitations in Sect. 10.2.4.

10.2.1 Symmetries of Guided Waves

Solutions for the electromagnetic and acoustic wave equations for quasi-1D and 2D geometries, i.e., beam/rod structures and slabs respectively, can be found in a variety of excellent references [21–26]. Here we primarily introduce a general overview of the results that are important for designing nanostructures. Symmetries of the structure can be immensely useful in characterizing the modes while avoiding the inevitable confusion surrounding historical nomenclature. They also aid in predicting the types of interactions that can occur between modes.

Consider first the reflection symmetry of a structure. A reflection operator, e.g., for reflections about the $y-z$ plane, is defined as σ_x with $\sigma_x(x, y, z) \equiv (-x, y, z)$. For a structure that is symmetric about reflections σ_n, i.e., one where $\sigma_n\{\bar{\bar{\varepsilon}}(\boldsymbol{r})\} = \bar{\bar{\varepsilon}}(\boldsymbol{r})$, vector fields representing physical solutions of the electromagnetic and acoustic modes of the structure can be characterized by their symmetry to be symmetric (+) or anti-symmetric (−), with $\sigma_n \boldsymbol{V}^{(\pm)}(\sigma_n \boldsymbol{r}) = \pm \boldsymbol{V}^{(\pm)}(\boldsymbol{r})$ where $\boldsymbol{V}(\boldsymbol{r})$ can be either $\boldsymbol{Q}(\boldsymbol{r})$ or $\boldsymbol{E}(\boldsymbol{r})$. Pseudo-vectors like the magnetic field, transform in the opposite way, i.e., for the magnetic field one has the vector relation $\sigma_n \boldsymbol{H}^{(\pm)}(\sigma_n \boldsymbol{r}) = \mp \boldsymbol{H}^{(\pm)}(\boldsymbol{r})$.

Of course, only some of the properties of the solutions of Eqs. (10.1) and (10.11) can be simply predicted from the symmetries alone. We assume that the reader has some familiarity with the solutions of these wave equations in a simple homogeneous isotropic material. In particular, for a given frequency and propagation direction (taken to be along y), Maxwell's equations permit two solutions with the same propagation vector $\boldsymbol{k}$ but with differing polarizations, where for both solutions the electric field $\boldsymbol{E}$, magnetic field $\boldsymbol{H}$, and $\boldsymbol{k}$ are all mutually orthogonal. These two waves propagate at the same speed, given by $c_{\text{diel.}} = 1/\sqrt{\mu_0 \varepsilon_{\text{diel.}}}$. The equations of elasticity in a homogeneous isotropic medium also permit a similar pair of solutions with $\boldsymbol{Q}$ perpendicular to the propagation vector, but there exists in addition a third type of elastic wave where $\boldsymbol{Q}$ is in the direction of propagation. Both types of waves have linear dispersion. The latter are called a dilatational, longitudinal, or pressure waves and have propagation velocity of $c_l = \sqrt{(\lambda + 2\mu)/\rho}$, while the two former solutions are often called transverse, or shear waves and have velocity $c_t = \sqrt{\mu/\rho}$. Here, μ and λ are Lamé constants, and ρ is the density of the material [22, 24]. It is easy to see from these relations that $2c_t^2 \leq c_l^2$. In what follows, unless otherwise stated, we consider structures formed from an elastic material with density $\rho = 2.329$ g/cm^3, Young's modulus $E = 170$ GPa, and Poisson's ratio $\nu = 0.28$, corresponding roughly to that of isotropic Si (i.e., neglecting anisotropy).

10.2.1.1 Photons in a Slab

Thin-film slabs possess continuous translational symmetry in the x and y directions, and a reflection symmetry in the z direction. More specifically, we take

$$\bar{\bar{\varepsilon}}(\boldsymbol{r}) = \begin{cases} \varepsilon_{\text{Si}} & |z| < t/2 \\ \varepsilon_0 & |z| \geq t/2 \end{cases}, \tag{10.17}$$

where $\varepsilon_{\text{Si}} = n_{\text{Si}}^2 \varepsilon_0$, and we take the refractive index of Si to be $n_{\text{Si}} = 3.48$, its value in the 1,500 nm wavelength band. This geometry, shown in Fig. 10.2, clearly satisfies both $\sigma_x\{\bar{\bar{\varepsilon}}(\boldsymbol{r})\} = \bar{\bar{\varepsilon}}(\boldsymbol{r})$ and $\sigma_z\{\bar{\bar{\varepsilon}}(\boldsymbol{r})\} = \bar{\bar{\varepsilon}}(\boldsymbol{r})$. The solutions of Maxwell's equations are often characterized to be symmetric or anti-symmetric about the mirror symmetry σ_x. For the symmetric solutions, it follows from $\sigma_x \boldsymbol{E}^{(+)}(\sigma_x \boldsymbol{r}) = +\boldsymbol{E}^{(+)}(\boldsymbol{r})$ and $\sigma_x \boldsymbol{H}^{(+)}(\sigma_x \boldsymbol{r}) = -\boldsymbol{H}^{(+)}(\boldsymbol{r})$, so that $E_x = 0$, and $H_z = H_y = 0$ on the $y - z$ plane. Due to the symmetry of the structure, these components of the electromagnetic fields

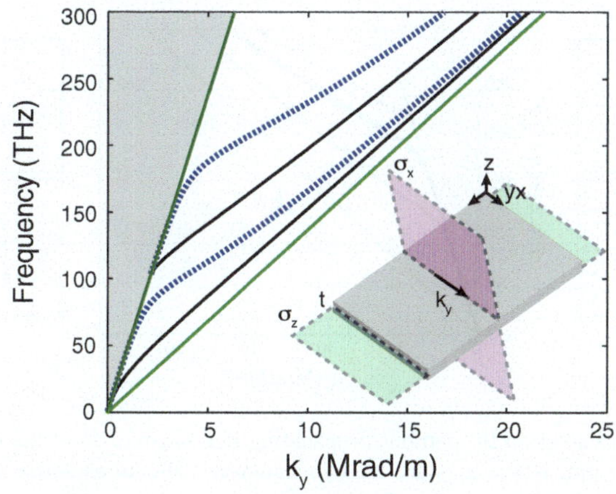

Fig. 10.2 Dispersion diagram for photons in a slab with thickness $t = 220$ nm and refractive index $n = 3.48$ (corresponding approximately to that of Si in the 1.5 μm wavelength band). The photonic bands were calculated by numerical solution of the determinantal equation [23]. The two lowest lying TM or p-mode bands are plotted as *blue dotted curves*, whereas the two lowest order TE or s-mode bands are plotted as *solid black curves*. The *upper* and *lower green curves* correspond to light propagation along $\boldsymbol{e}_y$ in vacuum and in a bulk material with refractive index of the slab, respectively. These are the *two light lines* of the slab, and the guided modes of the slab must lie in between them. The *shaded grey* region represents the continuum of modes which exist in the vacuum surrounding the slab

are also zero at all other points. Such solutions are often called transverse magnetic (TM) or p-modes in the literature, since there is no magnetic field component in the direction of propagation (y). Taking a propagation vector $\boldsymbol{k} = k_y\boldsymbol{e}_y$, and using the appropriate boundary conditions, the equations for the remaining three degrees of freedom, (E_z, E_y, H_x), can be solved to yield a determinantal equation relating ω and k_y [23]. The solutions anti-symmetric with respect to the transformation σ_x are called transverse electric (TE) or s-modes and similarly have non-zero field components (H_z, H_y, E_x). The two lowest lying bands shown in Fig. 10.2 are the fundamental TE (solid) and TM (dotted) modes. The fundamental TE mode is symmetric about σ_z, thus $E_z = H_x = H_y = 0$ on the $x - y$ plane. On the $x - y$ plane we have only non-zero H_z and E_x for the fundamental TE waves.[3] For an infinitely thick slab, the fundamental TE waves become the plane-wave solutions in a bulk dielectric that are linearly polarized in the x axis. Similarly, the fundamental TM waves are anti-symmetric about σ_z, and E_z and H_x become the only non-zero fields on the $x - y$ plane. For an infinitely thick slab, these waves reduce to the plane wave solution of orthogonal polarization (linearly polarized in the z axis). Bands for the next higher order TE (solid) and TM (dashed) waves are also plotted in Fig. 10.2. These waves

[3] The optical modes of structures in this chapter arise from engineering this TE mode, and thus in the following sections we will only plot the value of E_x on the $z = 0$ plane when representing optical mode profiles.

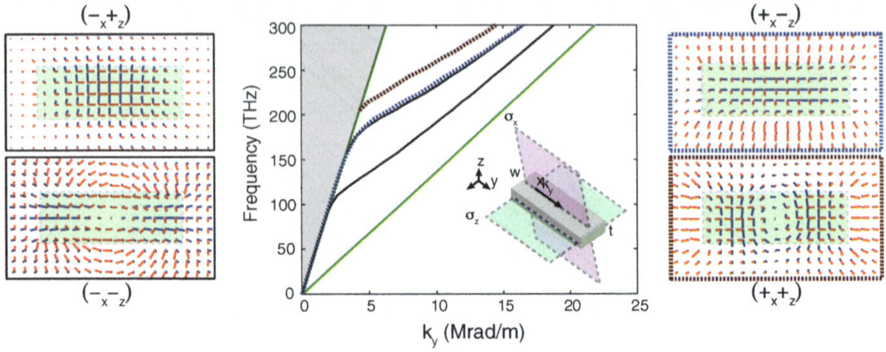

Fig. 10.3 Dispersion diagram for photons propagating in a beam of Si of thickness $t = 220$ nm and width $w = 600$ nm, where x, y are the two transverse directions and y is the long axis of the beam along which light propagates. Also shown are transverse vector mode profile plots for the four lowest lying guided mode optical bands (along with mode symmetries) at an excitation frequency of $\omega/2\pi = 235$ THz. Here the *green* region with *dashed grey outline* is the Si beam in cross-section, with the transverse electric field vector shown in *red* and the transverse magnetic field vector shown in *blue*. The *exterior outline color* and format of the mode field plots match that of the corresponding band in the band diagram. The band of interest in this work is the fundamental TE mode with symmetry $(-_x+_z)$, which as evident from the mode profile, is highly localized in the dielectric. The bands and mode profiles were calculated using a finite-difference frequency diagonalization method [27]

have the opposite symmetry about σ_z when compared to the fundamental modes, and thus have a node at the center of the slab.

10.2.1.2 Photons in a Beam

Beams or dielectric waveguides with rectangular cross sections have mode structures similar to slabs. The dielectric constant defining the structure is given by

$$\overline{\overline{\varepsilon}}(\boldsymbol{r}) = \begin{cases} \varepsilon_{\mathrm{Si}} & |z| < t/2, |x| < w/2 \\ \varepsilon_0 & |z| \geq t/2, |x| \geq w/2 \end{cases}. \tag{10.18}$$

When the beam is thicker than it is wide, i.e., $t < w$, the fundamental TM $(+_x)$ and TE $(-_x)$ guided modes of the structure are similar those of the slab. However, the field components (E_z, E_y, H_x) and (H_z, H_y, E_x), for the TM and TE waves respectively, are the *dominant* field components, as opposed to the only non-zero field components as is the case for the slab. In some literature, these guided modes are referred to as TM- or TE-like since the magnetic or electric field are no longer truly transverse to the direction of propagation throughout the structure. The four lowest guided mode bands of a Si beam are plotted in Fig. 10.3 (see caption for description).

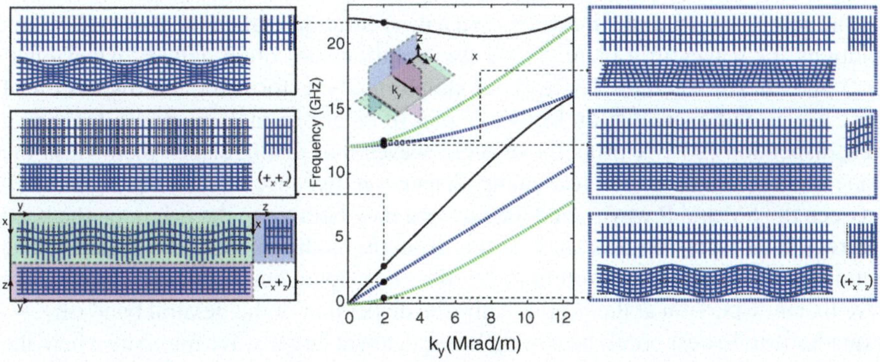

Fig. 10.4 Dispersion diagram for phonons propagating in a silicon slab, or plate, of thickness $t = 220$ nm. Here we assume Si to be an isotropic elastic material with density $\rho = 2.329$ g/cm^3, Young's modulus $E = 170$ GPa, and Poisson's ratio $\nu = 0.28$. The *solid black*, *dotted blue*, and *dotted green curves* correspond to bands with symmetry $(+_x+_z)$, $(-_x+_z)$, and $(+_x-_z)$, respectively. The displacement field profile for the six lowest lying phonon bands at $k_y = 2 \times 10^6$ rad/m are also presented. The mode field plots are represented by the distortion of a uniform Cartesian grid, and shown for the $x - y$ (*top left*), $y - z$ (*bottom left*), and $x - z$ (*top right*) planes of the structure. The vector symmetries for the three bands terminating at $\omega = 0$ are also shown. From lowest to highest frequency, these are the flexural, horizontal shear, and extensional waves described in the text, and have vector symmetries $(+_x-_z)$, $(+_x-_z)$, and $(+_x-_z)$, respectively. Note that the thickness of the slab (in the z direction) is exaggerated in the mode diagrams for clarity

10.2.1.3 Phonons in a Slab

Mechanical waves in a slab are superpositions of plane wave solutions of the bulk material, and thus share many characteristics with the three branches of acoustic waves in an isotropic elastic environment despite their strong modification by the boundary conditions. For an infinite isotropic elastic material, the three types of acoustic waves with propagation vector $\boldsymbol{k} = k_y \boldsymbol{e}_y$ can be taken to have polarizations in the $\boldsymbol{e}_x$, $\boldsymbol{e}_z$, and $\boldsymbol{e}_y$ directions. The former two are the horizontal shear (SH) and vertical shear (SV) waves, while the latter is called a pressure (P) wave. The vector symmetry of these modes with respect to σ_x and σ_z are $(-_x+_z)$, $(+_x-_z)$, and $(+_x+_z)$, respectively, as can be deduced simply by inspection of a constant vector field with the corresponding polarization.

At a boundary with normal $\boldsymbol{e}_z$, the symmetry σ_z is broken, while σ_x is preserved allowing for a coupling between SV and P modes. Thus, the modes of a slab, which possess two such boundaries, are either linear combinations of SH plane waves, or linear combinations of SV and P waves. The first few bands for waves in a plate are shown in Fig. 10.4. The modes related to the SH waves are easiest to understand as they share many properties with the corresponding plane wave solutions (including the symmetry $(-_x+_z)$). The dispersion of these modes is found to be $\omega_{\mathrm{SH},m}(k_y) = c_t\sqrt{k_y^2 + (m\pi/t)^2}$, where c_t is the velocity of SH (and SV) waves in the bulk, and m is the mode number.

The combined SV and P modes form a pair of solutions called the Rayleigh-Lamb solutions for waves in a plate. These two solutions are often called 'extensional' ($+_z$) and 'flexural' ($-_z$). There is no simple analytic form for the dispersion of these waves, but a calculation leads to a pair of determinantal equations describing their dispersion [24]. A property of these waves that distinguishes them from the plane wave solutions is their low energy dispersion. Because of the coupling induced between the SV and P waves, and the fact that they both go to the origin on the band diagram, the dispersion near $\omega, k \approx 0$ is strongly modified by level repulsion. The symmetric Rayleigh-Lamb solution, i.e., the extensional band ($+_x+_z$), continues to have linear dispersion at the origin, while the dispersion of the flexural band ($+_x-_z$) is quadratic to lowest order, becoming linear again at larger wave numbers where the frequencies of the two Rayleigh-Lamb waves diverge. The low-energy expansion for the dispersion can be calculated and for the fundamental extensional wave is given by $\omega_e(k) = c_e k$, with $c_e = c_t/c_l\sqrt{c_l^2 - c_t^2}$. For the flexural wave, we find that near $\omega, k = 0$, $\omega_f(k) = c_f k^2$, with $c_f = tc_t\sqrt{(c_l^2 - c_t^2)/3c_l^2}$ to lowest order.

10.2.1.4 Phonons in a Beam

The solutions for mechanical waves in a beam can be mostly understood by considering waves in a plate. Due to the addition of the boundaries normal to $\boldsymbol{e}_x$, the SH waves are also strongly modified, and will behave in a manner similar to the flexural bands of the plate. Therefore, there are two flexural modes of the beam, one polarized along the x direction with symmetry ($-_x+_z$), and the other along the z direction with symmetry ($+_x-_z$). Both exhibit quadratic dispersion at low energy as shown in Fig. 10.5. The 'extensional' wave in the plate corresponds to a similar wave in the beam with symmetry ($+_x+_z$) and also has linear dispersion at low energies.

There is however a fourth type of wave arising in a beam that is not analogous to any bulk or plate wave, but perhaps most closely related to SH waves in a plate. This solution to the equations of elasticity is called a 'torsional' wave and has symmetry ($-_x-_z$). It is well known that there are no rotational waves in bulk media [22]. An intuitive way to see this involves considering very low energy excitations of a beam and comparing to those of a plate. At a frequency close to zero, with the wavelength going to infinity, the three bands of plate waves simply reduce to displacements of the plate in the three directions. For a beam, the fourth band corresponds of a rotation of the beam about its longitudinal axis. This rotation can not constitute a low-energy degree of freedom for the plate, since for any finite rotation angle, the induced displacement vector $\boldsymbol{Q}(\boldsymbol{r})$ grows without bound as one moves away from the axis of rotation. Thus it is the finite transversal extent of the beam that allows torsional modes to exist—they are a product of the boundary conditions. The existence of torsional waves in beams, which are symmetry mismatched to bulk and plate waves has been used to create mechanical resonators with extremely low clamping

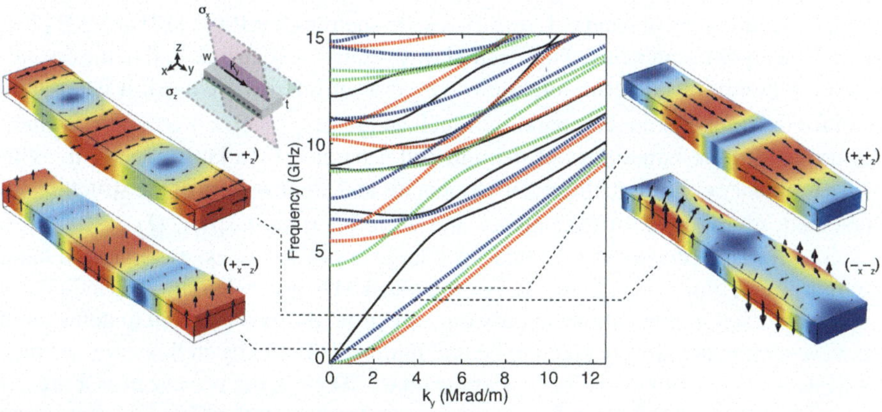

Fig. 10.5 Dispersion diagram for phonons propagating in a silicon beam of thickness $t = 220$ nm and width $w = 600$ nm. Here we assume Si to be an isotropic elastic material with density $\rho = 2.329$ g/cm^3, Young's modulus $E = 170$ GPa, and Poisson's ratio $\nu = 0.28$. The *solid black, dotted blue, dotted green,* and *dotted red curves* correspond to bands with symmetry $(+_x+_z)$, $(-_x-_z)$, $(-_x+_z)$, and $(+_x-_z)$, respectively. Mode plots for the four lowest lying bands which terminate at $\omega = 0$ are also shown for wavevector $k_y = 2 \times 10^6$ rad/m. In these mode plots the displacement of the beam from its reference form (*solid black outline*) is shown by the colored beam, and where *red* (*blue*) corresponds to regions of large (zero) displacement magnitude and the *black* vectors indicate the magnitude and direction of the displacement. The bands and mode profiles were calculated using the FEM software package COMSOL [12]

losses, since acoustic radiation away from the beam is prohibited by a symmetry mismatch [28–30].

10.2.2 Introducing Periodicity: Band Diagrams

Waves in periodic structures are best understood in terms of band theory. Bloch's theorem states that in a structure with periodic variation, the time-harmonic solutions to the optical and mechanical wave equations can be expressed as a product of a plane-wave solution ($e^{i\boldsymbol{k}\cdot\boldsymbol{r}}$) and a function ($\boldsymbol{e}_{\boldsymbol{k}}(\boldsymbol{r})$ or $\boldsymbol{Q}_{\boldsymbol{k}}(\boldsymbol{r})$) possessing the same periodicity as the crystal,[4]

$$\boldsymbol{e}(\boldsymbol{r}) = \boldsymbol{e}_{\boldsymbol{k}}(\boldsymbol{r})e^{i\boldsymbol{k}\cdot\boldsymbol{r}} \quad \text{and} \quad \boldsymbol{Q}(\boldsymbol{r}) = \boldsymbol{Q}_{\boldsymbol{k}}(\boldsymbol{r})e^{i\boldsymbol{k}\cdot\boldsymbol{r}}. \tag{10.19}$$

These solutions are for a given frequency, and the vector $\boldsymbol{k}$ is well-defined modulo a reciprocal lattice vector $\boldsymbol{G}$. For complex geometries, numerical methods are used to calculate the relation between $\boldsymbol{k}$, the wave vector, and $\omega_n(\boldsymbol{k})$, the frequency of the

[4] We refer the reader to a text on condensed matter physics [16] for a completely analogous treatment of these concepts for electrons.

Bloch wavefunctions of band n. Using a package such as COMSOL [12] or MPB [13], we find the eigenfrequencies of the desired wave equation for a unit-cell of the crystal. Boundary conditions determining the wavevector of interest are used. It is possible to set for example a boundary condition such as $\boldsymbol{Q}(\boldsymbol{r})|_{\partial_1} = \boldsymbol{Q}(\boldsymbol{r}+\boldsymbol{a})|_{\partial_2} e^{i\boldsymbol{k}\cdot\boldsymbol{a}}$ where $\partial_{1,2}$ are two of the boundaries of the unit-cell separated by a lattice vector $\boldsymbol{a}$. With such a boundary condition, the eigenfrequencies and eigenvectors of the structure are calculated, giving a set of frequencies and wavefunctions $\{\omega_n(\boldsymbol{k}), Q_{\boldsymbol{k},n} e^{i\boldsymbol{k}\cdot\boldsymbol{r}} : n = 1, 2, 3, \ldots\}$. This process is repeated for a large set of $\boldsymbol{k}$ values, generating a band diagram, i.e. a plot of $\omega_n(\boldsymbol{k})$ vs. $\boldsymbol{k}$. For the quasi-1D case, where the periodicity is defined by discrete translations in only one direction, the axis is trivial to define, and the wavevectors are simply taken to be real numbers. Typically in this case, we take $k \in [-G/2, G/2]$ with $G = 2\pi/a$. For the quasi-2D case, the wavevectors $\boldsymbol{k}$ are 2D vectors, and thus a 1D path in $\boldsymbol{k}$-space starting from $k = 0$ (the Γ-point), traversing the boundary of the First Brillouin Zone (FBZ), and coming back to the Γ-point is used.

10.2.2.1 Quasi-1D Nanobeam Optomechanical Crystals

Patterning an array of holes in a suspended beam realizes a quasi-1D crystal structure. Such dielectric crystals, sometimes referred to as photonic crystal wires, have been considered for the last two decades in optics [31, 32] due to their ease of design and fabrication. Figure 10.6 shows the calculated band diagrams for the propagation of light and sound in such a quasi-1D nanobeam crystal structure.

The band structure for photons at small values of k and at low frequencies is similar to that of photons propagating in an unpatterned beam studied in Sect. 10.2.1.2. In such a case, the wavelength is simply too large for the electromagnetic waves to sense the periodicity of the crystal. There are two bands terminating with linear dispersion at the origin of the $\omega - k$ diagram, corresponding to the fundamental TE-like and TM-like modes of an unpatterned beam. At close to the X-point however, strong backscattering due to Bragg reflection occurs, and the bands flatten. A splitting is induced by the periodic perturbation of the dielectric. For the geometry shown, this splitting produces a quasi-bandgap of nearly 50 THz centered around 190 THz for $(-_x+_z)$ symmetry guide modes. At the X-point, the upper and lower branches of the folded $(-_x+_z)$ guided mode band are called the conduction and valence bands, respectively, in analogy to the electronic bands of a semiconductor. Close to the X-point, these bands posses a quadratic dispersion.

The orange shaded region in the band diagram of Fig. 10.6b is not strictly a bandgap though we refer to it as such in this chapter. Rather, it is a quasi-bandgap region for the fundamental TE-like modes, protected by both mirror and discrete translation symmetries. Unlike a true bandgap, there are propagating modes that have frequencies in this region. These modes are of two types—those with a phase mismatch and those with a symmetry mismatch to the guided Bloch mode band of interest. The former can be found in the grey shaded region of Fig. 10.6b, where a

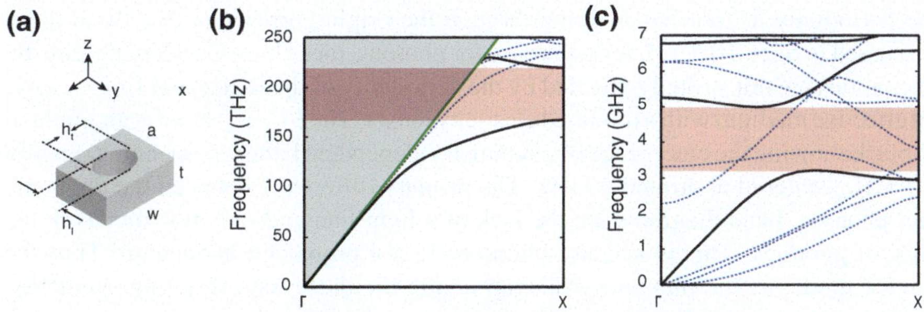

Fig. 10.6 **a** Unit cell of a quasi-1D crystal formed in a Si nanobeam. The unit-cell contains a hole shaped like an oval, with the transverse and longitudinal radii of the hole given by $h_t = 0.76a$ and $h_l = 0.54a$, respectively, and the lattice constant equal to $a = 448$ nm. **b** Photonic band diagram of the Si nanobeam. Here we take Si to be an isotropic dielectric with refractive index $n = 3.48$. The fundamental TE-like band with symmetry $(-_x+_z)$ is shown as a *solid black curve*, while the other bands are shown as *blue dotted curves*. The *solid green curve* is the vacuum *light line*, and the *gray shaded* region denotes the region with a continuum of radiation modes. The *orange shaded* region represents a quasi-bandgap for guided optical modes of symmetry $(-_x+_z)$. **c** Phononic band diagram of the Si nanobeam. Here we assume Si to be an isotropic elastic material with density $\rho = 2.329$ g/cm^3, Young's modulus $E = 170$ GPa, and Poisson's ratio $\nu = 0.28$. The $(+_x+_z)$ bands corresponding to the extensional modes of the beam are shown as *black curves*, with the bands of other symmetry shown as *blue dotted curves*. The *orange shaded* region corresponds to an acoustic mode bandgap for modes of symmetry $(+_x+_z)$. In both plots, the point Γ refers to $k = 0$ and X to $k = \pi/a$ in the y-direction

continuum of radiation modes exist. In a perfect and infinite structure, guided waves below the vacuum light line cannot leak into radiation modes due to a momentum or phase mismatch. However, if there are defects in the dielectric along the propagation direction y (whether designed or accidental) with sufficiently large Fourier components to compensate the phase mismatch, an excitation below the light line can couple to radiation modes above the light line. A structure with a highly localized perturbation (small Δy) can provide a large Δk_y which leads to "radiation losses" (see Sect. 10.2.3). The second set of modes that the Bloch mode of interest may couple to are those with differing symmetry found below the light line. In Fig. 10.6b, the dotted curve crossing through the bandgap corresponds to a guided mode with the same symmetry, $(+_x-_z)$, as the fundamental TM-like mode of the simple beam presented in Fig. 10.3. For this band to couple to the fundamental TE-like band requires breaking of both σ_x and σ_z mirror symmetries. In a perfect structure, one has essentially a perfect bandgap, while the introduction of nonidealities causes coupling to the out of symmetry modes and an increase in radiation or scattering losses. Introduction of engineered defects to localize the light in the bandgap will necessary yield radiation losses. These losses can be reduced below experimentally detectable limits by smoothing the defect (increasing Δy) as is studied in more detail below in Sect. 10.2.3.

A similar view emerges for the phonons in the 1D periodic structure with a few caveats. For small k and low energies, where the wavelengths are much longer than the periodicity a, four bands going through the origin correspond exactly to those discussed in Sect. 10.2.1.4. As is the case for photons, these long wavelength acoustic excitations are not strongly affected by the periodicity of the lattice and simply sense an effective medium with a modified elasticity tensor. The $(+_x+_z)$ band is highlighted in black showing the emergence of a symmetry-dependent band gap spanning roughly 1.5 GHz centered at around 4 GHz. The primary difference between the phononic and photonic band diagrams are the lack of a light line and radiation modes in the case of phonons. This is because phonons do not propagate in vacuum. Thus the guided modes in the phononic bandgap region are those with differing symmetry, and we call the quasi-bandgap in this case symmetry-dependent.

10.2.2.2 Quasi-2D Slab Optomechanical Crystals

Moving from one-dimensional to two-dimensional structures is highly desirable for certain experiments. First, there is the potential for more design freedom, as well as new opportunities in scaling and networking of systems. For example, two-dimensional arrays of optomechanical structures can give rise to novel collective phenomena not obtainable in 1D systems [5], as well as the ability to guide either light or sound by careful generation of line defects [1]. A major technical advantage to 2D slab structures is their enhanced power-handling ability. Heat generated by absorption of photons in the structure can escape more easily in a 2D setting than in 1D photonic structures [33, 34]. This becomes crucially important in low temperature experiments where quantum phenomena are more easily observed. In the mesoscopic systems under study here, and at temperatures below a few kelvin, phonon-mediated thermal transport is quantized [35]. This is a particularly limiting factor for 1D nanostructures when the cut-off energy of higher order propagating phonon modes $\hbar\Omega_{\mathrm{M}}$ surpasses $k_{\mathrm{B}}T_b$. In this case, thermal phonons can be carried away through at most four channels corresponding to the four polarizations of waves studied in Sect. 10.2.1.4. In an infinite slab, even if one dimension (for example, the thickness) is "small", there are still an infinite number of channels for guiding phonons, corresponding to waves propagating in different directions in the plane.[5] Though this argument holds only for phonons in thermal equilibrium, thermal conduction due to shorter wavelength "hot" phonons [36] is also enhanced from going to 2D structures.

Unfortunately, the phononic property of a structure essential for good heat management, i.e., a large number of phonon loss channels, is diametrically opposed to the central goal of creating long-lived mechanical resonances with low acoustic radiation losses. The quantization of thermal conductance and availability of a limited number of phonon loss channels, means that on one hand, the thermal conductance

[5] The band diagrams for phonons shown in Figs. 10.6, 10.7, and 10.8 look similar and hide this essential distinction. We remind the reader that the bands for the 2D structures represent modes along a path traversing the boundary of the FBZ, while in the band diagram for a 1D structure shown in Fig. 10.6, all of the points in the FBZ are represented.

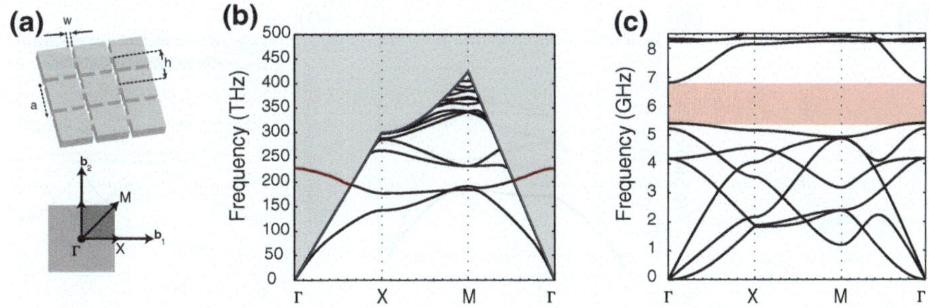

Fig. 10.7 **a** Schematic of the quasi-2D cross crystal slab structure (*upper*) along with its reciprocal lattice (*bottom*). The cross crystal structure considered here is formed from a Si thin-film of thickness $t = 220$ nm, and consists of a periodic array of cross-shaped holes etched into it with parameters $h = 0.8a$, $w = 0.2a$, and $a = 500$ nm. Here we assume the Si thin-film to be an isotropic dielectric with refractive index of $n = 3.48$ and an isotropic elastic material with $\rho = 2.329$ g/cm^3, $E = 170$ GPa, and $\nu = 0.28$. **b** Photonic band diagram, showing only the $(+_z)$ symmetry bands. The *gray shaded region* denotes the vacuum light *cone* region with a continuum of radiation modes. Leaky bands in the vacuum light *cone* are shown in *red*, and cause there to be no usable photonic bandgap. **c** Phononic band diagram, showing the acoustic bands of all symmetries. This structure exhibits a large bandgap centered at roughly 6 GHz for all possible acoustic waves (*orange shaded region*). The photonic and phononic band diagrams are both plotted along the $\Gamma \rightarrow X \rightarrow M \rightarrow \Gamma$ path in reciprocal space

of 1D structures is poor, while on the other hand, localized mechanical resonances in a 1D setting can be fairly robust against disorder. In fact, in 1D, symmetry-dependent bandgaps are sufficient for achieving large mechanical quality factors for modes of engineered defects [37, 38] because a localized mode at a specific frequency will at most couple to only a finite number of channels in the presence of disorder. In a 2D structure however, any symmetry-breaking perturbation couples a localized resonance to a much larger number of loss channels—the same channels that provide the desirably large thermal coupling to the environment.

A structure possessing a *full phononic bandgap*, i.e., a bandgap for all acoustic excitations regardless of their symmetry, provides the best of both worlds. A localized mechanical resonance with a frequency in the bandgap of such a crystal can be long-lived even under modest amounts of fabrication disorder. Additionally, since the thermal distribution of phonons is spectrally broad, phonon modes outside the bandgap provide a large thermal conductance to dissipate heat. Luckily, full phononic bandgaps are obtainable in thin-film nanostructures by patterning crystals with particular unit cell geometries. Perhaps even more surprisingly, it is possible to create simultaneously a full bandgap for phonons and a quasi-bandgap for guided mode photons as will be shown below for the snowflake lattice.

For a square lattice, an array of cross-shaped holes etched into a thin-film of Si can have a full phononic bandgap [39]. Each cross has a height h and a width w, which along with the lattice spacing a and slab thickness t, serve to fully define the geometry of the crystal. The reciprocal space representation of the lattice is shown

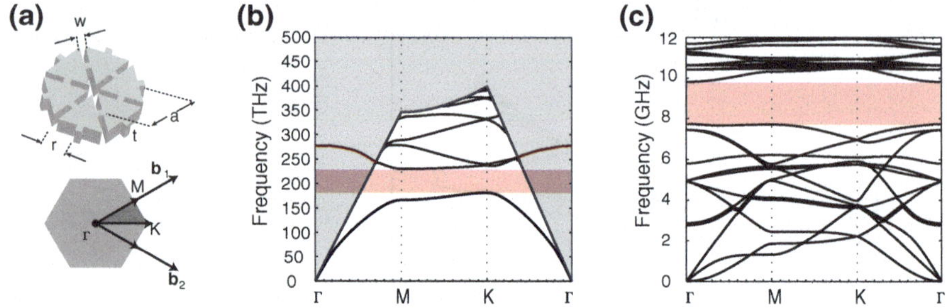

Fig. 10.8 **a** Schematic of the quasi-2D snowflake crystal slab structure (*upper*) along with its reciprocal lattice (*bottom*). The snowflake crystal structure considered here is formed from a Si thin-film of thickness $t = 220$ nm, and consists of a periodic array of snowflake-shaped holes with parameters $r = 0.4a$, $w = 0.15a$, and $a = 500$ nm. Here we assume the Si thin-film to be an isotropic dielectric with refractive index of $n = 3.48$ and an isotropic elastic material with $\rho = 2.329$ g/cm^3, $E = 170$ GPa, and $\nu = 0.28$. **b** Photonic band diagram, showing only the $(+_z)$ symmetry bands. The *gray shaded region* denotes the vacuum light *cone* region with a continuum of radiation modes. Leaky bands in the vacuum light *cone* are shown in *red*, and do not enter the region of the photonic quasi-bandgap. **c** Phononic band diagram, showing the acoustic bands of all symmetries. This structure exhibits a large bandgap centered at roughly 9 GHz for all possible acoustic waves (*orange shaded region*). The photonic and phononic band diagrams are both plotted along the $\Gamma \rightarrow M \rightarrow K \rightarrow \Gamma$ path in reciprocal space

in Fig. 10.7b, in which the common notation of the high symmetry points of the FBZ in a square lattice are used. The phononic bandstructure, including all symmetries of vibrational modes of the cross substrate is shown in Fig. 10.7c. For the plotted parameters, the Si structure has a bandgap between 5.3 and 6.8 GHz. This bandgap can be understood to arise due to the presence of small bridges between the connecting interconnected squares in the lattice [39]. For frequencies approaching the internal resonance frequencies of the squares, the wave propagation in the structure can be understood in a tight-binding picture. These tight-binding bands flatten as the nearest neighbor interaction is made smaller by reducing the width of the connecting bridges, giving rise to a bandgap.

Photons in this structure are better described in a plane-wave expansion approach as opposed to the tight-binding picture. The square lattice, with its low symmetry, behaves differently for plane waves propagating in different directions such as at the high symmetry X and M points of the FBZ boundary. This results in a much smaller in-plane photonic bandgap for the square lattice in comparison to a higher symmetry lattice such as the hexagonal lattice. The photonic bands for the cross structure are shown in Fig. 10.7c for $(+_z)$ symmetry (these modes include the fundamental TE-like bands). Unfortunately, due to the presence of a *leaky* guided mode resonance in the continuum of radiation modes, no photonic bandgap is present. The leaky resonance has a large local density of states in the slab, making localized photonic states highly susceptible to leakage in the presence of fabrication disorder. Thus, the cross substrate is ruled out as a suitable structure from which to form important photonic elements such as ultrahigh-Q optical cavities. Nonetheless, the simple design and full phononic bandgap of cross substrates has permitted their utilization as phonon "shields" in a variety of optomechanical experiments [7, 8, 10, 40, 41, 45].

The hexagonal lattice counterpart of the quasi-2D cross substrate is shown in Fig. 10.8, which we term the "snowflake" substrate. Each snowflake pattern has a radius r and a width w, which along with the lattice spacing a and slab thickness t, define the crystal geometry. Due to the higher symmetry of the hexagonal lattice, wavevectors at the high symmetry M and K points on the boundary of the FBZ sense a more similar medium preserving the photonic bandgap. In principle, using photonic cavity design techniques outlined in Sect. 10.2.3, the quasi-2D snowflake crystal substrate can support high-Q co-localized optical and mechanical resonances [42].

10.2.3 Breaking Periodicity: Localized States in Optomechanical Crystals

Any mechanical resonance with an eigenfrequency inside a full bandgap will be long-lived. Since phonons do not propagate in vacuum and there is no continuum of radiation modes outside of the thin-film, phononic cavities with high quality factors are easy to create, and can be defined by incorporating a defect into a quasi-2D phononic crystal slab. For example, one need only surround an area in the slab with a full bandgap phononic crystal like the cross structure presented above [40] to obtain long-lived phononic cavity resonances. The internal resonances of the surrounded region, cannot couple to any propagating leakage channel as long as the defect frequency is within the bandgap, and thus the quality factors of such resonators are not limited by clamping or radiation losses.

As explained in Sect. 10.2.2.1, the situation for photons is more complicated. The photonic bandgaps obtained in the thin-film systems are protected by phase matching and symmetries. By designing a structure so the vertical symmetry σ_z is conserved, coupling between TE-like and TM-like modes can be eliminated. On the other hand, to localize the field, a defect must be introduced breaking the translational symmetry of the structure and inducing coupling to the continuum of radiation modes, which leads to radiation losses. It was discovered nearly a decade ago that radiation losses can be greatly reduced in thin-film photonic crystal cavities by engineering defect modes with narrow momentum-space distributions [43, 44]. The essential idea is to generate a *smooth* defect. For these defects, modifications to the unit cell of the crystal are small enough that they can be described by a locally modified band picture. By slightly modifying a parameter, such as the radius or shape of a hole or the lattice constant, bands can be "pushed" into a bandgap, as shown in Fig. 10.9a–d. Surprisingly, even defects small enough to confine light in a nearly diffraction-limited volume can be designed to yield quality factors larger than 10^6 [45]. In Si, quality factors beyond 10^6 are often limited by other loss mechanisms [46].

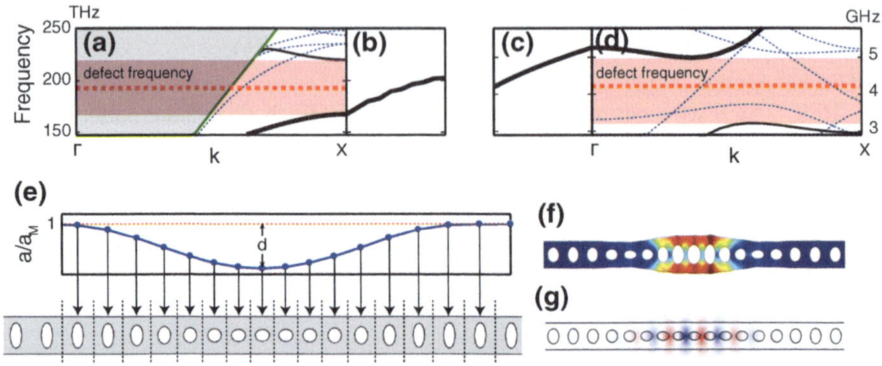

Fig. 10.9 Zoomed-in region of the **a** photonic and **d** phononic band diagrams of Fig. 10.6. The guided mode optical and acoustic bands from which the localized defect states are formed are shown as *thick black curves*. The resulting optical and acoustic defect mode frequency is shown as a *dashed red curve* in both diagrams. **e** Schematic view of the nanobeam defect cavity in the x–y plane, showing the variation of the unit cell along the length of the nanobeam. **b** The frequency of the X-point optical mode of interest as a function of unit cell perturbation. Moving from *left to right* on the plot corresponds to moving from the mirror region at the end of the nanobeam to the center of the defect cavity. **c** For the same perturbation, the frequency of the Γ-point mechanical mode of interest. **f**, **g** Plot of the FEM-simulated lowest order defect modes for the phonons ($\boldsymbol{Q}$) and photons (E_x) in the resulting defect cavity. Further details of the cavity design can be found in Ref. [47]

10.2.3.1 Nanobeam Optomechanical Crystal Cavities

Designs for photonic nanobeam cavity resonators with very large quality factors (exceeding 10^6) were realized in Ref. [48–50] with experimental demonstrations following shortly [51]. Quasi-1D nanobeam optomechanical crystal cavities were realized by Eichenfield et al. [37] following very similar principles. The extra complication in the design of optomechanical crystal cavities is that the perturbation of the crystal must push both the optical and mechanical bands edges into their respective bandgaps. The sections of the band diagrams of interest for the nanobeam geometry are shown in Fig. 10.9a and d for the optics and mechanics, respectively. The defect's perturbation must then increase the optical frequency as evident from Fig. 10.9a, since the optical defect state is formed from the valence band's X-point Bloch function. Γ-point optical resonances are unsuitable as they are embedded in a continuum of radiation modes and are usually leaky.

Since the mechanical field $\boldsymbol{Q}(\boldsymbol{r})$ enters the optomechanical coupling calculation linearly as opposed to quadratically (see Sect. 10.2.4), cancellations can occur in the overlap integral due to sign changes in the displacement field. Thus, the selection of the band edge for the mechanical resonance requires consideration of the translational and mirror symmetries of the mode. In particular, a mode taken from the Γ-point is preferred[6] since for X-point mechanical resonances the optomechanical coupling from consecutive cells for a mechanical mode with symmetry $(+_y)$ approximately cancels, leading to a reduced optomechanical coupling rate [39]. Additionally, the only mechanical mode symmetry giving a nonzero value for the integral in Eq. (10.23) is $(+_x+_z)$. The $(+_x +_y +_z)$ Γ-point phonon mode with the lowest frequency lies above a symmetry dependent bandgap. Hence the defect perturbation must reduce the frequency of this Γ-point phonon mode.

A perturbation of the nanobeam quasi-1D crystal that satisfies both the optical and mechanical conditions is one where both the radii and the inter-hole spacing is changed, as shown schematically in Fig. 10.9e. The exact form of this defect is described in Ref. [47]. Under such a perturbation, the valence band X-point photonic mode is pushed to higher frequencies in the center of the nanobeam (Fig. 10.9b) while the Γ-point phononic mode is reduced in frequency (Fig. 10.9c). The exact defect parameters are found through numerical optimization over hundreds of simulations used to optimize a fitness function approximately related to g_0^2/κ [47]. The co-localized mechanical and optical resonances of the resulting optimized nanobeam defect cavity are shown in Figs. 10.9f and g, respectively.

10.2.3.2 Snowflake Optomechanical Crystal Cavities

The design of cavities in a thin-film snowflake substrate follows generally the same principles as for the quasi-1D nanobeam cavity, with only a few caveats and complications. As a first step, we convert the quasi-2D crystal into an effective quasi-1D crystal by introduction of a line defect that possesses symmetry-protected optical and mechanical bandgaps. In addition to simplifying the design, this step embeds the band edge of interest in a region of frequency space with very low density of states, i.e. deep inside the bandgap region (essentially the same recipe has been used to create optical resonances of very high Q in a similar hexagonal lattice [45]). Next, we focus on the bands of the waveguide that provide the best optomechanical coupling estimated from an overlap integral of the optical and mechanical Bloch functions [39]. For a snowflake substrate with $(r, w) = (0.42, 0.15)a$, $t = 220$ nm, and $a = 540$ nm, a waveguide width of $W = 0.4a$ was found to produce strongly coupled optical and mechanical waveguide modes (see Fig. 10.10a). These bands are tuned into the bandgap regions in a manner analogous to that described above for the nanobeam structures. A local reduction of the radius of snowflakes causes both the mechanical and optical band edges to tune into their respective bandgaps, as elaborated in Ref. [39]. An exaggerated defect cavity in the snowflake substrate is shown in Fig. 10.10. In the actual cavity structure, the perturbation of the snowflake

[6] It is possible to generate mechanical modes with nonzero optomechanical coupling that do not have the "correct" Bloch function σ_y mirror symmetry (e.g, the 'accordion' modes demonstrated in Ref. [37]). However, a higher order resonance of the defect must be used. We do not consider these designs here as they generally have lower optomechanical coupling rates than fundamental modes formed from Γ-point Bloch functions with $(+_y)$ symmetry.

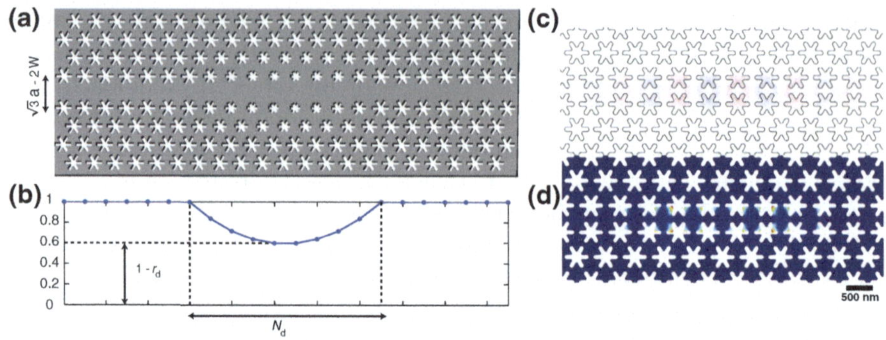

Fig. 10.10 **a** Schematic illustration (in the x–y plane) of a defect cavity in the quasi-2D snowflake substrate. A linear waveguide along the y-direction (Γ–K direction in reciprocal space) is first formed by removing a linear array of snowflake holes and shrinking the waveguide width by $2W$. Over $N_d^{\mathrm{WG}} = 7$ rows going out from the center of the waveguide, the snowflake radii are scaled quadratically from $r(1 - r_d)$ at the center to a nominal value of r. A cavity is formed from this waveguide by then varying the parameter r_d quadratically along the axis of the waveguide over N_d lattice periods. **b** A plot of the central waveguide hole radii along the length of the waveguide for $N_d = 10$ and a maximum r_d at the cavity center of 0.4. Plot of the FEM simulated **c** optical field ($E_x(\boldsymbol{r})$) and **d** magnitude of the mechanical displacement field ($\boldsymbol{Q}(\boldsymbol{r})$) for the co-localized resonances of a defect cavity formed in a Si snowflake substrate with parameters $(t, r, w, a) = (220, 210, 75, 500)$ nm, $N_d^{\mathrm{WG}} = 5$, $N_d = 14$, and $r_{d,\max} = 0.03$. For these parameters the mechanical resonance is at a frequency of $\nu_m = 9.50$ GHz and the optical mode at a wavelength of $\lambda_0 = 1.459\ \mu$m. The theoretical vacuum optomechanical coupling rate between these modes is $g_0/2\pi = 292$ kHz. See Ref. [39] for further details

radius due to the defect is only 3 %, and is spread over $N_d = 14$ unit cells. The snowflake defect cavity was not found through numerical optimization as in the nanobeam case due to the higher complexity of the structure and the impracticality of running a large number of simulations. Instead, the design of the quasi-2D snowflake cavity involved first optimizing the waveguide unit cell optomechanical coupling, and then forming the cavity with a simple harmonic defect perturbation [39]. The mode profiles of the strongly coupled optical and mechanical modes are shown in Fig. 10.10. The very small perturbation needed to localize the defect cavity modes has implications for how such structures are affected by fabrication imperfections as described in Sect. 10.3.3.2.

10.2.4 Optomechanical Coupling

Given a numerically computed optical and mechanical mode, the optomechanical coupling rate between the two modes can be calculated using perturbation theory. Physically, the optomechanical coupling arises from the shift in the optical cavity frequency by the mechanical deformation. In a dielectric structure characterized by $\overline{\overline{\varepsilon}}_0(\boldsymbol{r})$, modifications due to deformations can be taken into account with the expression

$$\overline{\overline{\varepsilon}}(\boldsymbol{r}) = \overline{\overline{\varepsilon}}_0(\boldsymbol{r}) + \overline{\overline{\delta\varepsilon}}(\boldsymbol{r}). \tag{10.20}$$

To first order, such a modification of the dielectric causes a shift in the optical resonance frequency of a mode with mode profile $\boldsymbol{e}(\boldsymbol{r})$ of

$$\omega^{(1)} = -\frac{\omega_0}{2}\frac{\langle \boldsymbol{e}|\overline{\overline{\delta\varepsilon}}|\boldsymbol{e}\rangle}{\langle \boldsymbol{e}|\overline{\overline{\varepsilon}}|\boldsymbol{e}\rangle}, \tag{10.21}$$

where

$$\langle \boldsymbol{a}|\overline{\overline{b}}|\boldsymbol{c}\rangle = \int \boldsymbol{a}(\boldsymbol{r}) \cdot \overline{\overline{b}}(\boldsymbol{r}) \cdot \boldsymbol{c}(\boldsymbol{r})\, \mathrm{d}^3\boldsymbol{r}. \tag{10.22}$$

Numerical evaluation of this integral can be simplified depending on the form of $\overline{\overline{\delta\varepsilon}}(\boldsymbol{r})$. Below we consider dielectric perturbations arising from moving boundaries and the photoelastic effect.

10.2.4.1 Boundary Perturbation

A deformation of the optical resonator affects the dielectric tensor at the boundaries between different materials. This is because the high-contrast step profile of $\overline{\overline{\varepsilon}}(\boldsymbol{r})$ across a boundary is shifted by deformations of the structure. By relating a deformation to a change in the dielectric constant, we can use Eq. (10.21) to calculate the optomechanical coupling. Johnson has derived a useful expression [52] for this shift in frequency, which when adapted to optomechanics [38], gives a frequency shift per unit displacement of

$$g_{\mathrm{OM,Bnd}} = -\frac{\omega_0}{2}\frac{\int (\boldsymbol{Q}(\boldsymbol{r}) \cdot \boldsymbol{n})\,(\Delta\overline{\overline{\varepsilon}}|\boldsymbol{e}^{\parallel}|^2 - \Delta(\overline{\overline{\varepsilon^{-1}}})|\boldsymbol{d}^{\perp}|^2)\mathrm{d}A}{\max(|\boldsymbol{Q}|)\int \overline{\overline{\varepsilon}}(\boldsymbol{r})|\boldsymbol{e}(\boldsymbol{r})|^2\mathrm{d}^3\boldsymbol{r}}, \tag{10.23}$$

for a mechanical vector displacement field $\boldsymbol{Q}(\boldsymbol{r})$.

10.2.4.2 Photoelastic Coupling

The photoelastic contribution to the optomechanical coupling arises from local changes in the refractive index due to strain in the structure. For a particular displacement vector $\boldsymbol{Q}(\boldsymbol{r})$ (and the corresponding strain tensor $\overline{\overline{S}}$, see Eq. (10.8)), the dielectric perturbation is given by

$$\overline{\overline{\delta\varepsilon}}(\boldsymbol{r}) = \overline{\overline{\varepsilon}} \cdot \frac{\overline{\overline{p}} \cdot \overline{\overline{S}}}{\varepsilon_0} \cdot \overline{\overline{\varepsilon}}, \tag{10.24}$$

which reduces to $\delta\varepsilon_{ij} = -\varepsilon_0 n^4 p_{ijkl} S_{kl}$ for an isotropic medium. The fourth-rank tensor $\overline{\overline{p}}$ with components p_{ijkl} is called the photoelastic tensor. Often, when considering the symmetries in the atomic structure of the material, a reduced tensor is used with elements p_{ij}. At this point, a simple volume integral, such as that shown in Eq. (10.21), can be used to find the frequency shift for a given displacement

$$g_{\text{OM,PE}} = -\frac{\omega_0}{2} \frac{\int \boldsymbol{e} \cdot \overline{\overline{\delta\varepsilon}} \cdot \boldsymbol{e} \, \mathrm{d}^3\boldsymbol{r}}{\max(|\boldsymbol{Q}|) \int \overline{\overline{\varepsilon}}(\boldsymbol{r}) |\boldsymbol{e}(\boldsymbol{r})|^2 \mathrm{d}^3\boldsymbol{r}}, \tag{10.25}$$

where for x and y axes aligned along the [100] and [010] crystal directions of Si, respectively, $\boldsymbol{e} \cdot \overline{\overline{\delta\varepsilon}} \cdot \boldsymbol{e}$ is given by [10],

$$\begin{aligned} \boldsymbol{e} \cdot \overline{\overline{\delta\varepsilon}} \cdot \boldsymbol{e} = -\varepsilon_0 n^4 \Big[& 2\Re\{E_x^* E_y\} p_{44} S_4 + 2\Re\{E_x^* E_z\} p_{44} S_5 + \Re\{E_y^* E_z\} p_{44} S_6 \\ & + |E_x|^2 (p_{11} S_1 + p_{12}(S_2 + S_3)) + |E_y|^2 (p_{11} S_2 + p_{12}(S_1 + S_3)) \\ & + |E_z|^2 (p_{11} S_3 + p_{12}(S_1 + S_2)) \Big]. \end{aligned} \tag{10.26}$$

The components of the photoelastic tensor are for such an orientation in Si, $(p_{11}, p_{12}, p_{44}) = (-0.094, 0.017, -0.051)$ [25].

10.2.4.3 Vacuum Coupling Rate

The expressions derived above give us the boundary and photoelastic components for a shift in the optical cavity frequency per unit displacement of the maximum deflection point of a deformation profile $\boldsymbol{Q}(\boldsymbol{r})$. A natural unit for displacement is the zero-point fluctuation amplitude. The vacuum coupling rate g_0 is defined as the shift in the optical cavity frequency due to a deformation equal in amplitude to the zero-point fluctuations of a mechanical mode $\boldsymbol{Q}_j(\boldsymbol{r})$. We can find $g_{0,\text{Bnd}}$ and $g_{0,\text{PE}}$ simply by multiplying the expressions (10.23) and (10.25) by the zero-point fluctuation length $x_{\text{zpf}} = \sqrt{\hbar/(2m_{\text{eff}}\Omega_{\text{M}})}$ (see Eq. 10.15). The total coupling rate is then simply their sum $g_0 = g_{0,\text{Bnd}} + g_{0,\text{PE}}$, and the corresponding optomechanical interaction Hamiltonian can be written as,

$$\begin{aligned} H_{\text{OM,int}} &= \hbar(g_{\text{OM,PE}} + g_{\text{OM,Bnd}})\hat{x}\hat{a}^\dagger\hat{a} \\ &= \hbar g_0(\hat{b}^\dagger + \hat{b})\hat{a}^\dagger\hat{a}. \end{aligned} \tag{10.27}$$

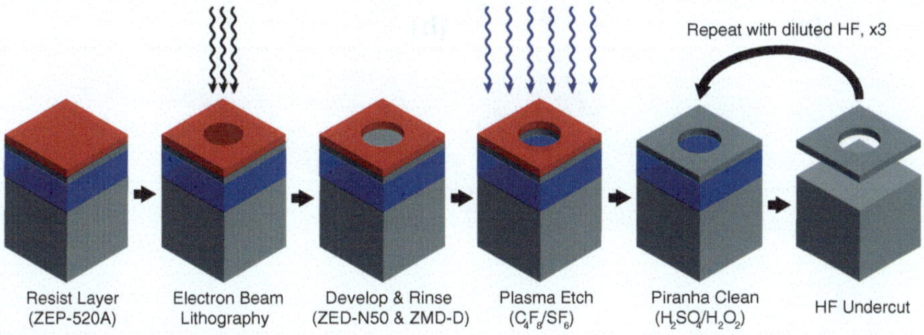

Fig. 10.11 Fabrication steps for a Si OMC device fabricated from SOI. See text for process description. Figure adapted from Ref. [47]

10.3 Fabrication and Characterization of Devices

10.3.1 Fabrication

The steps taken to fabricate the Si devices presented in this chapter are briefly summarized in Fig. 10.11. The starting point for device fabrication is a silicon-on-insulator (SOI) wafer, which has a thin top Si device layer (~200 nm) sitting on top of a buried silicon dioxide layer (3 μm) which itself is atop a Si handle wafer. The fabrication procedure involves one layer of electron-beam lithography (Vistec 100 kV lithography system with ZEP-520A resist), a dry plasma etch (SF_6/C_4F_8) to transfer the pattern into the top Si device layer, and a final wet etch (1:1 HF : H_2O) of the underlying silicon dioxide layer to create a free-standing structure. It was found that the mechanical and optical resonator quality factors were highly dependent on the final chemical etch steps in the fabrication in which the Si surfaces properties are determined. The recipe giving the best results was found to involve performing repeated Piranha (3:1 H_2SO_4 : H_2O_2) cleans (600 s, stirred at 250 rpm), followed by hydrogen termination using a weak 1:20 HF : H_2O solution (60 s) and a final DI water bath (30 s $\times$ 2). It is believed that such a preparation creates a stable hydrogen-terminated surface that is desirable for both the optical [46] and mechanical response of the structure.

Fabrication of the Si OMC cavities studied here benefit from highly optimized and established materials processing techniques that have been developed for the microelectronic industry and the availability of SOI wafers. Even though our focus here is on Si, similar techniques have recently been developed for materials such as diamond [53–55] and silicon carbide [56, 57], which are expected to have excellent optical and mechanical properties.

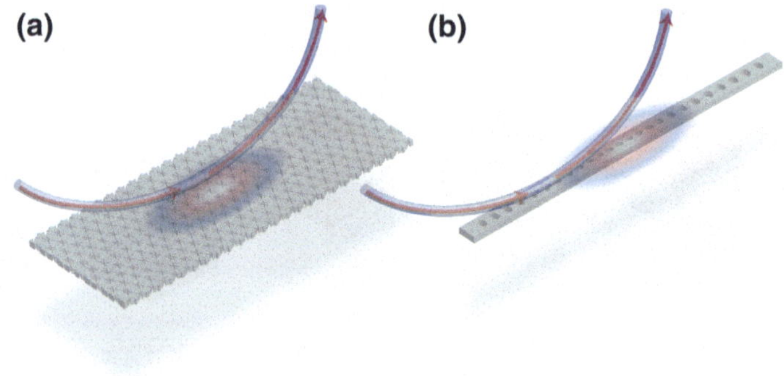

Fig. 10.12 Schematic showing the alignment and placing of a dimpled fiber taper in the near field of **a** a quasi-2D snowflake substrate cavity and **b** a quasi-1D nanobeam cavity. This sort of coupling does not rely on phase-matching and can provide efficient bi-directional coupling to such Si OMC cavities. Note that in many cases the taper is mechanically stabilized by bringing it into contact with the surface of the chip

10.3.2 Fiber-Taper Optical Coupling and Characterization

In order to rapidly test devices on a wafer scale, we have developed an optical probing technique for micro- and nano-photonic devices that utilizes a tapered optical fiber, heated and stretched down to a diameter on the order of a micron, and in which a small dimple has been flame polished [58, 59] to allow for local device probing. The first and most simple way to use a dimpled fiber is to directly couple to an optical resonator via the tapers evanescent field. Typically, a dimpled fiber taper is mounted in a "U"-shape with the dimpled region extending at the end of the "U", and brought into the near-field of a given optical resonator using electronically controlled stages with sub-micron-level positioning accuracy and visual feedback under a microscope. The taper is then placed in contact with the surface of the chip, close to the optical cavity, as shown in Fig. 10.12 for a snowflake and nanobeam OMC cavity. This induces some near field evanescent coupling between the resonator's modes and the guided modes of the fiber. The optical resonator can be then probed in both reflection and transmission using both directions of the fiber waveguide. Such fiber coupling systems have been demonstrated in room temperature ambient conditions as well as in cryogenic vacuum environments. In air, rapid characterization of a large number of devices on a chip is possible, facilitating a quick turnaround between design, fabrication, and characterization of devices.

The type of coupling achieved using this direct fiber taper coupling scheme is bi-directional in that light from the resonator can escape into directions of the fiber taper waveguide. This so-called "two-sided coupling" is schematically illustrated in Fig. 10.13a. There are two propagating modes in the fiber (L and R), both of which couple to the optical resonance at a rate $\kappa_e/2$. The total loading of the optical cavity is then $\kappa = \kappa_e + \kappa_i$, where κ is the optical energy decay rate of the loaded cavity and κ_i is the intrinsic energy loss rate that consists only of the undetected

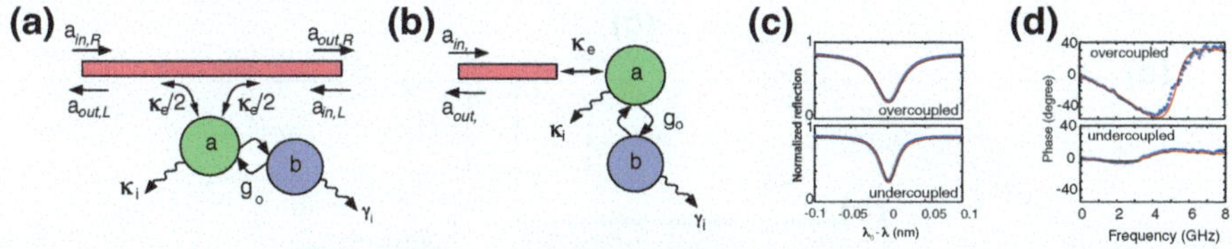

Fig. 10.13 **a** Bi-directional evanescent coupling geometry, in which the "transmitted" field goes into the forward a_{out} waveguide channel and the "reflected" field goes into the backward a'_{out} waveguide channel. This is the coupling geometry we will be focusing on in this work, in which the laser input channel is a_{in} and the detection channel is the forward waveguide channel, a_{out}. **b** Single-sided end-fire coupling geometry. This is the ideal measurement geometry since $\kappa = \kappa_{\mathrm{e}}$, in which all of the optical signal that is coupled into the cavity can be, in principle, collected and detected in the a_{out} channel. **c**, and **d** show the intensity and phase shift of the reflected laser light versus frequency for the over- ($\kappa_{\mathrm{e}} > \kappa$) and under- ($\kappa_{\mathrm{e}} < \kappa$) coupled cases of the single-sided cavity coupling scheme. Note that for the bi-directional coupling scheme, the transmitted light intensity and phase will be that of the undercoupled case shown in **c** and **d**

optical loss channels in the system (due to scattering, absorption, and coupling to higher order parasitic guided modes of the fiber). The situation, considering only the optical degree of freedom and neglecting any optomechanical coupling, is then well described by a Heisenberg-Langevin equation,

$$\frac{\mathrm{d}}{\mathrm{d}t}\hat{a} = -(i\Delta + \frac{\kappa}{2})\hat{a} - \sqrt{\kappa_{\mathrm{e}}/2}(\hat{a}_{\mathrm{in,L}} + \hat{a}_{\mathrm{in,R}}) - \sqrt{\kappa_{\mathrm{i}}}\,\hat{a}_{\mathrm{in,i}}, \tag{10.28}$$

and input–output boundary conditions,

$$\hat{a}_{\mathrm{out,k}} = \hat{a}_{\mathrm{in,k}} + \sqrt{\frac{\kappa_{\mathrm{e}}}{2}}\hat{a}, \qquad k = \mathrm{L, R}. \tag{10.29}$$

Assuming that a "good" coupling has been achieved, i.e., there is minimal insertion loss when the taper touches the chip and the observed transmission spectrum is flat away form the resonance, extracting the optical parameters of the system is straightforward. A measurement of the transmitted optical power versus laser frequency results in a (undercoupled) curve resembling that shown in Fig. 10.13c. The total optical cavity decay rate κ can be read from linewidth, while the transmission depth T_d can be used to find $\kappa_{\mathrm{e}}/\kappa = 1 - \sqrt{T_d}$.

The symmetric two-sided coupling achieved by direct taper coupling to the cavity has one major technical drawback when it comes to measurements in optomechanics. Considering that only one end of the waveguide is sent to the detection setup, 50 % of the photons leaving the cavity are automatically lost due to the presence of the alternate unmonitored propagation direction for light in the fiber. Though it is possible in principle to fully duplicate the measurement apparatus for both channels of the waveguide and combine the results to avoid loss of information for certain measurements, it is not obvious how to extend this to all experiments (e.g., generation of

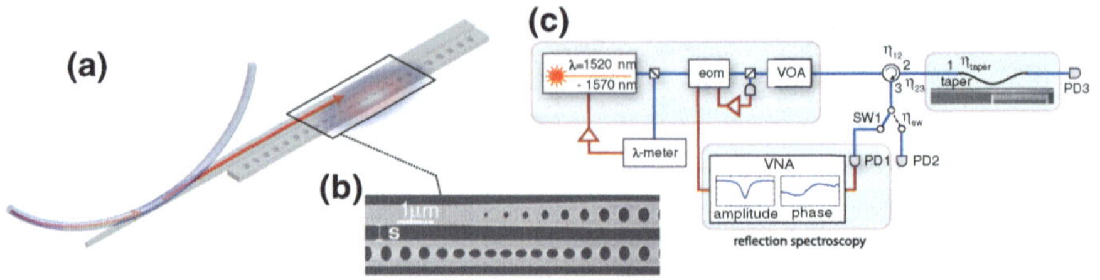

Fig. 10.14 **a** Schematic for a two-stage adiabatic taper coupler. The dimpled fiber taper is placed on a tapered Si waveguide with an effective index that crosses the fundamental mode of the fiber. **b** The Si waveguide is brought into the near field of the OMC cavity by lithographically defining the gap *S*. Terminating the Si coupling waveguide with a series of holes forming a photonic crystal mirror prevents leakage of light past the OMC cavity in the transmission channel. All the light coupled radiated by the OMC cavity is then captured in the reflection channel of the Si waveguide. **c** An experimental setup used to characterize optomechanical devices. Laser light goes through an electro-optic modulator (eom) before being coupled into the device under test, generating optical sidebands at a microwave modulation frequency. The reflected optical signal from the device is separated using a circulator and sent to a high-speed photodetector (PD1). The microwave signal from the photodetector corresponding to beating between the optical carrier and its optical sidebands is then sent into a vector network analyzer (VNA) which is also used to modulate the eom. This allows the measurement and characterization of the amplitude and phase response of an OMC cavity to a generated optical sideband near the optical cavity resonance

squeezed light [11]). An optimal scheme would consist of coupling a single-ended waveguide to the optical cavity, such that photons leaving the cavity can only enter a single propagating mode. Such a scheme is achievable using a dimpled fiber taper with the OMC cavities studied here through the use of a two-stage-coupler as shown in Fig. 10.14 and detailed in recent experiments [60, 61].

In the first stage, the adiabatic mode conversion stage, the propagating mode of the fiber taper is converted into the fundamental TE mode of a nanobeam waveguide. This is accomplished by creating a region of waveguide (roughly 30 μm) that has tapered profile, the details of which can be found in Ref. [62]. The second stage of the adiabatic coupler consists bringing this waveguide close to the cavity to achieve the desired near-field coupling. This distance, labeled S in Fig. 10.14b, can be continuously varied, giving fine control over the coupling. Moreover, there is no physical contact between the coupling waveguide and the nanobeam, better isolating the mechanical resonator from loss channels. The waveguide is then terminated in a way to minimize scattering into vacuum by introducing holes that gradually increase in size, as shown in Fig. 10.14b. This termination eliminates one of the two directions that light can propagate out of the system, forcing the light to leave in the backward propagating channel of the fiber. The Heisenberg-Langevin equations for the optical subsystem is

$$\frac{\mathrm{d}}{\mathrm{d}t}\hat{a} = -(i\Delta + \frac{\kappa}{2})\hat{a} - \sqrt{\kappa_\mathrm{e}}\,\hat{a}_{\mathrm{in,WG}} - \sqrt{\kappa_\mathrm{i}}\,\hat{a}_{\mathrm{in,i}}, \tag{10.30}$$

where

$$\hat{a}_{\text{out,WG}} = \hat{a}_{\text{in,WG}} + \sqrt{\kappa_e}\hat{a}. \tag{10.31}$$

It is not possible to distinguish between overcoupled ($\kappa_e > \kappa_i$) and undercoupled ($\kappa_e < \kappa_i$) devices by only looking at the reflected power (see Fig. 10.13c). It is therefore important to consider the phase response. Optical sidebands of the laser are generated using an electro-optic modulator (labeled 'eom') in Fig. 10.14c. This sideband is swept over the cavity by changing the modulation frequency, while its phase relative to the carrier is monitored using a vector network analyzer. The sideband transmission is found to be (in the absence of the coupled mechanical response)

$$t(\Delta_{\text{mod}}) = 1 - \frac{\kappa_e}{i(\Delta - \Delta_{\text{mod}}) + \frac{\kappa}{2}}. \tag{10.32}$$

The resulting phase response is shown in Fig. 10.14e. By normalizing the phase response to a detuning far from the cavity resonance, and then fitting the normalized phase response to the argument of $t(\Delta_{\text{mod}})/t(0)$, we obtain the linewidth κ and coupling rate κ_e unambiguously.

10.3.3 Characterization of Mechanical Properties

Having characterized the optical properties (ω_{opt}, κ, and κ_e) we turn our attention to the mechanical and optomechanical properties of the system: the mechanical frequency Ω_M, intrinsic mechanical linewidth γ_i, and optomechanical coupling rate g_0. These can be inferred by studying modifications to the optical cavity response due to the coupled phonon resonance. In particular, an effect akin to electromagnetically induced transparency (EIT) in 3-level atoms can be used to extract the optomechanical system parameters [45, 63, 64]. For a laser drive tuned to frequency $\omega_L = \omega_{\text{opt}} - \Delta$, with Δ close to the mechanical frequency, an effective beam-splitter interaction between the optical and mechanical cavities is induced with a characteristic rate of $G = |\alpha_0|g_0$ ($|\alpha_0|$ is the coherent state amplitude for the driven optical resonance). A modulated sideband at two-photon detuning Δ_{mod} will have a transmission coefficient [6]

$$t(\Delta_{\text{mod}}) = 1 - \frac{\kappa_e/2}{i(\Delta - \Delta_{\text{mod}}) + \frac{\kappa}{2} + \frac{|G|^2}{i(\Omega_M - \Delta_{\text{mod}}) + \gamma_i/2}}. \tag{10.33}$$

The experimental setup used to measure $t(\Delta_{\text{mod}})$ is similar to that shown in Fig. 10.14c. When the modulated tone is roughly equal to the mechanical frequency, the periodically oscillating radiation pressure force induced by the beating of the sideband and the carrier drives the mechanical motion to large amplitudes, modifying the transmission of the sideband. This is the physics captured in Eq. (10.33). This type of mechanical characterization, used in Refs. [6, 7, 10, 42], requires accurate calibration of the intracavity photon number since $|\alpha_0| = \sqrt{n_c}$ is required to convert G to g_0.

Alternatively, with no external modulation of the carrier, the mechanical motion due to thermal noise will generate a sideband on the carrier and the power spectral density of this thermal noise can be measured by heterodyne detection of the transmitted light. This detected spectrum is Lorentzian in shape and has a linewidth of approximately $\gamma_\pm = \gamma_\mathrm{i} \pm |\Gamma_\mathrm{opt}|$, $|\Gamma_\mathrm{opt}| \approx 4|G|^2/\kappa$ when $\Delta = \pm\Omega_\mathrm{M}$, with the $\Delta = -\Omega_\mathrm{M}$ measurement only achievable for optical powers where $\gamma_i > |\Gamma_\mathrm{opt}|$. By sweeping the optical power, and assuming that γ_i is constant, the value of G as a function of optical power can measured. Accurate calibration of $|\alpha_0|$ is again required for extraction of g_0. This method was used in conjunction with EIT spectroscopy in Refs. [7, 10].

Finally, in cases where g_0 is too small to accurately extract its value from optical modification of the mechanical linewidth, a fully calibrated thermometry procedure can be undertaken. The thermal Brownian motion power spectral density (PSD) is detected and the g_0 is extracted by assuming a bath and mode temperature. Such a procedure requires precise calibration of the inefficiencies of the optical train, the phase and intensity noise on the laser [11], as well as an accurate understanding of the gain and conversion ratios in the detector and microwave part of the setup [37, 40, 65].

10.3.3.1 Nanobeam in a Phononic Shield

The phononic bandgaps in the quasi-1D nanobeam OMCs studied here are symmetry-dependent, and as explained in Sect. 10.2.2.1, this leaves open channels for phonons to radiate away from the localized resonance.[7] By surrounding the nanobeam with a phononic crystal possessing a full phononic bandgap, such as the cross crystal introduced in Sect. 10.2.2.2, residual disorder-induced clamping losses can be eliminated from the system. Simulations taking into account disorder in the placement of the nanobeam holes show that given a certain amount of relative disorder, clamping losses can be reduced by six orders of magnitude through incorporation of a cross crystal incorporating six unit cells [47]. In experiments, we found that at cryogenic temperatures (~10 K), the mechanical mode quality factors of nanobeams with phononic shields were greater than those without phononic shields. The nanobeams without shields exhibited mechanical Qs not exceeding 3×10^4, while the shielded nanobeams had Q's approaching 10^6.

A shielded nanobeam OMC cavity, detailed in Ref. [10], is shown above in Fig. 10.15a. Five layers of crosses were used in the phononic shield in this experiment to help further isolate the localized cavity phonon resonance at frequency $\Omega_\mathrm{M}/2\pi = 5.1$ GHz from the environment. The fundamental optical resonance of the cavity at wavelength $\lambda = 1544.8$ nm is strongly coupled to localized phonon mode. Simulated mode profiles of the coupled optical and mechanical resonances are

[7] Recently 1D-OMC cavities with full phononic bandgaps have been demonstrated [66].

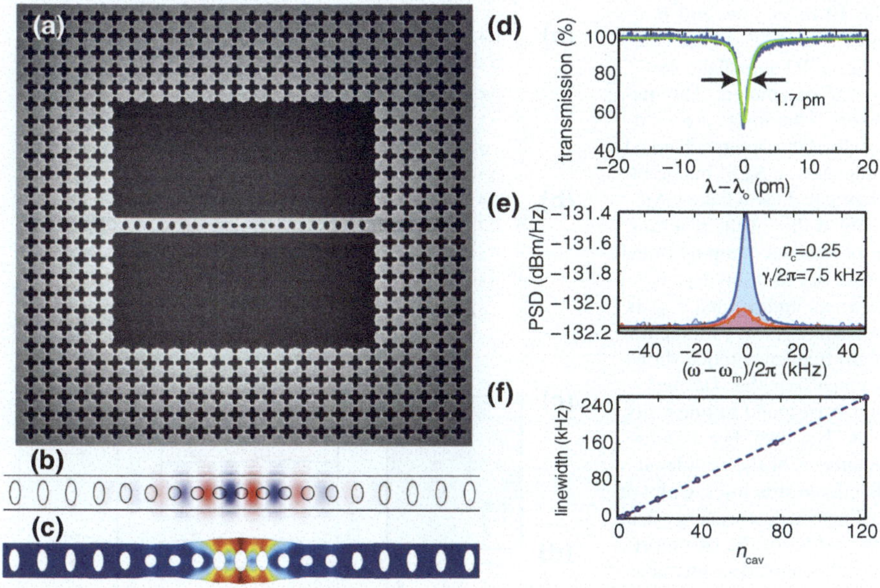

Fig. 10.15 **a** SEM image showing a cross crystal phononic shield surrounding a quasi-1D Si nanobeam OMC. FEM simulations of the strongly coupled **b**, optical and **c** mechanical resonance mode profiles. **d** Normalized optical transmission spectrum, centered at 1544.8 nm, showing the fundamental TE-like optical cavity mode of the nanobeam. **e** Optically transduced thermal noise power spectral density centered at the mechanical frequency, $\Omega_M/2\pi = 5.1$ GHz, with the input laser *red-detuned* (*red curve*) and *blue-detuned* (*blue curve*) from the cavity resonance by the mechanical frequency. **f** Plot of the measured optically-induced mechanical linewidth (Γ_{opt}) versus intra-cavity photon number (n_c). The slope of a linear fit to the measured data (dashed curve) is used to extract g_0

shown in Fig. 10.15b and c, respectively. The nanobeam was designed in accordance to the concepts described in Sect. 10.2.3.1, and in more detail elsewhere [47]. In this design, the nanobeam cavity was optimized in a way to minimize $g_{0,\mathrm{Bnd}}$ in favor of a large photoelastic contribution $g_{0,\mathrm{PE}}$ as it was found that the signs of the photoelastic and boundary perturbation optomechanical coupling oppose each other.

The optical characterization of the device was performed using direct fiber taper coupling. From the coupling depth and a the measured linewidth shown in Fig. 10.15d, an intrinsic optical quality factor of 1.2×10^6 was inferred. Mechanical spectroscopy data is shown in Fig. 10.15e, with the PSD of the transmitted laser light recorded at around the mechanical frequency for laser-cavity detunings of $\Delta = \pm\Omega_M$. An intrinsic mechanical linewidth of $\gamma_i/2\pi = 7.5$ kHz is inferred corresponding to a mechanical quality factor of $Q_m = 6.8 \times 10^5$. In addition the g_0 is measured by varying the optical power and recording the change in linewidth. From the linear fit shown in Fig. 10.15f, an optomechanical coupling rate of $g_0/2\pi = 1.1$ MHz is inferred.

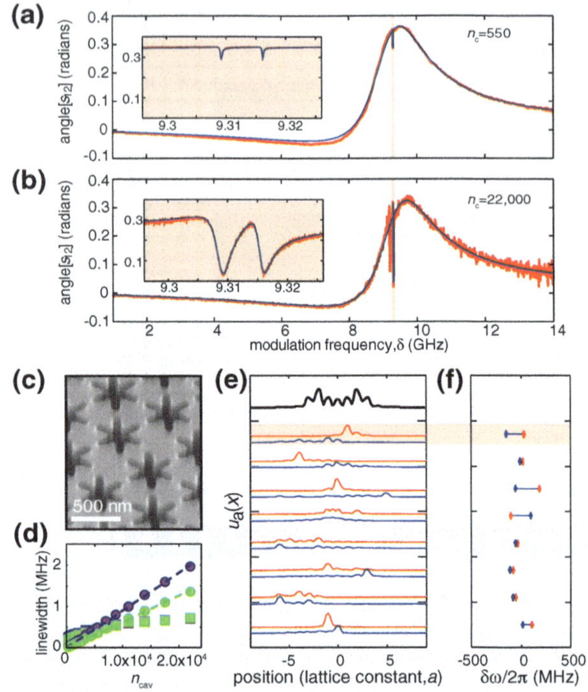

Fig. 10.16 **a** Low and **b** high power EIT spectra of a Si quasi-2D snowflake cavity with parameters. The fits shown in the insets are used to extract the optomechanical parameters. **c** SEM image of a fabricated snowflake OMC cavity. **d** Plot of the resulting fit mechanical damping rates versus n_c. Measured γ_i is shown as *squares* and Γ_{opt} is shown as *circles*, with the low (high) frequency mode shown in *green* (*purple*). *Dashed lines* correspond to linear fits to the Γ_{opt} data. For a representative 8 of the simulated disordered structures: **e** plot of the linear acoustic energy density profile for the two most strongly coupled mechanical resonances and **f** plot of the corresponding mechanical frequency difference for these mechanical resonances

10.3.3.2 Snowflake Cavities and Fabrication Disorder

The snowflake cavities presented in Sect. 10.2.3.2 were fabricated in a thin-film Si layer [42], and there properties were measured. Direct fiber taper coupling was used to determine the optical mode Q-factor. Afterwards, two-tone "EIT" spectroscopy was used to understand the mechanical mode properties using a set-up similar to that shown in Fig. 10.14c. In this scheme the VNA outputs a microwave tone that modulates the laser beam sent to the optical cavity, creating optical sidebands. The optomechanical interaction changes the phase between the sidebands and the carrier, and the extra phase imparted on the sidebands is measured by the VNA by taking the output of the high speed photodetector. The normalized phase response of the VNA (angle[$s_{12}(\delta)$]) for low and high optical input powers are shown in Figs. 10.16a, b respectively, for a laser-cavity detuning of $\Delta = \Omega_M$. The measured spectra show two mechanical modes with frequencies at 9.309 GHz and 9.316 GHz. By varying the optical power, and fitting these measured curves to the model described in Sect. 10.3.3, the intrinsic and damped mechanical linewidth can be measured. These measurements are plotted in Fig. 10.16d against optical power. The linear fits of Γ_{opt} are used to infer a value for $g_0/2\pi$, which is found to be 220 kHz for the upper frequency mode and 180 kHz for the lower frequency mode.

The presence of two nearly-degenerate and strongly coupled mechanical resonances likely arises from the effects of disorder in the fabricated structure. The flat dispersion (low group velocity) of the mechanical waveguide mode from which the snowflake cavity resonances are formed (see Ref. [39] for a detailed study of these guided modes) causes the mechanical mode spectrum to be highly sensitive to fabrication disorder. In fact, this flat dispersion allows an engineered defect consisting of only a 3 % modulation of the snowflake radius to be sufficient for localization of the mechanical mode, as described in Sect. 10.2.3.2. This extreme sensitivity to disorder makes exact determination of the mechanical mode structure from the measured spectra difficult. The simulated linear mechanical energy distribution along the length of the cavity for the two most strongly coupled mechanical modes of a series of disordered snowflake cavities is shown in Fig. 10.16e. In Fig. 10.16f we show the corresponding frequency splitting of these two modes. Here, disorder was introduced in the simulation by varying the snowflake hole position and dimensions with a normal distribution of 2 % (we expect disorder in the radii to be on the order of 1–2 % from fits to SEM images). The simulated disordered structures are found to yield similar mode couplings and mode frequency splitting as that in the measured snowflake device.

10.4 Outlook and Future Directions

A survey of a variety of thin-film OMC cavity structures, and their measured optical, mechanical, and optomechanical coupling properties are presented in Table 10.1. From this table it is clear that a variety of different geometries have been realized, utilizing 1D and 2D crystal structures with partial and full phononic bandgaps. New materials have also begun to be used to create OMC structures, including piezoelectrics such as AlN which offer the promise of integrating opto-mechanical and electro-mechanical functionality into the same device. For the most part, however, optomechanical crystals have thus far been experimentally realized only as individual cavity elements. Given the chip-scale nature of thin-film optomechanical crystal devices, and the ability to "wire up" different elements via photonic or phononic waveguides, an interesting future direction will be to explore the classical and quantum properties of arrays of coupled optomechanical crystal cavities [2, 4, 5, 8, 67–69].

The quantum nonlinear regime of cavity optomechanics, which requires both sideband resolution ($(\kappa/\Omega_{\mathrm{M}}) < 1$) and strong quantum coupling ($(g_0/\kappa) > 1$), remains completely unexplored experimentally. This regime would enable one to create quantum phononic gates for performing quantum information processing tasks, to generate non-Gaussian states of light through a photon-blockade-like effect, and to perform quantum non-demolition measurements of either photon or phonon number [70–73]. As such, it is interesting to consider the limits of optomechanical coupling in OMCs.

A back of the envelope estimate of the requirements needed to reach the strong quantum coupling limit can be made by considering the moving boundary contribu-

Table 10.1 Survey of experimentally realized thin-film optomechanical crystal devices

	T (K)	$\Omega_M/2\pi$ (GHz)	Q_m	$g_0/2\pi$ (kHz)	Phonon confinement	Material	Ref.
Eichenfield et al. (2009)	300	2	10^3–10^4	220	p1D	Si	[37]
Alegre et al. (2011)	30	1	10^4	60	f2D (+)	Si	[40]
Gavartin et al. (2011)	300	0.9	10^2–10^3	93	p2D	InP	[65]
Safavi-Naeini et al. (2011)	30	3.6	10^4	800	p1D and f2D (+)	Si	[6]
Chan et al. (2011)	20	3.7	10^5	800	p1D and f2D (+)	Si	[7]
Chan et al. (2012)	10	5.1	10^6	1,100	p1D and f2D (+)	Si	[10]
Bochmann et al. (2013)	300	4.2	10^3	30	p1D	AlN	[76]
Safavi-Naeini et al. (2014)	10	9.3	10^4–10^5	220	f2D (∗)	Si	[42]
Gomis-Bresco et al. (2014)	300	3-4	10^3	1,200	f1D	Si	[66]

Phonon confinement is obtained through full (f) and partial (p) bandgaps in quasi-1D and quasi-2D geometries. Systems using cross (+) and snowflake (∗) crystals are also indicated

tion to the optomechanical coupling. Assuming that the spatial extent of the photon cavity resonance can be limited to roughly λ_0/n (where λ_0 is the free-space optical wavelength for optical frequency $\omega_o = 2\pi c/\lambda_0$, and n is the refractive index of the material forming the OMC cavity), the optomechanical coupling will be roughly $g_0/2\pi \approx (n\omega_o/\lambda_0)x_{\text{zpf}}$. For typical solid state materials, with densities $\rho \sim 1$ g/cm^3 and Young's modulus E $\sim$ 100 GPa, the zero-point-motion amplitude of the fundamental mechanical resonances of a given structure is approximately $x_{\text{zpf}} \sim 1 - 10$ fm, even for the complex OMC structures studied here. For a free-space wavelength of 1.5 μm and a refractive index of $n = 3.5$, this yields a g_0 of approximately 2 MHz, very close to the maximum optomechanical coupling realized so far in OMCs. The corresponding ratio of vacuum coupling to optical linewidth is then, $g_0/\kappa \sim Q_o(nx_{\text{zpf}}/\lambda_0)$, where Q_o is the optical cavity resonance Q-factor and $\eta \equiv (nx_{\text{zpf}}/\lambda_0)$ is a Lamb-Dicke parameter, of order 10^{-8} in the currently realized OMC cavities.

For the optimized nanobeam OMC device of Ref. [10], the optical Q-factor was measured to be as high as $Q_o = 1.2 \times 10^6$ (limited by optical scattering and absorption), yielding a measured ratio of $g_0/\kappa = 0.0079$, still significantly far from the strong quantum coupling regime. One idea for increasing the optomechanical Lamb-Dicke parameter involves the use of nanoscale slots within OMC structures [74]. Similar structures have been proposed as a means in nanophotonic waveguides and cavities to generate large Purcell enhanced spontaneous emission of emitters embedded in the slot. In the case of optomechanics, nanoscale slots can be used to create extremely large electric field strengths at the material boundaries, greatly enhancing the moving boundary contribution to the optomechanical coupling. One may also seek to incorporate new materials with large photoelastic coefficients, such as GaAs, into OMCs so as to increase the attainable volume contribution to the optomechanical coupling. Further improvements in optical Q may also be expected [75], and looking to the future it will be interesting to see if this along with new clever techniques for increasing the optomechanical Lamb-Dicke parameter may be employed to reach the quantum nonlinear regime of cavity optomechanics.

Acknowledgments The authors would like to acknowledge the significant contributions to this work by Jasper Chan, Matt Eichenfield, Jeff Hill, Simon Gröblacher, Thiago Alegre, Alex Krause, Sean Meenehan, and Justin Cohen. The work was supported by the DARPA ORCHID and MESO programs, the Institute for Quantum Information and Matter, an NSF Physics Frontiers Center with support of the Gordon and Betty Moore Foundation, and the Kavli Nanoscience Institute at Caltech. ASN gratefully acknowledges support from NSERC.

References

1. A.H. Safavi-Naeini, O. Painter, New J. Phys. **13**, 013017 (2011)
2. D. Chang, A.H. Safavi-Naeini, M. Hafezi, O. Painter, New J. Phys. **13**, 023003 (2011)
3. G. Heinrich, M. Ludwig, J. Qian, B. Kubala, F. Marquardt, Phys. Rev. Lett. **107**(4), 043603 (2011)
4. M. Schmidt, M. Ludwig, F. Marquardt, New J. Phys. **14**(12), 125005 (2012)
5. M. Schmidt, V. Peano, F. Marquardt (2013), arXiv:1311.7095
6. A.H. Safavi-Naeini, T.P.M. Alegre, J. Chan, M. Eichenfield, M. Winger, Q. Lin, J.T. Hill, D. Chang, O. Painter, Nature **472**, 69 (2011)
7. J. Chan, T.P.M. Alegre, A.H. Safavi-Naeini, J.T. Hill, A. Krause, S. Gröblacher, M. Aspelmeyer, O. Painter, Nature **478**, 89 (2011)
8. A.H. Safavi-Naeini, J. Chan, J.T. Hill, T.P.M. Alegre, A. Krause, O. Painter, Phys. Rev. Lett. **108**(3), 033602 (2012)
9. J.T. Hill, A.H. Safavi-Naeini, J. Chan, O. Painter, Nat. Commun. **3**, 1196 (2012)
10. J. Chan, A.H. Safavi-Naeini, J.T. Hill, S. Meenehan, O. Painter, Appl. Phys. Lett. **101**(8), 081115 (2012)
11. A.H. Safavi-Naeini, J. Chan, J.T. Hill, S. Groeblacher, H. Miao, Y. Chen, M. Aspelmeyer, O. Painter, New J. Phys **15**(3), 035007 (2013)
12. COMSOL Multiphysics 3.5, http://www.comsol.com/
13. S.G. Johnson, J.D. Joannopoulos, Opt. Express **8**(3), 173 (2001)
14. J.V. Bladel, *Electromagnetic Fields*, 2nd edn. (IEEE Press, Wiley-Interscience, 2007)
15. S. Haroche, J.M. Raimond, *Exploring the Quantum: Atoms, Cavities, and Photons* (Oxford University Press, USA, 2006)
16. C. Kittel, *Introduction to Solid State Physics* (Wiley, New Jersey, 2005)
17. S. John, Phys. Rev. Lett. **58**(23), 2486 (1987)
18. E. Yablonovitch, T. Gmitter, J.P. Harbison, R. Bhat, Appl. Phys. Lett. **51**, 2222 (1987)
19. M.S. Kushwaha, P. Halevi, L. Dobrzynski, B. Djafari-Rouhani, Phys. Rev. Lett. **71**(13), 2022 (1993)
20. M. Sigalas, E. Economou, J. Sound Vib. **158**(2), 377 (1992)
21. J. Joannopoulos, S. Johnson, J. Winn, R. Meade, *Photonic Crystals: Molding the Flow of Light* (Princeton University Press, Princeton, 2008)
22. B.A. Auld, *Acoustic fields and waves in solids*, vol. 1 (Wiley, New York, 1973)
23. D. Marcuse, *Theory of dielectric optical waveguides* (Access Online via Elsevier, Amsterdam, 1974)
24. K. Graff, *Wave Motion in Elastic Solids* (Dover Publications, New York, 1975)
25. A. Yariv, P. Yeh, *Optical waves in crystals*, vol. 5 (Wiley, New York, 1984)
26. A.N. Cleland, *Foundations of Nanomechanics: from Solid-State Theory to Device Applications* (Springer, Heidelberg, 2003)
27. A. Fallahkhair, K. Li, T. Murphy, J. Lightwave Technol. **26**(11), 1423 (2008)

28. A.N. Cleland, M.L. Roukes, Nature **392**, 160 (1998)
29. M.D. Chabot, J. Moreland, L. Gao, S.H. Liou, C. Miller, J. Microelectromech. Syst. **14**(5), 1118 (2005)
30. P.H. Kim, C. Doolin, B.D. Hauer, A.J.R. MacDonald, M.R. Freeman, P.E. Barclay, J.P. Davis, Appl. Phys. Lett. **102**(5), 053102 (2013)
31. J.S. Foresi, P.R. Villeneuve, J. Ferrera, E.R. Thoen, G. Steinmeyer, S. Fan, J.D. Joannopoulos, L.C. Kimerling, H.I. Smith, E.P. Ippen, Nature **390**(6656), 143 (1997)
32. C. Sauvan, P. Lalanne, J.P. Hugonin, Phys. Rev. B **71**(16), 165118 (2005)
33. P. Barclay, K. Srinivasan, O. Painter, Opt. Express **13**, 801 (2005)
34. L.D. Haret, T. Tanabe, E. Kuramochi, M. Notomi, Opt. Express **17**(23), 21108 (2009)
35. K. Schwab, E.A. Henriksen, J.M. Worlock, M.L. Roukes, Nature **404**(6781), 974 (2000)
36. S.M. Meenehan et al. Thermalization properties at mK temperatures of a nanoscale optomechanical resonator with acoustic-bandgap shield (2014), arXiv:1403.3703
37. M. Eichenfield, J. Chan, R. Camacho, K. Vahala, O. Painter, Nature **462**(7269), 78 (2009)
38. M. Eichenfield, J. Chan, A.H. Safavi-Naeini, K.J. Vahala, O. Painter, Opt. Express **17**(22), 20078 (2009)
39. A.H. Safavi-Naeini, O. Painter, Opt. Express **18**(14), 14926 (2010)
40. T.P.M. Alegre, A. Safavi-Naeini, M. Winger, O. Painter, Opt. Express **19**, 5658 (2011)
41. P.L. Yu, K. Cicak, N. Kampel, Y. Tsaturyan, T. Purdy, R. Simmonds, C. Regal, (2013). arXiv:1312.0962
42. A.H. Safavi-Naeini, J.T. Hill, S. Meenehan, J. Chan, S. Groeblacher, O. Painter, (2014), arXiv:1401.1493
43. K. Srinivasan, O. Painter, Opt. Express **10**(15), 670 (2002)
44. Y. Akahane, T. Asano, B.S. Song, S. Noda, Nature **425**(6961), 944 (2003)
45. B.S. Song, S. Noda, T. Asano, Y. Akahane, Nat. Mater. **4**(3), 207 (2005)
46. M. Borselli, T.J. Johnson, O. Painter, App. Phys. Lett. **88**, 131114 (2006)
47. J. Chan, Laser cooling of an optomechanical crystal resonator to its quantum ground state of motion, Ph.D. thesis, California Institute of Technology 2012
48. A.R. Md Zain, N.P. Johnson, M. Sorel, R.M. De La Rue, Opt. Express **16**(16), 12084 (2008).
49. M. Notomi, E. Kuramochi, H. Taniyama, Opt. Express **16**(15), 11095 (2008)
50. J. Chan, M. Eichenfield, R. Camacho, O. Painter, Opt. Express **17**(5), 3802 (2009)
51. P.B. Deotare, M.W. McCutcheon, I.W. Frank, M. Khan, M. Loncar, Appl. Phys. Lett. **94**(12), 121106 (2009)
52. S.G. Johnson, M. Ibanescu, M.A. Skorobogatiy, O. Weisberg, J.D. Joannopoulos, Y. Fink, Phys. Rev. E **65**(6), 066611 (2002)
53. M.J. Burek, N.P. de Leon, B.J. Shields, B.J.M. Hausmann, Y. Chu, Q. Quan, A.S. Zibrov, H. Park, M.D. Lukin, M. Lonar, Nano Lett. **12**(12), 6084 (2012)
54. L. Li, M. Trusheim, O. Gaathon, K. Kisslinger, C.J. Cheng, M. Lu, D. Su, X. Yao, H.C. Huang, I. Bayn, J. Vac. Sci. Technol., B **31**(6), 06FF01 (2013)
55. P. Rath, S. Khasminskaya, C. Nebel, C. Wild, W.H. Pernice, Nat. Commun. **4**, 1690 (2013)
56. M. Radulaski, T.M. Babinec, S. Buckley, A. Rundquist, J. Provine, K. Alassaad, G. Ferro, J. Vuckovic, Opt. Express **21**(26), 32623 (2013)
57. J. Cardenas, M. Zhang, C.T. Phare, S.Y. Shah, C.B. Poitras, B. Guha, M. Lipson, Opt. Express **21**(14), 16882 (2013)
58. C.P. Michael, M. Borselli, T.J. Johnson, C. Chrystal, O. Painter, Opt. Express **15**, 4745 (2007)
59. B.D. Hauer, P.H. Kim, C. Doolin, A.J. MacDonald, H. Ramp, J.P. Davis, (2014), arXiv:1401.5482
60. J. Thompson, T. Tiecke, N. de Leon, J. Feist, A. Akimov, M. Gullans, A. Zibrov, V. Vuleti, M. Lukin, Science **340**(6137), 1202 (2013)
61. A.H. Safavi-Naeini, S. Groblacher, J.T. Hill, J. Chan, M. Aspelmeyer, O. Painter, Nature **500**(7461), 185 (2013)
62. S. Groeblacher, J.T. Hill, A.H. Safavi-Naeini, J. Chan, O. Painter, Appl. Phys. Lett. **103**(18), 181104 (2013)
63. M. Fleischhauer, A. Imamoglu, J.P. Marangos, Rev. Mod. Phys. **77**(2), 633 (2005)

64. S. Weis, R. Rivière, S. Deléglise, E. Gavartin, O. Arcizet, A. Schliesser, T.J. Kippenberg, Science **330**, 1520 (2010)
65. E. Gavartin, R. Braive, I. Sagnes, O. Arcizet, A. Beveratos, T.J. Kippenberg, I. Robert-Philip, Phys. Rev. Lett. **106**(20), 203902 (2011)
66. J. Gomis-Bresco, D. Navarro-Urrios, M. Oudich, S. El-Jallal, A. Griol, D. Puerto, E. Chavez, Y. Pennec, D. Djafari-Rouhani, F. Alzina, A. Martnez, C.M. Sotomayor Torres, (2014), arXiv:1401.1691
67. S.J.M. Habraken, K. Stannigel, M.D. Lukin, P. Zoller, P. Rabl, New J. Phys. **14**(11), 115004 (2012)
68. A. Tomadin, S. Diehl, M.D. Lukin, P. Rabl, P. Zoller, Phys. Rev. A **86**(3), 033821 (2012)
69. M. Ludwig, F. Marquardt, Phys. Rev. Lett. **111**, 073603 (2013)
70. P. Rabl, Phys. Rev. Lett. **107**(6), 63601 (2011)
71. A. Nunnenkamp, K. Borkje, S. Girvin, Phys. Rev. Lett. **107**(6), 63602 (2011)
72. M. Ludwig, A.H. Safavi-Naeini, O. Painter, F. Marquardt, Phys. Rev. Lett. **109**(6), 063601 (2012)
73. K. Stannigel, P. Komar, S.J.M. Habraken, S.D. Bennett, M.D. Lukin, P. Zoller, P. Rabl, Phys. Rev. Lett. **109**, 013603 (2012)
74. M. Davanco, J. Chan, A.H. Safavi-Naeini, O. Painter, K. Srinivasan, Opt. Express **20**(22), 24394 (2012)
75. H. Sekoguchi, Y. Takahashi, T. Asano, S. Noda, Opt. Express **22**(1), 916 (2014)
76. J. Bochmann, A. Vainsencher, D.D. Awschalom, A.N. Cleland, Nat. Phys. (2013)

Chapter 11
Introduction to Microwave Cavity Optomechanics

Konrad W. Lehnert

Abstract In this chapter, I introduce the concepts of electromechanical superconducting resonant circuits, a topic known as microwave cavity optomechanics. As part of that introduction, I will provide: a review of the field's development, a discussion of its relationship to "optical" cavity optomechanics, and a description of its current progress. In addition, I derive in pedagogical detail the classical dynamics of a mechanical oscillator parametrically coupled to a resonant circuit. Finally, I show that the cavity optomechanical Hamiltonian is the quantum description for a such an electromechanical circuit.

11.1 Introduction

Microwave cavity optomechanics is the part of the field of nano- and micro-electromechanics (NEMS and MEMS) that investigates the interaction between mechanical motion and electrical energy stored in microwave resonant circuits. It is one of the most promising strategies for observing and exploiting quantum behavior of macroscopic mechanical oscillators. The topic is the electrical analog of cavity optomechanics but has it roots in circuit quantum electrodynamics (cQED), the search for quantum effects in NEMS devices, and the effort to detect gravitational waves. The name microwave cavity optomechanics is a misnomer as the electrically resonant structures are not literally *cavities* but resonant circuits. The misnomer "cavity" persists to strengthen the analogy to cavity optomechanics and to resolve an ambiguity; the terms "resonator" and "oscillator" apply equally well to both the

K. W. Lehnert (✉)
Department of Physics, University of Colorado, Boulder, CO, USA
e-mail: konrad.lehnert@jila.colorado.edu

K. W. Lehnert
JILA, National Institute of Standards and Technology, University of Colorado,
Boulder, CO, USA

M. Aspelmeyer et al. (eds.), *Cavity Optomechanics*, Quantum Science and Technology,
DOI: 10.1007/978-3-642-55312-7_11,

mechanical and electrical degrees of freedom but "cavity" is understood to refer only to the electrical degree of freedom.

11.1.1 Relationship to Cavity Optomechanics

The reemergence of interest in radiation pressure in optical cavities occurred because advances in microfabrication enabled the creation of high finesse optical cavities incorporating light and floppy mechanical oscillators [1–3]. This advent of cavity optomechanics sparked a new and vigorous effort to observe the quantum motion of a mechanical oscillator and the quantum nature of radiation pressure [4, 5]. A few years earlier, experiments with superconducting circuits had beautifully demonstrated the quantum nature of microwave light [6]. These experiments exploited superconducting microwave circuits to create artificial atoms (qubits) and cavities (superconducting resonators) and achieved coherent coupling between a single microwave photon and a qubit. It was natural to consider if the same sort of superconducting circuits could be used to create a strong interaction between microwave light and mechanical motion, revealing the quantum nature of both mechanical motion and *microwave* radiation pressure. Indeed, a superconducting resonant circuit with a mechanically compliant element (whose motion alters the resonance frequency of that circuit) is described by the same optomechanical Hamiltonian as a Fabry-Perot cavity with a movable mirror.

Working with microwave light and superconducting circuits has a number of appealing features. The superconducting circuitry provides a highly flexible, engineerable platform for creating particular optomechanical interactions, including interaction between mechanical oscillators and superconducting qubits. Although all experiments must be performed at low temperatures, typically below 1 K, these experimental techniques are also immediately compatible with low temperatures. At low temperatures, the random thermal forces that decohere the quantum states of mechanical oscillators are greatly reduced. More practically, microwave cavity optomechanics is free from the experimental nuisances associated with aligning and stabilizing an optical cavity. The microwave cavity, which is formed from a lithographically patterned metal film, is absolutely passively stable and rigid except for the mechanically compliant element of interest.

Against these appealing features of microwave, rather than optical, light one must weigh two consequences of the lower frequency, lower energy photons. First, the scale for optomechanical forces is proportional to the photon energy, and is therefore 4–5 orders of magnitude smaller for microwave photons compared to optical photons. Second, the lower photon energy precludes using the quantum efficient photodetection technology of optics. Nevertheless, the experiments that have demonstrated interaction between micro- and nano-mechanical oscillators and microfabricated electrical circuits have been at the forefront of optomechanical physics. Initial work focused on coupling nanomechanical oscillators to superconducting cavities. These results progressed from observing the thermomechanical motion of a mechanical oscillator cooled below 40 mK [7], to detecting the dynamical backaction of microwave fields on mechanical oscillators [8], allowing them to be cooled by radiation pressure. This radiation pressure cooling has been used to bring a mechanical oscillator close to its motional ground state [9].

Recent work has focused on overcoming the poor quantum efficiency of the microwave measurement and the weak interaction between microwaves and mechanical motion. The poor measurement efficiency has been tackled either by creating quantum limited microwave measurement tools [10] or by implementing a backaction evading measurement of just one quadrature of mechanical motion [11]. The weak interaction can be enhanced by exploiting the ability to confine microwave photons into tiny volumes, much smaller than a cubic wavelength, using lumped element resonant circuits [12] or advanced nanofabriction [13]. In a parallel development, mechanical oscillators have been coupled to superconducting qubits either dispersively or directly. Respectively, these experiments demonstrated that it is possible to infer the state of the qubit through a modification of the mechanical oscillator's response [14], and to prepare the mechanical oscillator in a pure quantum state containing zero or one phonon through interaction with the qubit [15].

11.1.2 Emergence from Resonant Gravitational Wave Detectors and Quantum NEMS

Experimental work in microwave cavity optomechanics emerged from quantum nano-electromechanical systems [16], but was preceded by superconducting resonant-mass gravitational-wave detectors. Much of the early work in building gravitational wave detectors focused on meter-sized mechanical oscillators into which a gravitational wave would deposit energy (Ref. [17] has a good review of the experimental efforts). Energy deposited in the oscillator would be detected if it exceeded $k_B(T + T_N)$ where $k_B T$ is the thermal energy available to the oscillator and T_N is the noise added by the displacement measurement expressed as an apparent increase in temperature [18]. In order to detect the deflection of the mechanical oscillator various schemes were proposed and implemented. These schemes include transducing motion into a changing capacitance or inductance [19], which would then result in a changing current or voltage. A related scheme envisioned coupling motion to a nanometer-sized gap between electrodes, modulating a tunneling current [20].

Rather than detect directly the current or voltage created by a time varying capacitance, several groups pursue schemes where mechanical motion alters the resonance frequency of a literal microwave cavity [21, 22] (also see Ref. [23] and references therein). The microwave cavities are typically cylindrical Helmholtz resonators distorted so that bottom and top surfaces are brought close together forming something like a parallel plate capacitor. Strain in the cavity itself alters the size of the gap between the capacitor plates. In related work, the coupling of motion to microwave energy is studied in sapphire cylinders, which are simultaneously high-Q microwave cavities and acoustical oscillators [24].

But for matters of scale, these gravitational wave detectors contain the same physics as the microwave optomechanical structures that incorporate nano- or micro-mechanical oscillators. The meter-scale gravitational-wave detectors achieve excellent energy sensitivity [17], display dynamical backaction damping of the mechanical motion [24], and have even been used in a proof-of-principle demonstration of a backaction evading scheme [25].

Although work with resonant-mass gravitational-wave detectors continues, the dominant strategy for detecting gravitational waves now uses kilometer-sized optical interferometers. The theory of resonant-mass gravitational-wave detectors has found a new application in quantum NEMS devices. Describing the performance of NEMS devices in the language of quantum measurement occurred when micro- and nano- electromechanics had progressed to the point that it was plausible to imagine observing the quantum motion of these much smaller, higher-frequency mechanical oscillators. Initially, much of the effort was devoted to coupling nano-mechanical oscillators to ultrasensitive nano-electronic transistors, such as single-electron transistors (SETs) [26, 27], quantum point contacts (QPCs) [28], atomic point contacts (APCs) [29] and superconducting quantum interference devices (SQUIDs) [30]. Most of the insights gained from this work were in fact about the physics of the transistors, as both the imprecision and backaction of the displacement measurement are determined by the physics of the transistor itself [31].

In order to more directly pursue the goals of observing quantum behavior in macroscopic mechanical motion, nanomechanical oscillators have been embedded in microwave cavities [7]. The backaction on the mechanical oscillator is then determined by the the quality of the microwave source, which can be quantum-limited inside an ultralow temperature dilution refrigerator cryostat. That is, a random backaction force associated with the measurement arises from the amplitude or phase fluctuations of the microwave tone that excites the cavity. If the microwave tone is in a pure coherent state, the amplitude and phase fluctuations are associated only with the fundamental quantum noise of that state.

Detecting the motion of a nanomechanical oscillator via parametric, or dispersive, coupling to a microwave cavity recapitulated both the development of the resonant-mass gravitational-wave detectors and superconducting qubits. In fact, it was the development of cQED that inspired the development of microwave cavity optomechanics measurements of nanomechanical oscillators. In detecting the state of a superconducting qubit, particularly the Cooper-pair box, a readout of the qubit state using a microwave resonator provided dramatic advantages [6] over direct measurements with an SET [32]. Because the best measurements of nanomechanical motion were accomplished at that time with an SET [26, 27], it was natural to ask if the same benefits of cavity readout could be realized for detection of nanomechanical motion.

11.2 Superconducting Resonant Circuits

A true microwave cavity, meaning an empty volume enclosed by metal, is not well suited measuring nanomechanical motion. While cavities of the type described in [23] have a volume of approximately, 1 cm^3 set by the wavelength of microwave signals, a nanomechanical beam oscillator has a much smaller size, perhaps 10^{-10} cm^3 (Fig. 11.1). It is clear that only a tiny fraction of the energy stored in the cavity would interact with the oscillator. Instead, microwave cavity optomechanical structures use transmission line resonators or lumped element circuits comprising a separate inductive and capacitive element because they can have mode volumes much less than one cubic wavelength. A transmission line resonator has only one dimension on the order of a wavelength, while a lumped element resonator can be much smaller than a wavelength in all dimensions. For example, the cross-sectional dimension for a typical transmission line resonator used in these experiments is 5 μm, limited by lithography or other constraints. As such, these circuits or transmission lines have mode volumes only moderately larger than the size of the mechanical oscillator itself.

11.2.1 Circuit Design

In order for these small volume "cavities" to have any detectable resonance—let alone a high quality factor Q—they must be built from superconducting metals.[1] Consequently, the experiments only operate at or below liquid helium temperatures. Even lower temperatures are required if one would like use the interaction of the mechanical oscillator with the microwave field to cool the mechanical oscillator to its ground state. For this type of cooling, the microwave cavity must itself be in a pure quantum state [9]; therefore, the cavity should be in equilibrium with a microwave environment at temperature $T \ll \hbar\omega_r/k_B$, where ω_r is the cavity resonance frequency. An ultralow temperature cryostat with $T < 200$ mK is usually required to meet this condition because practical aspects of microwave technology are much simpler at frequencies $\omega/2\pi < 26$ GHz.

The physical and technical constraints of microwave cavity optomechanics are similar to those of cQED and a type of astrophysical detector known as a microwave kinetic inductance detector (MKID). The first cavity optomechanical structures [7] looked rather like cQED devices or MKID structures (Fig. 11.2, c.f. Ref. [33]).

In contrast to the cQED experiments in which the interaction between a qubit and a cavity is strong even with no photons in the cavity, the limiting factor in microwave optomechanics is the weak interaction between the motion of an oscillator and a cavity's resonance frequency. To increase this small coupling the microwave cavities have evolved into LC resonant circuits. These structures comprise separate

[1] Consider that the skin depth of a good normal metal at microwave frequencies is about 1 μm, indicating that a substantial fraction of the mode would be in the lossy metal itself.

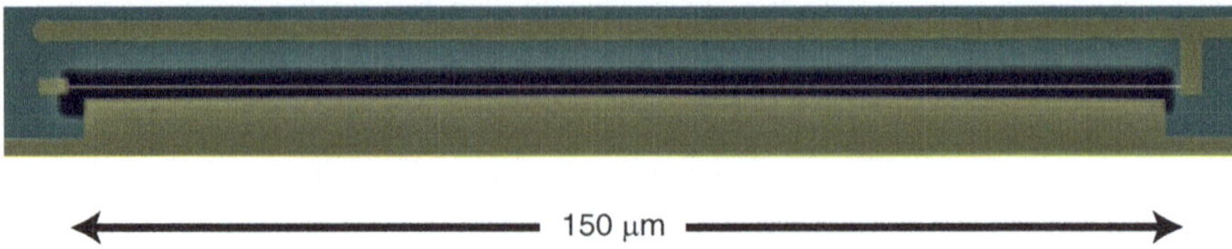

Fig. 11.1 A Nanomechanical oscillator embedded in a resonant circuit. In this device, a 150 μm long aluminum (*gold false color*) string is freely suspended over a hole etched into the silicon substrate (*green false color*). Motion of the string in the plane of the figure alters the capacitance between the two aluminum electrodes. Not visible in the figure is the inductor that completes the resonant circuit by connecting the two electrodes

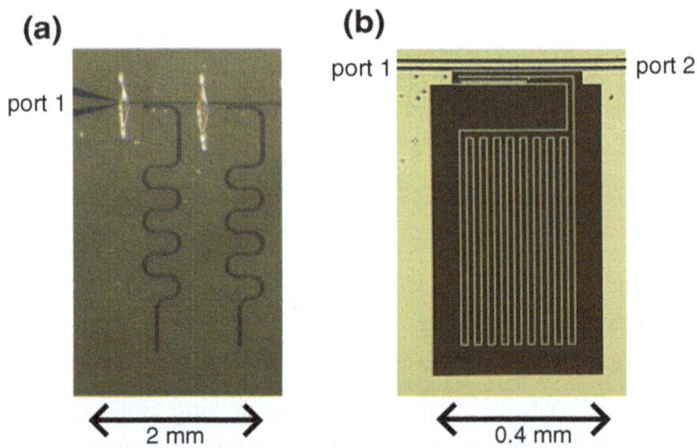

Fig. 11.2 Images of chips with microwave cavity optomechanical structures formed by patterning superconducting aluminum films. **a** The two meandering lines are coplanar waveguide (CPW) transmission line cavities. At the bottom of the meandering line the center of the coplanar waveguide structure is shorted to the adjacent ground planes, while the top is nearly an electrical open. The horizontal transmission line (called the feedline) passes near the electrically open ends of the cavities. Microwave power passing through the feedline excites the cavities and they resonate when the excitation frequency has a wavelength one-quarter of the cavity length. In this device, the two cavities have different lengths and consequently different resonance frequencies, allowing them to be studied separately. The wire-bonds visible in the image suppress the antisymmetric mode of the CPW feedline [34]. **b** This image shows a lumped element superconducting resonant circuit. The meandering wire forms the inductor, while the two ends of that wire fabricated close to one another form the capacitor. While this structures resonates at about the same frequency as in subfigure **a**, it is substantially smaller as indicated by the scale bars

inductive and capacitive elements and are more compact than a transmission line resonator with the same resonance frequency (Fig. 11.2b). Because they confine electromagnetic fields into even smaller volumes than their transmission line cousins, lumped element circuits can enhance the interaction between mechanical motion and electrical energy [12].

11.2.2 Circuit Analysis

Whether the resonant structure is a lumped element circuit or a transmission line cavity, it is useful to model the circuit as composed of separate inductor and capacitor elements. In particular, the interaction of a mechanical oscillator with the circuit is readily described as a capacitor whose capacitance varies with oscillator position. Although the experimental apparatus for studying these optomechanical structures is a full microwave network, when good microwave practice is followed one need only analyze electrical circuit models to describe the microwave resonance. In essence, good practice ensures that the left and right moving waves carrying microwave energy in a cable or transmission line do not couple to each other except at the ports as labeled in Fig. 11.3. Indeed, the simple one port network shown in Fig. 11.3a, is sufficient to understand the two-port microwave resonators used in the majority of microwave cavity optomechanics experiments [7–11]. Simple transformations extend the analysis of the one port network to either of the two types of two-port networks used, those represented by Fig. 11.3b and used in Refs. [7, 8, 10], or those represented by Fig. 11.3c and used in [9, 11].

Whether fabricated as lumped element circuits or as a transmission line resonators, all of the microwave cavities used in recent experiments [7–12] can be modeled as *parallel* LC circuits. The circuit shown in Fig. 11.2b is fabricated as a parallel LC circuit, while the quarter-wave transmission line cavities in Fig. 11.2a can be modeled as a parallel LC circuit. From standard microwave network analysis an electrical short at one end of a transmission line transforms to an open a quarter wavelength distant from the short. The lumped element model for such a quarter-wave resonator is then a parallel LC circuit because that circuit is also open on resonance. Finally, the transmission through a half-wave resonator, as implemented in [9, 11], can also be modeled as a parallel LC circuit, as it is essentially two quarter-wave resonators back-to-back. Loss in any of the resonators is modeled by including the R circuit element in Fig. 11.3a. A small capacitor C_c isolates the resonant circuit from power lost through to port 1 and dissipated in the source impedance Z_0.

For the circuit shown in Fig. 11.3a, straightforward circuit analysis yields the frequency-dependent impedance $Z_{\text{in}}(\omega)$ at port 1. One can relate the amplitude and phase of the voltage wave reflected from port 1 to the incident wave through the voltage reflection coefficient $\Gamma = (Z_{\text{in}}(\omega) - Z_0)/(Z_{\text{in}}(\omega) + Z_0)$, which can be written in terms of four dimensionless parameters as N/D with

$$N = i(\gamma_I - \gamma_E\varepsilon) + \gamma_I\gamma_E - 2z(1 - \gamma_I\gamma_E - i(\gamma_I + 2\gamma_E - 3\gamma_E\varepsilon)/2) + z^2(3i\gamma_E(1-\varepsilon) + \gamma_I\gamma_E - 1) + iz^3\gamma_E(1-\varepsilon) \tag{11.1}$$

$$D = i(\gamma_I + \gamma_E\varepsilon) - \gamma_I\gamma_E - 2z(1 + \gamma_I\gamma_E - i(\gamma_I - 2\gamma_E + 3\gamma_E\varepsilon)/2) + z^2(-3i\gamma_E(1-\varepsilon) - \gamma_I\gamma_E - 1) - iz^3\gamma_E(1-\varepsilon) \tag{11.2}$$

where $z = (\omega - \omega_r)/\omega_r$, $\omega_r = \sqrt{1/(L(C + C_c))}$, $\gamma_I = \omega_r L/R$, $\gamma_E = \omega_r C_c Z_0$, and $\varepsilon = C_c/(C + C_c)$. While this expression is simpler when written as function

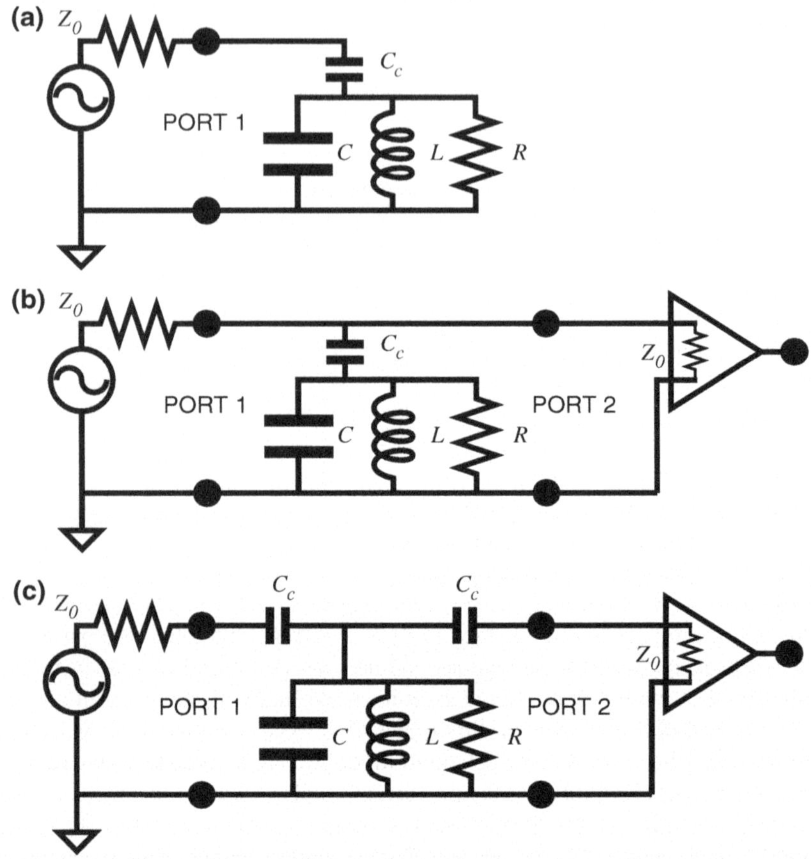

Fig. 11.3 Circuit diagrams for a lumped element model of a microwave cavity optomechanical structure. **a** The basic one-port network model of a resonant circuit. **b** The two-port model that describes the capacitively coupled feedline geometry of Fig. 11.2. **c** A two port network that describes the half wave resonators used for example in Ref. [35]

of (ω/ω_r), in this form each dimensionless parameter is small for a high-Q circuit probed near resonance. This exact expression captures more than is necessary to model a high-Q resonant circuit probed near its resonance frequency. From Eq. 11.1, one can find a simple approximate expression that describes the frequency dependence near a high-Q resonance in a systematic way

$$\Gamma(z) = \frac{i(\gamma_I - \gamma_E \varepsilon) - 2z}{i(\gamma_I + \gamma_E \varepsilon) - 2z}. \tag{11.3}$$

Although this expression accurately captures the magnitude of Γ and its phase dependence, it differs from the exact expression by an overall phase shift even in the high-Q

limit.[2] This deviation is of little consequence because the overall phase shift is usually not under experimental control. Ignoring the overall phase shift, Eq. 11.3 can be fit to a measured response extracting three parameters ω_r, γ_I, and $\gamma_E\varepsilon$. The rate of energy lost to dissipation in the resonant circuit itself is evidently $\omega_r\gamma_I = \kappa_I$. The rate at which energy is lost to port 1 is likewise $\omega_r\gamma_E\varepsilon = \kappa_E$.

Examining Eq. 11.3 one recognizes the reflection coefficient of a *series RLC* circuit probed near resonance. Namely, at resonance for a lossless resonator $\Gamma = -1$, the reflection from a short circuit. Even though the resonator itself is modeled as a parallel RLC circuit, when excited through C_c its behavior is that of a series RLC circuit with transformed values of R, L, and C. This transformation to a series circuit will simplify the analysis of electromechanics in Sect. 11.3.

From the reflection coefficient of Fig. 11.3a, the two-port parameters of networks B and C are determined. The dynamics of resonant circuits are the same in all three examples, only the interference of the excitation signal with the reflected signal changes. For Fig. 11.3b, the S-parameters for the network are $s_{11} = s_{22} = (\Gamma - 1)/(\Gamma + 3)$ and $s_{21} = s_{12} = (2\Gamma + 2)/(\Gamma + 3)$. For Fig. 11.3c, the expressions are a little more complex as they must explicitly contain the impedance of the coupling capacitor $Z_c = 1/i\omega C_c$. They are $s_{11} = s_{22} = (\Gamma - 1)[(Z_0 + Z_c)^2 + 4Z_0Z_c]/D_G$, and $s_{21} = s_{12} = 2Z_0[(\Gamma+1)Z_0+(\Gamma-1)Z_c]/D_G$, where $D_G = (Z_0+Z_c)[(\Gamma+3)Z_0 + (\Gamma - 1)Z_c]$.

Equation 11.3 is simple enough to allow compact analysis of a microwave optomechanical circuit, yet general enough to capture the behavior of the microwave networks used in recent experiments. In fact, approximating Eq. 11.1 with Eq. 11.3 is a version of a standard approximation for resonant systems where one replaces the full frequency dependence with a simpler Lorentzian form. In essence the frequency dependant impedance has been replaced with a response proportional to $1/(2zQ+i)$, a function whose magnitude is a Lorentzian. This simple form is used to model most resonant phenomena in cavity optomechanics, as both the cavities and mechanical oscillators are high-Q structures probed near resonance. Furthermore, the quantum equations of motion for a driven harmonic oscillator coupled to a simple environment are described in this form. (See for example Appendix E of Ref. [36]). In that sense, the high-Q and close to resonance approximation made in deriving Eq. 11.3 yield the same response function as the Markov approximation of the quantum input–output theory for cavities.

11.2.3 Electromechanical Coupling

The capacitors and inductors that determine the resonance frequency of an electrical circuit are structures that store either electrical or magnetic energy. The inductance and capacitance depend only on the geometry of the metal that makes up the circuit

[2] Even at $z = 0$ the exact and approximate expressions differ because the $\gamma_E\gamma_I$ terms in both the numerator and denominator have been dropped from the approximate expression.

and the permeability or permittivity of the insulating regions that contain the magnetic and electrical fields. Any change in that geometry should change the stored electrical energy and there will be an associated stress on the capacitor or inductor itself. If one of the elements is mechanically compliant, there will be an interaction between the motion of that object and the electrical energy stored in the circuit. Of course, microphones, speakers, relays and other electromechanical devices exploit exactly this type of interaction between motion and electricity. In the case of microwave cavity electromechanical structures mechanical motion alters the resonance frequency of the electrical circuit by a changing capacitance.[3] The mechanically compliant structure is itself a harmonic oscillator and in most cases the mechanical resonance frequency is far below the electrical resonance frequency. (Reference [15] is an important exception where the circuit and mechanical oscillator are nearly resonant with each other.) In these cases, the mechanical oscillator does not respond substantially to forces oscillating at the microwave frequency of the circuit resonance, but only to lower frequency variations in the microwave energy stored in the circuit.

In general, these electrical forces can reshape the very object whose motion defines the mechanical oscillator and one should include the electrical energy, as well as the elastic energy, when calculating the shapes of the resonant modes and their frequencies. In the analysis that follows, I will assume that the electrical forces are much smaller than the elastic forces of the mechanical oscillators, both because it is a great simplification and because it is accurate for the electromechanical experiments with superconducting circuits.

In this limit, each mode of a mechanically resonant structure can be modeled as an independent single degree of freedom, even though the mode shape is a full three-dimensional function of space. In the cavity optomechanical structures the motion of this mode changes the shape of a metal structure, thereby altering the capacitance C of this metal object to nearby metal structures. Referring to the coordinate of this single degree of freedom as x, the electro-mechanical coupling is determined by the function $C(x)$. As a matter of principle this function could be calculated by presuming that all metal objects are held at different fixed potentials and finding the change in electrical energy as function of x. In practice, the detailed geometry of the micro- or nano- fabricated mechanical oscillator is not known well enough for this to be a fruitful exercise. Estimates of $C(x)$ where the mechanical oscillator is treated as a thin wire [7] or a flat plate [12] translated by its motion towards or away from a nearby counter-electrode are accurate within a factor of two.

11.3 Electromechanics

Armed with the function $C(x)$, one can find the dynamics of the coupled electrical and mechanical system. In full generality this calculation is difficult because the interaction between motion and electricity is not linear. Nevertheless, microwave

[3] For microwave cavity optomechanics a mechanically compliant inductor has yet to be investigated.

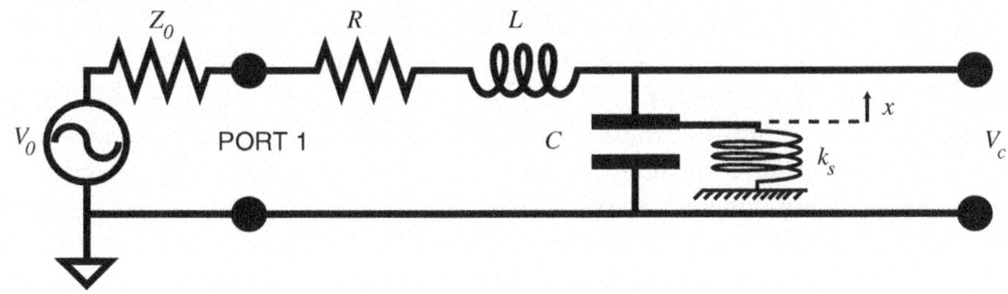

Fig. 11.4 A simple electromechanical circuit. The capacitor is assumed to be a mechanically compliant element whose capacitance depends on a coordinate x. That coordinate experiences a spring force with constant k_s

cavity optomechanical experiments have all been performed in a limit where the interaction can be linearized and the essential physics captured by a simple model.

11.3.1 Electromechanical Model

As the model for electromechanical effects, I assume a series RLC circuit driven with an ac-voltage source of source impedance Z_0, amplitude V_0, and angular frequency ω_L, as $V(t) = (V_0/2)(e^{i\omega_L t}) + c.c.$, where $c.c.$ is the complex conjugate (Fig. 11.4). If I identify $R\sqrt{C/L} \to \gamma_I$, $Z_0\sqrt{C/L} \to \gamma_E \varepsilon$, and $\omega\sqrt{LC} - 1 \to z$, this circuit has the same behavior close to resonance as Fig. 11.3a and has a reflection coefficient given by Eq. 11.3. The capacitor will be the electromechanical element; I model the mechanical coordinate x as a mass m subject to a spring force with constant k_s. For a distributed mechanical oscillator such as a beam or drumhead, there is not a unique definition of the mass m, spring constant k_s, or coordinate x. For microwave cavity optomechanical structures, in which no particular point on the distributed mechanical oscillator has special significance, the most sensible convention uses the total mass of the compliant structure as m. With this choice, x is the RMS deviation of the distributed oscillator away from its equilibrium position, and k_s is determined from the angular resonance frequency Ω_M as $k_s = m\Omega_M^2$.

11.3.2 Dynamics of Electromechanical Systems

Figure 11.4 is a simple enough model that it can be solved with a few approximations; yet, detailed enough to contain almost all of the phenomena observed in the recent experiments [7–12]. In order to solve for the dynamics of this electromechanical circuit, I will use a Lagrangian description. For electromechanical circuits, a Lagrangian description is preferable to standard ac circuit analysis because the electrical and mechanical degrees of freedom are treated equivalently in deriving the coupled equations of motion. Because the energy of the circuit $U = q^2/2C(x) + (L/2)\dot{q}^2$ is easily written in terms the charge q and its time derivative $\dot{q} = dq/dt$, I use q as the generalized coordinate of the circuit. The Lagrangian that describes the model is then $\mathscr{L} = [(1/2)L\dot{q}^2 + (1/2)m\dot{x}^2] - [(1/2)q^2/C(x) + (1/2)k_s x^2 - qV]$, where V is a generalized force acting on q. Rather than include dissipation into the Lagrangian formalism, I will introduce the dissipative forces directly into the equations of motion. The equations of motion with dissipation are

$$L\ddot{q} = -\frac{q}{C(x)} + V(t) - \dot{q}(R + Z_0) \tag{11.4}$$

$$m\ddot{x} = -k_s x + \frac{q^2}{2C(x)^2}\frac{\partial C(x)}{\partial x} - \Gamma_M m\dot{x}, \tag{11.5}$$

where the dissipative mechanical force is $-\Gamma_M m\dot{x}$.

Even numerically, these equations are too general to solve because the function $C(x)$ is unspecified. To make progress, one approximates the coupling between x and q^2 to linear order by expanding $1/C(x) \approx 1/C(\bar{x}) - (1/C(\bar{x})^2)(\partial C(\bar{x})/\partial x)(x - \bar{x})$, where $\bar{x}$ is the static value of the coordinate x. In order to determine $\bar{x}$ one must know the function $C(x)$ and specify the applied voltage $V(t)$. Equations 11.4 and 11.5 can then be solved for the static values of q^2 and x. If the applied voltage is sufficiently small that the change in $C(x)$ due to the application of V can be approximated linearly then one can find $\bar{x} = -\bar{F}/k_s$, where $\bar{F} = (\partial C/\partial x)(|V_0|^2/4)/[(1 - \omega_L LC)^2 + (\omega_L(R + Z_0))^2]$. In experimental practice, the technical limitations of micro- or nano- fabrication imply that quantities such as $C(x)$, k_s, and m that depend on the full 3-dimensional shape of an object are not known with better than 10 % accuracy. One still can proceed with the linearization without knowing $\bar{x}$; it will have some value that will determine $(1/C(\bar{x}))\partial C(\bar{x})/\partial x$ and that quantity can be determined experimentally. From the Lagrangian with linearized coupling we can write more tractable equations of motion as

$$L\ddot{q} = V - \frac{q}{C}\left(1 - \frac{1}{C}\frac{\partial C}{\partial x}x\right) - \dot{q}(R + Z_0) \tag{11.6}$$

$$m\ddot{x} = -k_s(x + \bar{x}) + \frac{q^2}{2C^2}\frac{\partial C}{\partial x} - \Gamma_M m\dot{x}, \tag{11.7}$$

where $C = C(\bar{x})$ and x now refers to displacement from $\bar{x}$ (i.e. $x - \bar{x} \rightarrow x$). Although the coupling no longer contains explicit x dependence, these equations of motion are themselves non-linear through the xq and q^2 terms. They can be solved either numerically or using approximate methods.

Let us find approximate solutions to Eqs. 11.6 and 11.7. In the presence of the drive $V(t)$, I will seek solutions of the form $q(t) = (1/2)(q_0 + q_1(t))e^{i\omega_L t} + c.c.$ and $x = x(t)$. In deriving linear equations of the motion for q_1 and x, I assume a strong drive $|q_1|/|q_0| \ll 1$ and a slowly varying response $|\dot{q}_1(t)|/|\omega_L q_1(t)| \ll 1$ and work to first order in both small quantities. In addition, I ignore terms oscillating near $2\omega_L$ as neither the circuit nor the oscillator will respond so far from resonance. The linear equations of motion are

$$-G\frac{q_0 q_1^*(t) + q_0^* q_1(t)}{2C\omega_{\rm opt}} + F_{\rm ext} = (\ddot{x} + \Gamma_m \dot{x} + \Omega_{\rm M}^2 x)m \tag{11.8}$$

$$2Gq_0\omega_{\rm opt}x(t) = q_1(t)(\omega_L^2 - \omega_{\rm opt}^2 - i\omega_L\kappa) + \dot{q_1}(t)(2i\omega_L + \kappa), \tag{11.9}$$

$$2Gq_0^*\omega_{\rm opt}x(t) = q_1^*(t)(\omega_L^2 - \omega_{\rm opt}^2 + i\omega_L\kappa) + \dot{q_1}^*(t)(-2i\omega_L + \kappa) \tag{11.10}$$

where the $k_s\bar{x}$ has vanished by construction, $\kappa = (\kappa_I + \kappa_E) = (R + Z_0)/L$ is the total rate at which energy is lost from the circuit and I have introduced a force $F_{\rm ext}$ applied to the mechanical oscillator by an external agent. In addition, I define the static resonance frequency $\omega_{\rm opt} = (LC(\bar{x}))^{-1/2}$ and the electromechanical coupling $G = -\omega_{\rm opt}(1/2C)(\partial C/\partial x)$. These equations are remarkable; they show that fluctuations in the charge act as a force that drives the oscillator and fluctuations in the position of the oscillator act as a voltage source that creates charge fluctuations on the capacitor. Implicitly the LHS of Eq. 11.8 contains the position of the oscillator. One could arrange such a condition by measuring the position of the oscillator and applying a force based on that measurement. Indeed literal feedback of this type is commonplace, but here the coupled electromechanics has reduced to a kind of endogenous feedback.

One can justify the assumptions made in deriving the equations of motion by noting the parameters used in the microwave cavity experiments to date [7–12]. Mechanical resonance frequencies range between $\Omega_{\rm M}/(2\pi) \approx 100$ kHz to 10 MHz, while typical cavity resonances are in the range 5 GHz $< \omega_{\rm opt}/(2\pi) <$ 8 GHz. The cavity decay rates are in the range 200 kHz $< \kappa/(2\pi) <$ 2 MHz, while the mechanical dissipation rates $\Gamma_M <$ 1 kHz are much smaller. For these parameters the mechanical degree of freedom is always slow compared to the electrical degree of freedom (this is not true for [15]) and both the microwave cavity and mechanical oscillator have $Q >$ 1,000. When discussing quantum effects in these systems (Sect. 11.4.2) it will also become clear that a strong drive is required for substantial coupling between electricity and motion.

These coupled equations can be readily solved via Fourier transform (or less formally by ansatz $x(t) \rightarrow x(\omega)e^{i\omega t}$ and $q_1(t) \rightarrow q_1(\omega)e^{i\omega t}$) for the purely mechanical susceptibility $x(\omega)/F_{\rm ext}(\omega)$ or for the electromechanical susceptibility $q_1(\omega)/F_{\rm ext}(\omega)$. The procedure requires a little more care than is usually necessary for linear equations of motion because $q_1(t)$ is complex. (I write both Eq. 11.9 and it complex conjugate Eq. 11.10 to emphasize that $q_1(t)$ is complex.) Furthermore $q_1^*(t) = q_1^*(\omega)e^{-i\omega t}$, thus; the frequency components rotating as $e^{i\omega t}$ are $q_1(\omega)$, $x(\omega)$, and $q_1^*(-\omega)$. The frequency domain equations are

$$x(\omega) = \chi_M(\omega)\left[F_{\rm ext}(\omega) - \frac{G(q_0 q_1^*(-\omega) + q_0^* q_1(\omega))}{2C\omega_{\rm opt}}\right] \tag{11.11}$$

$$q_1(\omega) = \chi_c(\omega)q_0 G x(\omega) \tag{11.12}$$

Fig. 11.5 A simple block diagram description of the effect of electromechanics on the susceptibility of a mechanical oscillator. As in the text, $Y(\omega) = A(\chi_c(\omega) + \chi_c^*(-\omega))$.

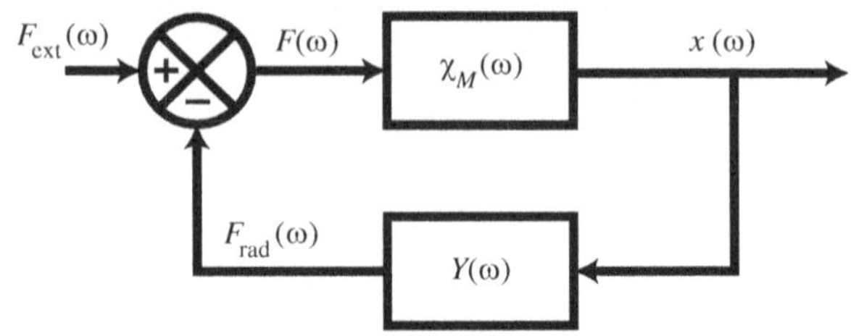

$$q_1^*(-\omega) = \chi_c^*(-\omega) q_0^* G x(\omega) \tag{11.13}$$

where I have defined the bare mechanical susceptibility $\chi_M^{-1} = m(\Omega_M^2 - \omega^2 + i\Gamma_M \omega)$, the circuit susceptibility $\chi_c(\omega) = [1 - (\Delta/2\omega_L)]/[(\Delta' - \omega) + i(\kappa/2)(1 - \omega/\omega_L)]$. I have written the circuit susceptibility in terms of the detuning of the drive from circuit resonance $\Delta = \omega_L - \omega_{opt}$ and defined $\Delta' = \Delta - (\Delta^2/2\omega_L)$.

These coupled equations constitute solutions to the electromechanical circuit linearized about a strong oscillatory drive. For example, they can be used to find the charge "sidebands" about a drive frequency ω_L generated by an oscillatory F_{ext}. In addition, they determine the mechanical susceptibility in the presence of electromechanical effects

$$\frac{x(\omega)}{F_{ext}} = \widetilde{\chi_M} = \frac{\chi_M}{1 + \chi_M A(\chi_c(\omega) + \chi_c^*(-\omega))}, \tag{11.14}$$

where $A = G^2|q_0|^2/(2C\omega_{opt})$. Written in this form, the description of electromechanical effects as endogenous feedback is evident (Fig. 11.5). The open loop response of the mechanical oscillator χ_M and the closed loop response is $\widetilde{\chi_M}$. In the recent literature, it is not control theory which has served as the intellectual touchstone for these electro- or opto-mechanical effects but rather quantum field theory. For example, the term $Y(\omega) = A(\chi_c(\omega) + \chi_c^*(-\omega))$ is known as the self-energy and Eq. 11.14 is a version of Dyson's equation [37]. This choice is natural when these equations of motion are derived directly from the quantum Hamiltonian and perturbation theory. Nevertheless, a classical feedback control offers a helpful alternative perspective. Most importantly, any effect contained within such a description is unambiguously classical. In addition, control theory has developed many powerful calculational techniques, all of which can be used once the electromechanical interaction is described in this language. For example, with these techniques it is simple to decide if the electromechanical circuit is stable as a function of A.

One of the most studied effects of electro- and opto- mechanics is the ability to damp, amplify, stiffen, or soften the motion of the mechanical oscillator by the radiation pressure. Such effects are easily calculated from Eq. 11.14. First I rewrite the closed loop response in a different form $\widetilde{\chi_M}(\omega)^{-1} = \chi_M(\omega)^{-1} + Y(\omega)$. For a high Q mechanical oscillator studied near resonance, I approximate $\chi_M(\omega)^{-1} \approx 2\Omega_M m[\Omega_M - \omega + i\Gamma/2]$. It is then easy to see that for a wide range of conditions $\widetilde{\chi_M}$

has the same form as χ_M but with a new resonance frequency and linewidth determined by the real and imaginary parts of $Y(\omega = \Omega_M)/(2\Omega_M m)$. Following through with this calculation, the additional "optical" damping and the "optical" frequency shift are

$$\Gamma_{\text{opt}} = \frac{2A}{\Omega_M m}\left[\frac{\kappa}{4(\Delta+\Omega_M)^2+\kappa^2} - \frac{\kappa}{4(\Delta-\Omega_M)^2+\kappa^2}\right] \tag{11.15}$$

$$\delta\Omega_{\text{M,opt}} = \frac{2A}{\Omega_M m}\left[\frac{\Delta-\Omega_M}{4(\Delta-\Omega_M)^2+\kappa^2} + \frac{\Delta+\Omega_M}{4(\Delta+\Omega_M)^2+\kappa^2}\right], \tag{11.16}$$

where I have assumed $\Omega_M \ll \omega_{\text{opt}}$ in order to write a simpler form for $\chi_c(\Omega_\text{M}) \approx [(\Delta - \Omega_\text{M}) + (i\kappa/2)]^{-1}$ and arrive at the form appearing in Refs. [8, 37]. In those papers, the prefactor $2A/\Omega_M m$ is written in a quantum form $4nG^2x^2_{\text{ZPF}}$, with an explicit dependence on the zeropoint motion of the oscillator $x^2_{\text{ZPF}} = \hbar/(2\Omega_M m)$ and on the number of photons in the cavity $n = E/\hbar\omega_{\text{opt}}$. By recognizing $E = q_0^2/2C$ as the average energy stored in the circuit, it is clear that ($4nG^2x^2_{\text{ZPF}} = 2A/\Omega_M m$) the prefactors are indeed the same and that x^2_{ZPF} appears if the energy stored in the circuit is expressed in units of energy quanta $n = E/\hbar\omega_{\text{opt}}$.

11.4 Quantum Effects in Electromechanics

By following the procedure of canonical quantization, the quantum optomechanical Hamiltonian can be derived from the classical circuit Lagrangian. Although the quantum description is widely used, the majority of microwave cavity optomechanical experiments are still contained within classical physics. With the recent demonstration of many photon strong coupling [12], microwave optomechanics is poised to escape a classical description.

11.4.1 The Cavity Optomechanical Hamiltonian

In the body of theory associated with cavity optomechanics, quantum equations of motion reminiscent of Eqs. 11.6 and 11.7 are usually derived by starting from a quantum cavity optomechanical Hamiltonian [37, 38]. One can show that the quantum version of the electromechanical circuit model is indeed this Hamiltonian. Returning to the Lagrangian with linearized coupling, one can recast this into a classical Hamiltonian using the formal definition, $H = \dot{q}\Phi + \dot{x}p - \mathscr{L}$, where the canonical momenta are $p \equiv \partial\mathscr{L}/\partial\dot{x} = m\dot{x}$ and $\Phi \equiv \partial\mathscr{L}/\partial\dot{q} = L\dot{q}$. Writing H as a function of q, x, p, and Φ gives

$$H = \frac{p^2}{2m} + \frac{k_s x^2}{2} + \frac{\Phi^2}{2L} + \frac{q^2}{2C}\left(1 - \frac{1}{C}\frac{\partial C}{\partial x}x\right) - qV. \tag{11.17}$$

Having written the classical Hamiltonian in terms of system coordinates and their canonical momenta, one simply writes the quantum Hamiltonian by replacing each coordinate or momentum by its operator (e.g. $q \rightarrow \hat{q}$) and requires that they obey canonical commutations relations $[\hat{x}, \hat{p}] = i\hbar$, and $[\hat{q}, \hat{\Phi}] = i\hbar$. By defining the creation and annihilation operators for the two harmonic oscillators

$$\hat{a} = \hat{q}\sqrt{Z_c/2\hbar} + i\hat{\Phi}\sqrt{1/2Z_c\hbar}$$

$$\hat{b} = \hat{x}\sqrt{Z_M/2\hbar} + i\hat{p}\sqrt{1/2Z_M\hbar},$$

one arrives at the celebrated optomechanical Hamiltonian

$$\hat{H} = \hbar\omega_{\text{opt}}\left(\hat{a}^\dagger\hat{a} + \frac{1}{2}\right) + \hbar\Omega_{\text{M}}\left(\hat{b}^\dagger\hat{b} + \frac{1}{2}\right) + \hbar G x_{\text{ZPF}}(\hat{b} + \hat{b}^\dagger)\hat{a}^\dagger\hat{a}$$

$$- (\hat{a} + \hat{a}^\dagger)\sqrt{\frac{\hbar}{2Z_c}}V + \frac{\hbar G}{2}\left[\hat{a}\hat{a} + \hat{a}^\dagger\hat{a}^\dagger + 1\right]x_{\text{ZPF}}(\hat{b} + \hat{b}^\dagger), \tag{11.18}$$

where $Z_c = \sqrt{L/C}$ and $Z_M = \sqrt{k_s m}$. The terms in square brackets of Eq. 11.18 are not usually included as part of the optomechanical Hamiltonian. Indeed there is a good physical argument to ignore them for the cavity optomechanical systems for which $\Omega_{\text{M}} \ll \omega_{\text{opt}}$. The force associated with these terms oscillates at about twice ω_{opt}, where the susceptibility of the mechanical oscillator is likely to be vanishingly small. Nevertheless, I have included them because it is possible [15] to build mechanical structures that responds at $2\omega_{\text{opt}}$ when $2\omega_{\text{opt}}$ is a microwave rather than optical frequency.

11.4.2 Observing Quantum Effects in Microwave Cavity Optomechanics

The quantum Hamiltonian of Eq. 11.18 has been studied in most of its possible limits [37, 38] and extended to include additional physical phenomena. In this work, the formalism of quantum optics and the descriptive language of optics has been dominant. Rather than recreate those calculations, I instead focus on the purely classical equations of motion. This choice is partly to acknowledge that the quantization of the cavity optomechanical Hamiltonian is an act more of aspiration than necessity. In addition, solving the classical equations of motion in the language of electromechanics, rather than quantum optics, establishes intuition for those more familiar with circuits.

From Eq. 11.18, one can understand why true quantum effects are elusive in electro-mechanics. The quantity $Gx_{ZPF} < 100$ Hz [12], which sets the rate for single photon-phonon processes in these circuits, is slower than the rate at which dissipative and decohering processes destroy the purity of quantum states of the cavity or oscillator ($\kappa > 100$ kHz). At the moment, strong coupling between photons and phonons requires the application of a strong cavity drive [12, 39], through the qV term in Eq. 11.17. In the presence of a strong drive, the effective Hamiltonian is bilinear in the photon $\hat{a}$ and phonon $\hat{b}$ operators. The phonon is just another boson; such linear interactions are analogous to the beam splitter and parametric down conversion interactions in quantum optics [39]. As a consequence of a theorem of quantum information [40], a system of harmonic oscillators: with only linear interactions, prepared in a Gaussian state, and measured with linear amplifiers has an average behavior that is classical, while the fluctuations away from the average remain Gaussian. In quantum optics, profoundly quantum behavior can be observed even for these linear interactions when the measurement process is nonlinear (e.g. photon detection). In electromechanics where there exists no readily available photon-counting technology, the predictions of quantum and classical equations of motion are very difficult to distinguish. In the linear measurement of microwave fields that have interacted with an electromechanical circuit, quantum effects are manifest primarily as an irreducible Gaussian noise process, identified as quantum noise. Indeed, many of the prominent results of this field constitute the measurement of a noisy quantity, such as the residual motion of a mechanical oscillator [9] upon cooling, or the added noise of measurement [10], which is then compared to a quantum-limited value. To date, there is one marvelous counter example [15] to the relative weakness of quantum effects in electro- or opto-mechanical structures. In this case, the mechanical oscillator inherits its nonlinearity through a strong interaction with a superconducting qubit with which it is nearly resonant. For the situation where low frequency mechanical motion alters the resonance of a high frequency cavity, the optomechanical nonlinearity is currently too weak for a single photon and phonon to influence one another. I am nevertheless optimistic that this may be overcome in the future. In the three years from the inception of the effort to measure and manipulate nanomechanical oscillators with microwave cavities, the coupling strength between motion and microwave electricity has increased more than 10,000 times (Fig. 11.6). Furthermore, the linewidth of microfabricated microwave cavities are decreasing as the loss processes in those structures are better understood [41–43].

In the meantime, the level of quantum interaction available is still quite useful for measurement and control. Returning to the feedback description of electromechanics, the performance of the endogenous electromechanical feedback can be compared to an active feedback scheme. For any feedback scheme, the quality of the oscillator control will be limited by any stochastic component in the feedback force. In an active feedback scheme, the dominant stochastic force may well arise from error in the position measurement. For electromechanial feedback, no explicit measurement is required. The stochastic force will arise from the random components in the applied voltage. For microwave signals applied inside a dilution refrigerator, these stochastic forces can be quantum-limited. From one point of view, the cooling of a mechanical

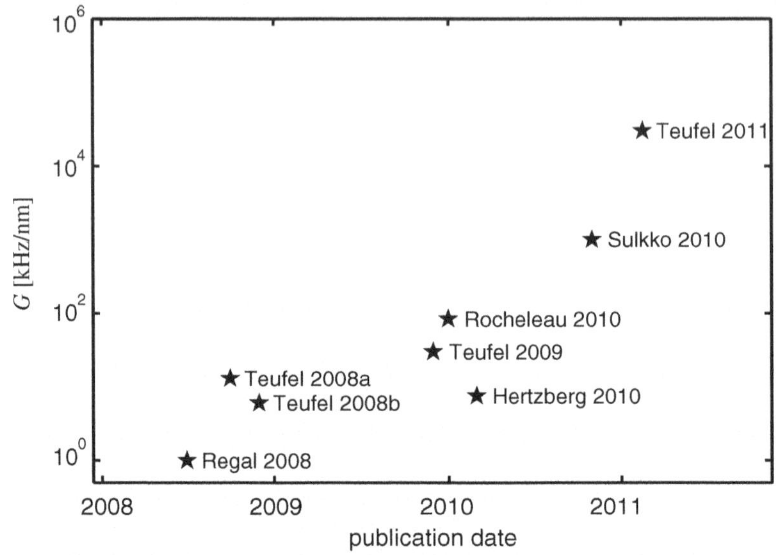

Fig. 11.6 The increase in G versus publication date. The cited works are Regal et al. [7], Teufel et al. [8, 10, 12, 44], Rocheleau et al. [9], Hertzberg et al. [11], Sulkko et al. [13]. A reasonable target for single photon strong coupling is about 10^7 kHz/nm

oscillator by radiation pressure is simply an exchange between the thermal state which characterizes the mechanical oscillator and the coherent state of the microwave field. Upon cooling the mechanical oscillator acquires the zero-entropy coherent state of the microwave field while the microwave field acquires the thermal state of the oscillator, carrying it away.

This point of view also demonstrates that it should be possible to use microwave cavity optomechanical structures to prepare mechanical oscillators in profoundly quantum non-Gaussian states before the single photon strong coupling regime is achieved. Non-Gaussian states of the microwave field are routinely created using superconducting qubits [45]. These in-principle can be transferred to the state of the mechanical oscillator using the already achieved many photon strong coupling [12, 39]. Whether using this strategy or direct coupling to a qubit [15], mechanical oscillators may soon be prepared in a quantum superposition of two places at once (so-called Schrodinger cat states), using already extant electromechanical technology.

Acknowledgments I thank John Price, Charles Rogers, Jennifer Harlow, Tauno Palomaki, Reed Andrews, Hsiang-Shen Ku, and William Kindel for helpful discussions and assistance in preparing this document.

References

1. S. Gigan, J.R. Böhm, M. Paternostro, F. Blaser, G. Langer, J.B. Hertzberg, K.C. Schwab, D. Bäuerle, M. Aspelmeyer, A. Zeilinger, Nature **444**, 67 (2006)
2. D. Kleckner, D. Bouwmeester, Nature **444**, 75 (2006)
3. O. Arcizet, P.F. Cohadon, T. Briant, M. Pinard, A. Heidmann, Nature **444**, 71 (2006)
4. T.J. Kippenberg, K.J. Vahala, Science **321**, 1172 (2008)
5. F. Marquardt, Nat. Phys. **4**, 513 (2008)
6. A. Wallraff, D. Schuster, A. Blais, L. Frunzio, R.S. Huang, J. Majer, S. Kumar, S.M. Girvin, R.J. Schoelkopf, Nature **431**, 162 (2004)
7. C.A. Regal, J.D. Teufel, K.W. Lehnert, Nat. Phys. **4**, 555 (2008)
8. J.D. Teufel, J.W. Harlow, C.A. Regal, K.W. Lehnert, Phys. Rev. Lett. **101**, 197203 (2008)
9. T. Rocheleau, T. Ndukum, C. Macklin, J.B. Hertzberg, A.A. Clerk, K.C. Schwab, Nature **463**, 72 (2010)
10. J.D. Teufel, T. Donner, M.A. Castellanos-Beltran, J.W. Harlow, K.W. Lehnert, Nat. Nanotech. **4**, 820 (2009)
11. J.B. Hertzberg, T. Rocheleau, T. Ndukum, M. Savva, A.A. Clerk, K.C. Schwab, Nat. Phys. **6**, 213 (2010)
12. J.D. Teufel, D. Li, M.S. Allman, K. Cicak, A.J. Sirois, J.D. Whittaker, R.W. Simmonds, Nature **471**, 204 (2011)
13. J. Sulkko, M.A. Sillanpaa, P. Hakkinen, L. Lechner, M. Helle, A. Fefferman, J. Parpia, P.J. Hakonen, Nano Lett. **10**, 4884 (2010)
14. M.D. LaHaye, J. Suh, P.M. Echternach, K.C. Schwab, M.L. Roukes, Nature **459**, 960 (2009)
15. A.D. O'Connell, M. Hofheinz, M. Ansmann, R.C. Bialczak, M. Lenander, E. Lucero, M. Neeley, D. Sank, H. Wang, M. Weides, J. Wenner, J.M. Martinis, A.N. Cleland, Nature **464**, 697 (2010)
16. K.C. Schwab, M.L. Roukes, Phys. Today **58**, 36 (2005)
17. M.F. Bocko, R. Onofrio, Rev. Mod. Phys. **68**, 755 (1996)
18. R.P. Giffard, Phys. Rev. D **14**, 2478 (1976)
19. P.F. Michelson, R.C. Taber, J. Appl. Phys. **52**, 4313 (1981)
20. M.F. Bocko, K.A. Stephenson, R.H. Koch, Phys. Rev. Lett. **61**, 726 (1988)
21. D.G. Blair, E.N. Ivanov, M.E. Tobar, P.J. Turner, F. van Kann, I.S. Heng, Phys. Rev. Lett. **74**, 1908 (1995)
22. V.B. Braginsky, *Systems with Small Dissipation* (University of Chicago Press, Chicago, 1985)
23. M.E. Tobar, E.N. Ivanov, D.G. Blair, Gen. Relativ. Gravit. **32**, 1799 (2000)
24. C.R. Locke, M.E. Tobar, E.N. Ivanov, D.G. Blair, J. Appl. Phys. **84**, 6523 (1998)
25. M.F. Bocko, W.W. Johnson, Phys. Rev. A **30**, 2135 (1984)
26. R. Knobel, A. Cleland, Nature **424**, 291 (2003)
27. M.D. LaHaye, O. Buu, B. Camarota, K.C. Schwab, Science **304**, 74 (2004)
28. A.N. Cleland, J.S. Aldridge, D.C. Driscoll, A.C. Gossard, Appl. Phys. Lett. **81**, 1699 (2002)
29. N.E. Flowers-Jacobs, D.R. Schmidt, K.W. Lehnert, Phys. Rev. Lett. **98**, 096804 (2007)
30. S. Etaki, M. Poot, I. Mahboob, K. Onomitsu, H. Yamaguchi, H.S.J. van der Zant, Nat. Phys. **4**, 785 (2008)
31. A.A. Clerk, Phys. Rev. B **70**, 245306 (2004)
32. K.W. Lehnert, K. Bladh, L.F. Spietz, D. Gunnarsson, D.I. Schuster, P. Delsing, R.J. Schoelkopf, Phys. Rev. Lett. **90**, 027002 (2003)
33. P.K. Day, H.G. LeDuc, B.A. Mazin, A. Vayonakis, J. Zmuidzinas, Nature **425**(6960), 817 (2003)
34. B.A. Mazin, Microwave kinetic inductance detectors. Ph.D. thesis, California Institute of Technology (2004)
35. K.R. Brown, J. Britton, R.J. Epstein, J. Chiaverini, D.L.D.J. Wineland, Phys. Rev. Lett. **99**, 137205 (2007)
36. A.A. Clerk, M.H. Devoret, S.M. Girvin, F. Marquardt, R.J. Schoelkopf, Rev. Mod. Phys. **82**, 1155 (2010)

37. F. Marquardt, J.P. Chen, A.A. Clerk, S.M. Girvin, Phys. Rev. Lett. **99**, 093902 (2007)
38. I. Wilson-Rae, N. Nooshi, W. Zwerger, T.J. Kippenberg, Phys. Rev. Lett. **99**, 093901 (2007)
39. S. Groblacher, K. Hammerer, M.R. Vanner, M. Aspelmeyer, Nature **460**, 724 (2009)
40. S.D. Bartlett, B.C. Sanders, S.L. Braunstein, K. Nemoto, Phys. Rev. Lett. **88**(9), 097904 (2002)
41. J. Gao, J. Zmuidzinas, B.A. Mazin, H.G. LeDuc, P.K. Day, Appl. Phys. Lett. **90**, 102507 (2007)
42. J.S. Gao, M. Daal, A. Vayonakis, S. Kumar, J. Zmuidzinas, B. Sadoulet, B.A. Mazin, P.K. Day, H.G. Leduc, Appl. Phys. Lett. **92**, 152505 (2008)
43. J. Gao, M. Daal, J.M. Martinis, A. Vayonakis, J. Zmuidzinas, B. Sadoulet, B.A. Mazin, P.K. Day, H.G. Leduc, Appl. Phys. Lett. **92**, 212504 (2008)
44. J.D. Teufel, C.A. Regal, K.W. Lehnert, New J. Phys. **10**, 095002 (2008)
45. M. Hofheinz, H. Wang, M. Ansmann, R.C. Bialczak, E. Lucero, M. Neeley, A.D. O'Connell, D. Sank, J. Wenner, J.M. Martinis, A.N. Cleland, Nature **459**, 546 (2009)

Chapter 12
Microwave-Frequency Mechanical Resonators Operated in the Quantum Limit

Aaron O'Connell and Andrew N. Cleland

Abstract In this chapter, we describe an experiment in which the quantum ground state of one vibrational mode of a mechanical resonator was reached when the structure was cooled in a dilution refrigerator to $T \sim 25$ mK. The resonator had a fundamental dilatational resonance frequency in excess of 6 GHz, so once cooled to this temperature, the number of thermal phonons at this frequency is vanishingly small. This achievement is a direct consequence of the high resonance frequencies obtainable with the class of mechanical resonator used in the experiment, which is known as a film bulk acoustic resonator, or FBAR. In this chapter, we begin by briefly describing the mechanics of bulk acoustic resonance and FBAR structures, and we present a simple electrical circuit model for the resonator. Experiments using this type of mechanical resonator in the classical regime are then described. We then introduce the Josephson phase quantum bit (qubit), a device which forms the heart of the measurement scheme used to probe the mechanical resonator in the quantum regime, and describe the coupling mechanism between the qubit and a mechanical resonator. Lastly, we present experimental measurements of the resonator in the quantum regime, where the qubit was used to both prepare and measure non-classical mechanical states in the resonator.

12.1 Film Bulk Acoustic Resonators

Film bulk acoustic resonators are dilatational-mode mechanical resonators fabricated using piezoelectric materials. Voltages applied to metal electrodes placed on the opposing surfaces of the resonator generate piezoelectric strain, either dilating or contracting the volume of the resonator. When the applied voltage oscillates at a frequency corresponding to a natural mechanical resonance of the structure, large

A. O'Connell · A. N. Cleland (✉)
University of California, Santa Barbara, CA, USA
e-mail: cleland@physics.ucsb.edu

M. Aspelmeyer et al. (eds.), *Cavity Optomechanics*, Quantum Science and Technology, DOI: 10.1007/978-3-642-55312-7_12,

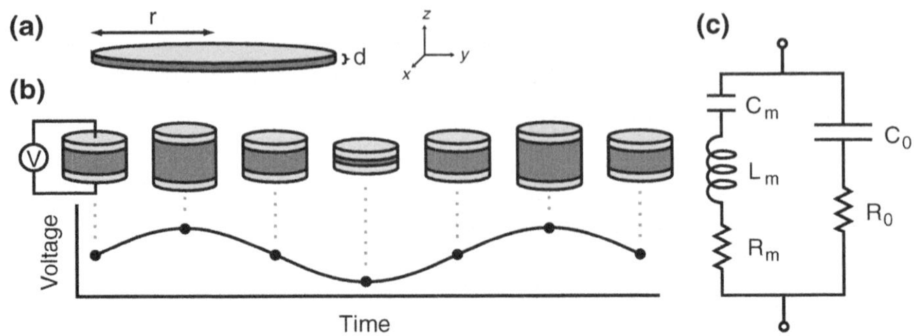

Fig. 12.1 Film bulk acoustic resonator geometry, piezoelectric response, and model circuit representation. **a** Idealized geometry of a mechanical resonator, comprising a piezoelectric material of thickness d with infinitely thin metal plates, or electrodes, on both the top and bottom surfaces. **b** Sketch illustrating the dilatational response of the piezoelectric structure to an externally-imposed voltage, which generates an electric field between the electrodes; *light gray* indicates metal electrodes, while *dark gray* represents piezoelectric material. The thickness dimension is grossly exaggerated for illustrative purposes, and the metal electrodes are shown here with non-zero thickness. **c** Equivalent electrical circuit model for mechanical response near the fundamental dilatational mechanical resonance. C_m, L_m, and R_m are the mechanical capacitance, mechanical inductance, and mechanical resistance (representing mechanical energy dissipation), respectively. The electrical branch is composed of a capacitor C_0 representing the geometric capacitance of the structure, and a resistor R_0 representing electrical dissipation with value R_0. The capacitance is given by $C_0 = \varepsilon_0 \varepsilon A/d$, where A is the area of an electrode, ε is the dielectric constant, and ε_0 is the permittivity of free space

and sustained oscillations can be generated. The symmetric nature of the piezoelectric response means that the converse is also true: A mechanical resonance will generate voltages on the same electrodes, which can be detected externally, allowing one to monitor the state of vibration of the resonator.

FBAR structures are of great interest in the telecommunications industry, where they are used to filter electrical signals, using the strong frequency dependence of the electromechanical response. Using mechanical resonators in this fashion provides filtering that is significantly superior to purely electronic filters. The mechanical structure is typically designed in a planar capacitor-like structure, with voltages driving a thickness-mode vibration, such that the top and bottom surfaces alternately approach and recede from one another. The fundamental resonance for this mode corresponds to a sound wavelength that is twice the thickness of the structure, so that the resonance frequency scales with the sound speed and inversely with the thickness of the structure. With typical sound speeds of ~10 km/s, resonance frequencies in the GHz band can be reached using structures with thicknesses of order 1 μm. For example, the structure depicted in Fig. 12.1 might have a radius r that is a few tens of micrometers, while the thickness of the piezoelectric layer d may be less than a micron. When a voltage is applied across the electrodes, a nearly uniform electric field is formed between the plates, directed along the z axis. The piezoelectric material between the plates reacts to the imposed electric field by polarizing, which due to the piezoelectric response generates strain (displacement) in the material. The electric field thus causes the piezoelectric to expand or contract proportional to the sign and magnitude of the electric field. This is illustrated in Fig. 12.1b for the quasi-static case of a slowly varying voltage.

We would like to examine the electrical response that occurs near a mechanical resonance. Consider a lossless piezoelectric slab sandwiched between infinitesimally thick resistance-less electrodes, vibrating at angular frequency ω close to a mechanical resonance frequency. The electrical impedance $Z(\omega)$ across the metal plates can be determined in the dissipation-free limit by solving the electro-mechanical equations of motion, yielding [1],

$$Z(\omega) \approx \frac{1}{i\omega C_0}\left(1 - \frac{\tan(\omega d/2v)}{\omega d/2v}k_{\text{eff}}^2\right) \tag{12.1}$$

where C_0 is the geometric capacitance of the structure, k_{eff}^2 is the piezoelectric coupling efficiency, typically of order a few percent, and v is the speed of sound in the piezoelectric medium. The impedance is a strong function of frequency for frequencies near the mechanical thickness resonances $\omega_n/2\pi = nv/2d$, where $n = 1, 3, 5, \ldots$, with the magnitude of the impedance $|Z|$ displaying a minimum and a maximum near these frequencies. The even-number modes do not appear because their strain symmetry does not generate an electrical signal for symmetrically placed electrodes, as in Fig. 12.1. In this model, the impedance maxima occur exactly at the mechanical resonance frequencies ω_n, which correspond to the surface-stress free thickness resonance modes of the structure. A local maximum of the impedance corresponds to a mode in which energy can be stored, as the voltage increases with the stored energy). The fundamental ($n = 1$) dilatational mode is given by

$$\frac{\omega_p}{2\pi} = \frac{v}{2d}. \tag{12.2}$$

The frequency of this resonance can be adjusted by changing the thickness d of the structure. For a piezoelectric material such as aluminum nitride (AlN), with a sound speed $v \approx 11$ km/s, resonant frequencies above 1 GHz are achievable for thicknesses less than 5 μm. The maximum in the impedance $Z(\omega)$ occurs at $\omega = \omega_p$, and is termed the "parallel resonance", for reasons that will be made clear below.

The impedance magnitude $|Z|$ has a second feature, a zero, at the frequency ω_s (the "series resonance" frequency, as explained below). This frequency is given by the transcendental equation

$$\frac{\tan(\omega_s d/2v)}{\omega_s d/2v}k_{\text{eff}}^2 = 1, \tag{12.3}$$

for which a closed-form solution does not exist. However, as the piezoelectric coupling efficiency k_{eff}^2 is much smaller than unity, the impedance can be approximated near its first pole by

$$Z \approx \frac{1}{i\omega C_0}\left(1 - \frac{8}{\pi^2 - (\omega d/v)^2} k_{\text{eff}}^2\right). \tag{12.4}$$

This yields the following approximate expression for the series-resonance frequency

$$\frac{\omega_s}{2\pi} \approx \frac{v}{2d}\left(1 - \frac{8}{\pi^2} k_{\text{eff}}^2\right)^{1/2}. \tag{12.5}$$

The frequency ω_s for the impedance minimum is slightly lower than the parallel-resonance frequency ω_p where the impedance is a maximum.

It is interesting to note that in the dissipation-free limit, the piezoelectric coupling efficiency can be determined from the two resonant frequencies ω_p and ω_s by combining Eqs. (12.2) and (12.3):

$$k_{\text{eff}}^2 \approx \frac{\pi\omega_s/2\omega_p}{\tan\left(\pi\omega_s/2\omega_p\right)} \approx \frac{\pi^2}{4}\left(\frac{\omega_p - \omega_s}{\omega_p}\right), \tag{12.6}$$

indicating that the frequency spacing between the two resonances is proportional to the coupling efficiency.

Equation (12.1) is useful for finding the resonance frequencies and evaluating their relation to the coupling efficiency. It also can be used to generate a lumped-element equivalent circuit model for the electromechanical response of the FBAR, one that gives a good approximation to the electrical impedance near the mechanical resonance. To the trained eye, Eq. (12.4) is the impedance of a static capacitor C_0, in parallel with a series-connected, mechanically-equivalent inductance and capacitance. This equivalent circuit model is termed the modified Butterworth-van Dyke (MBVD) model [2, 3]. In its full form it includes resistors that model the mechanical and electrical losses in the actual structure, as shown in Fig. 12.1c.

We will first examine the dissipation-free limit, where the resistances are set to zero. The dissipation-free circuit thus consists only of the static capacitor C_0, the electrical branch, in parallel with an electrical equivalent mechanical inductance L_m and capacitance C_m, which make up the mechanical motional branch of the circuit. Near the fundamental resonance frequency ω_p, the impedance of the lossless equivalent lumped circuit is given by

$$Z_{LE} = \frac{1}{i\omega C_0}\left(1 - \frac{C_m}{C_0 + C_m - C_0 C_m L_m \omega^2}\right), \tag{12.7}$$

which is clearly of the same form as Eq. (12.4). To relate the circuit elements to the mechanical properties, we can use the requirement that both the Eqs. (12.4) and (12.7) should exhibit resonances at the same frequencies. The equivalent circuit will have two resonance frequencies, the series resonance between L_m and C_m, and a parallel resonance between L_m with C_m and C_0 in parallel,

$$\omega_p = \frac{1}{\sqrt{L_m C_m \frac{C_0}{C_0+C_m}}}, \text{ and} \tag{12.8}$$

$$\omega_s = \frac{1}{\sqrt{L_m C_m}}. \tag{12.9}$$

This explains the origin of the terms identifying these frequencies. Equating these relations with those derived from Eq. (12.4), we find the relations

$$C_m = C_0 \frac{8}{\pi^2} k_{\text{eff}}^2 \left(1 - \frac{8}{\pi^2} k_{\text{eff}}^2\right), \text{ and} \tag{12.10}$$

$$L_m = \frac{d^2}{8 k_{\text{eff}}^2 v^2 C_0}. \tag{12.11}$$

Hence, we can easily find the electrical circuit element values from the electromechanical properties of the FBAR resonator (note the capacitance C_0 is determined from the geometry).

Thus far, we have ignored dissipation in our description. An actual FBAR resonator will of course exhibit both mechanical and electrical losses. As the motional branch of the circuit strongly dominates the impedance near the mechanical resonance, the series resistor R_m is included in the motional branch to account for mechanical dissipation. Far from the mechanical resonant frequency, the impedance of the FBAR approaches that of a capacitor with value C_0. Dielectric dissipation in this capacitor is included by adding the resistor R_0 to the electrical branch. The entire MBVD circuit model, including these dissipative elements, is shown in Fig. 12.1c. The electrical response of this circuit will exhibit resonances similar to those derived above, although the dissipation will suppress the amplitude of $|Z|$ near the resonance frequency, and slightly alter the exact resonance frequency.

12.2 Mechanical Resonator Characterization

We fabricated piezoelectric mechanical resonators of the FBAR design described above using aluminum metal electrodes and sputtered polycrystalline AlN as the piezoelectric material. The multi-layer structure was patterned on the surface of a high-resistivity (>10,000 Ω-cm) 100 mm diameter silicon wafer using interspersed lithographic processing steps of sputtered metal and piezoelectric deposition, optical lithography, and plasma etching; all patterning was done by etching previously-deposited layers, with no "lift-off" steps in the process. No electron beam lithography was needed to produce the structures. After processing the full wafer, individual dies were diamond-saw cut from the wafer, each containing one resonator structure. To mechanically isolate the mechanical FBAR structure from the substrate, the chip was then exposed to xenon difluoride gas in a custom-built vacuum system, a process

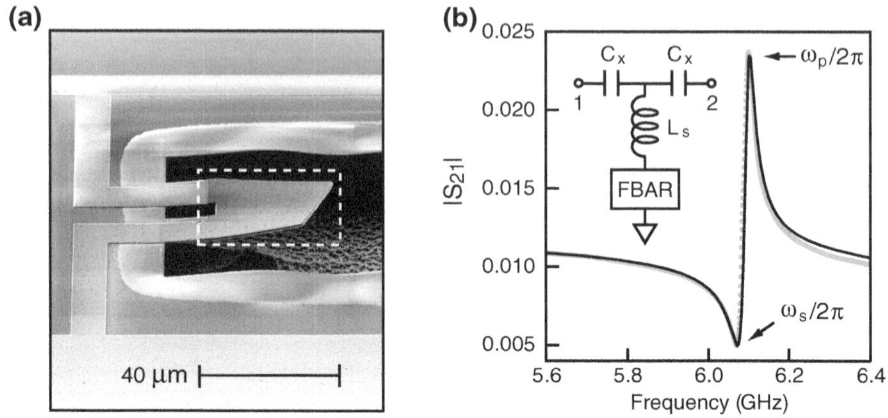

Fig. 12.2 Electron micrograph and electrical measurement of a mechanically-suspended resonator. **a** Mechanically active part of the resonator highlighted by white dashed rectangle. The resonator is supported on the left by the metal leads that form the two electrical connections, with their underlying AlN. **b** Electrical transmission measurement (*gray points*), with a fit to the equivalent circuit model (*black line*) using the (inset) circuit model. The transmission from port 1 to port 2 (inset) was measured with a calibrated network analyzer. C_x denotes the capacitance of external capacitors associated with the wiring between the measurement cables and the device, and L_S represents the stray inductance of the wiring leads [4]

that isotropically removes the silicon beneath the resonator structure, undercutting and releasing the mechanically-active part of the device. The resonator was left suspended over the remaining substrate, supported by the co-fabricated aluminum lead wires and underlying AlN. A scanning electron micrograph (SEM) of a completed device is shown in Fig. 12.2.

The mechanically-active metal/piezoelectric/metal stack is visible in the image, enclosed by the white dashed rectangle. The structure is supported by two metal leads joining the left edge of the stack to the bulk substrate and the external circuitry. In addition to providing mechanical support, the metal leads also serve as electrical connections to each electrode in the resonator stack. One of these leads was connected to the circuit electrical ground, while the other was connected to the microwave feed line shown at the top of the image. This microwave drive line was electrically coupled to wirebond pads using two on-chip interdigitated capacitors, C_x, which were used to isolate the resonant structure from the 50 Ω impedance of measurement lines (see inset circuit).

Electrical connections were made between the chip and its supporting aluminum microwave mount using 25 μm diameter wirebonds. The electromechanical response was measured using a microwave vector network analyzer. The normalized complex transmission S_{21} from port 1 to port 2 was measured as a function of frequency, producing the characteristic response illustrated in the main panel of Fig. 12.2b. There are two prominent features in the data: The dip in transmitted signal at $\omega_s/2\pi = 6.07$ GHz is attributed to the low-impedance series resonance. At the slightly higher frequency, $\Omega_M/2\pi = \omega_p/2\pi = 6.10$ GHz, the transmission takes on its maximum value due to the high-impedance, parallel mechanical resonance. This is the mechanical mode used in the quantum state experiments.

The equivalent circuit parameters are extracted from the spectroscopic data, using the electrical model shown inset to Fig. 12.2b. The capacitance C_0 is calculated using the lithographically defined area A of the electrodes, the estimated thickness of the AlN layer, $d = 330$ nm, and the approximate dielectric constant $\varepsilon_r \approx 10$ for AlN [5], yielding $C_0 = \varepsilon_0 \varepsilon A/d = 0.19$ pF. The stray inductance was estimated to be $L_s \approx 1$ nH, obtained by measuring the device response over a broad range of frequencies. The frequencies of the resonant responses, along with the equations linking the reactive elements in the circuit model, yield a mechanical capacitance $C_m = 0.655$ fF, a mechanical inductance $L_m = 1.043$ μH, and a piezoelectric coupling coefficient $k^2_{\text{eff}} \approx 1.2\,\%$ [3, 6]. The overall amplitude of the transmission indicates that the external capacitors were $C_x \approx 37$ fF, close to what is expected from the geometry, while the amplitude of the resonant response gives the resistive values $R_m = 146\,\Omega$ and $R_0 = 8\,\Omega$. The quality factor of the parallel resonant mode is correspondingly $Q \approx 260$.

We performed a number of experiments that ensured that the measured resonance feature was indeed mechanical in nature, and not a spurious electrical resonance. One such test was conducted by physically removing the mechanically-active part of the device and measuring the remaining circuitry. When measured in this configuration, no resonant response was seen. This result indicates that the response shown in Fig. 12.2b is due (in part, at least) to the presence of the suspended structure highlighted in Fig. 12.2a. We also fabricated an identical chip but using amorphous silicon nitride (SiN), a non-piezoelectric insulator material, in place of the piezoelectrically-active AlN. When this device was measured in the same manner, no resonant response was visible, an indication that the piezoelectric material is needed to generate an electrical response.

Two additional series of tests were performed, which more directly illustrate the mechanical nature of the resonance. The first series of tests was conducted to explore changes in the resonant response when the mechanically-active part of the resonator was not fully suspended. As previously mentioned, the active part of the resonator is released from the underlying Si substrate with XeF_2 gas, which isotropically removes silicon from around and eventually underneath the resonator. If instead of allowing the XeF_2 to fully undercut the resonator, a shorter exposure to the reactive gas is used, then the active part of the resonator will remain partially connected to the substrate by a pillar of silicon. If the chip is then exposed to additional doses of XeF2 gas, the diameter of the pillar will be reduced in size, freeing a greater area of the active part of the structure.

Using this technique, one particular sample was measured a number of times, each time after increasing the degree of mechanical release. The FBAR structure was first measured prior to any exposure to XeF_2, with the FBAR fully connected to the substrate. No resonant response was observed. Without removing the device from the microwave measurement mount, the resonator was then exposed to XeF_2 gas for a time long enough to release approximately 30 % of the area beneath the resonator.

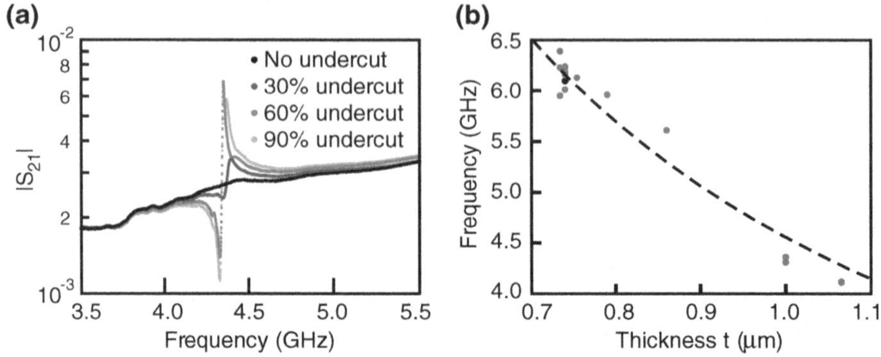

Fig. 12.3 Measured mechanical dissipation and frequency dependence of a FBAR structure. **a** Electrical resonant response of a resonator measured with varying degrees of mechanical release. A larger amplitude response indicates less mechanical dissipation. **b** Mechanical resonant frequencies $\Omega_M/2\pi$ obtained by measuring resonators with differing overall thicknesses t. The darker datum at 6.1 GHz corresponds to the resonator shown in Fig. 12.2

The sample was then re-measured and the characteristic resonant response began to emerge. The FBAR was then re-exposed to XeF_2 and measured two more times. Each time the resonant response grew in amplitude as illustrated in Fig. 12.3a. The trend toward greater amplitude indicates that the measured response is most likely mechanical in origin, and indicates that the structure was emitting less and less acoustic power into the substrate as its connection to the substrate was weakened.

To further test the nature of the electrical resonance, we fabricated a series of resonators with varying overall thicknesses t (including both the electrode thickness as well as the AlN thickness d). Equation (12.2) indicates that for greater thicknesses, the resonant frequency should decrease, approximately as $1/t$. This prediction is in stark contrast with what one would expect for a purely electrical resonator composed of the parallel connection of an inductor L and a capacitor C, where the capacitance is due to the geometric capacitance of the FBAR structure. In that case, one would have $C \propto 1/t$, so that the resonance frequency would scale as $1/\sqrt{LC} \propto \sqrt{t}$.

The spectroscopically determined mechanical resonance frequencies $\Omega_M/2\pi$ for a set of unique resonators with different thicknesses t are shown in Fig. 12.3b. The reported thickness correspond to the approximate total thickness of the resonator, where the thickness was varied from resonator to resonator by changing the deposition time of the AlN layer. It is clear from the data that the overall trend is toward lower resonant frequency with increased thickness, as expected. The data can be reasonably fit with the expected form $\Omega_M/2\pi \approx v/2t$, which produces the dashed line shown in Fig. 12.3b. The speed of sound was fit to $v \approx 9100$ m/s, and while not unreasonable, we emphasize that this oversimplifies the behavior of these composite structures. As a side note, the variation in the resonant frequency observed for a given thickness value was most likely due to variations in the actual thickness of the structure. As the deposited films tend to become thinner toward the edge of the substrate wafer, resonator dies taken from different locations on the same wafer will have varying thicknesses and thus resonant frequencies. The range of resonance frequencies we observe are within the range expected for this thickness variation.

This series of check experiments indicate quite strongly that we have realized FBAR structures with fundamental mechanical resonance frequencies in the few GHz band. These frequencies provide promise for the observation of quantum effects in such resonators when cooled to temperatures below 1 K. However, although the high resonance frequency of these structures avoids the typical limitation set by the presence of a strongly decohering thermal bath, these resonators do suffer from an inherent drawback: As the mechanical resonance frequency is increased, the quality factor of the resonance is seen to decrease correspondingly. A commonly used figure of merit that takes this effect into consideration is the f-Q product, the product of the resonance frequency and the quality factor. Here we find f-Q products of order 10^{12} Hz, within the range seen for most "high quality" mechanical resonators. However, as the FBARs measured here did not exhibit exceptionally high f-Q products, the high mechanical frequency comes at the expense of a lower quality factor.

The low quality factor has a direct implication for quantum measurements, as the lifetime T_1 for a single quantum of energy can be related to the classical quality factor, through $T_1 = Q/\omega_p$. Here we can estimate for our FBAR structures an energy lifetime of $T_1 = 6.7$ ns. A quantum operation on the mechanical resonator must be completed within this time before the quantum state is destroyed by dissipation. Creating and measuring quantum states in such a short time window requires careful planning, and in particular, a very strong coupling between the system used to create and measure the quantum state and the mechanical resonator. The approach we chose to use, to strongly couple a system that can quantum control and quantum measure a mechanical resonator, was to couple the resonator to an electrical equivalent of an atom: A superconducting Josephson phase qubit [7].

12.3 Josephson Junctions as Tunable Two Level Systems

A Josephson junction is created when a superconducting path is interrupted by a weak link that restricts the flow of electrons (or, in the superconducting state, Cooper pairs). For example, Josephson junctions are often formed by separating two superconducting metal electrodes with a thin insulating barrier, as depicted in Fig. 12.4, commonly referred to as a superconductor-insulator-superconductor (SIS) junction. The barrier completely separates the metal electrodes from one another, but is made thin enough that electrons (Cooper pairs) can tunnel through the barrier in the normal (superconducting) state. Although SIS junctions were used for all the experiments to be described below, Josephson junctions can be formed in other ways as well. Most notable are the superconductor-normal metal-superconductor (SNS) junction, where the superconductor is interrupted by a short normal metal section, sufficiently short that superconducting Cooper pairs can sometimes survive the passage through the

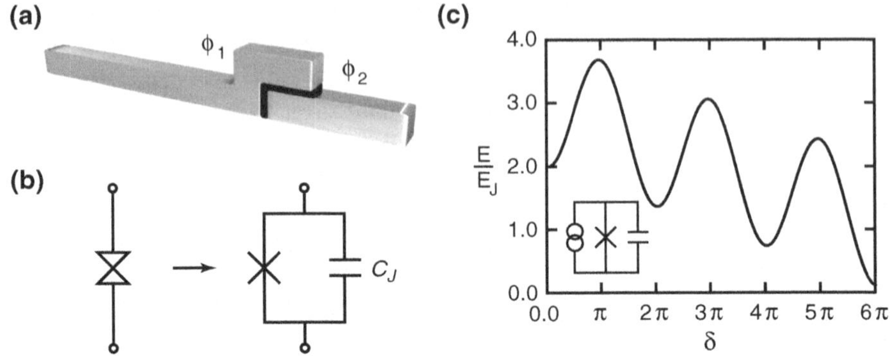

Fig. 12.4 A Josephson junction: SIS junction geometry, capacitively shunted circuit model, and "washboard" potential energy diagram. **a** Perspective drawing of an overlap Josephson junction. The two superconducting metal wires are separated by an insulating barrier, shown in *black*; this is typically a very thin (1 nm) metal oxide layer. The phase of the Ginzburg-Landau superconducting wavefunctions in each superconducting wire is represented by $\phi_{1,2}$. **b** Circuit representation of a Josephson junction, where the *hourglass symbol(left side)* represents the physical Josephson junction and is equivalent to an ideal Josephson junction shunted by a capacitor (*right*). **c** Potential energy model for a current-biased Josephson junction. The biasing current determines the overall tilt of the potential, while the oscillations are due to the dc Josephson effect. The local, periodic minima exist for bias currents I less than the critical current I_0

normal metal (proximity effect); and the microbridge junction, where a very narrow constriction in a superconducting wire limits the flow of supercurrent [8].

The physics of the Josephson junction has received much interest since its discovery in 1962 [9]. This is is part due to two remarkable characteristics of these unique devices, as described by the dc and ac Josephson relations. The dc Josephson relation describes the flow of a supercurrent through a junction, even in the absence of an applied voltage. The dc Josephson relation states that the supercurrent I_J through an SIS junction can be written as

$$I_J = I_0 \sin \delta, \tag{12.12}$$

where I_0 is the junction's critical current, and $\delta = \phi_2 - \phi_1$ is the gauge-invariant phase difference between the Ginzburg-Landau wavefunctions on either side of the junction's insulating barrier (in the absence of a magnetic field). If a voltage V is applied across the Josephson junction, the phase difference across the junction is predicted by the ac Josephson relation to evolve in time according to

$$\dot{\delta} = \frac{2e}{\hbar} V. \tag{12.13}$$

Combining these two relations implies that a voltage-biased junction will have a supercurrent that oscillates in time at a frequency $2eV/h$.

In order to exploit these novel properties, Josephson junctions have been incorporated into a wide array of superconducting circuits [10–24]. For illustrative purposes, we will examine a simple circuit, the current-biased Josephson junction. It is easiest to proceed by first generating an equivalent electrical circuit. Up to this point in our discussion, we have treated the junction in the quasi-static limit, where we have ignored the self-capacitance of the junction's geometric structure. As the junction is formed from two metal electrodes separated by a dielectric insulator, we expect that the resulting geometric capacitance will be electrically in parallel to the junction itself. A model circuit for a single junction is illustrated in Fig. 12.4b, where the junction has been split into an ideal junction, represented by an cross, and its parallel capacitance C_J. The ideal junction is assumed to follow exactly the two Josephson relations.

With this circuit model, we can analyze the effect of adding a time-varying current bias I to the circuit, as depicted in the insert of Fig. 12.4c. Kirchhoff's current law dictates that

$$I = C\dot{V} + I_J. \tag{12.14}$$

Using the dc Josephson effect to relate I_J to the phase difference δ, and recognizing that the voltage across the capacitor must be the same as that across the ideal junction, we can combine this equation with the ac Josephson relation to yield

$$\frac{\hbar}{2e}C\ddot{\delta} + I_0 \sin(\delta) - I = 0. \tag{12.15}$$

This is an equation of motion very similar to a classical particle moving in one dimension, with mass proportional to the capacitance C, interacting with a force proportional to $I - I_0 \sin\delta$. We can cast this equation of motion into the Lagrangian formalism and find the associated kinetic and potential energies

$$T = \left(\frac{\hbar}{2e}\right)^2 \frac{C}{2}\dot{\delta}^2 \tag{12.16}$$

$$U(\delta) = -E_J\left(\cos\delta + \frac{I}{I_0}\delta\right), \tag{12.17}$$

where we have introduced the parameter $E_J = \hbar I_0/2e$, the Josephson energy. Here we see that our fictitious particle has mass $M = (\hbar/2e)^2 C$, moving in the potential $U(\delta)$. A plot of the potential energy is shown in Fig. 12.4c for an applied current bias I far from the critical current I_0. The potential $U(\delta)$ is commonly referred to as the tilted washboard potential, with tilt proportional to the applied current. The potential displays 2π-periodic local minima for $I < I_0$, at which point the minima become inflection points. For bias currents $I \geq I_0$, the particle is free to run down the potential and the junction switches from the superconducting state to the "voltage state," where the junction develops a dc voltage.

To form the classical Hamiltonian, we first find the canonical momentum conjugate to the coordinate δ, [25]

$$p = \frac{\partial L}{\partial \dot{\delta}} = \left(\frac{\hbar}{2e}\right)\frac{\hbar C}{2e}\dot{\delta} = \frac{\hbar}{2e}Q, \tag{12.18}$$

where $Q = C\dot{\delta}$ is the charge on the capacitor. The classical Hamiltonian is then

$$H_{\text{cl}} = p\dot{\delta} - L = \frac{Q^2}{2C} - E_J\left(\cos\delta + \frac{I}{I_0}\delta\right). \tag{12.19}$$

If we now recognize that the capacitor has a Cooper pair charging energy $E_C = (2e)^2/2C$, then we can rewrite the Hamiltonian as [26] as

$$H_{\text{cl}} = E_C N^2 - E_J\left(\cos(\delta) + \frac{I}{I_0}\delta\right), \tag{12.20}$$

where $N = Q/2e$ is the number of Cooper pairs.

The classical Hamiltonian can now be quantized. Following the customary prescription, we define the quantum operators $\hat{p}$ and $\hat{\delta}$ and impose the commutation relation

$$\left[\hat{\delta}, \hat{p}\right] = i\hbar. \tag{12.21}$$

The coordinate representation of the momentum operator is then

$$\hat{p} \to -i\hbar\frac{\partial}{\partial \delta}, \tag{12.22}$$

which produces the coordinate representation for the Hamiltonian,

$$\hat{H} \to -E_C\frac{\partial^2}{\partial \hat{\delta}^2} - E_J\left(\cos\hat{\delta} + \frac{I}{I_0}\hat{\delta}\right). \tag{12.23}$$

The solution to the time-independent Schrödinger equation using this Hamiltonian will produce quantized energy levels (within the potential local minima) because we have ignored dissipation. Although these energy eigenstates can be solved for directly, we instead examine a harmonic approximation that captures the qualitative behavior of the lowest-lying energy levels, as illustrated in Fig. 12.5a. For small current bias, the shape of the potential minima can be accurately approximated by that of a harmonic oscillator. The energy eigenstates of the harmonic oscillator are determined by the curvature U'' of the potential function; in this approximation, this curvature is related to the plasma frequency ω_{plasma} through the defining relation [27]

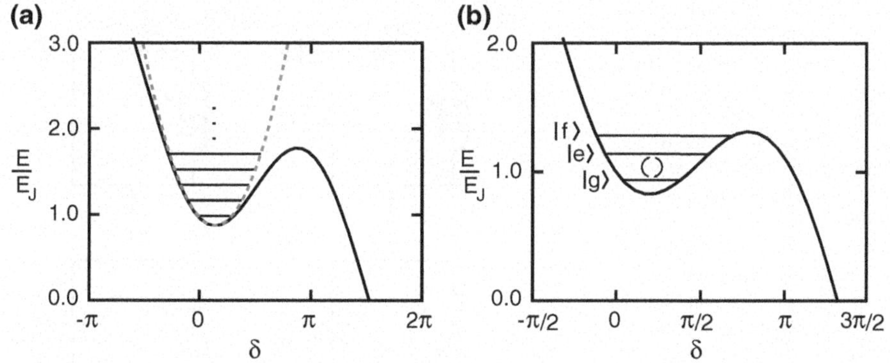

Fig. 12.5 Harmonic approximation and eigenstates of the washboard potential. **a** The *dashed line* overlay is a harmonic approximation to the local minimum. The evenly spaced energy levels are qualitative representations of the harmonic eigenstates. **b** Lowest energy levels of a local minimum (other eigenstates not shown). The nonuniform spacing arises from the nonlinearity of the potential. The two lowest-lying levels of a local minimum, the ground state $|g\rangle$ and the first excited state $|e\rangle$, form the qubit states

$$\omega_{\text{plasma}} \equiv \sqrt{\frac{U''\,|_{\min}}{M}} = \frac{\sqrt{2E_C E_J}}{\hbar}\left(1-\frac{I^2}{I_0^2}\right)^{1/4}. \tag{12.24}$$

The plasma frequency depends on the current bias I, with its maximum value at zero bias current. Equation (12.24) is a good approximation for the energy level spacing for a junction biased far from the critical current I_0. However, as the bias current approaches I_0, the shape of the potential becomes increasingly anharmonic. This anharmonicity results in a reduced number of energy levels, whose energy spacing, and thus transition frequency, decreases with level number, as shown in Fig. 12.5b. The lowest energy level we call the ground state $|g\rangle$, and the next higher level the excited state $|e\rangle$. The ground state $|g\rangle$ to excited state $|e\rangle$ transition can be addressed separately from the higher energy levels, due to the anharmonicity, as long as the microwave signal used to stimulate transitions is carefully selected in frequency and careful pulse shaping is used [28]. The two-state manifold $|g\rangle$ and $|e\rangle$ forms the qubit computational basis.

12.4 Coupling a Qubit to a Mechanical Resonator

A straightforward way to strongly couple a Josephson qubit to a piezoelectric mechanical resonator is to electrically connect the resonator in parallel with the Josephson junction. Using the modified Butterworth-van Dyke model for the FBAR resonator, we can design circuits as shown in Fig. 12.6. The two circuits shown there

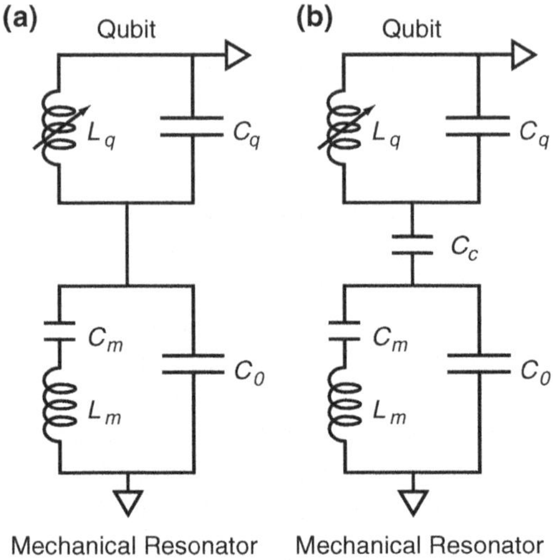

Fig. 12.6 Directly and capacitively coupled qubit-FBAR circuit representations. **a** The tunable inductor L_q and fixed capacitance C_q represent the qubit. The qubit is directly connected to a FBAR, which is modeled by the modified Butterworth-van Dyke circuit in the dissipation free limit. **b** A coupling capacitor can be used to decrease, and control, the electrical connection between qubit and FBAR

will support two resonant modes: One mode is primarily associated with the qubit itself (here we are implicitly approximating the qubit by its harmonic, that is linear, approximation), and the second mode is primarily attributed to the mechanical resonator. Figure 12.6a shows a circuit representation for the coupling scheme we have just described, where we have replaced the Josephson element with a tunable (linear) inductor, such that the L_q-C_q resonance reproduces the qubit $|g\rangle \leftrightarrow |e\rangle$ oscillation frequency. Note that while we are using linear circuit elements in this model, we must restrict excitations to the equivalent of a single photon in the system, maintaining the qubit in its lowest two levels, as otherwise the linear circuit element will not accurately reproduce the nonlinear response of the qubit.

The classical coupling strength between the two resonant modes in Fig. 12.6a can be found by isolating the qubit inductance L_q and working out the effective electrical admittance $Y(\omega)$ of the circuit in parallel with that inductance:

$$Y = \frac{1}{Z_{L_q}} + \frac{1}{Z_{C_q}} + \frac{1}{Z_{C_0}} + \frac{1}{Z_{C_m} + Z_{L_m}}, \tag{12.25}$$

where Z_x is the impedance of element x. The resonant modes are found by solving for the frequencies where the admittance goes to zero. The positive solutions to the resulting biquadratic equation are

$$\omega_\pm = \sqrt{\frac{-B \pm \sqrt{B^2 - 4AC}}{2A}}, \tag{12.26}$$

where $A = C_m(C_0 + C_q)L_q L_m$, $B = -C_m L_m - (C_0 + C_m + C_q)L_q$, and $C = 1$.

In the limit of small qubit inductance (thus higher qubit frequency), and for qubit and FBAR with quite different self-resonant frequencies, these solutions take on the limiting forms

$$\omega_+^L \to \frac{1}{\sqrt{L_q(C_0 + C_q)}}, \text{ and} \tag{12.27}$$

$$\omega_-^L \to \frac{1}{\sqrt{L_m C_m}}. \tag{12.28}$$

We can interpret the limiting form for ω_+^L as a qubit mode that resonates at the angular frequency $1/\sqrt{L_q C_{q,\mathrm{eff}}}$ with re-normalized capacitance $C_{q,\mathrm{eff}} = C_0 + C_q$. Likewise, the limiting form for ω_-^L can be attributed to the mechanical resonant frequency at $\omega_r = 1/\sqrt{L_m C_m}$. Now, using the fact that the qubit inductance L_q can be changed by varying the qubit bias, we can reduce the upper frequency ω_+^L, and thus tune the qubit frequency towards the mechanical resonance ω_-^L. Increasing L_q to make the two frequencies equal yields an on-resonance coupling between the qubit and resonator; this is analogous to a system of two different harmonic oscillators, coupled by a weak spring, and adjusting the mass of one of the oscillators to make its frequency equal to that of the other oscillator. This condition is colloquially referred to as the frequency for which the two modes are "on resonance".

For a given qubit inductance, we can calculate the frequency ω_+^L, and we define the detuning Δ as the difference between the qubit frequency ω_+^L and the resonator frequency ω_-^L, $\Delta \equiv \omega_+^L - \omega_-^L$. When the two modes are on resonance, $\Delta = 0$. However, the frequency difference between the full forms of the two modes, Eq. (12.26), is always nonzero, even at zero de-tuning, due to what in quantum mechanics is often called "level avoidance"; here we are seeing its classical representation. In fact, the frequency difference, $\Omega = (\omega_+ - \omega_-)_{\Delta=0}$, evaluated at $\Delta = 0$, is a measure of the coupling strength between the modes, $\Omega = 2g_0/\hbar$, where g_0 is the classical coupling strength, which corresponds in the quantum limit to vacuum Rabi coupling strength.

The coupling strength can be solved for analytically. The general solution for the difference between the positive roots of a biquadratic equation is given by

$$\omega_+ - \omega_- = \sqrt{-\frac{B + 2\sqrt{AC}}{A}} \tag{12.29}$$

For $\Delta = 0$ we find

$$\Omega = (\omega_+ - \omega_-)_{\Delta=0} = \sqrt{\frac{1}{L_m C_{q,\mathrm{eff}}}}, \tag{12.30}$$

where we have used the fact that $L_q = C_m L_m / C_{q,\mathrm{eff}}$ on resonance. Thus, if $C_{q,\mathrm{eff}}$ is held fixed, we see that the coupling strength depends only on the mechanical inductance L_m. This implies that larger-area FBARs with the same film thicknesses

t will be more strongly coupled to the qubit, and the area of the FBAR can be adjusted to obtain the desired coupling.

However, relying on the size of the resonator to set the coupling strength may prove inconvenient if the desired resonator size is difficult to realize in practice. To further control the coupling between qubit and mechanical resonator, a coupling capacitor may be added to the circuit, as shown in Fig. 12.6b. A similar analysis indicates that the two limiting modes are on resonance when $L_q = C_{m,\text{eff}} L_m / C_{q,\text{eff}}$, where $C_{q,\text{eff}} = C_q + C_0 C_c (C_0 + C_c)^{-1}$ and $C_{m,\text{eff}} = C_m (C_0 + C_c) / (C_0 + C_c + C_m)$. The frequency separation between the mode frequencies on resonance, and thus the coupling strength, is modified by the coupling capacitance to become

$$\Omega_C = \frac{C_c}{C_0 + C_c} \sqrt{\frac{1}{L_m C_{q,\text{eff}}}}. \tag{12.31}$$

We see that the general form is preserved, and we can now adjust the coupling strength via the coupling capacitor C_c as well as through the FBAR parameters.

12.5 Qubits as Quantum Transducers

Although cooling a mechanical resonator to its quantum ground state is not a trivial procedure, for the discussion that follows we will assume this has already been achieved. The next clear challenge is to excite the ground-state resonator into a non-classical state. However, the method needed to produce such states is not readily apparent. The use of classical excitation pulses to drive a harmonic oscillator will generate coherent states, whose subsequent behavior is indistinguishable from those of a purely classical oscillator: The simple harmonic oscillator, which is the appropriate model for a single harmonic resonance, is always in the correspondence limit between classical and quantum mechanics [29].

To illustrate this point, it is useful to outline the process that creates a coherent state. When a harmonic oscillator, initially in its ground state $|0\rangle$ (where here we represent the oscillator state in terms of its quantum number (Fock state) basis $|n\rangle$), is excited by an on-resonance classical force, the state of the system gradually increases its amplitude in the first excited $|1\rangle$ state. However, as soon as there is a non-zero amplitude in $|1\rangle$, the drive will begin to increase the amplitude of the next higher state $|2\rangle$, and so on. The state resulting from any classical pulse will thus be a superposition of number states $|0\rangle$, $|1\rangle$, $|2\rangle$ .... A Gaussian pulse yields a superposition such that the probability P_n to find the system in the state $|n\rangle$ after the pulse is given by the Poisson distribution,

$$\text{P}_n(a) = \frac{a^n e^{-a}}{n!}, \tag{12.32}$$

where a is the pulse amplitude, calibrated here to the average number of phonons excited by the pulse, and n is the index for the Fock state $|n\rangle$.

Instead of exciting the mechanical resonator directly, we chose to use the extraordinarily strong nonlinearity of our qubit, effectively using the qubit as a "classical-to-quantum transducer". As the qubit can be precisely manipulated using classical excitation pulses in a way that only involves its lowest two energy levels $|g\rangle$ and $|e\rangle$, a result of the anharmonicity of the qubit energy levels, purely classical excitations can be used to completely quantum-control the qubit state. For example, a qubit initially in $|g\rangle$ can be precisely excited, by an on-resonance, calibrated amplitude and duration classical microwave pulse, to the final state $|e\rangle$. At this point, further excitation will force the qubit to emit energy into the excitation field and return to its ground state $|g\rangle$. Indeed, the probability P_e of finding the qubit in the excited state $|e\rangle$ is described by the Rabi formula [30]

$$\mathrm{P}_e(t) = \sin^2\left(\frac{I_{\mathrm{rf}}\chi_{eg}E_J}{\hbar I_0}t\right), \tag{12.33}$$

where χ_{eg} is the dipole matrix element between states $|g\rangle$ and $|e\rangle$, I_{rf} the amplitude of the microwave current, E_J the qubit Josephson energy, and I_0 the qubit critical current [27]. In fact, by controlling the duration and phase of an on-resonant microwave pulse, the qubit can be accurately placed in any superposition state $|\psi\rangle = \alpha|g\rangle + \beta|e\rangle$ [31].

Due to the precision with which qubits can be quantum-controlled, they provide an ideal means for quantum control of a harmonic oscillator, such as a mechanical resonator. Producing a non-classical state in the mechanical resonator now becomes a two-step process. First, one creates an arbitrary qubit state using a classical excitation, with the qubit well de-tuned from the resonator to minimize interactions. Second, one transfers the quantum state to the mechanical resonator, by bringing the qubit into frequency resonance with the resonator in a carefully controlled fashion, and then waiting as the excitation transfers (Rabi swaps) from qubit to resonator. This is completely analogous to the classical transfer of energy that occurs between two coupled harmonic oscillators when one is excited: The classical "beating" described in elementary mechanics is the classical description of what here is the equivalent quantum process. As mentioned previously, the transfer time for the excitation (in both the quantum and classical systems) must be shorter or, at most, of the same order as the mechanical oscillator's energy relaxation time. For the FBAR resonators described above, this time is only several nanoseconds. Since the transfer time is inversely proportional to the coupling strength Ω_C, strong qubit-FBAR coupling is required.

The quantum dynamics of the capacitively-coupled qubit-FBAR system are most easily understood by casting the full qubit-resonator Hamiltonian into the Jaynes-Cummings approximate Hamiltonian (see [32]):

$$\frac{\hat{H}}{\hbar} = -\frac{\omega_q}{2}\hat{\sigma}_z + \Omega_{\mathrm{M}}\hat{a}^\dagger\hat{a} - i\frac{\Omega_C}{2}\left(\hat{a}\hat{\sigma}_- - \hat{a}^\dagger\hat{\sigma}_+\right). \tag{12.34}$$

In the Jaynes-Cummings Hamiltonian, written here in the rotating wave approximation, the qubit energy is given by the first term, which is that of a two-level system with tunable transition frequency $\omega_q/2\pi$, where $\hat{\sigma}_z$ is the Pauli z operator.

The second term corresponds to the mechanical resonator, which takes on the standard form of a harmonic oscillator with energy level transition frequency $\Omega_{\mathrm{M}}/2\pi$, and $\hat{a}^{\dagger}$ and $\hat{a}$ are the resonator phonon raising and lowering operators. The last term of the Hamiltonian expresses the qubit-mechanical resonator coupling, where $\hat{\sigma}_{-}$ and $\hat{\sigma}_{+}$ are the qubit raising and lower operators, respectively [27]. The rotating wave approximation is valid when the qubit-resonator coupling is much smaller than the mechanical transition frequency, $\Omega \ll \Omega_{\mathrm{M}}$ [33].

If a quantum state $\alpha|g\rangle + \beta|e\rangle$ is created in the qubit, it can be transferred to the mechanical resonator by tuning the qubit $|g\rangle \leftrightarrow |e\rangle$ transition frequency into resonance with the mechanical resonator frequency Ω_{M}. If we label the probability amplitudes C_{qr} for the basis state $|qr\rangle$ of the coupled system, i.e. a state is given by $|\psi\rangle = C_{g0}|g0\rangle + C_{g1}|g1\rangle + C_{e0}|e0\rangle + C_{e1}|e1\rangle$ (here including only the two lowest resonator Fock states $|0\rangle$ and $|1\rangle$), then the dynamics of the system are described by the set of equations [27]

$$c_{g0}(t) = \alpha, \tag{12.35}$$

$$c_{g1}(t) = \beta \frac{\Omega_C}{\Omega(\Delta)} \sin\left(\frac{\Omega(\Delta)t}{2}\right) e^{-i\Delta t/2}, \tag{12.36}$$

$$c_{e0}(t) = \beta \left[\cos\left(\frac{\Omega(\Delta)t}{2}\right) + i\frac{\Delta}{\Omega(\Delta)} \sin\left(\frac{\Omega(\Delta)t}{2}\right)\right] e^{i\Delta t/2}, \text{ and} \tag{12.37}$$

$$c_{e1}(t) = 0. \tag{12.38}$$

Here $\Delta = \omega_q - \Omega_{\mathrm{M}}$ is the qubit-resonator de-tuning and $\Omega(\Delta) = \sqrt{\Omega_C^2 + \Delta^2}$ is the vacuum Rabi frequency. In Fig. 12.7a we plot the time evolution for the initial qubit state $|e\rangle$ (i.e. $\alpha = 0, \beta = 1$), with the qubit transition frequency exactly on resonance with the mechanical resonator, $\Delta = 0$. The resonator is initially in the ground state, but after a quarter of a swap period, $\Omega_C t = \pi/2$, the qubit and mechanical resonator are maximally entangled, with an equal probability to find either in the excited state. The state transfer from qubit to mechanical resonator is complete after a time $t = \pi/\Omega_C$, which places the resonator in the quantum state $|1\rangle$ and leaves the qubit in the ground state. The energy quantum is returned to the qubit (up to a phase) if the interaction is allowed to proceed to $t = 2\pi/\Omega_C$. For a more general state, with both α and β non-zero, the same process occurs, with the resonator in the swapped state $\alpha|0\rangle + \beta|1\rangle$ after a time $t = \pi/\Omega_C$, and the state being completely returned to the qubit after a time $t = 2\pi/\Omega_C$.

If the qubit and mechanical resonator are not exactly on resonance, i.e. for non-zero de-tuning, then the state transfer will still take place, but with a higher frequency $\sqrt{\Omega_C^2 + \Delta^2}$, and with a reduced maximal transfer probability $\Omega_C^2/\left(\Omega_C^2 + \Delta^2\right)$. We have illustrated this dependence on de-tuning by plotting the probability to find the qubit in the excited state $P_e = |C_{e0}|^2$, shown in Fig. 12.7b. The light portions of the plot correspond to the quantum state being primarily found in the qubit, while the dark portions indicate a preferentially excited mechanical resonator $P_1 = 1 - P_e$,

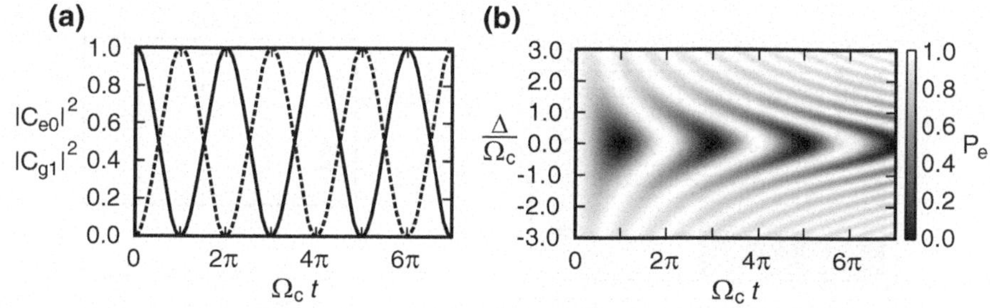

Fig. 12.7 Qubit-mechanical resonator probabilities on resonance and for variable de-tunings. **a** Shows the on-resonance time evolution of the excited state probability for the qubit, $|C_{e0}|^2$ (*solid line*), and the $|1\rangle$ state probability for the resonator $\left|C_{g1}\right|^2$ (*dashed line*), for the initial state $|e0\rangle$. **b** Shows the time evolution of the qubit $|e\rangle$ state probability P_e as for the same time evolution, but for different de-tunings Δ. The trace for $|C_{e0}|^2$ in **a** is the same as the evolution for P_e in **b** for $\Delta = 0$

from conservation of probability. Of course, in any real system, energy loss and phase decoherence will cause the system to relax, so the oscillations become less pronounced as time proceeds.

12.6 Quantum Ground State and Single Phonon Control

Now that we have constructed the theoretical tools needed to understand the quantum interaction between a qubit and a mechanical resonator, we can turn our attention to the results we obtained for our experimental implementation of this system. An image and circuit representation for the experimental device is shown in Fig. 12.8, where, for clarity, we have not included dissipative elements for the qubit. In this device, a flux-biased Josephson junction qubit (phase qubit) was used instead of the idealized current-biased Josephson qubit we discussed earlier. In the phase qubit, the effective bias current is produced using an on-chip flux bias coil, mutually coupled to the inductor L_B which is connected across the Josephson junction. This means that the inclined "washboard potential" of the current-biased junction is replaced by a larger parabolic shape with superposed washboard oscillations, but the local behavior (in one of the metastable wells) remains effectively identical. In addition, an additional shunting capacitor C_S is added to adjust the qubit resonant frequency to the desired microwave range of 4 GHz $< \omega_q/2\pi <$ 8 GHz.

There are slight changes in the dynamics of the flux-biased Josephson qubit in comparison with the current-biased version. Although the additional capacitance can be thought of as just an increased mass $M = (\hbar/2e)^2(C_J + C_S)$, the additional linear inductor changes the overall shape of the washboard potential by incorporating a term that is quadratic in the phase coordinate. This eliminates the "running" non-zero voltage states, and forms the double-well potential depicted in Fig. 12.8c. Properly biased, the left well includes the qubit state manifold $|g\rangle$ and $|e\rangle$, while the right well

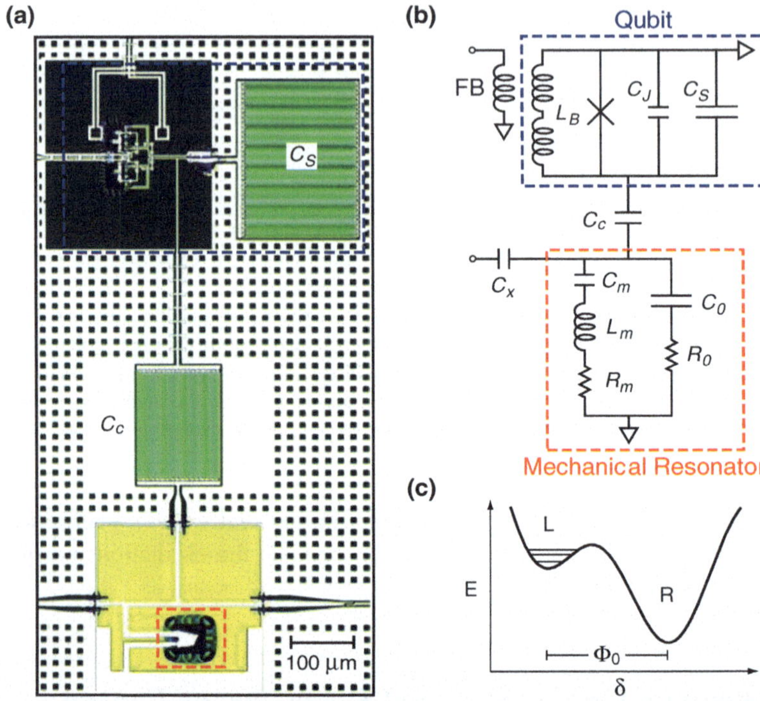

Fig. 12.8 Optical micrograph and circuit representation of coupled qubit-mechanical resonator. **a** The qubit at the top of the micrograph is contained within the *blue dashed line*. The Josephson junction is barely visible in the middle of the highlighted area and is shunted by an inductor to the *left* and an interdigitated capacitor to the *right*. A wire connects the qubit to an interdigitated coupling capacitor C_c, shown in the *middle* of the image. At the *bottom* of the image a mechanically-suspended FBAR is surrounded by the *red dashed box*. All circuit components are surrounded by an electrical ground plane. **b** Lumped-element circuit representation showing the capacitively-shunted Josephson junction in parallel with a gradiometric inductor L_B and a capacitance C_S resulting from the interdigitated inductor. The qubit is capacitively-coupled to the FBAR, which here is represented by the modified Butterworth-van Dyke model. **c** Double well potential energy of the phase qubit with left well, *L*, and right well, *R*. The three lowest lying energy levels of the left well are depicted schematically. The phase difference between left and right well corresponds to an approximate difference of one flux quanta threading the qubit loop

contains many lower-energy states. The two wells are separated by approximately 2π in the phase coordinate, which corresponds to approximately one flux quantum Φ_0 through the enclosing inductance.

The qubit is initialized by adjusting the flux bias until the phase "particle" is located in the left well, where in time it relaxes to the (metastable) ground state $|g\rangle$ of that well. This ground state, and the next higher energy level $|e\rangle$ in the same well, form the qubit manifold, with their energy level spacing dependent on the bias flux (instead of on the bias current). Microwave pulses resonant with the qubit $|g\rangle - |e\rangle$ transition can be applied through the flux bias line to prepare the qubit in any desired superposition of these states. In general, a sequence of these pulses, along with slower qubit frequency-tuning pulses, is used to control the qubit state, either by directly altering the state or by allowing the qubit to interact with the resonator. Once a pulse sequence is complete, a measurement of the qubit is performed by tilting the potential well with the external flux bias just enough so that the excited state $|e\rangle$ will preferentially tunnel from the left well into the right well, while the ground state $|g\rangle$ will remain in the left well. An on-chip superconducting quantum interference device (SQUID) reads out the measurement by detecting the flux threading the qubit loop, projectively differentiating between finding the phase particle in the left or right well. This procedure yields a 90–95 % visibility between the excited and ground state. The probability P_e that the qubit was in its $|e\rangle$ state just before measurement can be evaluated by repeating the entire process many ($\sim$ 1000) times, and averaging the results.

The coupled qubit-mechanical resonator sample shown in Fig. 12.8a was mounted in an aluminum box, attached to the mixing chamber of a dilution refrigerator, and cooled to the refrigerator base temperature of $T \approx 25$ mK. At this temperature, both the qubit and mechanical resonator should contain less than one energy quantum, as the inequality $\hbar\omega \gg k_B T$ is easily satisfied for typical qubit and resonator frequencies of a few GHz. Previous experiments on phase qubits have shown that the qubit is very reliably in its ground state at this temperature, and our aim in the next section is to demonstrate that the mechanical resonator is in its quantum ground state as well.

The first measurement is to determine the resonant frequencies of the coupled qubit-resonator system. This is performed using qubit spectroscopy, as illustrated in Fig. 12.9. For all measurements, the qubit flux is always first set to yield a $|g\rangle \leftrightarrow |e\rangle$ transition frequency $\omega_q/2\pi = 5.44$ GHz, which is what we term the "operating bias". Qubit spectroscopy is performed by pulsing the qubit dc flux away from the operating bias, and exciting the qubit with a microwave tone at some frequency for 1.0 μs. After tuning the (excited) qubit back to the operating bias, the qubit is measured, from which the excited state probability P_e can be determined. For a given microwave drive frequency, a substantial increase in qubit excited state probability is observed only when the qubit transition frequency is resonant with the microwave field. For non-resonant driving frequencies, the qubit remains in its ground state. Thus, by repeating the experiment while varying both the qubit flux bias and the microwave drive frequency, the resonant modes of the coupled system can be mapped out.

The measured spectroscopic response shows the qubit tuning as expected, with a strong avoided-level crossing, enclosed by the white dashed rectangle in Fig. 12.9b, appearing at the expected frequency. This is a characteristic feature of two interacting resonant systems, and arises here from the interaction of the qubit and resonator. A detailed view of the avoided crossing is shown in Fig. 12.9c, where the fixed mechanical resonant frequency is indicated by the dotted line at $\Omega_M/2\pi = 6.175$ GHz. The dashed white lines were produced by fitting the frequency response to the classical circuit model described above. The minimum frequency difference between the modes is $\Omega_C/2\pi = 124$ MHz and indicates an energy transfer time (Rabi swap time) between the qubit and mechanical resonator of approximately 4 ns.

We next exploited the strong qubit-resonator coupling to verify that the mechanical resonator was indeed in its quantum ground state. The general idea is to use the qubit

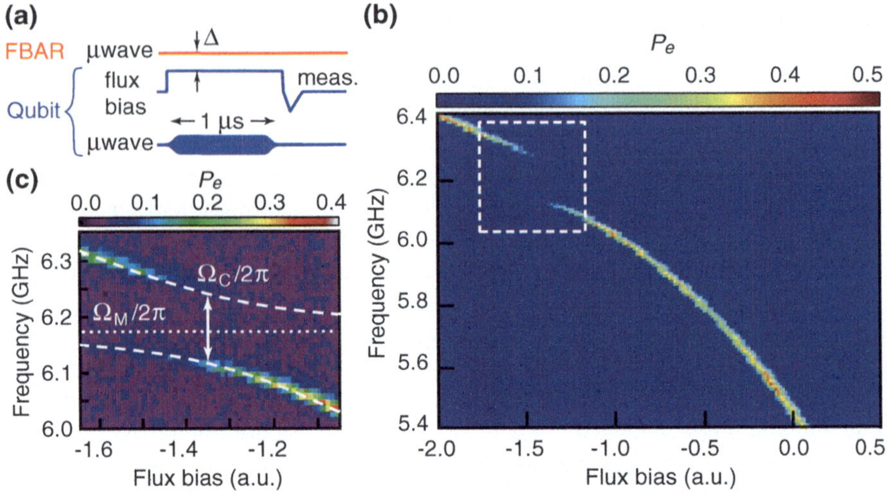

Fig. 12.9 Qubit spectroscopy: pulse sequence, qubit excited state probability, and detail of the avoided level crossing. **a** The qubit, initially in its ground state $|g\rangle$, is moved away from its operating bias and illuminated with microwaves for 1.0 μs. The qubit is then returned to the operating bias and measured. This experiment is repeated while varying both microwave frequency and flux bias tuning. **b** The probability for the qubit to be measured in the excited state, P_e, is maximal when the $|g\rangle \leftrightarrow |e\rangle$ energy level transition is resonant with the microwave drive. The distinct splitting enclosed in the *white dashed box* is due to the qubit-FBAR interaction. **c** Detail of **b** highlighting the avoided-level crossing with a fit to the model (*upper* and *lower dashed lines*). The *horizontal dotted line* indicates the frequency of the mechanical resonator $\Omega_M/2\pi$. The coupling strength $\Omega_C/2\pi$ = 124 MHz between qubit and resonator was determined from the minimum frequency difference between the two curves

as a "quantum thermometer," able to detect any non-zero state occupation in the system. If, for example, the mechanical resonator was not in the ground state, then some of that energy would transfer to the qubit when the qubit and resonator are brought into frequency resonance. Measuring that the qubit has a non-zero excited probability would imply that the mechanical resonator was not in the ground state, while a lack of qubit excitation would indicate that the FBAR was indeed in the quantum ground state.

We used two pulse sequences to measure the effective temperature of the resonator, as shown in Fig. 12.10a. As depicted in the upper pulse sequence, the qubit, initially in its ground state $|g\rangle$, was brought to a frequency de-tuning Δ from the mechanical resonance frequency and left there for 1.0 μs. The qubit was then returned to its operating bias and measured. The probability P_e of measuring the qubit in its excited state as a function of de-tuning Δ is shown in Fig. 12.10b (blue points). The lower pulse sequence shown in Fig. 12.10a, was used to establish consistency with the first measurement. In this pulse sequence, after interacting with the mechanical resonator, the qubit was returned to its operating bias, but just prior to measurement the ground and excited state probabilities were swapped by applying a microwave π-pulse. The results are shown in Fig. 12.10b by the black data points.

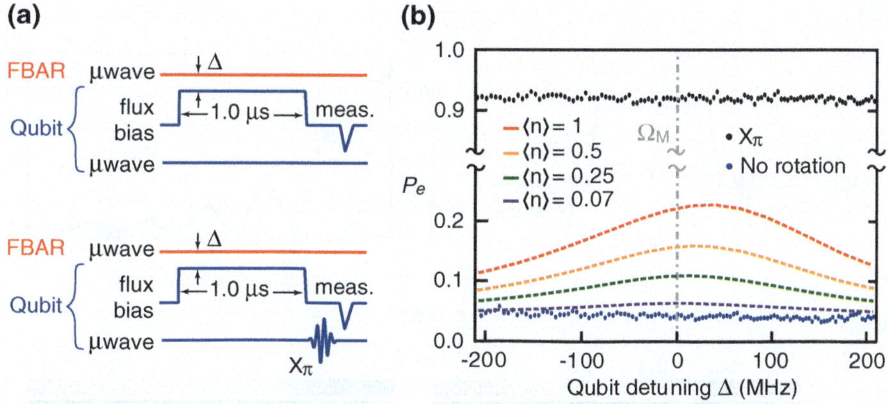

Fig. 12.10 Quantum ground state: pulse sequences and experimental data. **a** The qubit, initially in $|g\rangle$, was tuned to within Δ of the mechanical resonator and allowed to interact with the mechanical resonator for 1.0 µs. The qubit was then returned to its operating bias and measured. The experiment was repeated for a range of de-tunings. In the lower pulse sequence, an additional microwave π-pulse was used to swap the ground and excited state populations prior to measurement. **b** Qubit excited state probability without (*blue points*) and with (*black points*) the additional population-swapping π-pulse. The *dashed lines* are numerically simulated qubit probabilities, in which the mechanical resonator was maintained in a thermal state of $\langle n\rangle$ phonons during the qubit interaction process. The lack of a peak in the *blue data points* around zero de-tuning, and the lack of a dip in the *black data points*, which from the simulations would be evident even for a very small number of residual phonons $\langle n\rangle$, indicates that the mechanical resonator is with quite high probability (better than 93 %) in its quantum ground state $|0\rangle$

During the interaction, the qubit and mechanical resonator will come into energy equilibrium, and the qubit excited state population will then reflect any elevated population in the resonator, modulated by the strength of the interaction between the resonator and qubit, and by the qubit and resonator coupling to the thermal bath, as parameterized by their respective energy lifetimes. The numerical simulations include this physics. However, no change in qubit population was seen, even near zero de-tuning where the interaction is the strongest. The lack of a peak (dip) in the qubit excited state probability for the blue (black) data indicates that the mechanical resonator had no measurable additional energy to transfer to the qubit. The degree to which the resonator was in the quantum ground state was determined from dynamic quantum simulations to be $P_0 > 93\,\%$, with a corresponding average phonon number $\langle n\rangle < 0.07$.

With the assurance that the mechanical resonator was indeed in its quantum ground state, our focus shifted toward demonstrating quantum control of the mechanical resonator. We achieved this by using the qubit as a quantum transducer, generating and measuring a single phonon in the resonator, by applying the pulse sequence shown in Fig. 12.11a. The qubit, initially in $|g\rangle$, was promoted to its excited state with a microwave π-pulse (defined as the pulse that exchanges the populations of the $|g\rangle$ and $|e\rangle$ states, i.e. here exciting the qubit to its $|e\rangle$ state). The qubit was then biased to bring its transition frequency toward that of the mechanical resonator, and held at

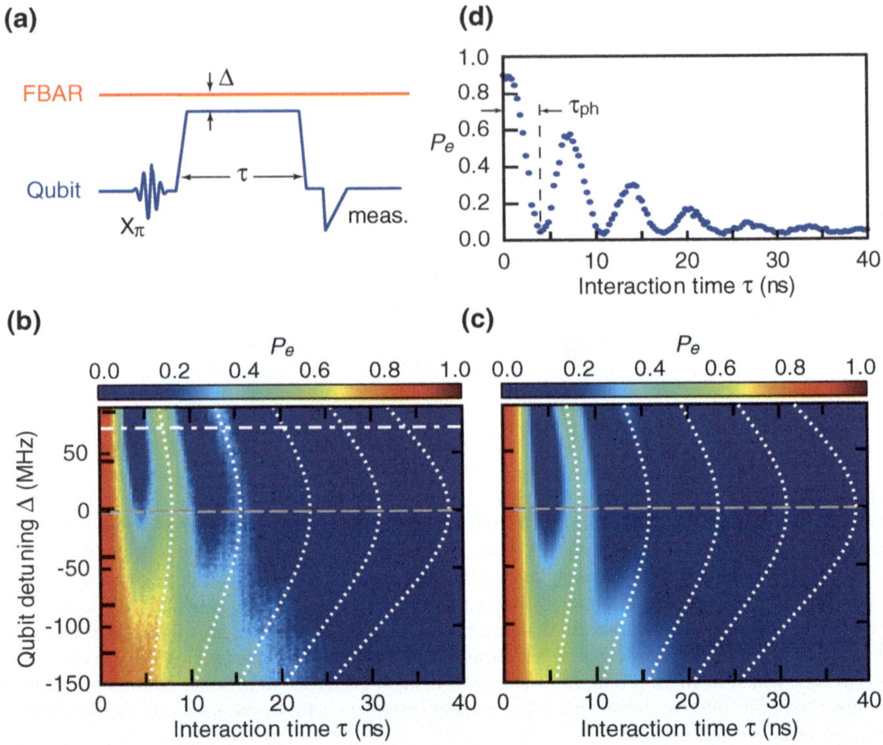

Fig. 12.11 Excited mechanical resonator states: pulse sequence, quantum simulation, and experimental data. **a** Pulse sequence in which the qubit is first promoted to $|e\rangle$ by a π-pulse, then allowed to interact with the mechanical resonator at a de-tuning Δ for a variable time τ, before being returned and measured at its operating bias. **b** Measurement of qubit excited state probability P_e as a function of de-tuning and interaction time, showing a single quantum excitation oscillating between the qubit and the mechanical resonator. The *dotted white lines* show the expected Lorentzian dependence. The *gray dashed line* indicates the mechanical resonance frequency. **c** Numerical simulation of the data in **b**, for a qubit initially in its $|e\rangle$ state interacting with a mechanical resonator, with the measured qubit and mechanical resonator parameters. **d** Qubit excited state probability for de-tuning $\Delta = 72$ MHz, indicated by the *white dashed line* slice of the experimental data shown in **b**. After a time $\tau = \tau_{ph}/2$, the qubit and mechanical resonator were maximally entangled. At interaction time τ_{ph}, the qubit was measured to be in its ground state, implying a single phonon excitation was created in the mechanical resonator. At time $2\tau_{ph}$ the excitation had been returned to the qubit, reduced in amplitude by dissipation (finite energy lifetime)

a de-tuning Δ for a variable time τ. Finally, we returned the qubit to its operating bias and measured its excited state probability P_e.

The expected oscillatory behavior, corresponding to the quantum state swapping between qubit and mechanical resonator, can be seen clearly in Fig. 12.11b. The oscillations occur at the correct Rabi swap frequency, and display the expected Lorentzian dependence on detuning. There is very rapid decay of the oscillations, due to the very limited energy lifetime of the resonator and the qubit. The asymmetry

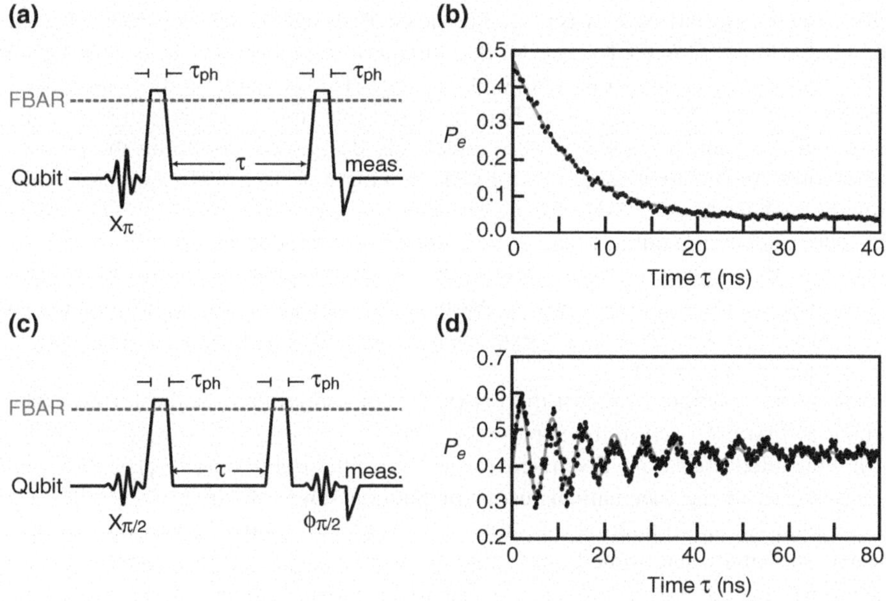

Fig. 12.12 Mechanical resonator energy relaxation and phase coherence times. **a** T_{1M} pulse sequence. The qubit was placed into $|e\rangle$ by a π-pulse, then tuned into resonance with the resonator for a duration sufficient to fully transfer the qubit excitation to the resonator, generating a single phonon. The qubit was then de-tuned from the resonator allowing the phonon to decay in the resonator for a time τ, after which the qubit was brought back into resonance with the resonator to transfer any remaining excitation to the qubit. **b** Measured qubit excited state probability $P_e(\tau)$. The fit line is a direct measure of the phonon decay time, yielding $T_{1M} = 6.1$ ns. **c** Resonator phase coherence pulse sequence. This is similar to **a**, but using a $\pi/2$-pulse to prepare the qubit in the superposed state $(|g\rangle + |e\rangle)/\sqrt{2}$. **d** The phase of the second $\pi/2$-pulse was swept in order to produce the oscillations seen in **d**, which is a measure of the resonator phase coherence. The fit line indicates $T_{2M} \approx 2T_{1M}$

in the magnitude of the probabilities around $\Delta = 0$ arise from the exact shape of the tuning pulse used to bring the qubit toward and away from the FBAR: The pulse is trapezoidal, due to the finite rise-time of the electronics that control the flux bias. Since the $\sim$1 ns rise and fall times of this trapezoidal shape are not fast compared to qubit-resonator swap time, setting the qubit frequency somewhat above the resonator frequency favors higher state transfer probabilities. This is borne out by the simulations presented in Fig. 12.11c. A line cut through the experimental data is shown in Fig. 12.11d, for a de-tuning $\Delta = 72$ MHz, corresponding to the white dashed line in panel (b). Guided by the solutions to the Jaynes-Cummings Hamiltonian presented above, the maxima of the oscillations were fit to a swap frequency $\Omega_C/2\pi = 132$ MHz, quite close to the value obtained from qubit spectroscopy.

We next used the ability to create a single phonon excitation in the resonator to measure its energy relaxation and phase coherence times. Fig. 12.12a shows the pulse sequence used to extract the resonator energy relaxation time T_{1M}. The qubit, initially in the ground state, was placed in the excited state by a π-pulse, then biased and held at $\Delta = 72$ MHz for τ_{ph}, so that the qubit excitation was fully transferred to the resonator, creating a one-phonon Fock state $|1\rangle$. The qubit, in its ground state, was then returned to its operating bias for a variable time τ. During this time the qubit and mechanical resonator were effectively decoupled, confining the phonon excitation to the mechanical resonator and allowing it to decay. The qubit was then brought back to $\Delta = 72$ MHz for τ_{ph}, transferring any excitation remaining in the resonator back to the qubit. Finally, the qubit was returned to its operating bias and measured. By varying the delay time τ, the measured qubit probability is a direct measure of the single-phonon energy decay in the resonator. The experimental data are shown in Fig. 12.12b, with a fit exponential corresponding to a mechanical energy decay time of $T_{1M} = 6.1$ ns.

The phase coherence time of the FBAR was measured in a similar fashion. The qubit was first prepared in the superposed state $(|g\rangle + |e\rangle)/\sqrt{2}$, using a $\pi/2$-pulse with controlled phase. As discussed earlier, an initial qubit state $\alpha|g\rangle + \beta|e\rangle$ can be transfered to the mechanical resonator phonon state $\alpha|0\rangle + \beta|1\rangle$ by allowing the coupled system to interact for a time $\tau = \pi/\Omega_C$; a length of time equivalent to the experimentally determined single phonon swap time τ_{ph}. Using this technique we transferred the initial superposed qubit state to the mechanical resonator, creating the superposed phonon state $(|0\rangle + |1\rangle)/\sqrt{2}$. After the state transfer was complete, the qubit, now in its ground state, was detuned from the resonator to decouple the system. The qubit was then held at its operating bias to allow the superposed phonon state to decay in the resonator. After a time τ, the qubit was brought back into resonance with the mechanical resonator and allowed to interact for τ_{ph} in order to transfer any remaining phonon superposition back to a superposition of qubit states. The qubit was then returned to the operating bias and measured. This experiment was repeated for varying hold times τ.

In order to obtain an accurate measure of the decay of the resulting superposed state, the measurement was conducted using a Ramsey fringe technique [34]. Just before the qubit excited state probability was read-out, a $\pi/2$-pulse was applied to the qubit. The phase of this pulse $\phi_{\pi/2}$ was then swept at an angular frequency ω_ϕ, producing the oscillation frequency seen in the data $\omega_\phi/2\pi$. The phase coherence time was determined from a fit to the resulting oscillation, $T_{2M} \simeq 20$ ns.

The fact that the dephasing time was measured to be more than twice the energy relaxation time (the theoretical limit) was most likely a result of the complexity of the pulse sequence, where small pulse shaping errors can lead to longer-than-expected phase coherence times. Note that during this measurement of the resonator phase coherence, the mechanical resonator was placed in a superposition of its ground and first excited states, a very non-classical quantum superposition.

We performed one final experiment to observe the effect of higher energy Fock states $|n\rangle$, with $n \geq 2$, in the mechanical resonator. These higher levels were populated by directly exciting the resonator with an on-resonance Gaussian microwave pulse coupled through the on-chip external capacitor C_x. During this microwave pulse, the qubit was kept off-resonance and thus not directly excited by the microwave signal. As previously discussed, the state of the resonator after this pulse will be a

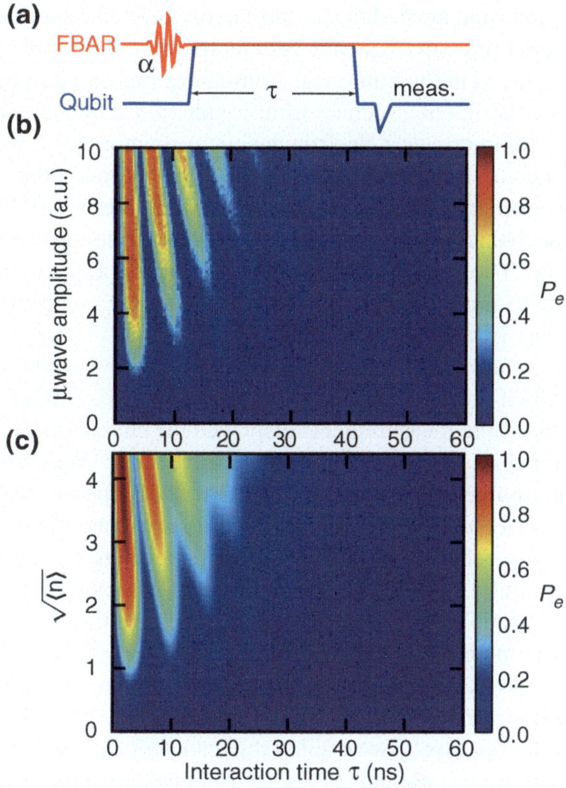

Fig. 12.13 Coherent state pulse sequence, data, and simulation. **a** A classical Gaussian microwave pulse was applied directly to the FBAR. After the microwave pulse was turned off, the qubit was tuned into resonance and allowed to interact with the excited FBAR for a time τ. The qubit, initially in its ground state $|g\rangle$, was measured and the excited state probability P_e recorded. The experiment was then repeated for varying initial microwave pulse amplitudes and interaction times. **b** Experimental data showing the expected increase in swap frequency when the qubit is allowed to interact with a more highly excited FBAR. **c** Quantum simulations show similar expected shortening of swap period for the interaction between qubit and mechanical resonator. For the simulations, the FBAR was initially placed in a coherent state with average phonon number $\langle n \rangle$

coherent state [35]. The qubit, initially in its ground state $|g\rangle$, was then brought into resonance with the resonator, and allowed to interact for a variable time τ at detuning $\Delta = 0$. The qubit was then returned to its operating bias and its excited state probability measured. The amplitude of the initial microwave pulse was varied while repeating this measurement. The pulse sequence and resulting data are shown in Fig. 12.13a, b. From the data presented there, the swap frequency can be seen to increase with higher microwave pulse amplitude. Larger microwave pulse amplitudes create initial coherent states with more highly populated Fock states $|n\rangle$. Since the qubit-resonator interaction strength scales as $\sqrt{n}g_0$, these more highly excited Fock states effectively cause the qubit and resonator to be more strongly coupled [35].

Correspondingly, the time needed to transfer energy between qubit and resonator is reduced. The increase in swap frequency with increasing excitation, and thus phonon number n, is reproduced in the numerical simulations presented in Fig. 12.13c.

More complex experiments, such as arbitrary phonon state generation, similar to those performed with microwave electromagnetic resonators [36], were not feasible with this system, due to the quite short energy relaxation time of the mechanical resonator. Future experiments, such as those involving Wigner tomography, to further verify that the mechanical quantum states were indeed non-classical, is an obvious future direction. The primary improvement needed to perform measurements in this area would be to increase the energy lifetime (i.e. the quality factor Q) of the resonator, preferably by a factor of ten or more.

In conclusion, we have created a novel system by coupling a qubit to a mechanical resonator, with which we were able to explore non-classical effects in a mechanical system. Spectroscopic measurements displayed the characteristic resonant mode splitting of a coupled system; an essentially classical result. We then used the qubit as a "quantum thermometer" to measure the thermal occupation of the mechanical resonator. Using this technique, the mechanical resonator was determined to be in the quantum ground state. To further reveal the quantum nature of the mechanical resonator, we used the qubit as a "quantum transducer" to prepare single phonon states, creating entangled qubit-phonon states in the process. We then measured the single phonon energy lifetime of the mechanical resonator $T_{1\mathrm{M}} = 6.1$ ns. A similar technique was employed to extract the resonator's phase coherence time $T_{2\mathrm{M}} \approx 2T_{1\mathrm{M}}$, which necessitated the creation of superposed phonon states. Finally, to explore the effects of higher phonon levels, we excited the resonator directly to create coherent phonon states. These measurements provide strong evidence that we have achieved elementary quantum control over a macroscopic mechanical system.

Acknowledgments We would like to thank J. M. Martinis for invaluable experimental support, M. Geller for numerous conversations and calculations, and A. Berube for assistance with resonator fabrication and measurements. The entire membership of the UC Santa Barbara phase qubit group assisted enormously in this experiment and in providing and maintaining the experimental infrastructure. This work was supported by the US National Science Foundation (NSF) under grant DMR-0605818 and by the Intelligence Advanced Research Projects Activity under grant W911NF-04-1-0204. Devices were made at the University of California, Santa Barbara, Nanofabrication Facility, which is part of the NSF-funded US National Nanotechnology Infrastructure Network.

References

1. J.F. Rosenbaum, Bulk Acoustic Wave Theory and Devices (Artech House, UK, 1988)
2. J.D.I. Larson, P.D. Bradley, S. Warternberg, R.C. Ruby (IEEE, 2000), pp. 863–868
3. R.C. Ruby, P. Bradley, Y. Oshmyansky, A. Chien, J.D.I. Larson (IEEE, 2001), pp. 813–821
4. A.D. O'Connell, M. Hofheinz, M. Ansmann, R.C. Bialczak, M. Lenander, E. Lucero, M. Neeley, D. Sank, H. Wang, M. Weides, J. Wenner, J.M. Martinis, A.N. Cleland, Nature **464**, 697 (2010)

5. M.A. Dubois, P. Muralt, Appl. Phys. Lett. **74**, 20 (1999)
6. K. Nam, Y. Park, B. Ha, D. Shim, I. Song, J. Pak, G. Park, J. Korean Phys. Soc. **47**, S309 (2005)
7. J.M. Martinis, Quant. Inf. Process. **8**, 81 (2009)
8. M. Tinkham, Introduction to Superconductivity (McGraw-Hill, US, 1996)
9. B.D. Josephson, Phys. Lett. **1**, 251 (1962)
10. A.O. Caldeira, A. Legget, Phys. Rev. Lett. **46**, 211 (1981)
11. M.H. Devoret, J.M. Martinis, J. Clarke, Phys. Rev. Lett. **55**, 1908 (1985)
12. A.J. Leggett, S. Chakravarty, A.T. Dorsey, M.P.A. Fisher, A. Garg, W. Zwerger, Rev. Mod. Phys. **59**, 1 (1987)
13. J. Clarke, A.N. Cleland, M.H. Devoret, D. Esteve, J.M. Martinis, Science **239**, 992 (1988)
14. J.E. Mooij, T.P. Orlando, L. Levitov, L. Tian, C.H. van der Wal, S. Lloyd, Science **285**, 1036 (1999)
15. C.H. van der Wal, A.C.J. ter Haar, F.K. Wilhelm, R.N. Schouten, C.J.P.M. Harmans, T.P. Orlando, S. Lloyd, J.E. Mooij, Science **290**, 773 (2000)
16. J.M. Martinis, S. Nam, J. Aumentado, C. Urbina, Phys. Rev. Lett. **89**, 117901 (2002)
17. I. Chiorescu, Y. Nakamura, C.J.P.M. Harmans, J.E. Mooij, Science **299**, 1869 (2003)
18. K.W. Lehnert, K. Bladh, L.F. Spietz, D. Gunnarsson, D.I. Schuster, P. Delsing, R.J. Schoelkopf, Phys. Rev. Lett. **90**, 027002 (2003)
19. Y.A. Pashkin, T. Yamamoto, O. Astafiev, Y. Nakamura, D.V. Averin, J.S. Tsai, Nature **421**, 823 (2003)
20. T. Yamamoto, Y.A. Pashkin, O. Astafiev, Y. Nakamura, J.S. Tsai, Nature **425**, 941 (2003)
21. I. Chiorescu, P. Bertet, K. Semba, Y. Nakamura, C.J.P.M. Harmans, J.E. Mooij, Nature **431**, 159 (2004)
22. K.B. Cooper, M. Steffen, R. McDermott, R.W. Simmonds, S. Oh, D.A. Hite, D.P. Pappas, J.M. Martinis, Phys. Rev. Lett. **93**, 180401 (2004)
23. R.W. Simmonds, K.M. Lang, D.A. Hite, S. Nam, D.P. Pappas, J.M. Martinis, Phys. Rev. Lett. **93**, 077003 (2004)
24. A. Wallraff, D.I. Schuster, A. Blais, L. Frunzio, R.S. Huang, J. Majer, S. Kumar, S.M. Girvin, R.J. Schoelkopf, Nature **431**, 162 (2004)
25. H. Goldstein, C. Poole, J. Safko, Classical Mechanics (Addison Wesley, Boston, 2002)
26. G. Wendin, V.S. Shumeiko (2005), arXiv:cond-mat/0508729v1
27. M.R. Geller, A.N. Cleland, Phys. Rev. A **71**, 032311 (2005)
28. E. Lucero, M. Hofheinz, M. Ansmann, R.C. Bialczak, N. Katz, M. Neeley, A.D. O'Connell, H. Wang, A.N. Cleland, J.M. Martinis, Phys. Rev. Lett. **100**(24), 247001 (2008)
29. E. Merzbacher, Quantum Mechanics (Wiley, New York, 1998)
30. J.J. Sakurai, Modern Quantum Mechanics (Addison Wesley, Boston, 1994)
31. M. Steffen, J.M. Martinis, I.L. Chuang, Phys. Rev. B. **68**, 224518 (2003)
32. A.N. Cleland, M.R. Geller, Phys. Rev. Lett. **93**, 070501 (2004)
33. A. Sornborger, A.N. Cleland, M.R. Geller, Phys. Rev. A **70**, 052315 (2004)
34. N.F. Ramsey, Phys. Rev. **78**, 695 (1950)
35. M. Hofheinz, E.M. Weig, M. Ansmann, R.C. Bialczak, E. Lucero, M. Neeley, A.D. O'Connell, H. Wang, J.M. Martinis, A.N. Cleland, Nature **454**, 310 (2008)
36. M. Hofheinz, H. Wang, M. Ansmann, R.C. Bialczak, E. Lucero, M. Neeley, A.D. O'Connell, D. Sank, J. Wenner, J.M. Martinis, A.N. Cleland, Nature **459**(7246), 546 (2009)

Chapter 13
Cavity Optomechanics with Cold Atoms

Dan M. Stamper-Kurn

Abstract The mechanical influence on objects due to their interaction with light has been a central topic in atomic physics for decades. Thus, not surprisingly, one finds that many concepts developed to describe cavity optomechanical systems with solid-state mechanical oscillators have also been developed in a parallel stream of scientific literature pertaining to cold atomic physics. In this chapter, I describe several of these ideas from atomic physics, including optical methods for detecting quantum states of single cold atoms and atomic ensembles, motional effects within single-atom cavity quantum electrodynamics, and collective optical effects such as superradiant Rayleigh scattering and cavity cooling of atomic ensembles. Against this background, I present several experimental realizations of cavity optomechanics in which an atomic ensemble serves as the mechanical element. These are divided between systems driven either by sending light onto the cavity input mirrors ("cavity pumped"), or by sending light onto the atomic ensemble ("side pumped"). The cavity-pumped systems clearly exhibit the key phenomena of cavity optomechanical systems, including cavity-aided position sensing, coherent back action effects such as the optical spring and cavity cooling, and optomechanical bistability; several of these effects have been detected not only for linear but also for quadratic optomechanical coupling. The extreme isolation of the atomic ensemble from mechanical disturbances, and its strong polarizability near the atomic resonance frequency, allow these optomechanical systems to be highly sensitive to quantum radiation pressure fluctuations. I describe several ways in which these fluctuations are observed experimentally. I conclude by considering the side-pumped cavity experiments in terms of cavity optomechanics, complementing recent treatments of these systems in terms of condensed-matter physics concepts such as quantum phase transitions and supersolidity.

D. M. Stamper-Kurn (✉)
Department of Physics, University of California, Berkeley, CA 94720, USA
e-mail: dmsk@berkeley.edu

Materials Sciences Division, Lawrence Berkeley National Laboratory,
Berkeley, CA 94720, USA

M. Aspelmeyer et al. (eds.), *Cavity Optomechanics*, Quantum Science and Technology,
DOI: 10.1007/978-3-642-55312-7_13,

13.1 Introduction

During the 1980s and 1990s, a major fraction of atomic physics research was focused on the mechanical effects of light-atom interactions (discussed nicely in the 1997 Nobel lectures [1–3]). This research area, the roots of which reach back much earlier, might now be called "atomic optomechanics," providing a second intellectual background, parallel to the studies related to gravity wave detection and quantum measurement limits, for present-day investigations of optomechanical interactions with massive solid-state objects.

Beyond this conceptual confluence, research on cavity optomechanics and on light-atom interactions has been united in recent years by experiments in which solid-state objects such as mirrors and membranes inside a cavity are replaced with gas-phase mechanical objects—non-degenerate ensembles of atoms, Bose-Einstein condensates, and even single trapped ions. These atomic cavity optomechanical systems circumvent the difficulties of preparing mechanical systems in the quantum regime by borrowing the methods of laser- and evaporative-cooling developed for the study of quantum gases. This capability has allowed experimentalists to explore quantum properties of both the "opto" [4] and "mechanical" [5] portions of cavity optomechanical systems, to achieve new regimes of optomechanical coupling, and to explore similarities between optomechanics and paradigmatic many-body Hamiltonians [6].

In this chapter, I attempt to summarize research on the optomechanics of atoms and atomic ensembles within optical cavities. The discussion begins with a review of basic single-atom optomechanical effects that are relevant to the ensuing discussion. I then summarize the single-atom-based view of cavity optomechanics that supplemented the study of single-atom cavity quantum electrodynamics (cQED) by considering mechanical effects such as diffusion, cooling and trapping. In that work, the motion of a single atom clearly represents a single mode of motion—or perhaps three modes if not only the axial motion is considered—and the connection to solid-state cavity optomechanics is readily apparent. In Sect. 13.4, the discussion turns from single atoms to atomic ensembles. I describe several investigations of optomechanical effects in continuous atomic media, highlighting situations were a small number of mechanical modes of the ensemble are at play; these include collective Rayleigh and Raman scattering from cold atomic gases, and the realizations of these phenomena within optical cavities. In Sects. 13.5 and 13.6, these precedents are combined to describe the realization of quantum cavity optomechanics with both spatially extended and confined cold-atom ensembles, summarizing recent research results and highlighting the new regimes accessible in these systems.

13.2 Basics of Light-Atom Interactions

The mechanical interactions between single atoms and an optical field may be differentiated according to whether the light-atom interaction is dissipative or dispersive. Dissipative interactions cause light to be absorbed, meaning that light power is transmitted with sub-unity efficiency due to the fact that photons are scattered out of the incident light beam. The scattered photons are emitted at random directions, according to an angular probability distribution function determined by the dipole emission pattern, and, disregarding antibunching over the very short lifetime of the atomic excited state, at random times. Dispersive interactions cause light to be phase shifted. As this phase shift is spatially varying, according to the position of the atom(s), the dispersive interaction also causes optical power to be redistributed, e.g. among the wavevectors of the incident light, in a deterministic manner. The dissipative and dispersive interactions can be treated on equal footing by considering the imaginary and real parts of the complex optical susceptibility (or, equivalently, the index of refraction), respectively. The two types of interactions can also be described as either spontaneous or stimulated emission.

The discussion above pertains both to Rayleigh (or Bragg) scattering, wherein an atom scatters photons and returns to the same internal state, and to Raman scattering, wherein light scattering shuffles an atom between internal states. In the former case, it is helpful to reduce the complexity of the atom to just two internal levels, the ground and excited internal states. This two-level atom approximation is valid when Raman scattering pathways are suppressed either by selection rules and the use of suitably polarized light, or when the light is sufficiently detuned from atomic resonances.

These two aspects of light-atom interactions lead to two types of optical forces on the atom. The mean force due to dissipative interactions is known as the scattering force or as radiation pressure, whereas that due to dispersive interactions is known as the optical dipole or gradient force. A unified derivation of both types of forces, making use of the optical Bloch equations to trace the internal-state evolution of an atom in a spatially inhomogeneous optical electric field $\mathbf{E}$, shows the scattering force to be proportional to the spontaneous scattering rate and the optical dipole force to be proportional to the Stark shift $\langle \mathbf{d} \cdot \mathbf{E} \rangle$, where $\mathbf{d}$ is the electric dipole operator [7]. In both cases, the forces arise from the redistribution of the momentum of the electromagnetic field due to the atom. The optical dipole force can also be envisioned via the dressed-state picture, in which an atom driven by a monochromatic light field is treated in a time-varying frame that rotates with the optical frequency. In the dressed-atom approach, the Hamiltonian is time-independent and the Hilbert space is spanned by products of atomic and optical quantum states. Approximating the atomic motion to be slow compared to the internal-state dynamics, the local eigenenergies of these dressed states define effective optical dipole potentials—the ac Stark shifts—from which optical dipole forces are derived [8]. Importantly, both treatments can be applied to describe velocity-dependent forces, which appear for both the scattering and the optical dipole force, and the effects of atomic saturation at high light intensity.

Owing to their small mass, atoms are strongly influenced by radiative forces, in contrast with solid-state mechanical systems for which radiative forces may be only a small perturbation atop other influences. The scattering force can be as large as $\hbar k \gamma$, where k is the photon wavevector and γ is the excited-state half-linewidth. For ^{87}Rb and light at the 780-nm-wavelength atomic resonance, this maximum force yields an acceleration of $10^5\ \mathrm{m/s^2}$. The optical dipole force is not limited by saturation,

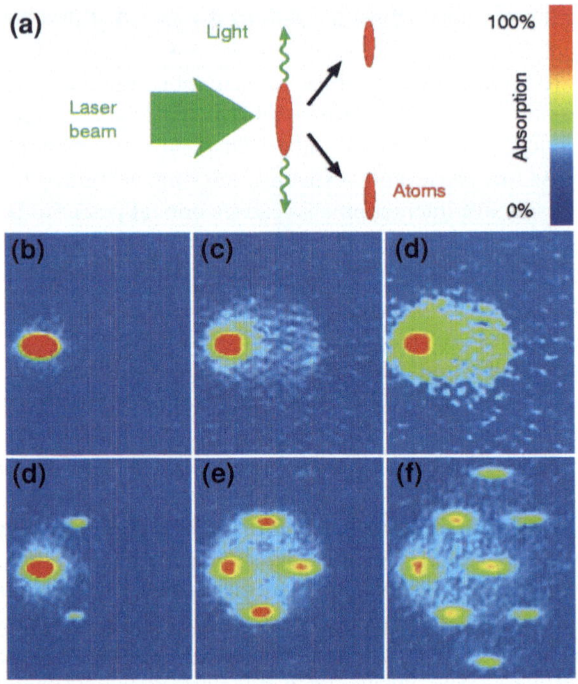

Fig. 13.1 Mechanical effects on atoms of single photon scattering. **a** A Bose-Einstein condensate is illumined with off-resonant light. **b–d** Choosing a polarization that inhibits light scattering down the long axis of the condensate suppresses collective scattering. After time of flight, a halo of scattered atoms is observed, indicating the momentum transferred to atoms by single photon scattering distributed according to a dipole emission pattern. **e–g** Superradiant Rayleigh scattering: allowing emission along the condensate axis allows for Brillouin instability. The coherent momentum populations emitted by light scattering indicate the wavevectors of the instability. At high probe fluence (increasing from *left* to *right* images), high-order Brillouin instability generates coherent populations at multiples of the recoil momentum. Figure reproduced from Ref. [9]

and can thus be even larger [8]. As illustrated in Fig. 13.1, optomechanical effects from the scattering of just a single photon are easily discernible given that the initial momentum of a neutral atom can be brought below the single-photon recoil, $\hbar k$, by cooling a gas to well below the recoil temperature $T = \hbar^2 k^2 / 2mk_B$ (where m is the atomic mass).

The optical forces on an atom fluctuate, leading to the diffusion of its momentum. The diffusion can be associated with two distinct processes. One process is the fluctuation of the atomic dipole, arising due to the quantum mechanical light-atom dynamics, which leads to fluctuations of the optical dipole force in the presence of an electric field gradient. A second process is the interaction of the atomic dipole with quantum fluctuations of the electromagnetic field. In the treatment of Gordon and Ashkin [7], for example, the atom interacts with quantum fluctuations of the electromagnetic field in free space, which are local both in time and space, i.e., the shot-noise spectrum of the light field is white and the atom interacts with fluctuations of the optical field at all incident angles.[1] This second process is modified within optical cavities as the quantum fluctuations of the optical field become modified by the cavity spectrum. The atom responds to the vacuum noise of the electromagnetic field by spontaneously emitting photons, and recoiling from each emission by a momentum $\hbar k$ directed counter to the photon emission direction. In the language of optomechanics, such recoil heating may be called the quantum fluctuations of the radiation pressure force, or radiation pressure shot noise. While such noise is challenging to observe for solid-state objects exposed to light, it is observed routinely in laser cooling experiments.

13.3 Optomechanics of Single Atoms in Cavities

The interaction between a single atom and light within an electromagnetic cavity, described by the theory of cQED, has been studied by atomic and optical physicists for decades. The cavity amplifies the influence of a single optical mode on the internal (i.e. electronic) dynamics of the atom so that coherent processes, such as the stimulated re-emission of photons into the cavity mode, can dominate the dissipative processes which typify light-atom interactions in free space. The deterministic exchange of energy between the electronic excitations of the atom and photons of the cavity field offers the prospect for quantum devices such as quantum memory registers, entangling quantum gates, and, for cascaded atom-cavity systems, quantum networks.

Early experiments on cQED in the optical domain were performed with high-velocity atomic beams transiting the optical resonator, so that the number and position of atoms within the cavity field was uncertain. While this condition was good enough to illustrate basic cQED phenomena, more ambitious experiments required that the nuisance of atomic motion within the cavity be controlled.

13.3.1 Sensing the Position of a Single Atom

The goal of position sensing for single atoms has been pursued both for atoms in free space and within optical resonators. Several free-space methods reminiscent of magnetic resonance imaging have been developed in which inhomogeneous magnetic or optical fields lead to strong spatial variation of optical absorption lines [10, 11] or (narrower) Raman transitions [12, 13]. In the former case, the atomic

[1] For a structured collection of atoms, as for solid-state mirrors and membranes used in cavity optomechanics experiments, light scattering becomes, of course, highly anisotropic. Force fluctuations due to the uncertain direction of light emission are reduced, but fluctuations due to the uncertain time of photon scattering remains.

position distribution is inferred by the shape of the optical absorption spectrum. In the latter, the atomic position becomes encoded in the atomic internal state which, in turn, can be detected efficiently. This Raman resonance imaging method was used to measure atomic distributions in beams with sub-optical-wavelength (200 nm) resolution [14, 15].

These free-space schemes provided some of the motivation for cavity-based schemes of single-atom imaging. Adopting the two-level atom approximation introduced above, the interaction of a single atom with a single cavity mode is quantified by the vacuum Rabi frequency, $2g(\mathbf{r}) \propto \mathbf{d} \cdot \mathbf{E}(\mathbf{r})$, which is the rate at which photons are cyclically emitted and reabsorbed by an excited atom placed in an empty, resonant cavity field. The interaction strength varies spatially according to the cavity mode's electric field $\mathbf{E}(\mathbf{r})$. In the common example of the TEM_{00} mode of a Fabry-Pérot cavity, at the cavity center, the interaction strength varies as[2]

$$g = g_0 \, e^{-\rho^2/w_0^2} \sin kz \tag{13.1}$$

where z denotes the position along the cavity axis and ρ the radial distance from that axis, k is the optical wavevector, and w_0 is the beam waist radius. Owing to this spatial dependence, the position of an atom within the resonator can be inferred by the resonator's optical properties; as in cavity optomechanical sensors employing solid-state objects, the cavity is regarded as an optical sensor of position.

Early works established that measurements on the optical output of a Fabry-Pérot cavity lead to a projective measurement of the position of a single atom within the standing-wave intracavity light field. The authors assumed the atom passes quickly through the light field (e.g. originating from a transversely oriented atomic beam), so that mechanical effects of the measurement on the subsequent evolution of the atom, i.e. the response to measurement back action, could be neglected. These works clarified how continuous homodyne measurement of the cavity-emitted light, treated as a quantum-optics measurement process, leads to an ever refined sensing of the atomic position [16–20]. A feasibility study of this approach, which accounted for realistic experimental parameters, indicated that high-temporal-bandwidth and high-spatial-resolution (below an optical wavelength) was achievable [21].

Subsequent work began taking into account the response of the atom to the continuous cavity-based measurement of its position. A quantum trajectory simulation showed that, even as the atom was disturbed by the measurement process, the cavity emission, specifically the phase quadrature of the near-cavity-resonant probe field, provided a continual record of its motion [22]. Subsequent works clarified that the momentum diffusion experienced by the atom under constant measurement is associated with vacuum fluctuations in the intracavity optical dipole potential experienced by the atom [23]. This diffusion serves as the back action of a quantum position measurement.

[2] For simplicity, we assume the cavity is near-planar and thus neglect the divergence of the cavity optical field beyond the Rayleigh range.

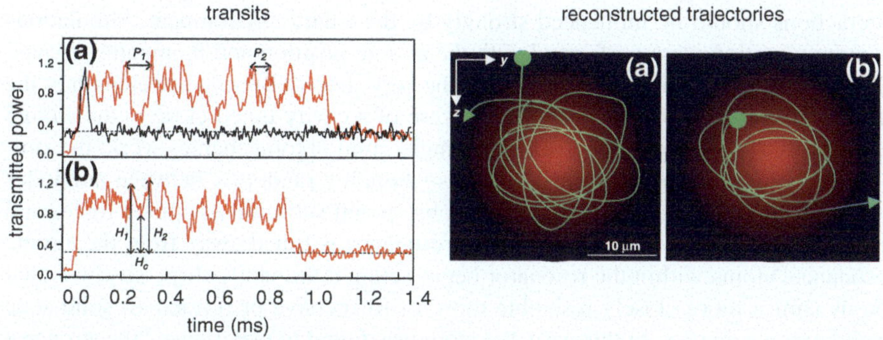

Fig. 13.2 Single atom transits and trajectory reconstructions. *Left* the transmitted power of a cavity probe red-detuned from the empty-cavity resonance is recorded on a heterodyne receiver, revealing two "transits" of single atoms passing through the cavity. The *black trace* shows reference data for an empty cavity. *Right* the motion of each atom in the radial plane is reconstructed. The atom enters the cavity mode, is captured and also perturbed by the optical potential of the cavity probe field, and is finally ejected from the cavity. Figures are reproduced from Ref. [26]

13.3.2 Single Atom Transits and the Atom-Cavity Microscope

Starting in 1996, these theoretical ideas became experimentally relevant with the observation of single-atom "transits." Cold atom sources replaced the atomic beams used previously, so that atoms traversing the cavity would spend enough time—just a few microseconds at first—within the cavity mode so that a cavity probe could detect their presence in real time (Fig. 13.2) [24]. Experiments were performed either with probe light resonant with the empty cavity, in which case a passing atom could shift the cavity resonances so that the cavity-transmitted light was variably extinguished, or off-resonant with the empty cavity, in which case the atom would shift the cavity temporarily into resonance and cause the transmitted signal to increase. The timing of these transits was probabilistic: on each repetition of the experiment, a small number of atoms would be launched toward the cavity, resulting in one or perhaps a few recorded transits. For medium-input-velocity atoms, the duration of these transits was found to correlate with the incident velocity [25]. For slower atoms, these transits showed characteristic ripples and wiggles that were taken to represent a real-time record of both internal-state transitions and also center-of-mass dynamics of an atom interacting with the cavity probe field. These experiments represent first light for atomic cavity optomechanics.

Numerical calculations soon showed that a single atom interacting *mechanically* with a strongly coupled single-mode cavity field is a complex dynamical system. Here, the single-atom strong coupling parameter is $C = g_0^2/2\gamma\kappa$, also known as the single-atom cooperativity, with κ being the half linewidth of the cavity resonance. The cooperativity quantifies the ratio of the emission rate of an atom into the cavity versus that into free space. The strong-coupling condition $C \gg 1$ implies that not only the internal dynamics but also the optomechanical effects of light-atom interactions should be influenced strongly by the cavity environment. Simulations of atomic motion were performed with the atomic position and momentum treated classically, an approximation justified by the very short deBroglie wavelength of the atoms in these first experiments. For the case of a cavity tuned close to the atomic resonance, which describes most of the single-atom optomechanics work, the simulated atom performed complex dynamics—hopping randomly between antinodes of the cavity field—resulting from the strong spatial variation of optical forces and force fluctuations [27]. When the cavity resonance is tuned away from the atomic resonance, atoms within the resonator begin acting dominantly dispersively so that the dynamics more closely resemble those of (dispersive or refractive) solid-state cavity optomechanics. In this case, the atom was found to experience "the strongest diffusion [at locations] where the output field contains the most information about the atomic position dynamics" [28]. In the language of cavity optomechanics, the measurement back action upon the atomic momentum is greatest where the measurement sensitivity to the atomic position is strongest.

Backed by these dynamical simulations, the temporal structure seen in the single-atom transit allowed researchers to surmise the real-space trajectory of the atom within the cavity mode (Fig. 13.2) [26, 29]. Interestingly, in contrast to typical cavity optomechanics experiments in which one detects the motion of a cantilever or membrane along the cavity axis, here, the single-atom transit was used to reconstruct the slower motion transverse to the cavity axis, a direction along which the cavity provides only quadratic ($\propto \rho^2$) sensitivity and no sensitivity to angle.

13.3.3 Cavity Cooling of Single Atoms and Ions

An important conceptual understanding of how mechanical light-atom interactions are affected by cQED was developed by Ritsch and colleagues, marked by the identification of cavity Doppler cooling and cavity-enhanced diffusive heating [30, 31]. Their work differs from previous derivations in the context of solid-state cavity optomechanics [32] in several respects. First, their treatment begins with the electric dipole coupling of the atom to the cavity field, explicitly including the atomic excited internal state so as to account for effects of saturation and spontaneous emission. As such, their work is applicable both to the case of single-atom experiments where the cavity probe field is nearly resonant with the atomic transition and is strongly coupled to the atom, and also to the case of many-atom experiments where the field is far off-resonant. Their treatment also closely parallels the Gordon and Ashkin treatment of free-space optical forces, cooling, and diffusion [7], allowing one to identify the new cavity optomechanical effects as resulting from the electrodynamics of the cavity. While the derivation specifically considers the motion of a two-level atom, the authors also recognize the generality of cavity Doppler cooling, stating that "As the only requirement on the particle is a strong coupling to the cavity mode, the results should also apply to small molecules or other more complex objects (as, e.g. a Bose condensate) with a sufficiently large dipole moment [30]."

Another valuable element of their work is the analogy between cavity cooling and Sisyphus laser cooling of atoms outside cavities [1, 2]. In free-space Sisyphus cooling, an atom moves in an internal-state and spatially dependent ac Stark shift produced at the intersection of several near-resonant light fields. Under proper conditions for cooling, the atom is preferentially optically pumped into the internal state for which the ac Stark potential has a local minimum. Because optical pumping is not instantaneous, but rather occurs only after a delay τ related inversely to the spontaneous scattering rate, the atom tends always to "climb up" the ac Stark potential energy landscape, always doing work against the optical field.

Similarly, in describing cavity cooling, Ritsch and colleagues consider the lowest lying atom-cavity excited dressed states, with spatially varying energies

$$E_{1,\pm}(\mathbf{r}) = \left(\frac{\omega_a + \omega_c}{2}\right) \pm \sqrt{\frac{\Delta_{\text{ca}}^2}{4} + g^2(\mathbf{r})} \tag{13.2}$$

Here ω_a (ω_c) is the atomic (cavity) resonance frequency, and $\Delta_{\text{ca}} = \omega_c - \omega_a$. If the cavity is driven with light tuned at or below the minimum value of $E_{1,-}$, an atom moving, for example, along the cavity axis is preferentially excited at a minimum of the optical-potential energy surface (left side of Fig. 13.3). The atom then climbs up the potential energy surface before relaxing back to the atom-cavity ground state, either by spontaneous emission or cavity decay, emitting a photon at higher energy than the probe and thus giving up some of its kinetic energy. Here, the delay time τ relates inversely to the cavity decay rate. The larger is the delay time, the longer the atom can work against the optical field and the greater is the cooling force; hence, narrower optical cavities cool the atom more strongly, allowing it to reach lower temperatures.

Vuletić and Chu provided an alternative, frequency-space description of cavity cooling [34]. Light scattering involves photon absorption followed by photon emission. Conventional Doppler cooling makes use of the frequency selectivity of *photon absorption*: Light tuned to the red of an atomic transition is absorbed preferentially by an atom moving counter to the photon wavevector, because the Doppler shift brings such photons closer to the atomic resonance. The cooling force, depending on the preference for absorbing light that is Doppler shifted toward rather than away from resonance, scales inversely with the linewidth of the transition. Given also that many scattered photons are needed to slow down an object appreciably, Doppler cooling is effective primarily for simple atoms.

In contrast, cavity Doppler cooling utilizes the frequency selectivity of *photon emission*: An atom inside an optical cavity, illuminated with light to the red of the cavity resonance, will preferentially emit blue-shifted photons, because the electromagnetic density of states is higher nearer to the cavity resonance (right side of Fig. 13.3). The vacuum-induced stimulation of high-energy photon emission was considered earlier by Mossberg et al. [35], although not within the specific context of optical cavities. Treatments following those of Ref. [34] have treated higher-order effects and clarified the potential velocity capture range of cavity cooling [36].

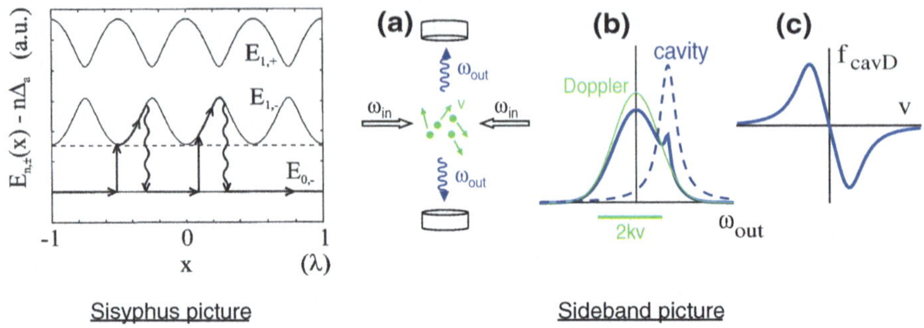

Fig. 13.3 Two conceptual pictures of cavity cooling provided by atomic physics. *Left* in the Sisyphus picture, an atom is excited by a red-detuned cavity probe to the first-excited atom-cavity state, at a position within the cavity where that excited state energy is a local minimum in space. Moving along the cavity axis, the atom remains in that excited state for a time $\sim\kappa^{-1}$, during which it does work against the optical dipole force, before decaying back down to the spatially uniform ground state. *Right* in a frequency-sideband picture, after an atom within a cavity absorbs a photon from a driving field that is red-detuned from the cavity resonance, it is stimulated by the cavity electromagnetic mode structure to re-emit nearer the cavity resonance frequency, cooling the atom. **b** Shows the overall spectrum of light scattered by an ensemble of rms velocity v: a sum of a Doppler profile of transversely scattered light, and a narrow spectrum of cavity-enhanced scattering into the cavity mode(s). **c** The net energy exchange provides a damping force $f_{\mathrm{cavD}}(v)$ on the atoms. Figures reproduced from Refs. [30] (*left*) and [33] (*right*)

Vuletić and Chu noted that since cavity-induced laser cooling does not rely on narrow atomic resonances, it can be applied to complex objects: molecules which, driven far from their resonances, need not undergo Raman transitions among ro-vibrational states; color centers, via which one can laser-cool solids; and even cantilevers and membranes.

The cavity cooling force is enhanced by making the cavity-stimulated emission stronger with respect to photon emission into free space. As such, the effectiveness of cavity Doppler cooling can be quantified by the ratio of the photon emission rate into the cavity versus that into free space. For a single-mode cavity, this ratio is simply the cooperativity C; in a multi-mode cavity, this ratio can be increased further [37].

In the demonstrations of cavity cooling of solid-state objects [38–41], the moving element is typically held by material supports, compared to which the optical forces affect the element weakly. Its motion within a narrow temporal frequency band is constrained along a single coordinate; correspondingly, the cavity emission spectrum within a narrow band serves as a (somewhat) direct record of its motion. The magnitude and width of this spectrum then reflect the damping of the object's motion under steady state conditions.

The motion of single atoms within cavities can be much more complex. Optical forces from the cavity-cooling light are often the dominant source of confinement, cooling, and diffusive heating. The atom transits the cavity, or resides briefly within it, and may not achieve steady state conditions. To avoid atomic saturation in strongly coupled cavities, the photon flux through the cavity is kept very low, corresponding to an average cavity photon number on the order of unity. The data collected from such experiments is thus noisy and transitory.

For these reasons, the early demonstrations of cavity cooling of single atoms relied on subtle data analysis and comparisons to complex numerical models. For instance, Münstermann et al. [42] examined the cavity spectrum produced with atoms transiting the cavity, and interpreted the spectrum as evidence for cavity-accentuated momentum diffusion and cavity cooling. In later experiments, cavity probes were applied to atoms trapped briefly within the cavity, and a slight increase in trapping lifetimes was taken as indicative of cavity cooling [43].

The evidence for cavity Doppler cooling was strengthened once schemes were developed to trap single atoms for as long as several seconds within intracavity optical traps [44]. Here, the atom is confined in far-detuned optical traps, providing an essentially conservative trapping potential, while different light fields closer to the atomic resonance produce the dissipative optical forces. This work also demonstrated a method to attain strong three-dimensional cooling. By illuminating atoms with light directed transverse to the cavity axis, the strong cavity Doppler cooling force acts in the two-dimensional plane defined by the pump-light and cavity-light propagation axes [34]. In the experiment of Nußman et al. [44], strong cooling in the third spatial dimension is obtained from the strong variation of the atomic resonance frequency in a far-detuned standing-wave optical potential.

Recently, sideband-resolved (mechanical frequency larger than the cavity linewidth) cavity cooling of a single trapped ion has also been demonstrated [45]. Such cavity-based cooling was proposed already in 1993 by Cirac et al. [46, 47], although their interpretation of the cooling mechanism led them to conclude erroneously that such cooling occurs exclusively in the Lamb-Dicke limit, in which the rms size of the atomic center-of-mass distribution is much smaller than the optical wavelength. Cavity cooling of tightly bound objects was also considered in Ref. [37], in which the conditions for ground-state cooling were spelled out; such analysis was later repeated in the context of cavity optomechanics with solid-state oscillators [48, 49]. In the experiment, the ion was only weakly coupled to the optical cavity ($C \ll 1$). Thus, even in the resolved sideband regime, the excess diffusive heating due to spontaneous emission outside the cavity prevented the ion from reaching the mechanical ground state.

13.3.4 Cavity-Enhanced Diffusion as Measurement Back Action

The enhanced position sensitivity provided by a high-finesse cavity must also yield enhanced momentum diffusion, as a form of measurement back action [23, 28, 50]. A quantum optical treatment of force fluctuations, for the case of a two-level atom, shows that the momentum diffusion constant D in the atom-cavity system can be cleanly separated into three components [51]:

$$D = (\hbar k)^2 \gamma P_e + |\hbar \nabla \langle \sigma^+ \rangle|^2 \gamma + |\hbar \nabla \langle a \rangle|^2 \kappa \tag{13.3}$$

The first term quantifies diffusion due to atomic spontaneous emission, similar to that in free space, where P_e is the excited state population. The second two terms quantify diffusion due to fluctuations of the optical dipole force. The first of these describes the interaction of a fluctuating atomic dipole (with the operator σ^+ proportional to the dipole moment operator) with the mean field within the cavity; the second describes the interaction of a fluctuating cavity field with the mean atomic dipole. The strength and spatial dependence of the atom-cavity coupling is implicit in the spatially varying expectation values, $\langle\sigma^+\rangle$ and $\langle a\rangle$. In free space, i.e. without the oscillating atomic dipole acting back on the driving optical field, the last term in Eq. 13.3 is absent, and one recovers the earlier results of Gordon and Ashkin [7]. In contrast, the diffusion rate inside a cavity is indeed enhanced, reaching a value that is up to $\sim C$ times larger than the free-space rate.

This enhanced diffusion is observed only indirectly in single-atom experiments, gleaned from measurements on normal-mode splitting for atoms transiting [42] or trapped within the cavity [52] and comparison with numerical Monte Carlo simulations [53]. More direct quantifications have been obtained from the many-atom cavity optomechanics system, as described in Sect. 13.5.

13.3.5 Feedback Cooling of a Single Atom

Finally, we discuss feedback control and cooling via cavity optomechanics. In solid-state optomechanics, such active feedback was used to cool mechanical objects by radiation pressure [54] before "passive" cavity cooling via the natural dynamics of the cavity field was demonstrated. Feedback control of the motion of atomic ensembles trapped in optical lattice potentials, but outside of a cavity, was implemented by Morrow et al. [55]. There, the intensity of the optical lattice beams provided a measurement of the force exerted by the lattice light onto the ensemble, since such forces can be understood as deriving from the coherent exchange of photons between the intersecting lattice beams [56]. This signal was fed back to the phase of one of the lattice beams, shifting the lattice potential spatially so as to amplify or damp the atomic motion.

The enhanced, and, ultimately, quantum limited position sensitivity attainable using a high-finesse optical cavity makes such feedback control schemes more powerful, in the sense that they can be effected with higher gain and may even cool a mechanical object into the lowest few quantum states regime of motion. The cavity enhances light scattering into a single electromagnetic mode, from which information can be extracted efficiently, thus minimizing the information lost to spontaneous emission into very many, usually unmeasured, emission modes. Several works have explored feedback control of the motion of a single atom within a cavity, in pursuit of the fundamental scientific goal of understanding and optimizing feedback control of an open dynamical quantum system.

Steck et al. [57, 58] considered controlling the one-dimensional motion along the axis of a Fabry-Pérot cavity. The atoms are assumed trapped within a red-detuned standing-wave potential formed by light that also serves as the measurement probe. In this case, the sensitivity to the atomic motion is quadratic in the atomic displacement, similar to that realized experimentally by positioning thin membranes [59, 60] or atomic gases [61] at the antinodes of the cavity field. The authors consider a "bang-bang" feedback scheme, a digital scheme in which the potential strength is toggled between high- and low-curvature settings, with the aim of effecting a Sisyphus-like cooling in which the atom moves away from the trap center against a strong confining force, and then returns to the center against a weaker force. When the "bang" amplitude is sufficiently high, such a scheme is predicted to bring the particle to the lowest parity-even or parity-odd states of the well (the feedback scheme conserves mirror symmetry about the trap center).

A different feedback scheme was considered by Vuletić et al. [62] for slowing down higher-velocity particles traveling along the standing-wave intensity pattern within a driven Fabry-Pérot resonator. This motion generates a cavity-field modulation at a frequency proportional to the particle velocity. Appropriate frequency-space conditioning of the feedback signal ensures that the cavity field is modulated so as to continually slow down the particle. Cooling rates and final-temperature limits were assessed taking into account the momentum diffusion of the particle and the limited signal-to-noise ratio of the measurement signal.

The "bang-bang" feedback scheme was successfully implemented by the Rempe group (Fig. 13.4) [63]. In this implementation, probe photons transmitted by the cavity were counted within consecutive short time bins, so as to estimate whether the atom was moving toward or away from the center of the cavity. The detection bandwidth was such that only the slower radial oscillatory motion was measured. The photon-number comparison was used to vary the intensity of light fields that resonated with higher-order transverse modes of the cavity, varying the radial optical confinement. The feedback provided active cooling, extracting energy from the atom and maintaining trap lifetimes on the order of 1 s [64].

13.4 Optomechanics of Continuous Atomic Media

We have seen how many of the phenomena of cavity optomechanics— cavity cooling, position sensing, measurement back action, and feedback control—apply to microscopic, single-atom mechanical oscillators just as they do to macroscopic solid-state objects. Whether the framework of cavity optomechanics could apply to atomic ensembles, bridging the gap between the microscopic and macroscopic realms, was initially unclear. For example, treatments of the motion of just two atoms interacting with a single cavity field suggested that their dynamics would be exceedingly complex, subject to cavity-mediated long-range interactions, "cross-friction" whereby the motion of one atom is damped due to the motion of the other, and diffusive heating that varies with the positions of each of the particles [65, 66]. An animation in Ref. [65] displays the simulated dynamics of two atoms dancing wildly within a cavity.

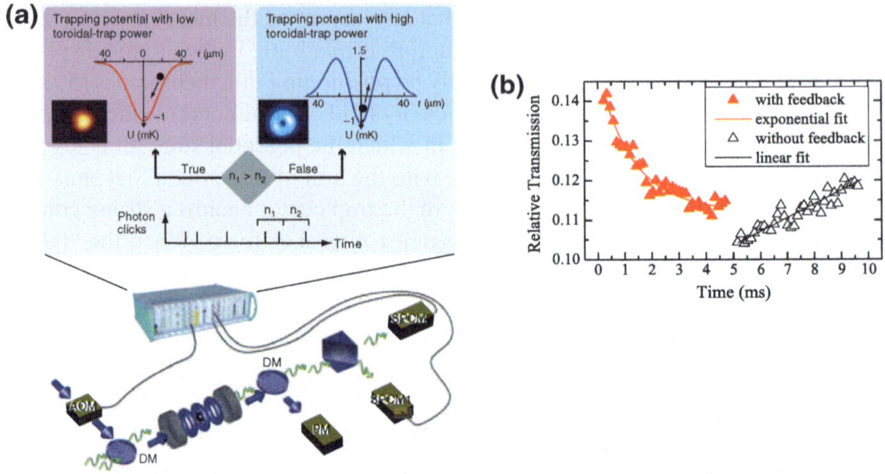

Fig. 13.4 Feedback cooling of a single atom within an optical cavity. **a** The rate of increase of ρ^2 is estimated by counting the number of cavity probe photons (using photon counters SPCM1 and SPCM2) in two consecutive time bins. Based on this difference, the power in a blue-detuned light field is varied between two settings, producing an optical potential on the atom with either low (*top left*) or high (*top right*) trap curvature. **b** With the feedback engaged, the average $\langle\rho^2\rangle$, proportional to the transmitted probe intensity, is decreased, demonstrating that energy is extracted from the atom by feedback. With the feedback disengaged, the atom heats up. **a** Reproduced from Ref. [63], and **b** from Ref. [64]

However, several works found evidence that cavity cooling could indeed apply not only to single atoms, but also to atomic ensembles. For example, the analogy of stochastic cooling of charged-particle beams suggested that dynamic back action of the cavity field could cool thermal fluctuations of an ensemble [34]. Of particular interest was the possibility that the cooling force on an atomic ensemble could be *collectively enhanced*, so that cavity-aided laser cooling could be not only more general, but also more powerful than free-space Doppler cooling [67, 68].

13.4.1 Collective Cavity Cooling and Self-organization via Brillouin Instability

This question was resolved experimentally by the MIT group [69]. An atomic gas, pre-cooled to moderate temperatures by conventional laser cooling, was launched at variable velocity within a (multimode) Fabry-Pérot cavity, and was then exposed to light aligned transverse to the cavity axis and detuned from the cavity resonance (Fig. 13.5). A time-of-flight analysis showed that a large portion of the gas was quickly decelerated, much faster than one would expect just from the theory of single-atom cavity cooling. The rapid deceleration could be ascribed to collective

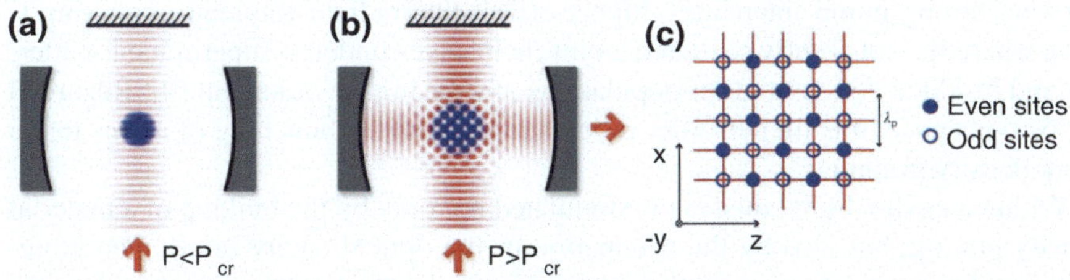

Fig. 13.5 **a** A near uniform gas placed within a Fabry-Pérot resonator is pumped with a standing-wave of light aligned orthogonal to the cavity axis, and with a frequency near the cavity resonance. **b** Above a threshold pump power, the gas organizes itself spatially so as to scatter pump light efficiently into the resonator. **c** Maximum collective scattering is achieved in two different spatial patterns of the gas, selecting either the "even" or "odd" sites of a checkerboard pattern. Figure adapted from Ref. [6]

light scattering by the ensemble into the cavity mode, as confirmed by the fact that although the single-atom cooperativity was low for this setup, the measured ratio of light scattering into the cavity versus outside the cavity was well above unity.

Such dynamics are now understood to result from the self-organization of the light-driven atoms into a spatially periodic configuration that enhances the collective scattering by the gas into the cavity field. That is, the continuous, deformable optomechanical medium organizes itself, in response to cavity-induced forces and damping, into compliance with the cavity field. This effect was predicted by Domokos and Ritsch [70], whose theoretical treatment demonstrated the collective nature of the cavity-induced cooling into a checkerboard spatial pattern.

The process by which a disordered medium spontaneously arranges itself into a collective coherent scattering state is known either as *superradiance* or *lasing*. The processes are loosely distinguished by whether the buildup of the optical field plays a major (lasing) or minor role (superradiance).

The gain mechanism for such self-organization is a form of Brillouin instability, exemplified first in observations of superradiant light scattering from Bose-Einstein condensates [9], and later also from nondegenerate Bose [71] and Fermi [72] gases. In these experiments, the cold gas is trapped in an elongated cigar-shaped geometry. Exposed to a plane wave of off-resonant pump light, the gas scatters photons, with each scattering event leaving behind a collective momentum excitation which conserves momentum in the scattering process. This excited residue establishes density modulations in the gas that persist for a time—very long in Bose-Einstein condensates and much shorter in nondegenerate or Fermi gases—which depends on the coherence properties of the gas. While they persist, these modulations preferentially scatter additional pump photons into the same output directions. This process represents a gain mechanism for density modulations. The rate constant for this density-grating amplification is proportional to the single-atom Rayleigh scattering rate and the optical depth of the gas along the optical emission direction; hence, for a prolate gas, the highest gain is seen for photons scattered along the long axis of the gas (Fig. 13.1).

For very strong pump intensities, higher-order superradiant Rayleigh scattering is seen, where the coherently scattered atoms themselves undergo superradiant scattering and produce coherent atom populations at ever higher momenta [9]. Regarded in position space, this higher-order process represents the bunching of atoms into a sharp density grating.

Within a cavity, light emission is stimulated not only by the buildup of a material density grating, but also by the occupation of the optical cavity mode. The aforementioned gain mechanism is then the basis of the "coherent atom recoil laser" [73, 74].

The relation between Rayleigh superradiance and the coherent atom recoil laser was studied experimentally by placing a cold atomic gas within a ring cavity (Fig. 13.6) [75]. The atoms-cavity system was pumped through one input port, exciting one running-wave mode of the cavity. Forward scattering by the atoms induces an atom-number-dependent frequency shift of both running-wave resonances of the cavity. Back scattering by the atoms couples the two modes. By monitoring the output of the second running-wave mode of the ring resonator, the spontaneous buildup of one-dimensional density modulations in the cold gas was observed. In addition, a time-of-flight imaging technique indicated the buildup of atom population at multiples of the back-scattering momentum recoil $2\hbar k$. By varying the temperature of the gas and parameters of the optical cavity, the relation between the lifetime of the matter-wave density modulations and of the cavity field was tuned between the "bad-cavity," or superradiant limit, and the "good-cavity" or "ringing-superradiance" limits of the coherent atom recoil laser [33, 76]. Depending on the experimental configuration, the cavity-field and matter-wave amplifications can either favor the same or different forms of the Brillouin instability; for example, a narrow frequency cavity can select the coherent emission of momentum populations different than those expected from free-space superradiant scattering [77].

A similar mechanism operates in the MIT (and Zürich) experiments [6, 69], where the atoms are pumped by an externally imposed standing-wave optical field with its axis orthogonal to the axis of a Fabry-Pérot cavity. The Brillouin instability organizes the gas into one of two distinct spatial patterns that maximize the collective scattering into the cavity mode, but that are distinguished by the relative phase, either 0 or π, between the probe and cavity fields (Fig. 13.5); the emergence of these checkerboard patterns was confirmed both by time-of-flight analysis [6] and by the observed bistability of the relative optical phase [69, 78].

In Sect. 13.6, we reconsider these experiments in terms of cavity optomechanics.

13.5 Quantum Cavity Optomechanics with Cold Atoms in a Driven Fabry-Pérot Resonator

As highlighted above, the full dynamics of a few or many particles, the motion of which is coupled by and to the dynamics of an optical resonator, is rich and complex. It was therefore surprising that a very simple, clarifying, and quantitative

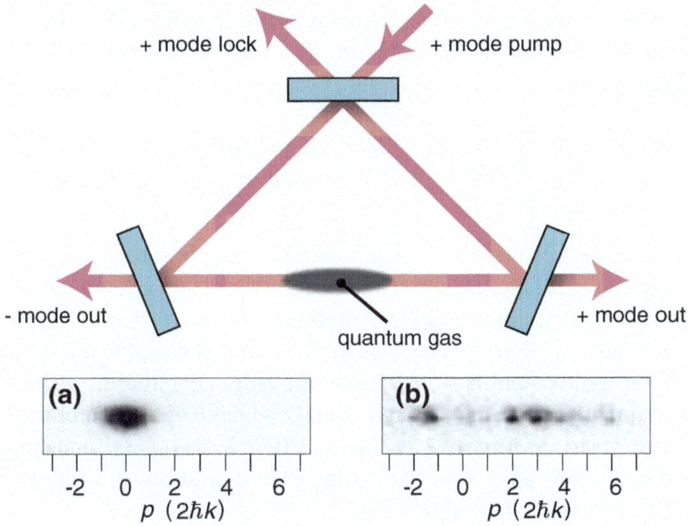

Fig. 13.6 Setup for pumped ring-cavity experiments. A quantum gas is placed within the mode volume of a ring cavity. The *right-going mode* (seen by the atoms, labeled with +) is pumped by a coherent-state input that is locked onto the cavity resonance. Monitoring the output of the *left-going mode* (labeled with −) reveals the coherent back scattering of light due to self-organization of the gas. Time-of-flight distributions of the gas **a** before and **b** after the coherent atom recoil lasing action show that coherent back scattering is accompanied by the distribution of the gas among several coherent momentum populations. These indicate the arrangement of the atoms into a periodic spatial structure. Panels **a** and **b** are adapted from Ref. [76]

treatment of such dynamics, using the paradigm of cavity optomechanics, could emerge. The key to this simplification was to create experimental situations where the mechanical dynamics of the atomic ensemble is restricted to a small perturbation atop a well characterized initial state. Such experimental starting points include near-uniform ultracold gases held within the resonator [6, 79] and atoms cooled near the ground state of axial motion within many [80] or just a few potential wells [61] of an intracavity optical lattice.

13.5.1 Collective Atomic Optomechanical Response

The derivation of the cavity optomechanics Hamiltonian as an approximation of the dispersive interaction between an atomic ensemble and a single-mode cavity has been outlined in several works [50, 79, 81, 82]. For the purpose of this discussion, we focus on forward scattering by the atom, a dispersive two-photon effect wherein an off-resonant cavity photon, detuned from the atomic resonance by the frequency Δ_{ca}, is absorbed and then re-emitted into the cavity field. As expected from second-order perturbation theory, the cavity energy shift due to the single atom, located at position $\mathbf{r}_i$, is given as $\hbar|g^2(\mathbf{r}_i)|/\Delta_{\text{ca}}$ (this relation is exhibited in Eq. 13.2 in the limit $|\Delta_{\text{ca}}| \gg |g(\mathbf{r})|$). For many atoms in the cavity, this energy shift is additive, giving the Hamiltonian

$$H = \hbar\omega_c \hat{a}^\dagger \hat{a} + H_{\text{mech}} + \sum_i \hbar \frac{|g^2(\mathbf{r}_i)|}{\Delta_{\text{ca}}} \hat{a}^\dagger \hat{a}, \tag{13.4}$$

where ω_c is the cavity resonance frequency, $\hat{a}$ the cavity photon annihilation operator, the sum is taken over all atoms, and we omit some constant energy terms. The term H_{mech} describes the dynamics of atomic motion.

We now consider two simple scenarios for the atomic ensemble. First, we consider the atoms to be tightly confined in a harmonic trap, centered at position z_0, with vibrational frequency ω_z along the cavity axis, and neglect motion in the transverse directions. The position of atom i is taken to be $z = z_0 + \delta z_i$ and thus $g(\mathbf{r}_i) = g_0 \sin(\phi_0 + 2k\delta z_i)$ with $\phi_0 = kz_0$. Expanding to first order in the small Lamb-Dicke parameters $k\delta z_i \ll 1$, we then obtain [61]

$$\begin{aligned} H \simeq \hbar \left(\omega_c + N\frac{g_0^2}{\Delta_{\text{ca}}} \sin^2 \phi_0 \right) \hat{a}^\dagger \hat{a} + \hbar\omega_z \sum_i \hat{b}_i^\dagger \hat{b}_i \\ + \hbar \frac{g_0^2}{\Delta_{\text{ca}}} \sin(2\phi_0)\, \hat{a}^\dagger \hat{a} \sum_i k\delta z_i, \end{aligned} \tag{13.5}$$

with $\hat{b}_i$ annihilating a phonon from the motion of atom i. From the sum of the cavity coupling to each of the individual atoms, we identify a single collective atomic variable with which the cavity interacts. Here, this collective variable is simply the center of mass $Z_{\text{cm}} = N^{-1} \sum \delta z_i$ of the ensemble. For this simple mechanical setup, the center of mass motion is a normal mode of the system: a harmonic oscillator of frequency ω_z, a mass $M = Nm$ equal to that of the entire N-atom ensemble, and a harmonic oscillator length $Z_{\text{ho}} = \sqrt{\hbar/2Nm\omega_z}$. Adjusting the cavity resonance frequency to $\omega_c' = \omega_c + N(g_0^2/\Delta_{\text{ca}}) \sin^2 \phi_0$, we thus obtain

$$H_{\text{om}} = \hbar\omega_c' \hat{a}^\dagger \hat{a} + \hbar\omega_z \hat{b}^\dagger \hat{b} + \hbar g \left(\hat{b}^\dagger + \hat{b} \right) \hat{a}^\dagger \hat{a} + H_{\text{bath}}. \tag{13.6}$$

Here, we write $H_{\text{mech}} = \hbar\omega_z \hat{b}^\dagger \hat{b} + H_{\text{bath}}$ where $\hat{b}$ annihilates a phonon from the center-of-mass collective mode, and H_{bath} describes the remaining normal modes of the system.

We thus identify the canonical cavity optomechanical Hamiltonian for linear optomechanical coupling. The single-photon/single-phonon optomechanical coupling strength is

$$g = N\frac{g_0^2}{\Delta_{\text{ca}}} \sin(2\phi_0) \times k\sqrt{\frac{\hbar}{2Nm\omega_z}}, \tag{13.7}$$

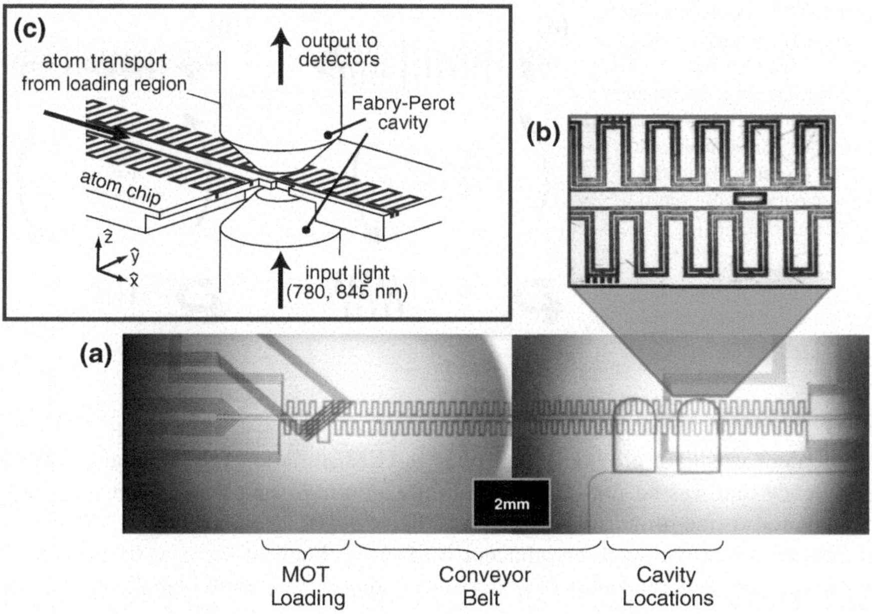

Fig. 13.7 The Berkeley atom chip experiment used to deliver an ultracold atomic ensemble into a high-finesse Fabry-Pérot cavity. **a** A transmission image of the silicon-substrate atom chip shows inlaid copper wires on front and back surfaces as opaque lines. These wires are used as electromagnets to generate the magnetic fields required to load atoms into a magneto-optical trap (*left*), transport them across the chip in a magnetic conveyor system (*middle*), and then confine them tightly and position them within the cavity region (*right*). The *tombstone-shaped shadows* indicate regions where the atom chip is etched down to a thickness of just 100 μm. **b** Within this thinned region, the atom chip is perforated to allow cavity light through the chip. **c** Two curved mirrors are positioned *above* and *below* this perforation to form the Fabry-Pérot cavity, with a mirror spacing of about 250 μm. Figure **c** is reproduced from Ref. [61]

an expression which exhibits the scaling with atom number, atom-cavity detuning, and mechanical frequency, all of which can be varied broadly via straightforward modifications of the experimental system, even during a single experiment.

Building upon this example, we consider an ensemble split into several harmonic traps. The cavity-chosen, linearly coupled, collective atomic variable is now a sum of the centers of mass of the sub-ensembles, weighted by their individual optomechanical coupling strengths [50, 80, 82]. This setup is akin to having several solid-state membranes within a single resonator, coupled to one another via the cavity field.

As with the membrane-based realizations of cavity optomechanics [59], the linear optomechanical coupling strength varies with the equilibrium position of the mechanical element within the cavity. This variation was demonstrated by the Berkeley group [61]. In these experiments, a cold atomic gas of several thousand atoms was produced and translated into an optical resonator by means of a microfabricated atom chip (Fig. 13.7). The chip provided for strong magnetic confinement of

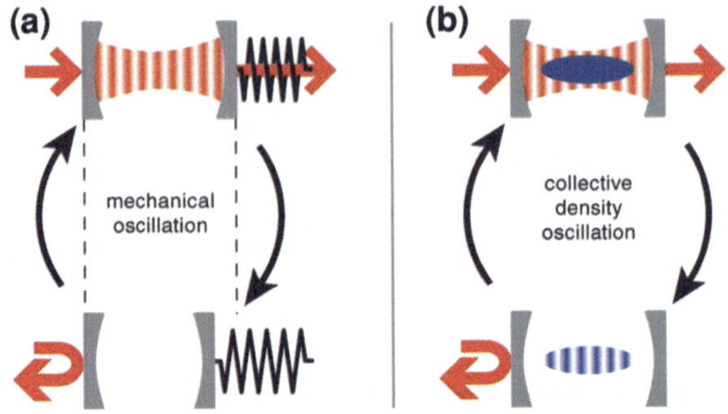

Fig. 13.8 The equivalence of cavity optomechanics realized with **a** moving mirrors or **b** cosine-mode density modulations within a Bose-Einstein condensate placed inside and coupled dispersively to the optical cavity. Figure reproduced from Ref. [79]

the gas, although not sufficiently strong to reach the Lamb-Dicke regime. Stronger confinement was then provided by driving a TEM_{00} mode of the optical cavity, at a frequency that was far red-detuned from the atomic resonance. This strong cavity field created a one-dimensional optical lattice potential, within which the atomic gas occupied just two or three adjacent wells [83]. Each of these wells provides a near-harmonic trap, relevant to the derivation above. Due to the small wavevector difference $q = k_p - k_t$ between the trapping light [$k_t = 2\pi/(850\,\mathrm{nm})$] and probe light [near the ^{87}Rb atomic resonance at $k_p = 2\pi/(780\,\mathrm{nm})$], the optomechanical coupling strength within neighboring wells of the optical lattice was similar, with ϕ_0 changing by 0.26 rad from well to well. Broad variations in ϕ_0 were achieved by varying the position of the magnetic trap. The tuning of the optomechanical interactions was indicated both by the strong variation in the atoms-induced cavity resonance shift $(\omega_c' - \omega_c)$, and by the variation in the optomechanical frequency shift (Sect. 13.5.3).

A second simple initial state is a uniform, stationary, non-interacting Bose-Einstein condensed gas (Fig. 13.8) [79]. We emphasize that the assumptions of the gas being Bose condensed and non-interacting are not essential. Bose-Einstein condensation ensures that density modulations at a fixed wavevector describe a normal mode of the system with a very sharp frequency response. In comparison, density modulations of a collisionally thin non-degenerate gas dephase more rapidly. Assuming the gas to be non-interacting allows us to neglect Bogoliubov transformations and the diminished structure factor for optically exciting phonons rather than free particles [84]. However, including such interaction effects presents no major difficulty.

With these assumptions, we recognize that, to lowest order, the spatial variation of $g^2(\mathbf{r}) \simeq g_0^2 \sin^2 kz = g_0^2\,(1 - \cos 2kz)\,/2$ couples the condensate to a cosine-function superposition of its $\pm 2\hbar k$ momentum excitations, with the excitation energy $\hbar\omega_{2k} = 2\hbar^2 k^2/m$ equal to four times the recoil energy. Using the Bose field operators $\hat{b}_0$ for the condensate and $\hat{b}_1 = (\hat{b}_{+2k} + \hat{b}_{-2k})/\sqrt{2}$ for the cosine-excitation mode where $\hat{b}_{\pm 2k}$ are momentum-space field operators, and substituting $\hat{b}_0 = \sqrt{N}$, we approximate [79]

$$H \simeq \hbar\left(\omega_c + N\frac{g_0^2}{2\Delta_{\mathrm{ca}}}\right)\hat{a}^\dagger\hat{a} + \hbar\omega_{2k}\hat{b}_1^\dagger\hat{b}_1 + \sqrt{N}\hbar\frac{g_0^2\sqrt{2}}{4\Delta_{\mathrm{ca}}}\left(\hat{b}_1^\dagger + \hat{b}_1\right)\hat{a}^\dagger\hat{a} \quad (13.8)$$

where now the single-photon/single-phonon optomechanical coupling strength is $g = \sqrt{N}(g_0^2\sqrt{2}/4\Delta_{\mathrm{ca}})$.

A similar analysis can be applied to the situation of a cold Fermi gas located within the cavity mode [85]. In this case, the cavity field may be coupled to zero-sound modes of the degenerate Fermi gas. This example illustrates the fact that neither Bose-Einstein condensation, nor quantum degeneracy, are necessary ingredients of most atoms-based cavity optomechanical systems.

13.5.2 Sensing Collective Atomic Motion

Now we review research findings on cavity optomechanics garnered with this atomic-ensemble approach. The experimental procedures utilized in this research differ from those of solid-state experiments. For instance, whereas the solid-state experiments generally persist in a steady state, the mechanical element used in the atoms-based experiments—the atomic gas—is short-lived. Thus, the gas must be prepared and positioned anew within the cavity before each measurement. This leads to experimental variability, e.g. in the atom number, temperature, and position, that must be controlled, or at least accounted for, in order to compile repeated measurements.

In the current Berkeley experiments, this variability is assessed and adjusted for by active feedback. For instance, variations in the atom number cause the atoms-shifted cavity resonance frequency ω_c' to vary between measurements, and also to vary *during* a measurement due to the ejection of atoms from their trap. To accommodate this variability, the cavity-probe light is actively stabilized to a fixed detuning from the cavity resonance.

The collective atomic motion is then sensed via the cavity field. For example, a single record of the cavity transmission collected with the probe locked onto the side of the cavity resonance shows a sharp frequency modulation at the mechanical resonance of a harmonically bound atomic ensemble (Fig. 13.9). This motional signal varies over the lifetime of the atomic ensemble, presumably due to light-induced displacements and heating of the ensemble, and finally disappears when the sample is fully depleted due to probe-induced heating.

13.5.3 Back Action onto the Mechanical Oscillator

The sensitivity of the cavity field to collective atomic motion can be thought of as one link in a closed-loop response connecting the optical and mechanical inputs and

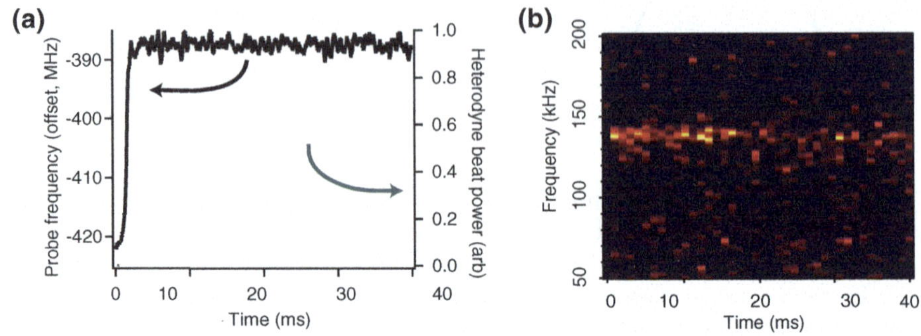

Fig. 13.9 Detecting motion in a newly prepared cold-atom optomechanics system. **a** Shot-to-shot variations in the number of atoms trapped within the cavity makes the atom-shifted cavity resonance frequency ω_c' uncertain at the start of an experiment. To probe the cavity at a reliable detuning from the cavity resonance, the probe light is switched on first at a frequency far from the cavity resonance, and then active stabilization is used to tune it to the side of the cavity resonance. The probe frequency (*gray line*, *left axis*) is seen to drift toward the resonance frequency and lock within about 1 ms. The transmitted probe intensity, seen here as the power on a heterodyne detector (*black line*, *right axis*), stabilizes to a near-constant value. The probe remains locked while atoms are slowly lost from the cavity, causing its frequency to drift slowly over a few MHz. **b** Demodulating the heterodyne signal, and taking the Fourier transform of the demodulated signal within 1 ms time bins, reveals a signal peak at around 140 kHz, recording the motion of the mechanical element within the cavity. The signal frequency and strength varies over 10s of ms as the atomic ensemble becomes increasingly perturbed by the constant probe

outputs, schematically represented in Fig. 13.10. A formal and quantitative representation of this closed-loop schematic is presented in Ref. [86].

Let us make the *linearization approximation,* wherein we take the cavity field to be $\hat{a} = e^{-i\omega_p t}\left(\bar{a} + \delta\hat{a}\right)$, where ω_p is the probe (or pump) optical frequency, $\bar{a}$ is the coherent-state amplitude of the driven cavity field, and we neglect terms quadratic in the field fluctuations $\delta\hat{a}$. This linearization approximation is valid in the non-granular regime where g is small compared to the cavity linewidth [50, 80]. With this approximation, the canonical cavity optomechanical Hamiltonian (Eq. 13.6) becomes

$$-\hbar\Delta\hat{a}^\dagger\hat{a} + \hbar\omega_z\hat{b}^\dagger\hat{b} + \hbar g|\bar{a}|^2\left(\hat{b}^\dagger + \hat{b}\right) + \hbar\bar{a}g\left(\hat{b}^\dagger + \hat{b}\right)\left(\delta\hat{a}^\dagger + \delta\hat{a}\right) \tag{13.9}$$

where we have assumed without loss of generality that $\bar{a}$ is real. The energy of the cavity is now measured with respect to the probe frequency, with $\Delta = \omega_p - \omega_c$. The optomechanical interaction term proportional to $|\bar{a}|^2$ describes a constant force on the oscillator, and may be absorbed by shifting its equilibrium position. The linearized optomechanical coupling strength is now $g\bar{a}$, and the interaction contains only bilinear terms.

Within this linearized picture, we state that the motion of the oscillator $z(\omega)$ translates to a modulation of the cavity field (generally in both its amplitude and phase quadratures) via a transfer function $\mathbf{F}_a\mathbf{T}$ that is a product of two

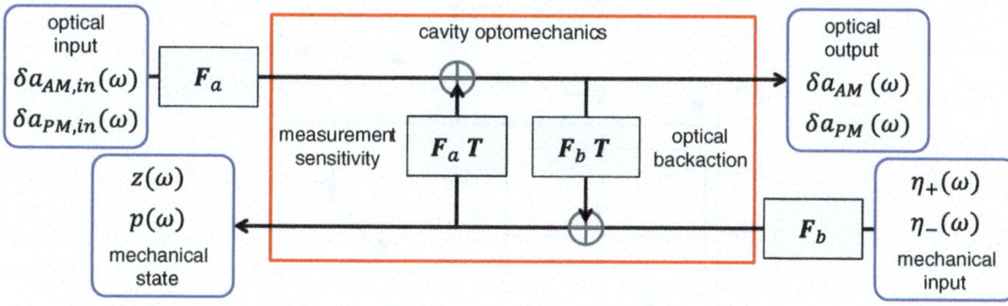

Fig. 13.10 The interactions between the optical field and the mechanical oscillator in cavity optomechanics are represented as a closed-loop feedback system, with the approximation that optical field is linearized about a steady coherent-state field with amplitude $\bar{a}$. The system is driven by optical inputs, the amplitude (AM) and phase (PM) modulation at the frequency ω, and mechanical inputs, $\eta_\pm(\omega)$ representing force modulations and fluctuations due to mechanical dissipation. The feedback system relates these inputs to the optical and mechanical outputs according to the closed-loop, ponderomotive gain spectrum. Elements in the feedback system include the cavity field and mechanical dynamical responses, summarized by transfer matrices $\mathbf{F}_a$ and $\mathbf{F}_b$, respectively, and the linear optomechanical coupling $\mathbf{T}$, which is proportional to $g\bar{a}$. For example, the maximum closed-loop optical-to-optical response is proportional to the optomechanical cooperativity $4g^2\bar{n}/\kappa\gamma$, with $\bar{n} = |\bar{a}|^2$, and the linewidths quantify the strength of the optical cavity (via $\mathbf{F}_a$) and mechanical oscillator (via $\mathbf{F}_b$) resonant responses. The figure is adapted from Ref. [86]

matrices: an optomechanical coupling matrix $\mathbf{T}$ that is proportional to $g\bar{a}$, and a cavity conditioning matrix $\mathbf{F}_a$ that describes the evolution of amplitude and phase modulations in the cavity. The latter matrix depends on the detuning Δ of the cavity probe from the cavity resonance, and captures the fact that the motion is recorded on the phase quadrature if the cavity is probed on its resonance, and in a combination of amplitude and phase quadratures if probed off resonance.

Simultaneously, the cavity field influences the motion of the mechanical oscillator. This influence is summarized by the transfer function $\mathbf{F}_b\mathbf{T}$ connecting the two quadratures of field fluctuations to the two quadratures of mechanical fluctuations. This connection emphasizes the fact that, in quantum mechanics, the act of measurement is necessarily influential on the object being measured.

The effects of such back action may be described as either *coherent* or *incoherent*, based on whether or not quantum fluctuations play a role. One form of coherent back action is the optical spring effect, in which the mechanical oscillation frequency is shifted by the adiabatic response of radiation pressure to the moving oscillator [87–89]. The optomechanical frequency shift was analyzed by the Berkeley group [61]. Positioning the harmonically trapped atoms at the location of maximum linear optomechanical coupling, the frequency shift was measured as a function of probe-cavity detuning Δ, and found to be in quantitative agreement with a no-free-parameter theoretical prediction (Fig. 13.11), demonstrating the appropriateness of using cavity optomechanics to describe mechanical effects of atomic ensembles in cavities.

The optomechanical frequency shift describes the adiabatic dynamic back action of the cavity field upon the mechanical oscillator in the limit of small oscillation amplitude, where the radiation pressure force varies linearly with the position of the

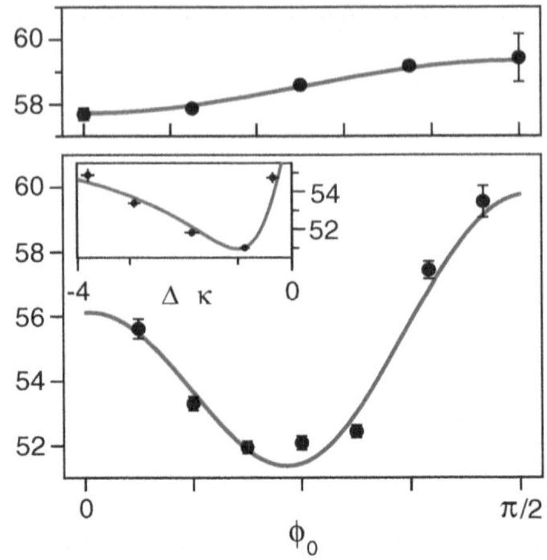

Fig. 13.11 Optomechanical frequency shifts measured with an atomic ensemble trapped at various positions along the cavity axis. *Top* the mechanical oscillation frequency ω_z determined via parametric heating reveals the static frequency shift from quadratic optomechanical coupling; here, $\Delta_{ca}/2\pi = 20.1$ GHz, $\omega_z/2\pi = 58.5$ kHz and $\bar{n} = 0.8$. *Bottom* the oscillation frequency for the linearly coupled mode is derived from the cavity transmission intensity spectrum. The deviation between the frequencies in the top and bottom figures quantifies the optomechanical frequency shift due to linear optomechanical coupling. This frequency shift is maximal in the case of maximal linear coupling ($\phi_0 = \pi/4$). Here, $\Delta_{ca}/2\pi = 40$ GHz, $\omega_z/2\pi = 58.9$ kHz, $\bar{n} = 3.5$ and $N = 3{,}750$. Inset shows this frequency at $\phi_0 = \pi/4$ for varying detuning Δ from the cavity resonance. *Solid lines* show predictions calculated with no free parameters. Figure reproduced from Ref. [61]

oscillator. For larger mechanical displacements, when the cavity resonance frequency is made to vary by an amount comparable to the cavity linewidth, the oscillatory mechanical dynamics become more complex. Such large-scale motion, represented by periodic spikes in the power transmitted through the driven optical cavity, was observed by the Zürich group (Fig. 13.12) [79]. Their data also display strong optomechanical frequency shifts [90].

A second coherent back action effect is cavity cooling. Experimental studies of cavity cooling with atomic ensembles were carried out with gases at the 100-μK-range temperatures reached by laser cooling. To observe directly the coherent damping of collective atomic motion, the researchers excited large scale motion, using cavity-induced amplification from a cavity probe that was blue-detuned from the cavity resonance, before switching the probe frequency to the red of the cavity resonance (Fig. 13.13). It was confirmed that the mechanical damping rate from cavity cooling, proportional to g^2, indeed scaled linearly with the atom number (see Eq. 13.7), demonstrating that cavity cooling is a collective optical effect [82].

It is important to point out that, in contrast with solid-state optomechanics for which cavity cooling is an essential means to bring solid-state objects toward their mechanical ground state (except for the highest-frequency oscillators that reach the ground state at dilution-refrigerator temperatures [91]), atomic ensembles can be cooled to the quantum regime of motion by evaporative cooling to micro- or

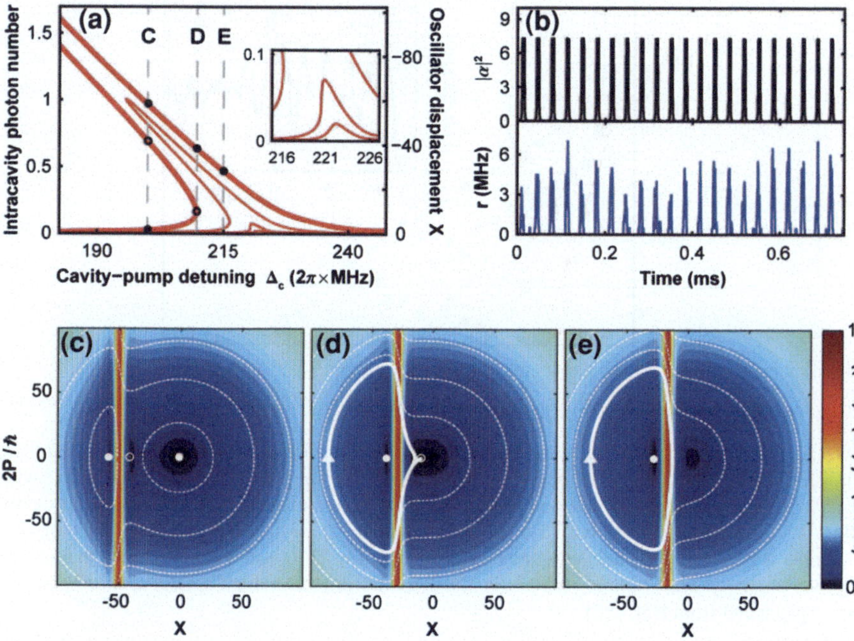

Fig. 13.12 Large scale motion of the mechanical oscillator leads to complex mechanical and optical dynamics. **a** The cavity optomechanical system is operated at strong input probe power, reaching deep into the regime of optomechanical bistability. **b** In this regime, the cavity output (*top* theory, *bottom* experiment) is a nearly periodic sequence of sharp emission bursts. **c**–**e** The mechanical trajectory corresponding to these bursts is shown on a phase-space plot. The mechanics evolve in a dark cavity until the cavity is tuned briefly into resonance with the probe, providing a strong radiation pressure impulse to the mechanical element and a burst of light through the cavity. This pulsed force occurs twice per mechanical cycle. Figure reproduced from Ref. [79]

nanokelvin-range temperatures. Indeed, for such ensembles, cavity-induced diffusive heating generally dominates cavity cooling, since the equilibrium temperatures reached by cavity cooling, on the order of $\hbar\kappa/k_B$, are typically higher than the "base temperatures" reached by evaporative cooling. Nevertheless, cavity cooling may be advantageous for gases that cannot be cooled effectively by evaporative cooling, for example, gases of complex atoms or of molecules which suffer from large inelastic collision rates.

13.5.4 Observations of Radiation Pressure Shot Noise

Incoherent back action is represented by the quantum fluctuations of the radiation pressure force, representing the back action of a continuous quantum measurement of the position of the mechanical oscillator. Considering the linear optomechanical

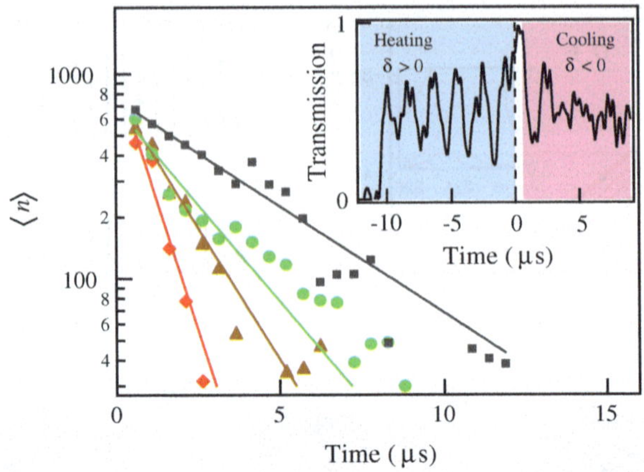

Fig. 13.13 Real-time observation of optomechanical cooling of an atomic ensemble. *Inset* the cavity is driven first with a *blue*-detuned probe, to excite large-amplitude motion of the intracavity ensemble, and then with a red-detuned probe, to observe the decay of this motion due to cavity cooling. A single record of the cavity transmission is shown. *Main figure* the mean phonon occupation number is derived from the cavity transmission spectrum. The cooling rate is shown to increase with probe intensity (*black squares* to *red diamonds* trend from low to high optical scattering rates). Figure reproduced from Ref. [82]

coupling Hamiltonian H_{om} (Eq. 13.6), the force on the mechanical oscillator can be taken as

$$\hat{f} = -\frac{\partial}{\partial \hat{Z}} H_{\mathrm{om}} = -\frac{\hbar g}{Z_{\mathrm{ho}}} \hat{a}^\dagger \hat{a} \tag{13.10}$$

Fluctuations in this force lead to diffusion, leading to an increase in the energy of the mechanical object. For an undamped harmonic oscillator, the energy increases at a rate proportional to the spectrum of force fluctuations—or equivalently, according to the above expression, the spectrum of photon number fluctuations—at the mechanical frequency.

In the absence of technical fluctuations, a linear cavity driven by a coherent-state input, detuned by Δ from the cavity resonance, has a (two-sided) spectrum of photon number fluctuations given as

$$S_{nn}(\omega) = 2\bar{n}\frac{\kappa}{\kappa^2 + (\Delta + \omega)^2} \tag{13.11}$$

where $\bar{n}$ is the average intracavity photon number. Unlike in free-space, where the shot-noise spectrum is white, within a cavity the photon shot noise is accentuated at the frequency $\omega = -\Delta$ by the cavity resonance. These intracavity fluctuations can be regarded as coming from the white field fluctuations of the cavity input. At the cavity resonance, this input noise enters the cavity where it beats with the coherent-state field to produce strong photon number fluctuations. Away from the cavity resonance, this input noise is predominantly reflected and does not enter the cavity, causing photon number fluctuations to be relatively suppressed.

It is no accident that the photon shot-noise spectrum also quantifies the sensitivity of a cavity-based position measurement of the mechanical oscillator. Variations δZ in the position are seen via variations $g\delta Z/Z_{\text{ho}}$ in the cavity resonance frequency. To first order, these lead to variations in the cavity electric field by the amount

$$\delta E = E_0 \frac{ig}{\kappa - i\Delta} \frac{\delta Z}{Z_{\text{ho}}} \tag{13.12}$$

where E_0 is the cavity field strength within the cavity with the oscillator at rest (so that $\bar{n} \propto |E_0|^2$), and we assume $\omega_z \rightarrow 0$ for simplicity. In this limit, the signal strength, taken as $|\delta E|^2$, is indeed proportional to the photon shot noise spectrum S_{nn}.

The cold-atoms approach has provided the first direct observations of radiation pressure shot noise in cavity optomechanics. These observations are enabled by the fact that the atoms are so well isolated from their environment, that radiation pressure fluctuations cause large accelerations of light mechanical objects, and that these forces are accentuated by the strong polarizability of atoms near their resonance frequency.

Radiation pressure shot noise was observed in four separate measurements by the Berkeley group: (M1) bolometric detection measured by atom loss, (M2) the observation of ponderomotive squeezing, (M3) the observation of ponderomotively amplified shot-noise fluctuations, and (M4) measurement of the heat flux onto the mechanical element due to force fluctuations.

In measurement (M1), these shot-noise driven fluctuations were observed by the bolometric quantification of the diffusive heating rate of the atomic ensemble. Here, one makes use of the finite depth of the trapping potential, which allows the heating rate to be quantified by the atom loss rate. This loss rate was measured as a function of the probe-cavity detuning Δ. At constant intracavity probe power, the diffusive heating rate showed the line shape predicted in Eq. 13.11, and matched the magnitude predicted for shot-noise-driven heating within about 20 % [50].

13.5.5 Mechanically Induced Nonlinear Optics: Bistability, Ponderomotive Amplification, and Ponderomotive Squeezing

Returning to our schematic representation of cavity optomechanics in Fig. 13.10, we understand that optical signals entering the cavity act upon the mechanical element, which then acts back onto the light. As such, the mechanical oscillator mediates optical self-interactions, providing the cavity with a nonlinear optical response.

Consider the dc effect of this self-interaction. A weak probe of the cavity reveals the cavity resonance frequency ω_c'. At higher probe-light power, radiation pressure shifts the mechanical oscillator's equilibrium position, shifting the cavity resonance.

The cavity is therefore a non-linear optical element, showing a cavity resonance that varies with the light intensity.

For sufficiently strong probe light, the light-induced shift of the mechanical oscillator is sufficient to shift the cavity resonance by more than its linewidth. Under these conditions, the optical cavity becomes dispersively bistable. Bistability induced by optical radiation pressure was observed in early experiments on solid-state optomechanics, using a mg-scale mass, a low-finesse optical resonator, and an optical power of several Watts [92]. In experiments using atomic ensembles, the higher cavity finesse and lower mechanical mass leads to cavity bistability at much lower powers, observed experimentally as low as 100 fW corresponding to an average intracavity photon number $\bar{n} = 0.05$ far below unity [80, 90]. In side-pumped cavity optomechanical systems, it is also possible to observe regimes where no stable atom-cavity configurations exist; rather, the dynamics may be described by limit cycles [93].

The nonlinear response of the cavity optomechanical system varies with frequency, in the standard sense of four-wave mixing where a monochromatic cavity drive field is regarded as the pump and weak sidebands about the pump frequency are regarded as signal and idler beams. For weak optical signals and in the nongranular regime (see Sect. 13.5.8 for discussion of the granular regime), the linearized cavity optomechanical system represented in Fig. 13.10 allows one to define a closed loop optical gain relating the cavity input signals to the intracavity field [86], where the frequency variation of this closed-loop gain stems from the dynamical response of the cavity and mechanical systems. The cavity is thus regarded as a ponderomotive amplifier. The amplification spectrum has been measured in solid-state optomechanical systems using input optical drives that greatly exceed the optical shot noise level, so that the response to the deliberately modulated radiation pressure dominates over that of other mechanical and optical perturbations [94, 95]. The same closed-loop optical response is observed also in experiments on optomechanically induced transparency [96, 97], albeit by focusing on different optical inputs and outputs [86].

The ponderomotive amplification spectrum was measured using atomic ensembles (Fig. 13.14). The cavity was pumped by a monochromatic coherent drive with the input power of around 36 pW, tuned below the cavity resonance, and then probed by placing single-tone amplitude modulation (AM) on the pump beam. A heterodyne receiver at the cavity output was used to quantify the complex-valued gain for transducing the input AM tone to AM and phase modulation (PM) at the cavity output. At and below the optomechanically shifted mechanical frequency, the system showed strong amplification in both the AM and PM output quadratures, with a power gain as high as 20 dB. This amplification results from the in-phase response of the mechanical oscillator to the radiation pressure modulation, which then feeds back so as to further amplify the cavity field modulations. At frequencies above the mechanical frequency, where the oscillator responds out of phase with its force drive, the AM signal is suppressed by as much as 26 dB. The remarkable quantitative match between measurements and the predictions of cavity optomechanics are, again, a testament to the utility of regarding collective atomic motion as equivalent to that of a solid-state mechanical oscillator [4].

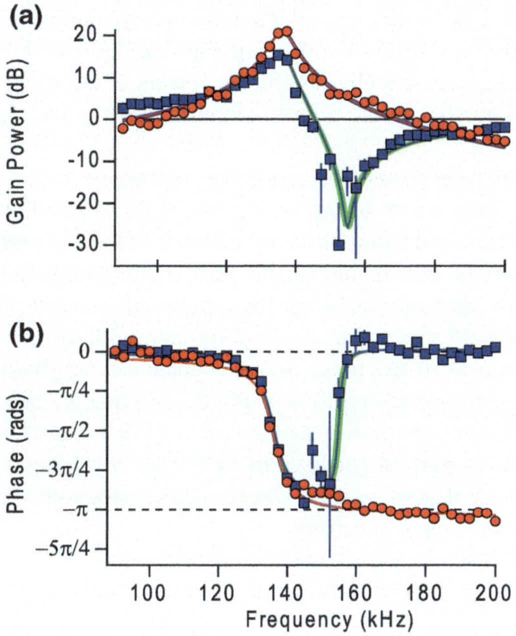

Fig. 13.14 Strong ponderomotive amplification is observed in a cavity optomechanical system pumped with just 36 pW of input power. The gains for converting an AM input to either AM (*blue squares*) or PM (*red circles*) at the cavity output are complex numbers. Here, **a** the power gains (square magnitude of the complex-valued gain), and **b** the phase shifts (arguments of the complex-valued gain) are plotted versus the input modulation frequency. The peak AM → PM gain occurs at the optically shifted mechanical resonance frequency, reaching a maximum power gain of 26 dB. Above the mechanical resonance frequency, the out-of-phase mechanical response leads to a strong suppression of the AM output. In the frequency region of strong suppression, one expects and indeed observes ponderomotive squeezing when the system is driven by a shot-noise-dominated probe field. The agreement between measurement and theory (*solid lines*, with *shaded region* indicating systematic uncertainty) is remarkable. Figure reproduced from Ref. [4]

Acting on the quantum fluctuations of the input field, the nonlinear suppression of input signals yields inhomogeneously squeezed light [98, 99]. Such ponderomotive squeezing differs from other methods of squeezing light in that it originates not from electronic motion, but rather from the collective motion of more massive objects. Ponderomotive squeezing has important consequences for quantum-limited and "sub-quantum-limited" detection of motion and forces, for example in gravity-wave observatories where it will modify the nature of squeezed light injected into the cavity optomechanical detector, and will require proper conditioning at the output of the detector in order to sense the squeezed, rather than the anti-squeezed, field quadratures [100].

Ponderomotive squeezing in an atoms-based cavity optomechanical system was observed following the aforementioned measurements of ponderomotive amplification simply by extinguishing the deliberate AM tone and allowing the system to be driven by shot-noise-dominated fluctuations. In regions of ponderomotive suppression, the observed level of squeezing was small (about 1.5 % of shot noise) but statistically and systematically significant (errors at the level of 0.1 % of shot noise), and limited mostly by the poor quantum efficiency for photon detection in this system.

The detection of ponderomotive squeezing represents a direct observation of radiation pressure shot noise acting on the mechanical oscillator [measurement (M2)]. That is, light emitted from the cavity is a sum of the vacuum field fluctuations reflecting off the cavity mirror and of the intracavity optical field. The reduction of optical quadrature fluctuations below their standard quantum limit confirms that the cavity field fluctuations are negatively correlated with those of the cavity input field by the back action of the noise-driven mechanical oscillator onto the cavity field. In addition, in frequency regions of ponderomotive amplification, the cavity output spectrum was in quantitative agreement with that expected for a shot-noise driven optomechanical system [measurement (M3)]. Mechanical influences other than radiation pressure shot noise were constrained to have contributed no more than 3 % of the mechanical power spectrum.

13.5.6 Sideband Thermometry and Calorimetry in Quantum Cavity Optomechanics

Ponderomotive squeezing represents a clear quantum signature of the "opto" portion of the cavity optomechanical system. One signature for the quantum nature of the "mechanical" portion is the observation of asymmetric motional sidebands in light scattered off the mechanical oscillator. This sideband asymmetry represents the fact that a quantum mechanical oscillator will more readily gain than lose energy. That is, the sideband asymmetry reflects the fact that, exposed to classical (optical) force fluctuations (equal spectral densities for positive and negative frequencies), the ν-phonon state will be excited by one phonon at a rate $R_+ \propto |\langle \nu + 1|Z|\nu\rangle|^2 \propto \nu + 1$ and de-excited at the smaller rate $R_- \propto |\langle \nu - 1|Z|\nu\rangle|^2 \propto \nu$. Correspondingly, the powers in the emitted Stokes (red-shifted, $\propto \nu + 1$) and anti-Stokes (blue-shifted, $\propto \nu$) sidebands are unequal. The asymmetry is most pronounced for a ground-state harmonic oscillator: Since the oscillator cannot emit energy, the anti-Stokes sideband produced by its motion vanishes altogether.

The asymmetry between Stokes and anti-Stokes sidebands is readily observed for microscopic quantum systems, for example in the Raman spectra of molecular gases. Sideband asymmetry is also observed in light scattered by single trapped ions [101], single neutral atoms [102], and also by chains of as many as 14 ions [103, 104] on the modes of collective motion cooled to near the ground state. This telltale sideband asymmetry has been observed recently in two cavity optomechanical experiments: one involving a microfabricated solid-state mechanical oscillator [105], and another involving the collective motion of an atomic ensemble [5].

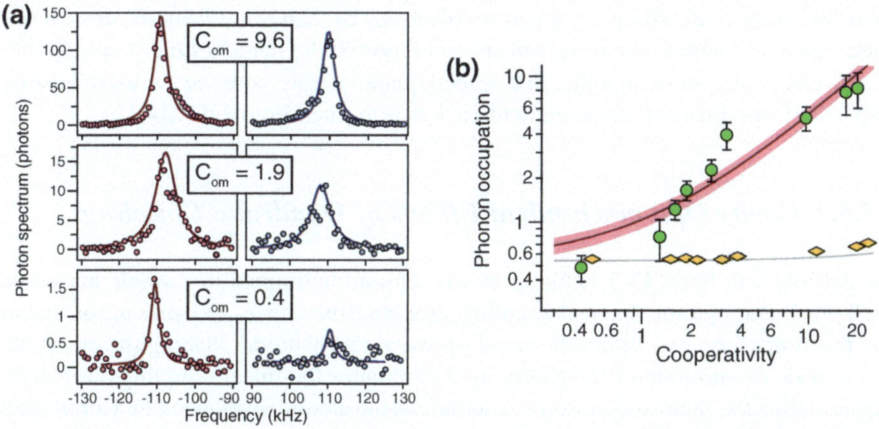

Fig. 13.15 Asymmetry between the Stokes (*red*) and anti-Stokes (*blue*) sidebands placed by the mechanical element on the cavity probe field reveals the quantization of collective atomic motion. **a** Sideband spectra are presented for probe power increasing from *bottom* to *top*. **b** The mean phonon occupation $\bar{\nu}$ determined by the sideband asymmetry (*circles* data, *shaded line* theory) is shown as a function of increasing probe power (quantified by the cooperativity $C_{\mathrm{om}} = 4g^2\bar{n}/\kappa\gamma$), and compared with the phonon occupation number expected at the measured temperature of the atomic ensemble (*diamonds*). At low probe power, a 3:1 asymmetry is observed, corresponding to $\bar{\nu} = 0.5$ and agreeing with the expected value at thermal equilibrium. With increasing probe power, the sideband asymmetry diminishes, showing the collective atomic modes probed by the cavity to be selectively heated by measurement back action, while the remainder of the mechanical modes, acting as a large thermal bath, remain at nearly constant temperature. Figure reproduced from Ref. [5]

In the atomic-ensemble work, the cavity was probed with light at the cavity resonance and monitored in transmission by a heterodyne measurement. Demodulating the heterodyne signal provided the sideband spectrum of the cavity field (Fig. 13.15). For a weak cavity probe, the sidebands showed a 3:1 power ratio, indicating the mechanical system to have a steady-state average phonon occupation of $\bar{\nu} = 0.5$. This was the value expected given the temperature at which the atomic ensemble was prepared by evaporative cooling, indicating that the collective mode (or a few modes) probed by the measuring the cavity-selected collective atomic variable was at equilibrium with the many other modes of motion within the ensemble.

For stronger cavity probes, the relative asymmetry of the sidebands diminished, indicating that $\bar{\nu}$ increased in response to the measurement. In contrast, the overall temperature of the atomic gas, probed by time-of-flight methods, remained nearly constant, supporting the assumption that the measurement directly disturbs only the collective mode(s) being probed.

Further information is gleaned from the absolute, rather than the relative, asymmetry of the Stokes and anti-Stokes sidebands. Every detected photon that is red-shifted by a frequency ω records the addition of an energy $\hbar\omega$ into the mechanical system, while blue-shifted photons record a similar subtraction of mechanical energy. The cavity spectrum thus serves not only as a thermometer, but also as a heat-flux sensor, recording the spectrum of energy exchange between the optical and mechanical elements. The observed spectra reported the net addition of energy into the mechanical system, at a flux that agreed quantitatively with the diffusive heating expected from quantum measurement back action [measurement (M4)].

13.5.7 Cavity Optomechanical Effects of Quadratic Coupling

As described in Sect. 13.5.1, for an atomic ensemble trapped in a single harmonic well, the linear optomechanical coupling strength (linear in Z_{cm}) varies according to the position of the gas within the standing-wave cavity mode. Placing the ensemble at the node or antinode of the cavity field eliminates the linear coupling altogether, necessitating the inclusion of terms quadratic in the atomic displacement. Continuing the expansion of Eq. 13.5 to second order in the Lamb-Dicke parameters, we obtain a quadratic optomechanical coupling term of the following form:

$$H_{\mathrm{quad}} = \hbar g\, k Z_{\mathrm{ho}} \cos(2\phi_0) \left(\frac{1}{N} \sum_i \frac{\delta z_i^2}{Z_{\mathrm{ho}}^2} \right) \hat{a}^\dagger \hat{a} \tag{13.13}$$

$$= \hbar g\, k Z_{\mathrm{ho}} \cos(2\phi_0) \left(\frac{Z_{\mathrm{cm}}^2}{Z_{\mathrm{ho}}^2} + \frac{\sigma^2}{Z_{\mathrm{ho}}} \right) \hat{a}^\dagger \hat{a}. \tag{13.14}$$

The cavity frequency displays a quadratic response to the center-of-mass displacement, and also (dominantly) to the position variance σ^2 of the compressible atomic medium. All normal modes of axial motion contribute to the position variance; hence, the quadratic optomechanical coupling senses, acts back upon, and potentially cools (by cavity cooling or by active feedback), all modes of motion in the atomic gas.

Quadratic optomechanical sensitivity is also achieved in membrane-based cavity optomechanics, although effects of this quadratic coupling have only been observed in the atomic-ensemble realization [61]. One such effect is cavity nonlinearity and bistability induced by variation in the strain of the atomic gas: Cavity probe light either adds or detracts from the trap curvature, depending on the sign of the quadratic coupling term, causing the gas to either contract or expand, and thereby shifting the cavity resonance frequency. The variation of the trap curvature with probe power also leads to a quadratic optomechanical frequency shift (Fig. 13.11). Future work may explore phenomena such as ponderomotive amplification and squeezing induced by the quadratic coupling term.

13.5.8 The Granular Regime of Cavity Optomechanics

One of the most important contributions from the work on cavity optomechanics with atomic ensembles is the identification of a new, "granular" regime of cavity optomechanics, one in which the coupling between photons and phonons is significant at the single quantum level [80]. Quantitatively, the granular regime is reached when $\varepsilon = g/\kappa > 1$.

To illustrate the implication of this relation, let us consider the strength by which an optical interrogation of the cavity measures the position of the mechanical element. The transmission of a single photon through the cavity determines the cavity resonance frequency to within the cavity half-linewidth κ. In turn, this knowledge determines the position of the mechanical element to within a position uncertainty of $\delta Z = Z_{\text{ho}}(\kappa/g)$. In the granular regime, $\delta Z < Z_{\text{ho}}$, so that a single photon measurement is sufficient to collapse the position of the mechanical oscillator to an uncertainty smaller that the range of its zero-point motion. The granular regime is thus an inherently strong quantum measurement regime.

Concomitantly, the granular regime is one in which the back action of a measurement is strong. A single photon within the cavity exerts a radiation pressure force $f = -\hbar g/Z_{\text{ho}}$ on the mechanical element. Its residence time within the cavity is uncertain to within the cavity ring-down time, $\sim\kappa^{-1}$. Thus, the impulse exerted by a single photon onto the mechanical element has an uncertainty of $\delta P = (g/\kappa)(\hbar/Z_{\text{ho}})$. In the granular regime, $\delta P > \hbar/Z_{\text{ho}}$, i.e. the measurement back action increases the uncertainty in the momentum of the mechanical element by an amount greater than its zero-point uncertainty. Conversely, viewing the mechanical element as a detector for intracavity photons, the granular regime is one in which single photons passing through the cavity are strongly measured by imparting distinguishable changes in the mechanical state.

The reported experiments on cavity optomechanics with atomic ensembles have approached the granular regime, with the highest reported value of the granularity parameter being $\varepsilon = 1.5$ in the work of Gupta et al. [80]. In contrast, in solid-state cavity optomechanical systems, the granularity parameter is typically much smaller, below 10^{-2} in the recent work on "optomechanical crystals," [106], and several orders of magnitude lower still in other experiments.

It is unclear how the range of optomechanical effects obtained in the nongranular regime, such as those discussed previously in this Section, will be modified in the regime of stronger optomechanical coupling. The common linearization approximation that is applied to describe those effects is clearly invalid in the granular regime. Several theoretical works have begun to consider this regime, pointing to new features such as strong photon antibunching [107], cavity absorption and emission spectra dressed by several mechanical sidebands [107–109], unstable dynamics driven by strong radiation pressure fluctuations [110] and with distinct quantum signatures [111], photon bursts related to optomechanical bistability [112], and entanglement generated in an optical interferometer in which photons on one arm are effectively measured by the displacement of a mechanical element [113]. The last work relates to earlier proposals to measure gravity-induced decoherence of massive objects by entangling their motion with single photon states [114], demonstrating that the granular regime is required for such investigations. Many of these theoretical predictions rely on realizing the granular and resolved-sideband regime simultaneously, posing a clear, but attainable, challenge for future experiments.

13.6 Cavity Optomechanics of a Side-Pumped Ensemble

Now we reconsider two experimental configurations discussed in Sect. 13.4.1: an atomic ensemble within a Fabry-Pérot cavity pumped with a transversely oriented standing-wave field (Fig. 13.5), and within a ring-cavity pumped on one of its running-wave modes (Fig. 13.6). Unlike in the single-mode cavity-driven optomechanical systems discussed in the previous Section, here, the system is driven by a field external to an otherwise undriven cavity mode. In the ring-cavity case, the undriven cavity mode (say, the right-going mode, as seen by the atoms) is one of the running wave modes, while the pump field is itself resonant within a second cavity mode (the left-going mode). However, for the case of a strong pump field and only a weak occupation of the undriven cavity modes, we may neglect the cavity dynamics of the pumped mode and consider it simply as an externally imposed field. My aim in this Section is to treat these situations in the language of cavity optomechanics, both to supplement other descriptions of these systems in terms of, for example, realizations of quantum phase transitions, supersolidity, superradiance, and Brillouin instabilities, and also to set a precedent for similar experimental pursuits with solid-state cavity optomechanics.

13.6.1 Cavity Optomechanics of a Side-Pumped Fabry-Pérot Resonator

Consider the side-pumped Fabry-Pérot resonator studied by the MIT and Zürich groups, shown schematically in Fig. 13.5. We treat the standing-wave pump field initially as a finite-volume cavity mode with atom-cavity coupling strength $g_p(\mathbf{r}) = g_p \sin kx$, photon annihilation operator $\hat{a}_p$, and resonant frequency ω_p. The coupling strength to the Fabry-Pérot cavity is $g(\mathbf{r}) = g_0 \sin kz$. In both cases we neglect the transverse variation of the field, assuming the atoms to be well confined at the intersection of the two fields. There are three forms of coherent scattering: forward scattering into the pump field, forward scattering into the Fabry-Pérot field, and scattering of photons between the two fields. Assuming a large optical detuning from the excited atomic state, the optomechanics of a *single atom* in the cavity is given by the following Hamiltonian:

$$
\begin{aligned}
H = &\hbar\omega_c \hat{a}^\dagger \hat{a} + \hbar\omega_p \hat{a}_p^\dagger \hat{a}_p + H_{\text{mech}} \\
&+ \hbar \frac{g_0^2}{\Delta_{\text{ca}}} \sin^2(kz)\ \hat{a}^\dagger \hat{a} + \hbar \frac{g_p^2}{\Delta_{\text{ca}}} \sin^2(kx)\ \hat{a}_p^\dagger \hat{a}_p \\
&+ \hbar \frac{g_p g_0}{\Delta_{\text{ca}}} \sin(kx)\ \sin(kz) \left(\hat{a}^\dagger \hat{a}_p + \hat{a}_p^\dagger \hat{a} \right)
\end{aligned}
\tag{13.15}
$$

Now, we assume the pump field is a coherent state, allowing us to substitute $\hat{a}_p \rightarrow e^{-i\omega_p t}\bar{a}_p$. Evaluating the cavity field in the probe's rotating frame, i.e. $\hat{a} \rightarrow e^{-i\omega_p t}\hat{a}$, we find

$$H = -\hbar\Delta\hat{a}^\dagger\hat{a} + H_{\text{mech}} + \hbar\frac{g_0^2}{\Delta_{\text{ca}}}\sin^2(kz)\ \hat{a}^\dagger\hat{a}$$
$$+\hbar\frac{G_p^2}{\Delta_{\text{ca}}}\sin^2(kx) + \hbar\frac{G_p g_0}{\Delta_{\text{ca}}}\sin(kx)\sin(kz)\left(\hat{a}^\dagger + \hat{a}\right) \tag{13.16}$$

where $G_p = g_p\bar{a}_p$ and we dispense with the pump-field cavity altogether.

As in Sect. 13.5.1, to simplify the mechanical description of the atomic gas, let us specify the initial state of the gas and consider only small perturbations from that initial state. Specifically, let us treat the case that the initial state is a non-interacting Bose-Einstein condensate. Strictly speaking, with the cavity field empty, this condensate resides in the lowest Bloch state of the x-oriented optical lattice formed by the standing-wave probe field (orientations are shown in Fig. 13.5). However, assuming this lattice to be weak, let us neglect it and consider the condensate to be spatially uniform; adapting our treatment to include this optical lattice is straightforward. To match the spatial structure of the photon-exchange coupling term (last term in Eqs. 13.15 and 13.17), we consider an excitation of the condensate with an annihilation operator of the form

$$\hat{b} = \frac{\hat{b}_{k(-\mathbf{x}+\mathbf{z})} + \hat{b}_{k(\mathbf{x}-\mathbf{z})} - \hat{b}_{k(\mathbf{x}+\mathbf{z})} - \hat{b}_{k(-\mathbf{x}-\mathbf{z})}}{2} \tag{13.17}$$

where numerator contains momentum-space field operators, with $\mathbf{x}$ and $\mathbf{z}$ being unit vectors. These momentum modes are all degenerate, with the excitation energy $\hbar\omega_M = \hbar k^2/m$.

Let us also neglect the lattice potential $\hbar(g_0^2/\Delta_{\text{ca}})\sin^2(kz)\hat{a}^\dagger\hat{a}$ formed by the cavity field itself, under the approximation that the pump field is far stronger than the cavity field. For weak excitations atop the condensate, we thus obtain

$$H = -\hbar\Delta\hat{a}^\dagger\hat{a} + \hbar\omega_M\hat{b}^\dagger\hat{b} + \hbar\frac{\sqrt{N}G_p g_0}{2\Delta_{\text{ca}}}\left(\hat{b}^\dagger + \hat{b}\right)\left(\hat{a}^\dagger + \hat{a}\right) \tag{13.18}$$

This expression matches the canonical optomechanical Hamiltonian under the linearization approximation (Eq. 13.9), except in that, here, the mean cavity field is zero. This connection between intracavity Brillouin instabilities and cavity optomechanics is discussed also in Ref. [115], focusing on aspects of ponderomotive amplification.

We are left with a very simple description of two harmonic oscillators coupled by a spring, a system whose normal modes are simply obtained from classical mechanics. Let us focus on the optomechanical frequency shift in the limit of a weak cavity field. If we approximate that the coupling frequency $\lambda = (\sqrt{N}G_p g_0)/(2\Delta_{\text{ca}})$ is small compared to the magnitude of the frequency difference $|\Delta| - |\omega_M|$, a condition satisfied in the experiments, we may apply second-order perturbation theory to determine the energy difference between the lowest phonon states, and obtain a frequency shift from ω_M by the amount

$$\delta\omega_M = \frac{\lambda^2}{\omega_M + \Delta}. \tag{13.19}$$

As in the cavity-driven situation, the mechanical frequency is shifted upward for pump light that is blue-detuned from the cavity resonance, and shifted downward for red-detuned pump light.

Such optomechanical spring effects were observed recently by the Zürich group, who describe the cavity-modified collective excitations of the Bose condensate as "roton-like" modes [116]. In the case of red detuning, as these modes are "softened" by the downward optomechanical frequency shift, indicating, as in liquid helium, the propensity of the fluid to solidify. For red detuning and a sufficiently strong pump field ($|\lambda| > \sqrt{\omega_M(|\Delta| - \omega_M)}$), the mechanical spring constant changes sign, and the gas undergoes a dynamical instability. As for an inverted harmonic oscillator, the "position" $\hat{b}^\dagger + \hat{b}$ grows exponentially toward either large positive or large negative values; these correspond to the even or odd checkerboard ("solid") patterns discussed in Sect. 13.4.1 and shown in Fig. 13.8. This instability thus breaks a $\mathbb{Z}_2$ symmetry of the initial mechanical state. The strength of the density modulation is stabilized by the energy terms ignored in our various approximations.

This optomechanical system also shows a strong analogy to the quantum Dicke model [117], which describes a collection of two-level atoms with equal electric-dipole coupling to a single cavity mode. In this analogy, developed in Refs. [6, 118], the collective mechanical position operator $\hat{b}^\dagger + \hat{b}$ plays the role of the dipole moment operator, coupled to the electric field $(\hat{a}^\dagger + \hat{a})$. The analogy breaks down for very large mechanical excitation, where rather than exhibiting saturation as one expects in the Dicke model, the mechanical system undergoes higher order Brillouin instability [9, 75].

The connection between cavity-mediated dynamics of a Bose-Einstein condensate and the Dicke model was anticipated by Refs. [119, 120], which treated the more general case of two bosonic modes coupled by light scattering involving a cavity-field photon and a driving-field (or a second cavity-field) photon. The model of Ref. [119] included also effects of self- and cross-interactions between the two bosonic modes, as well as a classical drive coupling the modes, as could be achieved in the cavity optomechanical system by illuminating the atoms both with a transverse pump field and also with a coherent cavity field. The interesting variation between first and second order transitions in their work may, therefore, be observed in optomechanics experiments similar to those of the MIT and Zürich groups.

One may ask whether the optomechanical instability described above and observed in experiments [6, 69, 78] is a realization of the quantum phase transition exhibited by the zero-temperature quantum Dicke model, or of, perhaps, a classical phase transition exhibited in a finite-temperature system. These situations are distinguished by whether quantum-mechanical or thermal fluctuations dominate near the phase transition. Drawing on lessons from cavity optomechanics, one suspects that the mechanical system is beset by thermal fluctuations that originate from measurement-induced mechanical diffusion. That is, the Fabry-Pérot field provides a constant measurement of the position of the mechanical oscillator. This constant measurement implies the presence of force fluctuations; here, the relevant fluctuations originate from the radiation-pressure shot noise coming from the interference of vacuum fluctuations of the cavity field with the coherent-state field of the pump. Indeed, recent measurements by the Zürich group indicate that the system shows mechanical fluctuations much larger than the predicted zero-point fluctuations, with the disparity growing larger as the mechanical mode is softened when one approaches the phase transition [121].

Whether or not this system displays quantum phase transitions, it is highly significant as an example of an *open* quantum system undergoing dynamical transitions under constant perturbation and measurement. Predicted phenomena include complex long-term dynamics and dynamical multistability [122, 123], and an alteration of the critical exponents for the phase transition in an open system [124].

13.6.2 Cavity Optomechanics in a Pumped Ring Cavity

For the ring-cavity setup (Fig. 13.6), let the atom-cavity coupling strength to the running wave modes be $g_\pm(z) \propto e^{\pm ikz}$, retaining only the spatial variation along the cavity axis. We focus on the process whereby an atom backscatters light from one running wave mode to the other, giving a single-atom coupling strength of the form

$$\frac{g_-^*(z) g_+(z)}{\Delta_{\mathrm{ca}}} \hat{a}_-^\dagger \hat{a}_+ + h.c. \tag{13.20}$$

where $\hat{a}_\pm$ are the optical field operators for the respective cavity modes. For the situation where one running-wave cavity mode (say the $+$ mode) is strongly driven and treated as a classical pump field with amplitude $e^{-i\omega_p t}\bar{a}_+$, this single-atom interaction term becomes

$$\frac{g_0^2 \bar{a}_+}{\Delta_{\mathrm{ca}}} \left(e^{2ikz} \hat{a}_- + e^{-2ikz} \hat{a}_-^\dagger \right), \tag{13.21}$$

where z is the atomic position, and $\bar{a}_+$ is real.

Let us again specify the initial state of the gas to be a uniform, stationary, and non-interacting Bose-Einstein condensate. The spatially dependent optomechanical interaction term (Eq. 13.21) couples this initial state to excitations at momenta $\pm 2\hbar k$, with the well-defined energy $\hbar\omega_{2k}$. Summing over all N atoms in the system and including the atomic excitation energy yields the following Hamiltonian:

$$H \simeq -\hbar\Delta\hat{a}^\dagger\hat{a} + \hbar\omega_{2k}\left(\hat{b}^\dagger_{+2k}\hat{b}_{+2k} + \hat{b}^\dagger_{-2k}\hat{b}_{-2k}\right) + \hbar\sqrt{N}\frac{g_0^2\bar{a}_+}{\Delta_{\rm ca}}\left[\left(\hat{b}^\dagger_{+2k} + \hat{b}_{-2k}\right)\hat{a}_- + \left(\hat{b}_{+2k} + \hat{b}^\dagger_{-2k}\right)\hat{a}^\dagger_-\right] \tag{13.22}$$

In the case of the side-pumped Fabry-Pérot cavity, the checkerboard patterns favored by collective scattering emerge at fixed positions determined by the overlap of the two standing-wave fields. In the case of the ring cavity, the position of the emergent standing wave pattern is not fixed. To exhibit this fact, let us rewrite the optomechanical coupling term above in terms of the cosine and sine spatial modulation patterns; that is, we define $\hat{b}_{\cos} = (\hat{b}_{2k} + \hat{b}_{-2k})/\sqrt{2}$ and $\hat{b}_{\sin} = (\hat{b}_{2k} - \hat{b}_{-2k})/(i\sqrt{2})$ and obtain the optomechanical coupling term as

$$\hbar\sqrt{2N}\frac{g_0^2\bar{a}_+}{\Delta_{\rm ca}}\left[\left(\hat{b}^\dagger_{\cos} + \hat{b}_{\cos}\right)\left(\hat{a} + \hat{a}^\dagger\right) + \left(\hat{b}^\dagger_{\sin} + \hat{b}_{\sin}\right)\left(\frac{\hat{a} - \hat{a}^\dagger}{i}\right)\right] \tag{13.23}$$

The two independent quadratures of the cavity field are coupled to two independent harmonic oscillators. The optomechanical frequency shift and the condition for optomechanical instability are treated similarly as for the Fabry-Pérot cavity setup discussed above. The instability besetting the two mechanical modes is similar to that of a particle in a two-dimensional degenerate inverted harmonic oscillator potential. The mechanical system spontaneously breaks a $U(1)$ symmetry by forming a density modulation at an offset position defined to within the modulation wavelength. This symmetry breaking also selects the $U(1)$ phase of the backscattered running-wave cavity mode.

In the symmetry-broken phase, one cannot ignore the dynamical back action on the pumped cavity mode. The phenomenology of nonlinear cavity optics in this system is quite rich, including cavity bistability, as observed by Elsässer et al. [125], and predicted regions of multi-stability and disconnected regions ("isolas") of stable cavity operation that are flanked by unstable modes of operation [126, 127].

13.6.3 Optomechanics in Side-Pumped Multi-mode Cavities

These experiments have given rise to theoretical proposals for continuous optomechanical systems with greater dynamical complexity. Notably, Gopalakrishnan et al. [128] considered the low-energy states of an atomic gas that is placed within an optical cavity with a high degeneracy of cavity modes, such as a concentric or confocal Fabry-Pérot resonator, and illuminated by a standing wave of light. The gas is susceptible to Brillouin instabilities into several cavity modes, each of which is supported by a distinct spatial pattern of the gas medium. In this case, the gas may spontaneously adopt one of several distinct spatial (potentially "supersolid") distributions, as would be evident in the spatial mode of light emitted by the cavity. In this sense, the "crystallinity" of the gas, rather than being externally imposed upon the atoms by an optical lattice of fixed geometry, now emerges due to the internal dynamics of the atoms-cavity system. Alternately, different portions of the gas may support collective scattering into different cavity modes, producing a multi-mode optical output and a spatially modulated density pattern interrupted by dislocations or domain walls. Applying this idea to the internal (spin) rather than the external (center-of-mass motion) degrees of freedom of the gas, this realization of the multi-mode Dicke model may produce spin-glass phases in a highly tunable, and essentially open, many-body quantum system [129, 130].

In the language of cavity optomechanics, these proposals relate to a system in which several mechanical modes of the same medium are coupled to several optical modes. Studies of multi-mode solid-state cavity optomechanics have only recently begun. One expects both the atomic- and solid-state-based investigations of multi-mode optomechanical systems to yield new insight on optomechanical effects such as phonon lasing, amplification, synchronization and gain saturation; coherent energy transfer between mechanical modes; cavity-induced cooling; ponderomotive amplification and squeezing; etc.

13.7 Future Directions

The clearest imperative for future research on cold-atomic cavity optomechanics is the exploration of optomechanics in the granular regime. As discussed in Sect. 13.5.8, current experimental setups are sufficient to reach this regime. Moderate increases in the mechanical trapping frequency and the cavity decay time will allow one to reach also the resolved sideband regime, so that distinctly nonlinear single-photon effects, such as photon antibunching [107], and other departures from weak-coupling cavity optomechanics may be observed.

Beyond this, theoretical studies indicate that novel optomechanical phenomena may arise by enriching the optomechanical system by adding multiple mechanical or optical modes. Additionally, one may consider situations where the physical state of the mechanical medium, now considered as an interacting many-body system, affects or is affected by the cavity optomechanical interaction. For instance, the equation of state of the gaseous mechanical medium changes when the gas undergoes a metal/superfluid to insulator transition. This phase change may influence the cavity field, e.g. due to a change in the compressibility of the medium. In turn, the cavity field, by changing the optical potential in which the gas resides, may itself drive the phase transition of the gas. The self-consistent phase diagram of this atoms-cavity system has been considered for the case of the single-band Bose-Hubbard model, showing features that are distinct from those observed for atoms in free-space lattices [131, 132]. Another interesting intracavity medium to consider is the one-dimensional strongly interacting Bose gas, which shows a strong susceptibility to being "pinned" by weak periodic potentials, thereby enhancing the optomechanical response [133]. It will be interesting to examine how such many-body phases and phase transitions influence the dynamics of optomechanical systems.

Finally, several works have explored the analogy between the motional dynamics and the internal-state dynamics of atoms in a cavity. By this analogy, new phenomena have been predicted to arise from the parametric coupling between spin ensembles and a single-mode cavity field, such as cavity magneto-optical bistability and spontaneous cavity birefringence, coherent amplification and damping of the collective spin, and the generation of inhomogeneously squeezed light [134]. It will be interesting to explore such phenomena experimentally with both atomic and solid-state spin ensembles.

Acknowledgments The author is deeply grateful to his co-researchers on cQED with cold atoms, whose persistence, curiosity and keen insight led to the development of the optomechanics picture for describing the interactions of trapped gases with single-mode optical cavities. This team includes Thierry Botter, Nathaniel Brahms, Daniel Brooks, Subhadeep Gupta, Zhao-Yuan Ma, Kevin Moore, Kater Murch, Sydney Schreppler, and Thomas Purdy. Additional contributions to the development of the experimental apparatus were made by Kevin Brown, Keshav Dani, Marilena LoVerde, and Guilherme Miranda. I am thankful to T. Esslinger, H.J. Kimble, G. Rempe, H. Ritsch, V. Vuletić, and C. Zimmermann for permission to use figures from their work, and also to A. Nunnenkamp and to P. Rabl for critical readings of the manuscript. Financial support for our research was provided by the DARPA QuIST program, the NSF, the David and Lucile Packard Foundation, a critical seedling grant from DARPA through the AFOSR, and the AFOSR directly.

References

1. S. Chu, Rev. Mod. Phys. **70**, 685 (1998)
2. C. Cohen-Tannoudji, Rev. Mod. Phys. **70**, 707 (1998)
3. W. Phillips, Rev. Mod. Phys. **70**, 721 (1998)
4. D.W.C. Brooks, T. Botter, S. Schreppler, T.P. Purdy, N. Brahms, D.M. Stamper-Kurn, Nature **488**, 476 (2012)
5. N. Brahms, T. Botter, S. Schreppler, D.W.C. Brooks, D.M. Stamper-Kurn, Phys. Rev. Lett. **108**, 133601 (2012)
6. K. Baumann, C. Guerlin, F. Brennecke, T. Esslinger, Nature **464**, 1301 (2010)
7. J.P. Gordon, A. Ashkin, Phys. Rev. A **21**, 1606 (1980)
8. J. Dalibard, C. Cohen-Tannoudji, J. Opt. Sci. Am. B **2**, 1707 (1985)
9. S. Inouye, A. Chikkatur, D. Stamper-Kurn, J. Stenger, W. Ketterle, Science **285**, 571 (1999)
10. C. Salomon, J. Dalibard, A. Aspect, H. Metcalf, C. Cohen-Tannoudji, Phys. Rev. Lett. **59**, 1659 (1987)
11. J.H. Thywissen, M. Prentiss, New J. Phys. **7**, 47 (2005)
12. J.E. Thomas, Opt. Lett. **14**, 1186 (1989)
13. S. Kunze, G. Rempe, M. Wilkens, Europhys. Lett. **27**, 115 (1994)
14. K.D. Stokes, C. Schnurr, J.R. Gardner, M. Marable, G.R. Welch, J.E. Thomas, Phys. Rev. Lett. **67**, 1997 (1991)
15. J.R. Gardner, M.L. Marable, G.R. Welch, J.E. Thomas, Phys. Rev. Lett. **70**, 3404 (1993)
16. P. Storey, M. Collett, D. Walls, Phys. Rev. Lett. **68**, 472 (1992)
17. M.A.M. Marte, P. Zoller, Appl. Phys. B: Photophys. Laser Chem. **54**, 477 (1992)
18. P. Storey, M. Collett, D. Walls, Phys. Rev. A **47**, 405 (1993)
19. A.M. Herkommer, H.J. Carmichael, W.P. Schleich, Quantum and semiclassical optics. J. Eur. Opt. Soc. Part B **8**, 189 (1996)
20. H. Mabuchi, Quantum and semiclassical optics. J. Eur. Opt. Soc. Part B **8**, 1103 (1996)
21. G. Rempe, App. Phys. B **60**, 233 (1995)

22. R. Quadt, M. Collett, D.F. Walls, Phys. Rev. Lett. **74**, 351 (1995)
23. J.A. Dunningham, H.M. Wiseman, D.F. Walls, Phys. Rev. A **55**, 1398 (1997)
24. H. Mabuchi, Q.A. Turchette, M.S. Champan, H.J. Kimble, Opt. Lett. **21**, 1393 (1996)
25. P. Münstermann, T. Fischer, P.W.H. Pinkse, G. Rempe, Opt. Commun. **159**, 63 (1999)
26. C. Hood, T. Lynn, A. Doherty, A. Parkins, H. Kimble, Science **287**, 1447 (2000)
27. A.C. Doherty, A.S. Parkins, S.M. Tan, D.F. Walls, Phys. Rev. A **56**, 833 (1997)
28. A.C. Doherty, A.S. Parkins, S.M. Tan, D.F. Walls, Phys. Rev. A **57**, 4804 (1998)
29. A.C. Doherty, T.W. Lynn, C.J. Hood, H.J. Kimble, Phys. Rev. A **63**, 013401 (2001)
30. P. Horak, G. Hechenblaikner, K.M. Gheri, H. Stecher, H. Ritsch, Phys. Rev. Lett. **79**, 4974 (1997)
31. G. Hechenblaikner, M. Gangl, P. Horak, H. Ritsch, Phys. Rev. A **58**, 3030 (1998)
32. V. Braginsky, A. Manukin, *Measurement of Weak Forces in Physics Experiments* (University of Chicago Press, Chicago and London, 1977)
33. A.T. Black, J.K. Thompson, V. Vuletic, J. Phys. B **38**, S605 (2005)
34. V. Vuletić, S. Chu, Phys. Rev. Lett. **84**, 3787 (2000)
35. T.W. Mossberg, M. Lewenstein, D.J. Gauthier, Phys. Rev. Lett. **67**, 1723 (1991)
36. K. Murr, Phys. Rev. Lett. **96**, 253001 (2006)
37. V. Vuletić, H.W. Chan, A.T. Black, Phys. Rev. A **64**, 033405 (2001)
38. O. Arcizet, P.F. Cohadon, T. Briant, M. Pinard, A. Heidmann, Nature **444**, 71 (2006)
39. S. Gigan, H.R. Bohm, M. Paternostro, F. Blaser, G. Langer, J.B. Hertzberg, K.C. Schwab, D. Bauerle, M. Aspelmeyer, A. Zeilinger, Nature **444**, 67 (2006)
40. A. Schliesser, P. Del'Haye, N. Nooshi, K.J. Vahala, T.J. Kippenberg, Phys. Rev. Lett. **97**, 243905 (2006)
41. A. Naik, O. Buu, M.D. LaHaye, A.D. Armour, A.A. Clerk, M.P. Blencowe, K.C. Schwab, Nature **443**, 193 (2006)
42. P. Münstermann, T. Fischer, P. Maunz, P. Pinkse, G. Rempe, Phys. Rev. Lett. **82**, 3791 (1999)
43. P. Maunz, T. Puppe, I. Schuster, N. Syassen, P.W.H. Pinkse, G. Rempe, Nature **428**, 50 (2004)
44. S. Nuszmann, K. Murr, M. Hijlkema, B. Weber, A. Kuhn, G. Rempe, Nat. Phys. **1**, 122 (2005)
45. D.R. Leibrandt, J. Labaziewicz, V. Vuletic, I.L. Chuang, Phys. Rev. Lett. **103**, 103001 (2009)
46. J.I. Cirac, A.S. Parkins, R. Blatt, P. Zoller, Opt. Commun. **97**, 353 (1993)
47. J.I. Cirac, M. Lewenstein, P. Zoller, Phys. Rev. A **51**, 1650 (1995)
48. I. Wilson-Rae, N. Nooshi, W. Zwerger, T.J. Kippenberg, Phys. Rev. Lett. **99**, 093901 (2007)
49. F. Marquardt, J.P. Chen, A.A. Clerk, S.M. Girvin, Phys. Rev. Lett. **99**, 093902 (2007)
50. K.W. Murch, K.L. Moore, S. Gupta, D.M. Stamper-Kurn, Nat. Phys. **4**, 561 (2008)
51. K. Murr, P. Maunz, P.W.H. Pinkse, T. Puppe, I. Schuster, D. Vitali, G. Rempe, Phys. Rev. A **74**, 043412 (2006)
52. P. Maunz, T. Puppe, I. Schuster, N. Syassen, P. Pinkse, G. Rempe, Phys. Rev. Lett. **94**, 033002 (2005)
53. T. Puppe, I. Schuster, P. Maunz, K. Murr, P.W.H. Pinkse, G. Rempe, J. Mod. Opt. **54**, 1927 (2007)
54. P.F. Cohadon, A. Heidmann, M. Pinard, Phys. Rev. Lett. **83**, 3174 (1999)
55. N.V. Morrow, S.K. Dutta, G. Raithel, Phys. Rev. Lett. **88**, 093003 (2002)
56. G. Raithel, W.D. Phillips, S.L. Rolston, Phys. Rev. Lett. **81**, 3615 (1998)
57. D.A. Steck, K. Jacobs, H. Mabuchi, T. Bhattacharya, S. Habib, Phys. Rev. Lett. **92**, 223004 (2004)
58. D.A. Steck, K. Jacobs, H. Mabuchi, S. Habib, T. Bhattacharya, Phys. Rev. A **74**, 012322 (2006)
59. J.D. Thompson, B.M. Zwickl, A.M. Jayich, F. Marquardt, S.M. Girvin, J.G.E. Harris, Nature **452**, 72 (2008)
60. J.C. Sankey, C. Yang, B.M. Zwickl, A.M. Jayich, J.G.E. Harris, Nat. Phys. **6**, 707 (2010)
61. T. Purdy, D. Brooks, T. Botter, N. Brahms, Z.Y. Ma, D. Stamper-Kurn, Phys. Rev. Lett. **105**, 133602 (2010)
62. V. Vuletić, J.K. Thompson, A.T. Black, J. Simon, Phys. Rev. A **75**, 051405(R) (2007)

63. A. Kubanek, M. Koch, C. Sames, A. Ourjoumtsev, P.W.H. Pinkse, K. Murr, G. Rempe, Nature **462**, 898 (2009)
64. M. Koch, C. Sames, A. Kubanek, M. Apel, M. Balbach, A. Ourjoumtsev, P.W.H. Pinkse, G. Rempe, Phys. Rev. Lett. **105**, 173003 (2010)
65. T. Fischer, P. Maunz, T. Puppe, P.W.H. Pinkse, G. Rempe, New J. Phys. **11**, 1367 (2001)
66. J.K. Asboth, P. Domokos, H. Ritsch, Phys. Rev. A **70**, 013414 (2004)
67. M. Gangl, H. Ritsch, Phys. Rev. A **61**, 011402 (1999)
68. P. Horak, H. Ritsch, Phys. Rev. A **64**, 033422 (2001)
69. A.T. Black, H.W. Chan, V. Vuletić, Phys. Rev. Lett. **91**, 203001 (2003)
70. P. Domokos, H. Ritsch, Phys. Rev. Lett. **89**, 253003 (2002)
71. Y. Yoshikawa, Y. Torii, T. Kuga, Phys. Rev. Lett. **94**, 083602 (2005)
72. P. Wang, L. Deng, E.W. Hagley, Z. Fu, S. Chai, J. Zhang, Phys. Rev. Lett. **106**, 210401 (2011)
73. R. Bonifacio, L. De Salvo, Nucl. Instrum. Methods Phys. Res. Sect. A **341**, 360 (1994)
74. R. Bonifacio, L. De Salvo, L.M. Narducci, E.J. DAngelo, Phys. Rev. A **50**, 1716 (1994)
75. S. Slama, S. Bux, G. Krenz, C. Zimmermann, P.W. Courteille, Phys. Rev. Lett. **98**, 053603 (2007)
76. S. Slama, G. Krenz, S. Bux, C. Zimmermann, P.W. Courteille, Phys. Rev. A **75**, 063620 (2007)
77. S. Bux, C. Gnahm, R.A.W. Maier, C. Zimmermann, P.W. Courteille, Phys. Rev. Lett. **106**, 203601 (2011)
78. K. Baumann, R. Mottl, F. Brennecke, T. Esslinger, Phys. Rev. Lett. **107**, 140402 (2011)
79. F. Brennecke, S. Ritter, T. Donner, T. Esslinger, Science **322**, 235 (2008)
80. S. Gupta, K.L. Moore, K.W. Murch, D.M. Stamper-Kurn, Phys. Rev. Lett. **99**, 213601 (2007)
81. T. Botter, D. Brooks, S. Gupta, Z.Y. Ma, K.L. Moore, K.W. Murch, T.P. Purdy, D.M. Stamper-Kurn, in *Quantum Micro-mechanics with Ultracold Atoms* (World Scientific, Singapore, 2009), pp. 117–130
82. M.H. Schleier-Smith, I.D. Leroux, H. Zhang, M.A. Van Camp, V. Vuletic, Phys. Rev. Lett. **107**, 143005 (2011)
83. N. Brahms, T.P. Purdy, D.W.C. Brooks, T. Botter, D.M. Stamper-Kurn, Nat. Phys. **7**, 604 (2011)
84. D. Stamper-Kurn, A. Chikkatur, A. Görlitz, S. Inouye, S. Gupta, D. Pritchard, W. Ketterle, Phys. Rev. Lett. **83**, 2876 (1999)
85. R. Kanamoto, P. Meystre, Phys. Rev. Lett. **104**, 063601 (2010)
86. T. Botter, D.W.C. Brooks, N. Brahms, S. Schreppler, D.M. Stamper-Kurn, Phys. Rev. A **85**, 013812 (2012)
87. B.S. Sheard, M.B. Gray, C.M. Mow-Lowry, D.E. McClelland, S.E. Whitcomb, Phys. Rev. A **69**, 051801 (2004)
88. T. Corbitt, D. Ottaway, E. Innerhofer, J. Pelc, N. Mavalvala, Phys. Rev. A **74**, 021802 (2006)
89. T. Corbitt, Y.B. Chen, E. Innerhofer, H. Muller-Ebhardt, D. Ottaway, H. Rehbein, D. Sigg, S. Whitcomb, C. Wipf, N. Mavalvala, Phys. Rev. Lett. **98**, 150802 (2007)
90. S. Ritter, F. Brennecke, K. Baumann, T. Donner, C. Guerlin, T. Esslinger, App. Phys. B **95**, 213 (2009)
91. A.D. OConnell, M. Hofheinz, M. Ansmann, R.C. Bialczak, M. Lenander, E. Lucero, M. Neeley, D. Sank, H. Wang, M. Weides, J. Wenner, J.M. Martinis, A.N. Cleland, Nature **464**, 697 (2010)
92. A. Dorsel, J.D. Mccullen, P. Meystre, E. Vignes, H. Walther, Phys. Rev. Lett. **51**, 1550 (1983)
93. T. Griesser, H. Ritsch, Opt. Express **19**, 11242 (2011)
94. F. Marino, F.S. Cataliotti, A. Farsi, M.S. de Cumis, F. Marin, Phys. Rev. Lett. **104**, 073601 (2010)
95. P. Verlot, A. Tavernarakis, T. Briant, P.F. Cohadon, A. Heidmann, Phys. Rev. Lett. **104**, 133602 (2010)
96. S. Weis, R. Rivière, S. Deléglise, E. Gavartin, O. Arcizet, A. Schliesser, T.J. Kippenberg, Science **330**, 1520 (2010)
97. A.H. Safavi-Naeini, T.P.M. Alegre, J. Chan, M. Eichenfield, M. Winger, Q. Lin, J.T. Hill, D.E. Chang, O. Painter, Nature **472**, 69 (2011)

98. C. Fabre, M. Pinard, S. Bourzeix, A. Heidmann, E. Giacobino, S. Reynaud, Phys. Rev. A **49**, 1337 (1994)
99. S. Mancini, P. Tombesi, Phys. Rev. A **49**, 4055 (1994)
100. H.J. Kimble, Y. Levin, A.B. Matsko, K.S. Thorne, S.P. Vyatchanin, Phys. Rev. D **65**, 022002 (2001)
101. F. Diedrich, J.C. Bergquist, W.M. Itano, D.J. Wineland, Phys. Rev. Lett. **62**, 403 (1989)
102. A.D. Boozer, A. Boca, R. Miller, T.E. Northup, H.J. Kimble, Phys. Rev. Lett. **97**, 083602 (2006)
103. R. Islam, E.E. Edwards, K. Kim, S. Korenblit, C. Noh, H. Carmichael, G.D. Lin, L.M. Duan, C.C.J. Wang, J.K. Freericks, C. Monroe, Nat. Commun. **2**, 1374 (2011)
104. T. Monz, P. Schindler, J.T. Barreiro, M. Chwalla, D. Nigg, W.A. Coish, M. Harlander, W. Hansel, M. Hennrich, R. Blatt, Phys. Rev. Lett. **106**, 130506 (2011)
105. A.H. Safavi-Naeini, J. Chan, J.T. Hill, T.P.M. Alegre, A. Krause, O. Painter, Phys. Rev. Lett. **108**, 033602 (2012)
106. M. Eichenfield, J. Chan, R.M. Camacho, K.J. Vahala, O. Painter, Nature **462**, 78 (2009)
107. P. Rabl, Phys. Rev. Lett. **107**, 063601 (2011)
108. A. Nunnenkamp, K. Borkje, S.M. Girvin, Phys. Rev. Lett. **107**, 063602 (2011)
109. J.Q. Liao, H.K. Cheung, C.K. Law, Phys. Rev. A **85**, 025803 (2012)
110. M. Ludwig, B. Kubala, F. Marquardt, New J. Phys. **10**, 095013 (2008)
111. J. Qian, A.A. Clerk, K. Hammerer, F. Marquardt, Phys. Rev. Lett. **109**, 253601 (2012)
112. A. Kronwald, M. Ludwig, F. Marquardt, Phys. Rev. A **87**, 013847 (2013)
113. T. Hong, H. Yang, H. Miao, Y. Chen, Phys. Rev. A **88**, 023812 (2013)
114. W. Marshall, C. Simon, R. Penrose, D. Bouwmeester, Phys. Rev. Lett. **91**, 130401 (2003)
115. G. Szirmai, D. Nagy, P. Domokos, Phys. Rev. A **81**, 043639 (2010)
116. R. Mottl, F. Brennecke, K. Baumann, R. Landig, T. Donner, T. Esslinger, Science **336**, 1570 (2012)
117. R.H. Dicke, Phys. Rev. **93**, 99 (1954)
118. D. Nagy, G. Kónya, G. Szirmai, P. Domokos, Phys. Rev. Lett. **104**, 130401 (2010)
119. G. Chen, X.G. Wang, J.Q. Liang, Z.D. Wang, Phys. Rev. A **78**, 023634 (2008)
120. S. Morrison, A.S. Parkins, Phys. Rev. Lett. **100**, 040403 (2008)
121. F. Brennecke, R. Mottl, K. Baumann, R. Landig, T. Donner, T. Esslinger, PNAS **110**, 11763 (2013)
122. J. Keeling, M.J. Bhaseen, B.D. Simons, Phys. Rev. Lett. **105**, 043001 (2010)
123. M.J. Bhaseen, J. Mayoh, B.D. Simons, J. Keeling, Phys. Rev. A **85**, 013817 (2012)
124. D. Nagy, G. Szirmai, P. Domokos, Phys. Rev. A **84**, 043637 (2011)
125. T. Elsasser, B. Nagorny, A. Hemmerich, Phys. Rev. A **69**, 033403 (2004)
126. W. Chen, D.S. Goldbaum, M. Bhattacharya, P. Meystre, Phys. Rev. A **81**, 053833 (2010)
127. S.K. Steinke, P. Meystre, Phys. Rev. A **84**, 023834 (2011)
128. S. Gopalakrishnan, B.L. Lev, P.M. Goldbart, Nat. Phys. **5**, 845 (2009)
129. S. Gopalakrishnan, B.L. Lev, P.M. Goldbart, Phys. Rev. Lett. **107**, 277201 (2011)
130. P. Strack, S. Sachdev, Phys. Rev. Lett. **107**, 277202 (2011)
131. W. Chen, K. Zhang, D.S. Goldbaum, M. Bhattacharya, P. Meystre, Phys. Rev. A **80**, 011801 (2009)
132. S. Fernandez-Vidal, G. De Chiara, J. Larson, G. Morigi, Phys. Rev. A **81**, 043407 (2010)
133. Q. Sun, X.H. Hu, W.M. Liu, X.C. Xie, A.C. Ji, Phys. Rev. A **84**, 023822 (2011)
134. N. Brahms, D. Stamper-Kurn, Phys. Rev. A **82**, 041804(R) (2010)

Chapter 14
Hybrid Mechanical Systems

Philipp Treutlein, Claudiu Genes, Klemens Hammerer, Martino Poggio and Peter Rabl

Abstract We discuss hybrid systems in which a mechanical oscillator is coupled to another (microscopic) quantum system, such as trapped atoms or ions, solid-state spin qubits, or superconducting devices. We summarize and compare different coupling schemes and describe first experimental implementations. Hybrid mechanical systems enable new approaches to quantum control of mechanical objects, precision sensing, and quantum information processing.

14.1 Introduction

The ability of functionalized mechanical systems to respond to electric, magnetic and optical forces has in the past led to widespread applications of mechanical resonators as sensitive force detectors. With improved technology the same principle will apply for resonators in the quantum regime and allow the integration of mechanical

P. Treutlein (✉) · M. Poggio
Department of Physics, University of Basel, Basel, Switzerland
e-mail: philipp.treutlein@unibas.ch

M. Poggio
e-mail: martino.poggio@unibas.ch

C. Genes
Institute for Theoretical Physics, University of Innsbruck, Innsbruck, Austria
e-mail: Claudiu.Genes@uibk.ac.at

K. Hammerer
Institute for Theoretical Physics and Institute for Gravitational Physics, University of Hannover, Hannover, Germany
e-mail: Klemens.Hammerer@itp.uni-hannover.de

P. Rabl
Institute of Atomic and Subatomic Physics, Vienna University of Technology, Vienna, Austria
e-mail: peter.rabl@ati.ac.at

M. Aspelmeyer et al. (eds.), *Cavity Optomechanics*, Quantum Science and Technology, DOI: 10.1007/978-3-642-55312-7_14, oscillators with a large variety of other (microscopic) quantum systems such as atoms and ions, electronic spins, or quantized charge degrees of freedom. The benefits of such hybrid quantum systems are quite diverse. On the one hand, the motion of the resonator can be used as a sensitive probe and readout device for static and dynamic properties of the quantum system. On the other hand, coupling the resonator to a coherent and fully controllable two-level system provides a way to prepare and detect non-classical states of mechanical motion. Finally, the mechanical system can serve as a quantum transducer to mediate interactions between physically quite distinct quantum systems. This can be used to coherently couple e.g. an electronic spin to charge or optical degrees of freedom with various potential applications in the context of (hybrid) quantum information processing.

From a practical point of view the combination of mechanical resonators with microscopic quantum systems faces considerable challenges. Often the functionalization of mechanical resonators with electrodes, magnets, or mirrors competes with the requirement of a small mass to achieve a sufficient coupling strength on a single-quantum level. At the same time both the resonator and the other quantum system must be exceptionally well isolated from the environment to avoid decoherence. Various hybrid setups have been proposed that address those challenges, and some have already been implemented in experiments. This includes solid-state systems such as spin qubits, quantum dots, and superconducting devices, as well as atomic systems such as trapped atoms, ions, and molecules. In this chapter we give a brief overview of the different approaches towards mechanical hybrid quantum systems and discuss some basic examples from the fields of solid-state and atomic physics.

14.2 Solid-State Quantum Systems Coupled to Mechanics

Within the field of solid-state physics, a large variety of microscopic two- or few-level systems have been identified that are well isolated from the environment and allow for a coherent manipulation of their quantum state. Examples range from electronic or nuclear spin states associated with naturally occurring defect centers [1] to electronic states of so-called artificial atoms such as quantum dots [2] or superconducting Josephson devices [3]. Nanomechanical systems are naturally integrated with such solid-state quantum systems by fabricating them on the same chip, where they may interact with spins or charges via strong magnetic or electric forces. In contrast to most of the atomic implementations described below, the system dimensions are usually not limited by optical properties or trapping requirements, and without those restrictions strong interactions between an individual two-level system and a mechanical mode can be achieved more easily. On the other hand, it is more challenging to achieve long coherence times in the solid state. In combination with cryogenic temperatures the solid-state approach to mechanical hybrid systems offers a promising route towards manipulating mechanical motion on a single-phonon level.

We first present a brief overview of different physical mechanisms that have been suggested for achieving strong coupling between solid-state systems and mechanical

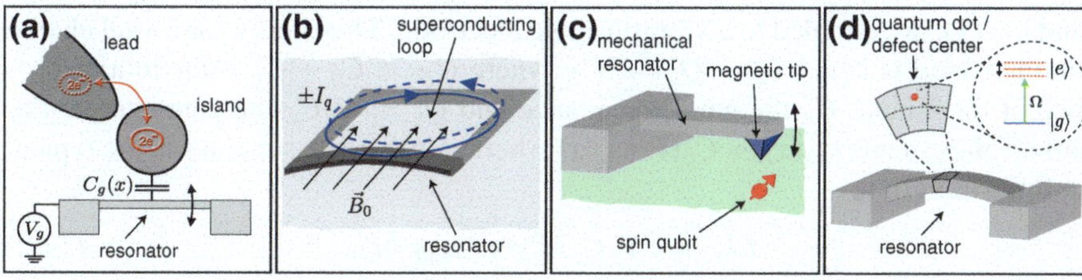

Fig. 14.1 Different schemes for coupling solid-state qubits and mechanical resonators. **a** Electrostatic coupling to charge qubits. **b** Lorentz force interactions with current states of a flux qubit. **c** Magnetic coupling to spins. **d** Deformation potential coupling to quantum dots or defect centers

motion. For the specific examples of superconducting charge qubits and electronic spin qubits we then describe mechanical sensing techniques and quantum control schemes as basic applications of these systems in the weak and strong interaction regime.

14.2.1 Overview of Systems and Coupling Mechanisms

The coupling of mechanical motion to other harmonic oscillators has been treated in other chapters of this book and we restrict the discussion in this section to microscopic two-level systems with a ground and excited state $|g\rangle$ and $|e\rangle$. In solid-state systems the energy separation E_{eg} between the two states is strongly dependent on the local electrostatic and magnetic fields, which also provides a way to couple to mechanical motion. For example, by fabricating an oscillating electrode or a vibrating magnetic tip, the system energy $E_{eg}(\hat{x}) = E_{eg}^0 + \partial_x E_{eg}\hat{x} + (1/2)\partial_x^2 E_{eg}\hat{x}^2 + \cdots$ now explicitly depends on the resonator displacement $\hat{x} = x_{\text{ZPF}}(\hat{b}+\hat{b}^\dagger)$. Even for nanoscale devices the zero point motion $x_{\text{ZPF}} \approx 10^{-13}$ m is still much smaller than other system dimensions and corrections beyond the linear coupling term are usually negligible. Therefore, the generic Hamiltonian for the qubit-resonator system is given by

$$\hat{H}(t) = \hat{H}_q(t) + \hbar\Omega_{\text{M}}\hat{b}^\dagger\hat{b} + \hbar\lambda(\hat{b} + \hat{b}^\dagger)\hat{\sigma}_z, \tag{14.1}$$

where $\hat{\sigma}_z = |e\rangle\langle e| - |g\rangle\langle g|$ is the Pauli operator and $\hat{H}_q(t)$ denotes the unperturbed Hamiltonian for the solid-state qubit. The relevant parameter in Eq. (14.1) is the coupling strength $\lambda = \partial_x E_{eg} x_{\text{ZPF}}/2\hbar$, which is the frequency shift per vibrational quantum.

While the basic form of the interaction in Eq. (14.1) has been derived from quite general considerations, the origin and the magnitude of the qubit-resonator coupling λ depends on the specific physical implementation. Figure 14.1 illustrates four basic mechanisms for coupling different charge and spin qubits to mechanical motion. In Fig. 14.1a two states encoded in quantized charge degrees of freedom, e.g. an electron on a quantum dot [4–7] or a Cooper pair on a small superconducting island [8–11], are coupled to a vibrating gate electrode. The energy for a total charge Q on the island is $E_Q = (Q - Q_g)^2/2C_\Sigma$ where $C_\Sigma = C_0 + C_g$ is the total capacitance of the island, C_g the gate capacitance and $Q_g = V_g C_g$ the gate charge. For small displacements $C_g(x) \approx C_g(1 - x/d)$ where d is the gate separation, the typical coupling strength is

$$\hbar\lambda_{el} \approx eV_g C_g/C_\Sigma \times x_{\rm ZPF}/d. \tag{14.2}$$

For $d \approx 100$ nm and voltages up to $V_g = 10$ V this coupling is quite substantial and can reach values in the range of $\lambda_{el}/2\pi \approx 5 - 50$ MHz [8].

Instead of using charge states, a two-level system can alternatively be encoded in clockwise and anti-clockwise circulating currents in a superconducting loop [12–17] as shown in Fig. 14.1b. Here an interaction with a freely suspended arm of the loop can arise from the Lorentz force created by a magnetic field B_0 perpendicular to the bending motion [14–16]. For circulating currents of magnitude I_q and a length l of the resonator we obtain

$$\hbar\lambda_{Lor} \approx B_0 I_q l x_{\rm ZPF}. \tag{14.3}$$

Although the applied magnetic field is limited by the critical field of the superconductor, $B_0 \leq 10\,$mT, typical values of $I_q \approx 100$ nA and $l = 5\ \mu$m still result in a coupling strength of $\lambda_{Lor}/2\pi \approx 0.1 - 1\,$MHz [14–16].

Qubits encoded in electronic or nuclear spin states can be coupled to the motion of a magnetized tip [18–22] as shown in Fig. 14.1c. Here strong magnetic field gradients ∇B lead to a position dependent Zeeman splitting of the spin states and for an electron spin,

$$\hbar\lambda_{mag} \approx g_s \mu_B x_{\rm ZPF} \nabla B/2, \tag{14.4}$$

where μ_B is the Bohr magneton and $g_s \approx 2$. On the scale of a few nanometers, magnetic field gradients can be as high as $\nabla B \sim 10^7$ T/m [23, 24], which corresponds to a coupling strength of $\lambda_{mag}/2\pi \approx 10 - 100$ kHz [22]. Due to a smaller magnetic moment, the coupling to a single nuclear spin is reduced by a factor $\sim 10^{-3}$, but is partially compensated by the much longer coherence times of nuclear spin qubits.

Mechanical resonators cannot only modulate the configuration of externally applied fields, but for example also couple to quantum dots or defect centers by changing the local lattice configuration of the host material [25]. This deformation potential coupling is illustrated in Fig. 14.1d where flexural vibrations of the resonator induce a local stress $\sigma \sim z_0 x_{\rm ZPF}/l^2$, where l is the resonator length and z_0 the distance of the defect from the middle of the beam. The corresponding level shift is given by

$$\hbar\lambda_{def} \approx (D_e - D_g) z_0 x_{\rm ZPF}/l^2, \tag{14.5}$$

where D_e and D_g are deformation potential constants for the ground and excited electronic states. For quantum dots a coupling strength of $\lambda_{def}/2\pi \approx 1 - 10$ MHz can be achieved [25], but competes with radiative decay processes of the same order.

In summary, this brief overview shows that various different mechanisms lead to interactions between solid-state two-level systems and mechanical resonators. In many cases the single-phonon coupling strength λ can be comparable to or even exceed the typical decoherence rate of the qubit T_2^{-1} as well as the mechanical heating rate $\Gamma_{\rm th} = k_B T/\hbar Q$. As we describe now in more detail this enables various applications ranging from measurement and ground state cooling schemes for weak coupling to quantum control techniques in the strong coupling regime.

14.2.2 Superconducting Devices and Mechanics

Solid-state qubits which are encoded in a quantized charge degree of freedom can be coupled to mechanical motion via electrostatic interactions. A prototype example is the Cooper Pair Box (CPB), i.e. a small superconducting island where Cooper pairs are coherently coupled to a large reservoir via a Josephson tunnel junction (see Fig. 14.1a). The CPB belongs to a larger class of superconducting qubits [3, 26] and is—in its simplest realization—described by the number N of excess Cooper pairs on the island and its conjugate variable δ which is the difference of the superconducting phase across the Josephson junction. As discussed in more detail in the chapter by Konrad Lehnert and the chapter by Aaron O'Connell and Andrew Cleland, the corresponding quantum operators obey the standard commutation relations $[\hat{N}, \hat{\delta}] = i$ and the Hamiltonian operator for the CPB is

$$\hat{H}_{\rm CPB} = E_C(\hat{N} - N_g)^2 - E_J \cos(\hat{\delta}). \tag{14.6}$$

Here $E_C = 4e^2/2C_\Sigma$ is the charging energy for a total island capacitance C_Σ, E_J is the Josephson energy, and $N_g = C_g V_g/(2e)$ the dimensionless gate charge which can be adjusted by the voltage V_g applied across the gate capacitance C_g. The CPB is usually operated in a regime where the charging energy $E_C/h \sim 50$ GHz is the dominant energy scale and the dynamics of the CPB is restricted to the two energetically lowest charge states. For example, by setting $N_g = n + 1/2 + \Delta N_g$ with integer n and $\Delta N_g < 1$, these two states are $|g\rangle = |N = n\rangle$ and $|e\rangle = |N = n+1\rangle$ and form the basis of a so-called 'charge qubit' [26]. The states $|g\rangle$ and $|e\rangle$ are separated by an adjustable charging energy $E_{eg} = 2E_C \Delta N_g$ and coupled by the Josephson tunneling term $\langle g|\hat{H}_{\rm CPB}|e\rangle = -E_J/2$.

When the gate electrode is replaced by a vibrating mechanical beam the capacitance $C_g(x) \approx C_g(1 - x/d)$ varies with the beam displacement x and the resulting change in the charging energy $E_{eg}(x)$ introduces a coupling between the qubit states and the mechanical resonator. The Hamiltonian for the combined system is then given by [8, 9]

$$\hat{H} = E_c \Delta N_g \hat{\sigma}_z - \frac{E_J}{2}\hat{\sigma}_x + \hbar\Omega_{\rm M}\hat{b}^\dagger\hat{b} + \hbar\lambda(\hat{b} + \hat{b}^\dagger)\hat{\sigma}_z\,, \tag{14.7}$$

and we recover the general form of Eq. (14.1) of the qubit-resonator coupling with a single phonon coupling constant $\lambda \equiv \lambda_{el}$ as defined in Eq. (14.2). Due to the electrostatic nature of the interaction the achievable coupling strength $\lambda_{el}/2\pi \sim 10$ MHz between a charge qubit and a mechanical resonator can be substantially larger than the corresponding magnetic interactions with spin qubits discussed in Sect. 14.2.3. However, for the same reason charge states are also more susceptible to random interactions with the environment and typical dephasing times T_2 for charge superposition states are in the pico- to nanosecond regime. An exception to this rule occurs when the CPB is operated at the charge degeneracy point $\Delta N_g = 0$. Here the eigenstates of $\hat{H}_{\mathrm{CPB}}$, namely $|\tilde{g}\rangle = (|g\rangle + |e\rangle)/\sqrt{2}$ and $|\tilde{e}\rangle = (|g\rangle - |e\rangle)/\sqrt{2}$, are combinations of different charge states and therefore highly insensitive to ubiquitous sources of low frequency electric noise [27]. By assuming $\Delta N_g = 0$ and re-expressing Eq. (14.7) in terms of the Pauli operators $\tilde{\sigma}_j$ for the rotated basis states $|\tilde{g}\rangle$, $|\tilde{e}\rangle$ we obtain

$$\hat{H} = \frac{E_J}{2}\tilde{\sigma}_z + \hbar\Omega_{\mathrm{M}}\hat{b}^\dagger\hat{b} + \hbar\lambda(\hat{b} + \hat{b}^\dagger)\tilde{\sigma}_x\,, \tag{14.8}$$

as our final model for the coupled resonator charge-qubit system. Indeed, by using optimized charge qubit designs dephasing times $T_2 > 1\,\mu$s have been demonstrated [27, 28]. This makes the CPB a promising candidate to achieve strong coupling $\lambda T_2 > 1$ with a mechanical resonator.

Equation (14.8) is familiar from related models studied in the context of cavity QED [29] or trapped ions [30], where usually a resonant exchange of excitations between the qubit and the resonator is used for cooling or quantum control of the resonator mode. However, with the exception of the high frequency dilatation modes described in the chapter by Cleland et al., mechanical frequencies of μm sized beams are typically in the range of 10–100 MHz and Eq. (14.8) describes a highly non-resonant coupling to a qubit with a transition frequency $E_J/h \sim 5$ GHz. Therefore, in the following we briefly outline two possibly strategies for potential quantum applications in the present system. First, we remark that in the relevant regime $\lambda \ll \Omega_{\mathrm{M}} < E_J$ second order perturbation theory can be used to approximate Eq. (14.8) by an effective Hamiltonian [9, 11]

$$\hat{H} \simeq \frac{E_J}{2}\tilde{\sigma}_z + \hbar\Omega_{\mathrm{M}}\hat{b}^\dagger\hat{b} + \hbar\chi(\hat{b}^\dagger\hat{b} + 1/2)\tilde{\sigma}_z\,. \tag{14.9}$$

The resulting coupling term with a strength $\hbar\chi = (\hbar\lambda)^2 \times 2E_J/(E_J^2 - (\hbar\Omega_{\mathrm{M}})^2)$ can be interpreted as a shift of the qubit frequency proportional to the phonon number $\hat{b}^\dagger\hat{b}$. Under the condition $\chi T_2 > 1$ this frequency shift can in principle be detected and the charge qubit can be used to implement a quantum non-demolition measurement of the number of vibrational quanta of the mechanical mode. To resolve a single vibrational level in time the coupling χ must also exceed the rate Γ_{th} at which the environment induces jumps between different vibrational states of the resonator. Estimates show that in this setting the combined condition $\chi > \Gamma_{\mathrm{th}}, T_2^{-1}$ for a phonon resolved measurement is experimentally feasible [9].

To go beyond passive measurement applications a second strategy is to realize effective resonance conditions by applying an oscillating gate voltage $V_g(t) \sim \cos(\omega_0 t)$ such that the microwave frequency $\omega_0 = E_J/\hbar - \Omega_M$ is used to gap the energy between vibrational and qubit excitations [10, 31]. In the interaction picture with respect to the free evolution $\hat{H}_0 = E_J/2\tilde{\sigma}_z + \hbar\Omega_M \hat{b}^\dagger \hat{b}$ the resulting Hamiltonian is then of the form

$$\hat{H} \simeq \lambda(\tilde{\sigma}_+ \hat{b} + \tilde{\sigma}_- \hat{b}^\dagger) + \mathcal{O}\left(e^{\pm i2\Omega_M t}, e^{\pm i2\omega_0 t}\right), \tag{14.10}$$

where for $\lambda \ll \Omega_M, \omega_0$ the oscillating terms can be neglected by using a rotating wave approximation. Hamiltonian (14.10) reduces to the resonant Jaynes-Cummings model which allows a coherent exchange of qubit and vibrational excitations. Applications of this model such as sideband cooling, state preparation and detection have been discussed in different areas of quantum optics [32]. The driven CPB provides the tool to implement similar applications [10, 31] for the vibrational modes of macroscopic mechanical resonators. We close this section by noting that strong coupling of a superconducting phase qubit to a mechanical oscillator has been observed experimentally [33], as discussed in more detail in the chapter by O'Connell and Cleland.

14.2.3 Spin Qubits and Mechanics

As discussed in Sect. 14.2.1, electronic and nuclear spin states can be coupled to mechanical motion by way of a magnetic field gradient. In the solid state, this situation is realized by positioning a spin qubit—typically residing in the lattice of some material—in close proximity to a strongly magnetized tip. One of the two elements, the tip or the qubit, is then rigidly affixed to a cantilever or other mechanical resonator. The most prominent examples of such experiments include mechanically detected magnetic resonance and optical experiments on nitrogen-vacancy defects in diamond.

14.2.3.1 Mechanical Detection of a Single Electron Spin

The first experiments demonstrating coupling between a nanomechanical cantilever and the spin of an isolated single electron appeared in 2004. In a landmark experiment, Rugar et al. measured the force of flipping a single unpaired electron spin contained in a silicon dangling bond (commonly known as an E' center) using a NEMS cantilever [19]. This achievement concluded a decade of development of a technique known as magnetic resonance force microscopy (MRFM) and stands out as one of the first single-spin measurements in a solid-state system.

The principle behind MRFM is simple (see Fig. 14.2). Magnetic moments—such as those associated with single electron or single nuclear spins—produce a force when in a magnetic field gradient: $F = \mu \nabla B$, where μ is the spin's magnetic moment and

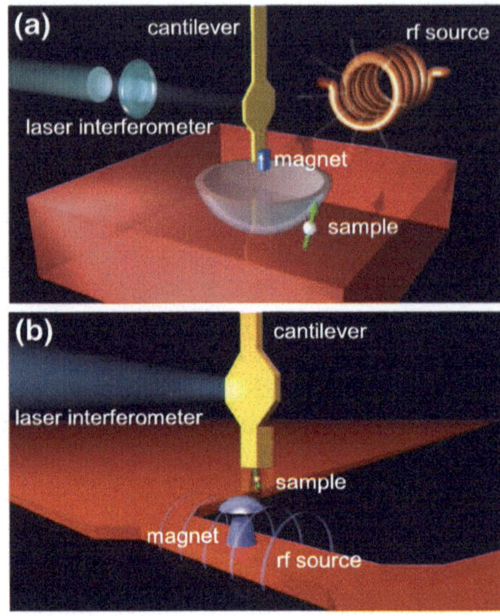

Fig. 14.2 Schematics of an MRFM apparatus. **a** "Tip-on-cantilever" arrangement, such as used in the single electron MRFM experiment of 2004 [19]. **b** "Sample-on-cantilever" arrangement, like the one used for the nanoscale virus imaging experiment in 2009 [20]. In both cases the hemispherical region around the magnetic tip is the region where the spin resonance condition is met—the so-called "resonant slice"

∇B is the spatial field gradient. This force can couple the deflection of a compliant cantilever to the spin if either the spin or the gradient source, such as a small magnet, are fixed to the cantilever. If the magnetic field gradient is large enough (i.e. the coupling is strong enough), the spin polarization and the cantilever's motion will be coupled.

Most MRFM techniques utilize this coupling to make measurements of spin density on the micro- or nanometer scale [34]. They employ extremely compliant cantilevers capable of detecting forces as small as 1 $\mathrm{aN}/\sqrt{\mathrm{Hz}}$ and optical interferometers that can measure the cantilever's displacement to resolutions of 1 $\mathrm{pm}/\sqrt{\mathrm{Hz}}$. Furthermore, magnetic resonance techniques are used in order to selectively address ensembles of spins in a sample, allowing for both spatial and chemical selectivity. For example, a pulse sequence known as a rapid adiabatic passage can be applied using a radio-frequency (rf) source. Each adiabatic passage pulse causes only the spins on resonance with the rf carrier frequency to flip. Using a periodic pulse sequence, a small ensemble of spins—at a particular region in space—can be made to flip at nearly any desired frequency. By choosing the flip frequency to be at the cantilever's mechanical resonance, the periodic spin oscillations will in turn drive the cantilever into oscillation. Because this spin force is made to occur at the cantilever's resonant frequency and at a particular phase, it can be distinguished from all other random electro-static and thermal forces disturbing the cantilever's motion.

We should note that the force sensitivity required for the single electron spin measurement imposed several limitations. Since the single-shot signal-to-noise ratio (SNR) was 0.06, up to 12 h of averaging per data point were required [19]. In addition to making useful imaging prohibitively time-consuming, this small SNR precludes the technique from following the dynamics of the single electron spin. Large SNRs would allow for shorter measurement times. If the measurement time could be reduced below the correlation time of the electron spin, which is related to the relaxation time of the electron spin in its rotating frame, real-time readout of the spin quantum state would become possible. Such readout would enable a wide variety of quantum measurement experiments. In the case of the electron spin E' centers, the correlation time was measured to be 760 ms. Until the SNR improves dramatically, real-time readout of spins will only be possible for small ensembles of electrons rather than for single electrons. One example is the real-time measurement of the direction of spin polarization for an ensemble of spins (~70 electron spins) [35]. In contrast, due to low SNR, the single-spin measurement [19] could not discern the direction of the measured spin; it could only ascertain its position.

14.2.3.2 Mechanical Detection of Nuclear Spins

Using this technique to couple and detect a single nuclear spin is far more challenging than to detect a single electron spin. The magnetic moment of a nucleus is much smaller than that of an electron: a ^{1}H nucleus (proton), for example, possesses a magnetic moment that is only $\sim 1/650$ of an electron spin moment. Since the measured force, F, is directly proportional to the magnetic moment of the spin, a significantly higher force resolution is required for nuclear spin experiments than for electron spin experiments. Other important nuclei, such as ^{13}C or a variety of isotopes present in semiconductors, have even weaker magnetic moments than ^{1}H. In order to observe single nuclear spins, it is necessary to improve the state-of-the-art sensitivity by another two to three orders of magnitude. While not out of the question, this is a daunting task that requires significant advances to all aspects of the MRFM technique.

14.2.3.3 Strong Magnetic Coupling

The strong magnetic coupling achieved between spins and the cantilever enables the high sensitivity of MRFM. This coupling is mediated by field gradients that can exceed 5×10^6 T/m [23, 24]. For the cantilevers and magnetic tips used in these experiments this corresponds to $\lambda_{mag}/2\pi \approx 10$ kHz for a single electron spin and 10 Hz for a proton. Such high gradients have been achieved using micro-fabricated Dy or FeCo magnetic tips and by the ability to make stable measurements with the sample positioned less than 50 nm from the apex of the tip. The strong interaction between spins and the mechanical sensor has been the subject of a number of theoretical studies, and is predicted to lead to a host of intriguing effects. These range from shortening of spin lifetimes by "back action" [36, 37], to spin alignment by specific mechanical modes either at the Larmor frequency or in the rotating frame [38], to a mechanical analog of a laser [18], and to long-range mediation of spin couplings using charged resonator arrays [39].

The first direct experimental evidence for accelerated nuclear spin relaxation induced by a single, low-frequency mechanical mode was reported in 2008 [40].

In these experiments the slight thermal vibration of the cantilever generated enough magnetic noise to destabilize the spin. Enhanced relaxation was found when one of the cantilever's upper modes (in particular the third mode with a frequency of about 120 kHz) coincided with the Rabi frequency of the ^{19}F spins in CaF_2. In this regime, the spins are more tightly coupled to one mechanical resonator mode than to the continuum of phonons that are normally responsible for spin-lattice relaxation. Interestingly, these initial experiments showed a scaling behavior of the spin relaxation rate with important parameters, including magnetic field gradient and temperature, that is substantially smaller than predicted by theory.

14.2.3.4 Nano-MRI and Potential Practical Applications

The coupling of small nuclear spin ensembles to a compliant mechanical oscillator through strong magnetic tips has resulted in the highest magnetic resonance imaging (MRI) resolution achieved by any method. In 2009, Degen et al. demonstrated three-dimensional (3D) MRI of ^{1}H nuclear spins in a biological specimen (tobacco mosaic virus) with a spatial resolution down below 10 nm [20]. This resolution represents a 100 million-fold improvement in volume resolution over conventional MRI and shows the potential of MRFM as a tool for elementally selective imaging on the nanometer scale. If the development of such techniques continues, these results indicate that force-detected spin resonance has the potential to become a significant tool for structural biologists.

14.2.3.5 Nitrogen Vacancy Centers and Mechanics

Recently, single nitrogen vacancy (NV) centers hosted in diamond have been proposed as solid-state qubits amenable to mechanical coupling [22]. Again, as in the case of MRFM, the coupling rests on bringing the qubit—in this case a single NV—in close proximity to a strongly magnetized tip. Then, either the NV or the tip must be affixed to a mechanical oscillator. NV centers appear especially attractive qubits due to their excellent optical and electronic properties. A single NV spin can be readily initialized and measured by optical means, manipulated using resonant rf pulses, and excellent coherence times up to a few milliseconds persist even in ambient conditions [41, 42]. As a result NVs have been proposed and used as ultra-sensitive scanning magnetic sensors [43–46].

First experiments have recently demonstrated the coupling of the NV spin to mechanical motion. Arcizet et al. coupled an NV to the motion of a SiC nanowire using field gradients around 7×10^3 T/m [47]. The NV was fixed to the tip of the nanowire while the magnet was placed near to it. Nanowire vibrations of a few tens of nanometers in amplitude were detected through a change in the lineshape of the NV spin resonance. More recently, Kolkowitz et al. used an NV spin to sense the vibrations of a cantilever resonatorwith a magnetic tip [48]. A sequence of coherent manipulation pulses was applied to the spin in order to enhance its sensitivity to the resonator vibrations while suppressing noise from other sources. In this way, mechanical vibrations down to a few picometers in amplitude were detected without phase locking NV spin dynamics and resonator vibrations. In these initial experiments, the spin-resonator coupling strength was $\lambda_{mag}/2\pi = 70$ Hz [47] or lower [48]. Coupling strengths in the kHz range could be reached by combining a strong magnet with a nanoscale oscillator with large zero-point motion [48].

14.2.3.6 Increasing the magnetic coupling strength

The prospects of improving hybrid mechanical systems based on spin qubits depend on progress in increasing the magnetic coupling strength λ_{mag}. First, the magnetic tips can and must be improved with the use of cleaner materials and lithographic processing techniques. Second, the development of experimental techniques designed to bring the spin qubit and the gradient source as close together as possible without destroying either qubit coherence or introducing mechanical dissipation should also yield significant gains in coupling strength.

14.3 Atoms, Ions, and Molecules Coupled to Mechanics

Atoms, ions, and molecules are quantum systems par excellence, and a sophisticated toolbox exists for coherent manipulation of their electronic, spin, and motional degrees of freedom [49, 50]. It is therefore natural to ask whether such atomic quantum systems can be coupled to mechanical oscillators. Through the coupling, the tools of atomic physics could become available for quantum control of mechanical devices. On the other hand, mechanical oscillators could find new applications in atomic physics experiments, such as optical lattices with vibrating mirrors.

Compared to the solid-state based approaches discussed in the previous sections, coupling atomic systems to mechanical oscillators creates a qualitatively different setting. Atoms in a trap can be regarded as mechanical oscillators themselves. Acting as a dispersive medium inside an optical cavity, an interesting variant of cavity optomechanics in the quantum regime can be realized (see the chapter by Stamper Kurn). Atomic systems offer both the continuous degree of freedom of their motion in a trap as well as a discrete set of internal electronic and spin states that can be reduced to two-level systems. Both the internal and motional state can be initialized, coherently manipulated, and detected on the quantum level with high fidelity, using techniques that have been developed in experiments on atomic clocks and interferometers [51, 52], Bose-Einstein condensation [53–55], and quantum information processing [56]. Coherence times of atomic systems are typically in the range of milliseconds up to many seconds [57], and thus much longer than those of most solid-state quantum systems discussed in the previous sections. Moreover, many properties of atomic systems can be widely tuned in-situ with external fields, including trapping frequencies, laser cooling rates, and even the strength of atom–atom interactions.

While the good isolation of atoms trapped in a vacuum chamber enables long coherence times, it renders coupling to mechanical oscillators more challenging. Various coupling mechanisms have been proposed, such as electrostatic coupling to the motion of trapped ions [58–60] and molecules [61], magnetic coupling to atomic spins [62–66], and optomechanical coupling to atoms in free space [67–69] and in optical cavities [70–80]. Remarkably, some of these schemes predict strong atom-oscillator coupling even for a single atom [73, 74]. First experimental implementations of hybrid atom-oscillator systems have recently been reported [81–84]. In the following, we discuss several of these proposals and experiments (see also the review in [85]).

14.3.1 Direct Mechanical Coupling

The conceptually most straightforward approach is to directly couple the vibrations of a mechanical oscillator to the vibrations of an atom or ion in a trap with the help of a "spring", i.e. a distance-dependent force between the two systems [85]. We consider an atom of mass m_{at} in a trap of frequency Ω_{at} and a mechanical oscillator of mass $m_{\rm eff}$ and frequency $\Omega_{\rm M}$. The coupling force derives from a potential $U_c(d)$ that depends on the distance d between the oscillator and the atom. For small displacements $\hat{x}, \hat{x}_{at} \ll d$, the resulting coupling Hamiltonian is of the form $\hat{H}_c = U_c''(d)\hat{x}\,\hat{x}_{at}$, where $U_c''(d)$ is the curvature of U_c evaluated at the mean atom-oscillator distance d. The oscillator displacement can be written as $\hat{x} = x_{\rm ZPF}(\hat{b} + \hat{b}^\dagger)$ in terms of creation/annihilation operators $[\hat{b}, \hat{b}^\dagger] = 1$ and the zero-point amplitude $x_{\rm ZPF} = \sqrt{\hbar/2m_{\rm eff}\Omega_{\rm M}}$. Similarly, the atomic displacement is $\hat{x}_{at} = x_{at,0}(\hat{a}_{at} + \hat{a}_{at}^\dagger)$ with $[\hat{a}_{at}, \hat{a}_{at}^\dagger] = 1$ and $x_{at,0} = \sqrt{\hbar/2m_{at}\Omega_{at}}$. The Hamiltonian of the coupled system is then given by

$$\hat{H} = \hbar\Omega_{at}\hat{a}_{at}^\dagger\hat{a}_{at} + \hbar\Omega_{\rm M}\hat{b}^\dagger\hat{b} + \hbar\lambda(\hat{a}_{at} + \hat{a}_{at}^\dagger)(\hat{b} + \hat{b}^\dagger), \tag{14.11}$$

with a single-phonon atom-oscillator coupling constant

$$\hbar\lambda = U_c''(d)x_{\rm ZPF}\,x_{at,0} \simeq \varepsilon\frac{\hbar\Omega_{at}}{2}\sqrt{\frac{m_{at}}{m_{\rm eff}}}, \tag{14.12}$$

for near-resonant coupling $\Omega_{at} \simeq \Omega_{\rm M}$. It is important to note that U_c also modifies the atomic trapping potential by contributing a term of order $U_c''(d)\hat{x}_{at}^2$ [85]. We therefore introduce the dimensionless parameter $\varepsilon = U_c''(d)/(m_{at}\Omega_{at}^2)$, which compares $U_c''(d)$ to the curvature of the atom trap. To avoid strong trap distortion, we typically have $\varepsilon \ll 1$. In the special case where U_c itself provides the atom trap, we have $\varepsilon = 1$. To achieve $\varepsilon > 1$, the effect of U_c on the trap has to be partially compensated, requiring sophisticated trap engineering. For direct mechanical coupling, we thus find that λ scales with Ω_{at} but is reduced by the atom-oscillator mass ratio, which is typically very small ($\sqrt{m_{at}/m_{\text{eff}}} \sim 10^{-8} - 10^{-4}$). To achieve significant coupling strength, oscillators with small m_{eff} are advantageous.

Several theoretical proposals consider direct mechanical coupling between a trapped ion and an oscillator with a metallic electrode on its tip [58–60]. In this case, $U_c = eq/(4\pi\varepsilon_0 d)$ is the Coulomb interaction between the ion of elementary charge e and the charge $q = C_q V_q$ on the oscillator tip. For a nanoscale oscillator with $m_{\text{eff}} = 10^{-15}$ kg coupled to a single $^9\text{Be}^+$ ion in a trap with $\Omega_{at}/2\pi = 70$ MHz, we obtain $\lambda/2\pi = \varepsilon \times 150$ Hz assuming $\Omega_{\text{M}} = \Omega_{at}$ [85]. A value of $\varepsilon = 1$ can be achieved e.g. with $d = 10\,\mu\text{m}$ and a metallic tip with a capacitance $C_q = 10^{-17}$ F and an applied voltage $V_q = 90$ V.

Stronger coupling is possible if $N \gg 1$ atoms are simultaneously coupled to the mechanical oscillator. In this case, $\hat{a}_{at}$ and $\hat{a}_{at}^{\dagger}$ refer to the atomic center-of-mass (COM) motion. The coupling is collectively enhanced by a factor $\sqrt{N}$, so that

$$\lambda_N = \lambda\sqrt{N} = \varepsilon\frac{\Omega_{at}}{2}\sqrt{\frac{Nm_{at}}{m_{\text{eff}}}}. \tag{14.13}$$

This result can be intuitively obtained by replacing $m_{at} \rightarrow Nm_{at}$ in Eq. (14.12), for a derivation see [68]. An example of such collective coupling where λ_N can reach several kHz is given in Sect. 14.3.3.

In the experiment of [82], a direct mechanical coupling between a cantilever oscillator and ultracold atoms was demonstrated for the first time. An atomic Bose-Einstein condensate (BEC) of $N = 2 \times 10^3$ atoms was placed at about one micrometer distance from the surface of the cantilever and used as a probe for cantilever oscillations (see Fig. 14.3a). The coupling potential U_c is due to attractive atom-surface interactions, which substantially modify the magnetic trapping potential U_m at such small distance. One effect of the surface force is to reduce the potential depth (see Fig. 14.3b). In addition, it shifts the trap frequency and minimum position. When the cantilever oscillates, the trapping potential is modulated at the cantilever frequency Ω_{M}, resulting in mechanical coupling to the atoms as described above.

In the experiment, the vibrating cantilever induced large-amplitude atomic motion that was detected simply via atom loss across the barrier U_0, see Fig. 14.3c. The observed atom-cantilever coupling depends strongly on the trap parameters and shows resonant behavior if $\Omega_{\text{M}} = \Omega_{at}$. Coupling to collective mechanical modes of the BEC other than the COM mode was observed as well [82]. While it was possible to detect the cantilever motion with the atoms, the backaction of atoms onto the cantilever was negligible in this experiment, mainly because the relatively large $m_{\text{eff}} = 5$ ng results in a small coupling constant $\lambda_N/2\pi \simeq 10^{-2}$ Hz. Much stronger coupling could be achieved by miniaturizing the cantilever. Since coupling via surface forces does not require functionalization of the cantilever, it could be used to couple atoms to molecular-scale oscillators such as carbon nanotubes [86]. In this case, a coupling constant of a few hundred Hz could be achieved [85].

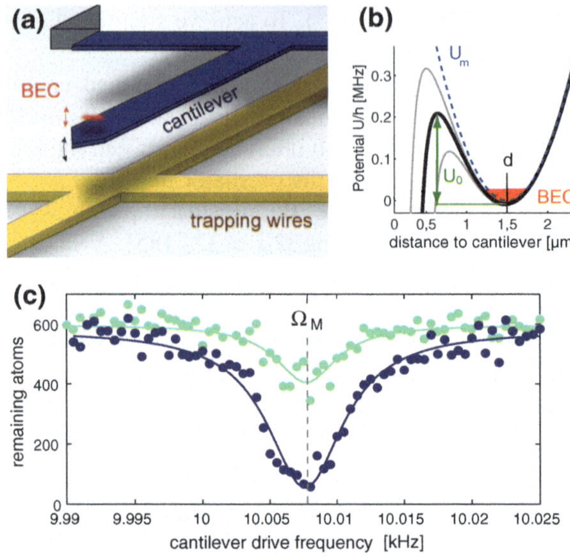

Fig. 14.3 Coupling of a mechanical cantilever and an atomic BEC via atom-surface forces [82]. **a** Atom chip with cantilever oscillator (length 200 µm, $\Omega_M/2\pi = 10\,\text{kHz}$, $m_{\text{eff}} = 5\,\text{ng}$, $Q = 3{,}200$). The atoms can be trapped and positioned near the cantilever with magnetic fields from wire currents. **b** Combined magnetic trapping and surface potential. The surface potential reduces the trap depth to U_0. Cantilever oscillations modulate the potential, thereby coupling to atomic motion. **c** Cantilever resonance detected with the atoms, for two different driving strengths of the cantilever

14.3.2 Magnetic Coupling to Atomic Spin

The vibrations of a mechanical oscillator can also be coupled to the spin of the atoms. This has several advantages compared to coupling to atomic motion. First, the atomic spin can be manipulated with higher fidelity. For hyperfine spins, coherence times T_2 of many seconds have been achieved [57]. Second, it is easier to isolate a two-level system among the internal states, providing a way to the preparation of non-classical quantum states of the oscillator. Third, hyperfine spin transition frequencies lie in the MHz to GHz range, significantly higher than typical trap frequencies. This enables coupling to high-frequency mechanical oscillators, which are easier to cool to the ground state.

To couple to the spin of the atoms, the oscillator is functionalized with a small magnet that generates a field gradient ∇B, see Fig. 14.4a. Mechanical oscillations $x(t)$ are transduced into an oscillating magnetic field $B_r(t) = \nabla B\, x(t)$ that couples to the spin. If $B_r(t)$ is perpendicular to the static field B_0 in the trap, the Hamiltonian is

$$\hat{H} = \frac{\hbar\omega_L}{2}\hat{\sigma}_z + \hbar\Omega_{\text{M}}\hat{b}^\dagger\hat{b} + \hbar\lambda_{mag}(\hat{b} + \hat{b}^\dagger)\hat{\sigma}_x \qquad (14.14)$$

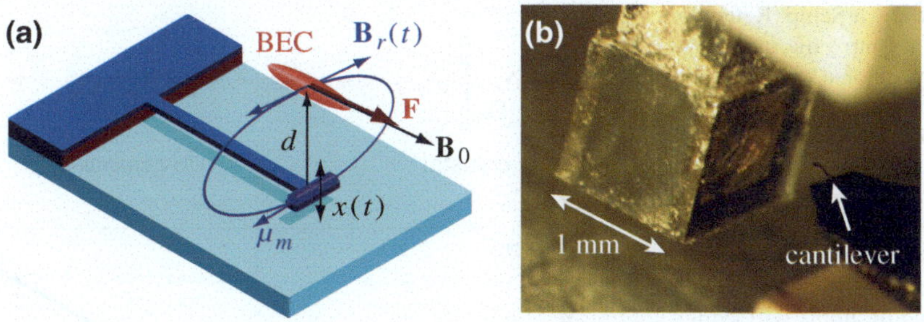

Fig. 14.4 **a** Schematic setup for coupling the spin of an atomic BEC to a nanoscale cantilever with a magnetic tip [62]. Cantilever oscillations $x(t)$ are transduced by the magnet into an oscillating field $B_r(t)$, which couples to the spin F of the atoms. **b** Experimental setup of [81], where a vibrating cantilever with a magnetic tip induces atomic spin resonance in a room-temperature Rb vapor cell

with a coupling constant $\lambda_{mag} = g\mu_B x_{\mathrm{ZPF}} \nabla B/2\hbar$ similar to Eq. (14.4). The prefactor g (of order unity) accounts for the matrix element of the atomic hyperfine transition considered. For transitions between Zeeman sublevels, the Larmor frequency is $\omega_L = g_F \mu_B B_0/\hbar$, with the hyperfine Landé factor g_F. It can be widely tuned by adjusting B_0 in order to achieve resonance $\omega_L = \Omega_{\mathrm{M}}$. The coupled system realizes a mechanical analog of the Jaynes-Cummings model in cavity quantum electrodynamics [29], with the *phonons* of the mechanical oscillator playing the role of the *photons* of the electromagnetic field.

To achieve large λ_{mag} a strong gradient ∇B is required. Approximating the magnet by a dipole of magnetic moment μ_m, we have $\nabla B = 3\mu_0\mu_m/4\pi d^4$. It is thus essential to trap and position the atoms at very small distance d from the oscillator tip. At the same time, care has to be taken that ∇B does not significantly distort the atomic trapping potential. With neutral atoms in magnetic microtraps, d can be as small as a few hundred nanometers [82]. Compared with the solid-state implementations discussed in Sect. 14.2.3, where d can be in the tens of nanometers range, it is thus more difficult to achieve large ∇B. On the other hand, the spin decoherence rates T_2^{-1} of trapped atoms are exceptionally small. The main challenge is the thermal decoherence rate $\Gamma_{\mathrm{th}} = k_B T/\hbar Q$ of the mechanical oscillator. The single-phonon single-atom strong coupling regime requires $\lambda_{mag} > \Gamma_{\mathrm{th}}, T_2^{-1}$. As in the previous section, collective coupling to $N \gg 1$ atoms is a possible strategy to enhance the coupling strength. In this case, the collective strong-coupling regime requires $\lambda_N = \lambda_{mag}\sqrt{N} > \Gamma_{\mathrm{th}}, T_2^{-1}$. Several theory papers investigate how to achieve the strong coupling regime by coupling nanoscale cantilevers to the spin of ultracold neutral atoms [62–66]. The predicted coupling constants lie in the range of $\lambda/2\pi \approx 10 - 10^3$ Hz, so that very high Q and low T are required for strong coupling.

In [81], a first experiment was reported where a cantilever with a magnetic tip was coupled to atoms in a room-temperature vapor cell, see Fig. 14.4b. The cantilever was piezo-driven and induced spin resonance in the atomic vapor, which was recorded with a laser. Besides demonstrating spin-oscillator coupling, such a setup is of interest for applications in magnetic field sensing, where the cantilever is essentially used as a tool for spectroscopy of the atomic transition frequency.

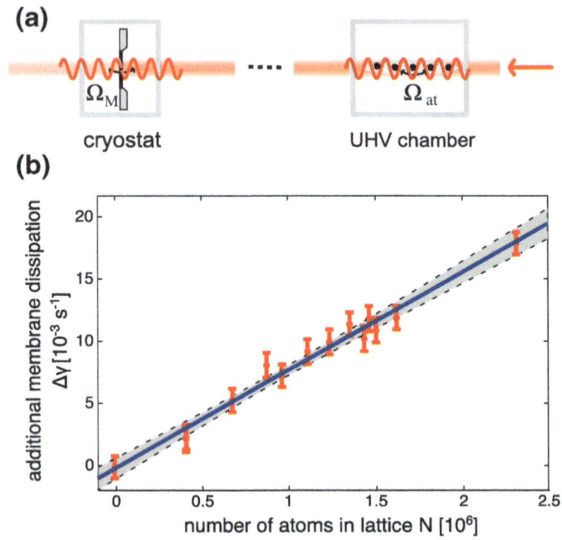

Fig. 14.5 **a** Setup: Reflection of light from a micromechanical membrane results in a standing wave which provides an optical lattice potential for ultracold atoms. This configuration gives rise to a coupling of the membrane vibrations and the center of mass motion of the cloud of atoms. Laser cooling of atoms will sympathetically cool along the membrane [68]. **b** Measured change in the mechanical damping rate due to the sympathetic cooling effect in its dependence on atom number in the lattice [83]

14.3.3 Optomechanical Coupling in Free Space

The coupling mechanisms discussed in the preceding sections require to position atoms close to the mechanical oscillator. Combining trapping and cooling of atoms in ultra-high vacuum (UHV) with a cryogenic environment as required for minimizing decoherence of the micromechanical system is a demanding task. In contrast, an indirect coupling mechanism acting over some distance would allow to keep the atomic and the micromechanical system in separate environments. Such a scheme was suggested in [68, 69] and experimentally implemented as described in [83, 84].

Consider the setup shown in Fig. 14.5a. Laser light is retroreflected from a partially reflective membrane, resulting in a standing wave light field, which in turn provides an optical lattice potential for a cloud of cold atoms. When the membrane supports a mechanical degree of freedom its position fluctuations will move the optical lattice, and thereby shake along the atoms. Conversely, position fluctuations of atoms in the potential will couple to the membrane's motion: When an atom is displaced from its potential minimum it will experience a restoring force which is due to transfer of photon momentum to the atom. This change in photon momentum is caused by an unbalancing of power between left and right propagating beams, which ultimately changes also the radiation pressure force on the membrane. Thus, also the membrane vibrates along with the atoms.

A more quantitative, semi-classical consideration along these lines reveals that the (dimensionless) position and momentum fluctuations of the membrane and the atomic COM motion, $\hat{q} = (\hat{b} + \hat{b}^\dagger)/\sqrt{2}$, $\hat{p} = i(\hat{b}^\dagger - \hat{b})/\sqrt{2}$ and $\hat{q}_{at} = (\hat{a}_{at} + \hat{a}_{at}^\dagger)/\sqrt{2}$, $\hat{p}_{at} = i(\hat{a}_{at}^\dagger - \hat{a}_{at})/\sqrt{2}$, respectively, *on average* obey the equations of motion

$$\langle \dot{\hat{p}} \rangle = -\Omega_{\mathrm{M}} \langle \hat{q} \rangle - 2r\lambda_N \langle \hat{q}_{at} \rangle, \quad \langle \dot{\hat{q}} \rangle = \Omega_{\mathrm{M}} \langle \hat{p} \rangle,$$
$$\langle \dot{\hat{p}}_{at} \rangle = -\Omega_{at} \langle \hat{q}_{at} \rangle - 2\lambda_N \langle \hat{q} \rangle, \quad \langle \dot{\hat{q}}_{at} \rangle = \Omega_{at} \langle \hat{p}_{at} \rangle. \tag{14.15}$$

Here Ω_{at} denotes the trap frequency for atoms provided by the optical potential, λ_N is the coupling strength between the COM motion of atoms and the membrane, and r is the power reflectivity of the membrane. The semiclassical calculation yields a coupling strength of $\lambda_N = (\Omega_{at}/2)\sqrt{N m_{at}/m_{\mathrm{eff}}}$ as in Eq. (14.13), assuming resonance between the two systems $\Omega_{\mathrm{M}} = \Omega_{at}$. For mechanical frequencies on the order of several 100 kHz this condition is routinely met in state of the art optical lattices. As expected from the discussion in Sect. 14.3.1 the coupling scales with the mass ratio between an atom and the membrane m_{at}/m_{eff}, but it is also collectively enhanced by the number of atoms N. Therefore even for a mass ratio $m_{at}/m_{\mathrm{eff}} \simeq 10^{-14}$, a large but feasible atom number $N = 10^8$ will still give rise to an appreciable coupling λ_N on the order of kHz for a trap frequency around one MHz. Moreover, it was recently shown that λ_N can be further increased by placing the membrane inside an optical cavity [69]. Since the atoms are still trapped in the optical lattice forming *outside* the cavity, the long-distance nature of the coupling is maintained.

Curiously, the semiclassical consideration outlined above predicts a coupling between atoms and membrane that is stronger in one direction than in the other by a factor given by the power reflectivity r. This scaling is in fact confirmed and explained by a full quantum treatment of this system. Starting from a complete Hamiltonian description including the motional degress of freedom of the membrane and atoms, as well as the quantized electromagentic field, it is possible to derive a master equation for the density matrix $\hat{\rho}$ of the membrane and atomic COM motion [68]. It has the form

$$\dot{\hat{\rho}} = -i[\hat{H}_{sys} - 2\lambda_N \hat{q}_{at}\hat{q}, \hat{\rho}] + C\hat{\rho} + L_m\hat{\rho} + L_{at}\hat{\rho}. \tag{14.16}$$

and implies the equations of motion (14.15) for the mean values. The term $C\hat{\rho} = -i(1-r)\lambda_N\big([\hat{q}, \hat{q}_{at}\hat{\rho}] - [\hat{\rho}\hat{q}_{at}, \hat{q}]\big)$ is responsible for the asymmetric coupling. Its form is well known in the theory of *cascaded quantum systems* [87, 88], and arises here due to the finite reflectivity of the membrane. The Lindblad terms $L_m\hat{\rho}$ and $L_{at}\hat{\rho}$ correspond to momentum diffusion of, respectively, the membrane and the atomic COM motion. They arise due to vacuum fluctuations of the radiation field giving rise to fluctuations of the radiation pressure force on the membrane and the dipole force on atoms. The full quantum treatment of the system correctly reproduces these well known effects, and shows that the corresponding diffusion rates are well below the rates of other relevant decoherence processes in this system: For the membrane mode this is thermal heating at a rate Γ_{th} due to clamping losses or absorption of laser light, which will heat the mechanical mode to thermal occupation $\bar{n}_{\mathrm{th}}$ in equilibrium. Atoms on the other hand can be *laser cooled* to the motional ground state by well established techniques such as Raman sideband cooling [83]. The corresponding cooling rate $\gamma_{at,cool}$ is widely tunable and can in fact be significantly larger than Γ_{th}. Note that in contrast to the normal optomechanical situation the cooling rate of atoms can be *switched off* giving rise to a regime of coherent coupling between atoms and the membrane.

This opens up the interesting possibility to *sympathetically cool* the membrane motion via laser cooling of atoms in the lattice. Adding a corresponding heating term for the membrane and a cooling term for atoms to the master equation (14.16), and solving for the steady state it is possible to determine the effect of the atom-membrane coupling and the associated sympathetic cooling. In [68] it was shown that ground state cooling might in fact be within reach. In the weak coupling regime $\lambda_N \ll \gamma_{at,cool}$ we expect in analogy to the treatment of optomechanical sideband cooling that laser cooling of atoms results in an increased effective mechanical damping rate $\Gamma_{\text{eff}} = \Gamma_{\text{M}} + 4r\lambda_N^2/\gamma_{at,cool}$, where $\Gamma_{\text{M}} = \Omega_{\text{M}}/Q$ is the intrinsic damping rate of the membrane. Such an increase was observed in the experiment reported in [83]. Figure 14.5b shows the results for the change in mechanical damping $\Delta\gamma = \Gamma_{\text{eff}} - \Gamma_{\text{M}}$ for various experiments with different number of atoms N in the lattice. The linear scaling as expected from the model discussed here is clearly confirmed. Also the magnitude of $\Delta\gamma$ in the experiment agrees well with the prediction by this model.

Overall this setup provides exciting first results and perspectives for interfacing micromechanical oscillators with ultracold atoms. The interface works at a distance, easing experimental requirements, and enables sympathetic cooling towards the ground state, as well as coherent dynamics for quantum state preparation and measurement of the mechanical mode via coupling to ultracold atoms.

14.3.4 Cavity-Optomechanical Coupling Schemes

The optomechanical coupling discussed in the previous section can be enhanced by placing the atoms and the oscillator inside a high-finesse optical cavity. In such a system, the cavity field can mediate an interaction between the internal or motional degrees of freedom of the atoms and the vibrations of the oscillator. In the following, we first discuss a scheme where the oscillator interacts with the internal state of an atomic ensemble. Subsequently, we present a system where the motion of a single trapped atom is strongly coupled to a membrane oscillator.

14.3.4.1 Coupling of Atomic Internal Levels to Resonator Motion

The presence of an ensemble of N two-level atoms inside a driven optical cavity modifies the cavity response. For example when the atomic resonance is far from the cavity mode frequency, i.e. in the dispersive limit, the effect of the atoms onto the field is a phase shift. Similarly a vibrating cavity end-mirror also leads to a phase shift of the intracavity field. The field can then be used as a mediator between atoms and the mechanical resonator to either allow an exchange of quantum states, entangle the two systems, or lead to enhanced optical cooling of the mirror. The last motive will be explored in the following and it can be seen as an atom induced effect of spectral filtering of the mirror scattered optical sidebands. The upshot is that low finesse

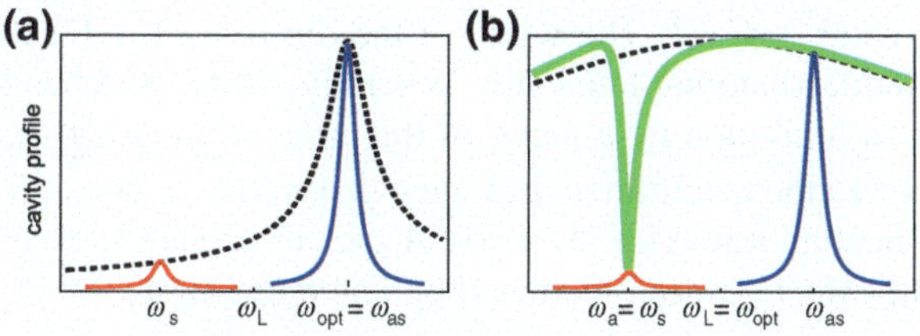

Fig. 14.6 Resolved sideband cooling of an end-mirror **a** using a sharp resonance optical cavity and **b** using a bad-cavity with an ensemble of atoms inside. The *dotted black lines* show the empty cavity response while the *red* and *blue curves* represent optical sidebands. The *green line* shows the modification of the cavity response in the presence of atoms

cavities can provide resolved sideband cooling when supplemented with filtering ensembles of atoms [76].

The total Hamiltonian of the system can be split as $\hat{H} = \hat{H}_0 + \hat{H}_I + \hat{H}_{dis}$, where $\hat{H}_0$ and $\hat{H}_I$ are the free part and the interaction term, while $\hat{H}_{dis}$ describes dissipation. The mirror quadratures are defined as above, $\hat{q} = (\hat{b} + \hat{b}^\dagger)/\sqrt{2}$ and $\hat{p} = i(\hat{b}^\dagger - \hat{b})/\sqrt{2}$, and the atomic ensemble of frequency splitting ω_a is described by creation/annihilation operators $[\hat{c}, \hat{c}^\dagger] = 1$. The atom-cavity coupling $G_a = g_a\sqrt{N}$ is collectively enhanced by $\sqrt{N}$ compared to the single atom-single photon coupling strength g_a. The harmonic oscillator description for the atomic cloud is accurate when the cavity photons are much less numerous than atoms $\bar{n}_{\text{cav}} \ll N$. The cavity resonance is ω_{opt} and the laser driving shows up in $\hat{H}_I$ as a displacement term of amplitude $\mathscr{E} = \sqrt{2P_{\text{in}}\kappa/\hbar\omega_L}$. One can derive equations of motion from the Hamiltonian dynamics and then perform a linearization of fluctuations around steady state values that leads to a set of quantum linearized Langevin equations

$$\dot{\hat{q}} = \Omega_{\text{M}}\hat{p}, \tag{14.17}$$

$$\dot{\hat{p}} = -\Gamma_{\text{M}}\hat{p} - \Omega_{\text{M}}\hat{q} + g(\hat{a} + \hat{a}^\dagger) + \xi, \tag{14.18}$$

$$\dot{\hat{a}} = -(\kappa + i\Delta_f)\hat{a} + ig\hat{q} - iG_a\hat{c} + \hat{a}_{in}, \tag{14.19}$$

$$\dot{\hat{c}} = -(\gamma_a + i\Delta_a)\hat{c} - iG_a\hat{a} + \hat{c}_{in}, \tag{14.20}$$

where $\Delta_f = \omega_{\text{opt}} - \omega_L - g^2/\Omega_{\text{M}}$ is the effective cavity detuning, κ is the cavity decay rate, γ_a and $\Delta_a = \omega_a - \omega_L$ are the atom decay rate and detuning respectively, and $\xi, \hat{a}_{in}, \hat{c}_{in}$ is quantum noise describing the effect of $\hat{H}_{dis}$ in the Langevin approach.

Setting $G_a = 0$ we recover the typical resolved sideband regime in optomechanics where under the conditions $\kappa \ll \Omega_{\text{M}}$ and $\Delta_f = \Omega_{\text{M}}$, optimal cooling of the mirror via the field is obtained as illustrated in Fig. 14.6a. The scattered sidebands are at $\omega_{as,s} = \omega_L \pm \Omega_{\text{M}}$ and the sharp response of the cavity around ω_{opt} leads to a suppression of the heating sideband. Assuming a bad cavity that cannot resolve sidebands and $G_a > 0$, we are instead in the situation depicted in Fig. 14.6b where the atoms, placed at $\Delta_a = -\Omega_{\text{M}}$, induce a dip in the cavity profile at ω_s that inhibits scattering into the Stokes sideband. Defining atomic cooperativity $C = G_a^2/\kappa\gamma_a$, the dip at ω_s scales as $(1 + C)^{-1}$. The width of the dip is $\gamma_a(1 + C)$ representing the enhanced light-induced atomic linewidth. When the Stokes sideband fits inside the dip one expects an inhibition by a factor of the order of $(1 + C)^{-1}$. For a rigorous analysis one analyzes the spectrum of the Langevin force $\hat{F} = g(\hat{a}_{in}+\hat{a}_{in}^{\dagger})$ which gives the cooling and heating rates (for $\Delta_f = 0$): $A_{as} \simeq g^2/\kappa$ and $A_s \simeq g^2/[\kappa(1 + C)]$. Subsequently, the effective atom-mediated optical damping is

$$\Gamma_{\text{opt}} = \frac{g^2}{\kappa}\frac{C}{1+C}. \tag{14.21}$$

The residual occupancy is given by $n^{res} = A_s/(\Gamma_{\text{M}}+\Gamma_{\text{opt}}) \to C^{-1}$ (in the limit of large C) and can be compared with $n^{res} = (\kappa/2\Omega_{\text{M}})^2$ for the purely optomechanical system; an immediate advantage of this hybrid system comes from the scaling of n^{res} with the controllable parameter N.

While in the above the atomic system has been viewed rather as a high-Q system that acts as a spectral filter for light to relax the resolved sideband limit requirements, we nevertheless stress that one can as well take a different stand by seeing the cavity mode rather as a mediator between atoms and mechanical resonator. From here, following the elimination of the cavity field as a fast variable, one can also derive the exact form of the implicit atom-mirror interaction and show effects such as a quantum state swap or entanglement [71].

14.3.4.2 Coupling of Atomic Motion to Resonator Motion

In a situation as that depicted in Fig. 14.7a, where a single atom is trapped inside a cavity that surrounds a vibrating membrane, the effect of light on atomic motion can be exploited to generate motion-motion coupling between membrane and atom [73, 74]. To this purpose we choose two fields of slightly different wavelengths and opposite detunings $\pm\Delta$ with respect to two cavity resonances (see Fig. 14.7b) that provide harmonic trapping of the atom. The two fields have frequencies ω_1, ω_2 equally far-detuned by δ from two internal transitions of the atom as shown in Fig. 14.7c. Around the atomic equilibrium position $\bar{x}_{at}$ a Lamb-Dicke expansion leads to a linear atom-field interaction

$$\hat{H}_{at,f} = g_{at,f}[(\hat{a}_1 + \hat{a}_1^{\dagger}) - (\hat{a}_2 + \hat{a}_2^{\dagger})](\hat{a}_{at} + \hat{a}_{at}^{\dagger}), \tag{14.22}$$

where the atom-field coupling $g_{at,f}$ as well as the atomic trapping frequency Ω_{at} inside the two-field optical trap are quantities depending on the single-photon Stark shift, cavity photon number and the geometry of the fields at position $\bar{x}_{at}$. This interaction can be interpreted as follows (Fig. 14.7d, e): fluctuations in the amplitudes of the two cavity fields exert oppositely oriented forces on the atom. Conversely, fluctuations of the atom around its mean position cause changes of opposite sign

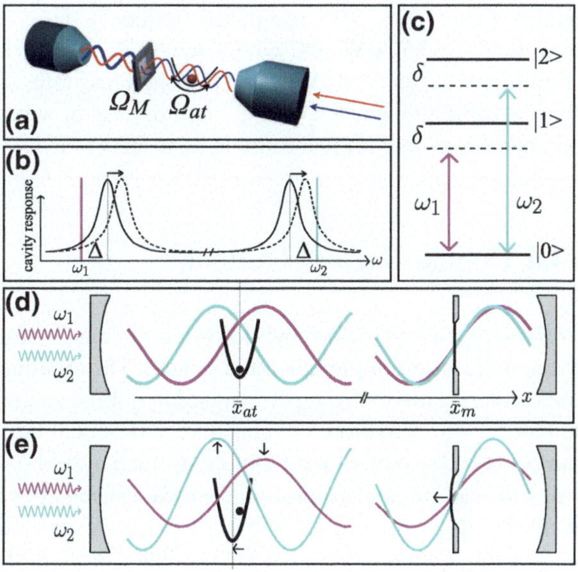

Fig. 14.7 Linear atom-membrane coupling mediated by two driven cavity modes. **a** Schematics of the setup with atom, membrane and two cavity fields shown. **b** Two cavity resonances and the frequency position of the two driving lasers. **c** Internal structure of the atom. **d** Static illustration of optical potentials. **e** Dynamical illustration of potentials showing the modification induced by motion of atom and membrane

in the amplitudes of the two cavity fields. The membrane-field interaction takes a similar form

$$\hat{H}_{m,f} = g_{m,f}[(\hat{a}_1 + \hat{a}_1^\dagger) - (\hat{a}_2 + \hat{a}_2^\dagger)](\hat{b} + \hat{b}^\dagger), \tag{14.23}$$

with a similar interpretation. To the reversible Hamiltonian dynamics one has of course to add the dissipation channels: cavity decay, momentum diffusion owing to spontaneous scattering of the atom and thermal decoherence of the membrane.

The elimination of the fast varying cavity fields can be done in the limit $|\Delta| \gg g_{at,f}, g_{m,f}$, and the reduced atom-membrane dynamics is governed by a linear two-mode Hamiltonian

$$\hat{H}_{at,m} = \Omega_{\mathrm{M}}\hat{b}^\dagger\hat{b} + \Omega_{at}\hat{a}_{at}^\dagger\hat{a}_{at} - \lambda(\hat{a}_{at} + \hat{a}_{at}^\dagger)(\hat{b} + \hat{b}^\dagger), \tag{14.24}$$

with atom-membrane coupling strength

$$\lambda = \frac{2g_{at,f}g_{m,f}(\Delta + \Omega_{\mathrm{M}})}{\kappa^2 + (\Delta + \Omega_{\mathrm{M}})^2} + \frac{2g_{at,f}g_{m,f}(\Delta - \Omega_{\mathrm{M}})}{\kappa^2 + (\Delta - \Omega_{\mathrm{M}})^2}, \tag{14.25}$$

to which decoherence at rates Γ_c, Γ_{at} and Γ_{th} adds irreversible dynamics.

The goal is to obtain a coupling λ much larger than the rates of decoherence. For a demonstrative example we consider a single Cs atom and a SiN membrane of small effective mass $m_{\mathrm{eff}} = 0.4$ ng inside a cavity of finesse of $\mathscr{F} \simeq 2 \times 10^5$. A small cavity waist of $w_0 = 10\ \mu$m results in a cooperativity parameter of 140. With a mechanical quality factor $Q = 10^7$, resonance frequency $\Omega_M = 2\pi \times 1.3$ MHz, circulating power $P_c \simeq 850\,\mu$W and cavity length $L = 50\,\mu$m we find a cavity mediated coupling $\lambda \simeq 2\pi \times 45$ kHz and decoherence rates Γ_c, Γ_{th}, $\Gamma_{at} \simeq 0.1 \times \lambda$. It is thus possible to enter the strong coupling regime with state-of-the-art experimental parameters, even with just a single atom in the cavity.

14.4 Conclusion and Outlook

We have discussed various hybrid systems in which a mechanical oscillator is coupled to another (microscopic) quantum system. The approaches that are being pursued are quite diverse, involving superconducting qubits, single spins in the solid state, quantum dots, ultracold atoms in magnetic and optical traps, as well as trapped ions and molecules. One motivation for building such hybrid systems is that they enable novel ways to read out and control mechanical objects. For example, a switchable, linear coupling of a mechanical oscillator to a two-level system allows for the preparation of arbitrary quantum states of the oscillator through the Law-Eberly protocol [89].

Experimentally, the coupling of superconducting two-level systems to mechanical oscillators is most advanced. First experiments have already reached the strong-coupling regime (see [33] and the chapter by O'Connell and Cleland). However, the coherence time of the involved qubits is very short (nano- to microseconds), and it is thus highly desirable to develop and implement strategies for strong coupling of mechanical oscillators to long-lived qubits such as spins in the solid-state or ultracold atoms in a trap. Strong coupling of mechanical oscillators to solid-state spins can build on the impressive achievements of magnetic resonance force microscopy, which has reached single-spin detection sensitivity already some time ago [19]. Recently, a novel system was realized in which the spin of a nitrogen vacancy center in diamond was used to sense mechanical motion [47, 48]. In another recent experiment, an ensemble of ultracold atoms in an optical lattice was optically coupled to vibrations of a micromechanical membrane [83, 84], enabling sympathetic cooling of the membrane through laser-cooled atoms. By enhancing the coupling with a high-finesse cavity, strong coupling could be achieved even for a single atom [73].

The fact that very different microscopic quantum systems are investigated as potential candidates for strong coupling to mechanical oscillators points to one of the big strengths of mechanical quantum systems: the oscillator can be functionalized with electrodes, magnets, or mirrors while maintaining high mechanical quality factor. Mechanical oscillators are thus particularly well suited to serve as quantum transducers [39] for precision sensing or hybrid quantum information processing [90]. Through the mechanical vibrations, spin dynamics can e.g. be transduced into electric or optical signals, and one can envision scenarios where atomic quantum memories are interfaced with superconducting quantum processors. Another important application is the transduction of microwave or radio-frequency signals into optical signals [91, 92], ultimately at the level of single quanta [93, 94]. The versatility of mechanical devices makes them a fascinating toy in the playground of quantum science and technology and we expect many exciting developments in the future.

Acknowledgments P.T. acknowledges fruitful discussions with D. Hunger and support by the EU project AQUTE and the Swiss National Science Foundation. K.H. acknowledges support through the cluster of excellence QUEST at the University of Hannover. C.G. acknowledges support from the Austrian Science Fund (FWF).

References

1. R. Hanson, D.D. Awschalom, Nature **453**(7198), 1043 (2008)
2. A.J. Shields, Nat. Photonics **1**(4), 215 (2007)
3. J. Clarke, F.K. Wilhelm, Nature **453**, 1031 (2008)
4. M.P. Blencowe, M.N. Wybourne, Appl. Phys. Lett. **77**, 3845 (2000)
5. R.G. Knobel, A.N. Cleland, Nature **424**, 291 (2003)
6. S. Zippilli, G. Morigi, A. Bachtold, Phys. Rev. Lett. **102**, 096804 (2009)
7. S.D. Bennett, L. Cockins, Y. Miyahara, P. Grütter, A.A. Clerk, Phys. Rev. Lett. **104**, 017203 (2010)
8. A.D. Armour, M.P. Blencowe, K.C. Schwab, Phys. Rev. Lett. **88**, 148301 (2002)
9. E.K. Irish, K. Schwab, Phys. Rev. B **68**, 155311 (2003)
10. I. Martin, A. Shnirman, L. Tian, P. Zoller, Phys. Rev. B **69**, 125339 (2004)
11. M.D. LaHaye, J. Suh, P.M. Echternach, K.C. Schwab, M.L. Roukes, Nature **459**, 960 (2009)
12. X. Zhou, A. Mizel, Phys. Rev. Lett. **97**, 267201 (2006)
13. E. Buks, M.P. Blencowe, Phys. Rev. B **74**, 174504 (2006)
14. F. Xue, Y.D. Wang, C.P. Sun, H. Okamoto, H. Yamaguchi, K. Semba, New J. Phys. **9**, 35 (2007)
15. Y.D. Wang, K. Semba, H. Yamaguchi, New J. Phys. **10**, 043015 (2008)
16. K. Jähne, K. Hammerer, M. Wallquist, New J. Phys. **10**, 095019 (2008)
17. S. Etaki, M. Poot, I. Mahboob, K. Onomitsu, H. Yamaguchi, H.S.J. van der Zant, Nat. Phys. **4**, 785 (2008)
18. I. Bargatin, M. Roukes, Phys. Rev. Lett. **91** (2003)
19. D. Rugar, R. Budakian, H. Mamin, B. Chui, Nature **430**, 329 (2004)
20. C.L. Degen, M. Poggio, H.J. Mamin, C.T. Rettner, D. Rugar, Proc. Natl. Acad. Sci. U.S.A. **106**, 1313 (2009)
21. F. Xue, L. Zhong, Y. Li, C.P. Sun, Phys. Rev. B **75**, 033407 (2007)
22. P. Rabl, P. Cappellaro, M.V.G. Dutt, L. Jiang, J. Maze, M.D. Lukin, Phys. Rev. B **79**, 041302 (2009)
23. M. Poggio, C. Degen, C. Rettner, H. Mamin, D. Rugar, Appl. Phys. Lett. **90**, 263111 (2007)
24. H. Mamin, C. Rettner, M. Sherwood, L. Gao, D. Rugar, Appl. Phys. Lett. **100**, 013102 (2012)
25. I. Wilson-Rae, P. Zoller, A. Imamoglu, Phys. Rev. Lett. **92**, 075507 (2004)
26. Yu. Makhlin, G. Schön, A. Shnirman, Rev. Mod. Phys. **73**, 357 (2001)
27. D. Vion, A. Aassime, A. Cottet, P. Joyez, H. Pothier, C. Urbina, D. Esteve, M.H. Devoret, Science **296**, 886 (2002)
28. J.A. Schreier, A.A. Houck, J. Koch, D.I. Schuster, B.R. Johnson, J.M. Chow, J.M. Gambetta, J. Majer, L. Frunzio, M.H. Devoret, S.M. Girvin, R.J. Schoelkopf, Phys. Rev. B **77**, 180502 (2008)
29. J.M. Raimond, M. Brune, S. Haroche, Rev. Mod. Phys. **73**, 565 (2001)
30. D. Leibfried, R. Blatt, C. Monroe, D. Wineland, Rev. Mod. Phys. **75**, 281 (2003)
31. P. Rabl, A. Shnirman, P. Zoller, Phys. Rev. B **70**, 205304 (2004)

32. S. Haroche, J.M. Raimond (eds.), *Exploring the Quantum* (Oxford University Press, New York, 2006)
33. A.D. O'Connell, M. Hofheinz, M. Ansmann, R.C. Bialczak, M. Lenander, E. Lucero, M. Neeley, D. Sank, H. Wang, M. Weides, J. Wenner, J.M. Martinis, A.N. Cleland, Nature **464**, 697 (2010)
34. M. Poggio, C.L. Degen, Nanotechnology **21**, 342001 (2010)
35. R. Budakian, H. Mamin, B. Chui, D. Rugar, Science **307**, 408 (2005)
36. D. Mozyrsky, I. Martin, D. Pelekhov, P. Hammel, Appl. Phys. Lett. **82**, 1278 (2003)
37. G. Berman, V. Gorshkov, D. Rugar, V. Tsifrinovich, Phys. Rev. B **68** (2003)
38. P. Magusin, W. Veeman, J. Magn. Reson. **143**, 243 (2000)
39. P. Rabl, S.J. Kolkowitz, F.H.L. Koppens, J.G.E. Harris, P. Zoller, M.D. Lukin, Nat. Phys. **6**, 602 (2010)
40. C.L. Degen, M. Poggio, H.J. Mamin, D. Rugar, Phys. Rev. Lett. **100** (2008)
41. F. Jelezko, T. Gaebel, I. Popa, A. Gruber, J. Wrachtrup, Phys. Rev. Lett. **92**, 076401 (2004)
42. G. Balasubramanian, P. Neumann, D. Twitchen, M. Markham, R. Kolesov, N. Mizuochi, J. Isoya, J. Achard, J. Beck, J. Tissler, V. Jacques, P. Hemmer, F. Jelezko, J. Wrachtrup, Nat. Mater. **8**, 383 (2009)
43. C. Degen, Appl. Phys. Lett. **92**, 243111 (2008)
44. J. Maze, P. Stanwix, J. Hodges, S. Hong, J. Taylor, P. Cappellaro, L. Jiang, M. Dutt, E. Togan, A. Zibrov, A. Yacoby, R. Walsworth, M. Lukin, Nature **455**, 644 (2008)
45. G. Balasubramanian, I. Chan, R. Kolesov, M. Al-Hmoud, J. Tisler, C. Shin, C. Kim, A. Wojcik, P. Hemmer, A. Krueger, T. Hanke, A. Leitenstorfer, R. Bratschitsch, F. Jelezko, J. Wrachtrup, Nature **455**, 648 (2008)
46. M. Grinolds, P. Maletinsky, S. Hong, M. Lukin, R. Walsworth, A. Yacoby, Nat. Phys. **7**, 687 (2011)
47. O. Arcizet, V. Jacques, A. Siria, P. Poncharal, P. Vincent, S. Seidelin, Nat. Phys. **7**, 879 (2011)
48. S. Kolkowitz, A.C.B. Jayich, Q. Unterreithmeier, S.D. Bennett, P. Rabl, J.G.E. Harris, M.D. Lukin, Science **335**, 1603 (2012)
49. S. Chu, Nature **416**, 206 (2002)
50. M.A. Kasevich, Science **298**, 1363 (2002)
51. R. Wynands, S. Weyers, Metrologia **42**, S64 (2005)
52. A.D. Cronin, J. Schmiedmayer, D.E. Pritchard, Rev. Mod. Phys. **81**, 1052 (2009)
53. W. Ketterle, D.S. Durfee, D.M. Stamper-Kurn, Bose-Einstein condensation in atomic gases, in *Proceedings of the International School of Physics "Enrico Fermi", Course CXL*, ed. by M. Inguscio, S. Stringari, C.E. Wieman (IOS Press, Amsterdam, 1999), pp. 67–176
54. I. Bloch, Science **319**, 1202 (2008)
55. J. Reichel, V. Vuletić (eds.), *Atom Chips* (Wiley-VCH, Weinheim, 2011)
56. D. Wineland, Phys. Scr. **T137**, 014007 (2009)
57. C. Deutsch, F. Ramirez-Martinez, C. Lacroûte, F. Reinhard, T. Schneider, J.N. Fuchs, F. Pichon, F. Laloë, J. Reichel, P. Rosenbusch, Phys. Rev. Lett. **105**, 020401 (2010)
58. D. Wineland, C. Monroe, W.M. Itano, D. Leibfried, B. King, D. Meekhof, J. Res. Natl. Inst. Stand. Technol. **103**, 259 (1998)
59. L. Tian, P. Zoller, Phys. Rev. Lett. **93**, 266403 (2004)
60. W.K. Hensinger, D.W. Utami, H.S. Goan, K. Schwab, C. Monroe, G.J. Milburn, Phys. Rev. A **72**, 041405(R) (2005)
61. S. Singh, M. Bhattacharya, O. Dutta, P. Meystre, Phys. Rev. Lett. **101**, 263603 (2008)
62. P. Treutlein, D. Hunger, S. Camerer, T.W. Hänsch, J. Reichel, Phys. Rev. Lett. **99**, 140403 (2007)
63. A.A. Geraci, J. Kitching, Phys. Rev. A **80**, 032317 (2009)
64. S. Singh, P. Meystre, Phys. Rev. A **81**, 041804 (2010)
65. C. Joshi, A. Hutter, F. Zimmer, M. Jonson, E. Andersson, P. Ohberg, Phys. Rev. A **82**, 043846 (2010)
66. S. Steinke, S. Singh, M. Tasgin, P. Meystre, K. Schwab, M. Vengalattore, Phys. Rev. A **84**, 023841 (2011)

67. K. Hammerer, M. Aspelmeyer, E.S. Polzik, P. Zoller, Phys. Rev. Lett. **102**, 020501 (2009)
68. K. Hammerer, K. Stannigel, C. Genes, P. Zoller, P. Treutlein, S. Camerer, D. Hunger, T.W. Hänsch, Phys. Rev. A **82**, 021803 (2010)
69. B. Vogell, K. Stannigel, P. Zoller, K. Hammerer, M.T. Rakher, M. Korppi, A. Jöckel, P. Treutlein, Phys. Rev. A **87**, 023816 (2013)
70. D. Meiser, P. Meystre, Phys. Rev. A **73**, 033417 (2006)
71. C. Genes, D. Vitali, P. Tombesi, Phys. Rev. A **77**, 050307 (2008)
72. H. Ian, Z.R. Gong, Y. Liu, C.P. Sun, F. Nori, Phys. Rev. A **78**, 013824 (2008)
73. K. Hammerer, M. Wallquist, C. Genes, M. Ludwig, F. Marquardt, P. Treutlein, P. Zoller, J. Ye, H.J. Kimble, Phys. Rev. Lett. **103**, 063005 (2009)
74. M. Wallquist, K. Hammerer, P. Zoller, C. Genes, M. Ludwig, F. Marquardt, P. Treutlein, J. Ye, H.J. Kimble, Phys. Rev. A **81**, 023816 (2010)
75. A. Bhattacherjee, Phys. Rev. A **80**, 043607 (2009)
76. C. Genes, H. Ritsch, D. Vitali, Phys. Rev. A **80**, 061803 (2009)
77. K. Zhang, W. Chen, M. Bhattacharya, P. Meystre, Phys. Rev. A **81**, 013802 (2010)
78. M. Paternostro, G. De Chiara, G. Palma, Phys. Rev. Lett. **104**, 243602 (2010)
79. G. De Chiara, M. Paternostro, G. Palma, Phys. Rev. A **83**, 052324 (2011)
80. Y. Chang, T. Shi, Y.x. Liu, C. Sun, F. Nori. Phys. Rev. A **83**, 063826 (2011)
81. Y.J. Wang, M. Eardley, S. Knappe, J. Moreland, L. Hollberg, J. Kitching, Phys. Rev. Lett. **97**, 227602 (2006)
82. D. Hunger, S. Camerer, T.W. Hänsch, D. König, J.P. Kotthaus, J. Reichel, P. Treutlein, Phys. Rev. Lett. **104**, 143002 (2010)
83. S. Camerer, M. Korppi, A. Jöckel, D. Hunger, T.W. Hänsch, P. Treutlein, Phys. Rev. Lett. **107**, 223001 (2011)
84. M. Korppi, A. Jöckel, M.T. Rakher, S. Camerer, D. Hunger, T.W. Hänsch, P. Treutlein, EPJ Web Conf. **57**, 03006 (2013)
85. D. Hunger, S. Camerer, M. Korppi, A. Jöckel, T.W. Hänsch, P. Treutlein, C. R. Phys. **12**, 871 (2011)
86. M. Gierling, P. Schneeweiss, G. Visanescu, P. Federsel, M. Häffner, D.P. Kern, T.E. Judd, A. Günther, J. Fortagh, Nat. Nanotech. **6**, 446 (2011)
87. C. Gardiner, Phys. Rev. Lett. **70**, 2269 (1993)
88. H. Carmichael, Phys. Rev. Lett. **70**, 2273 (1993)
89. C.K. Law, J.H. Eberly, Phys. Rev. Lett. **76**, 1055 (1996)
90. M. Wallquist, K. Hammerer, P. Rabl, M. Lukin, P. Zoller, Phys. Scr. **T137**, 014001 (2009)
91. T. Bagci, A. Simonsen, S. Schmid, L.G. Villanueva, E. Zeuthen, J. Appel, J.M. Taylor, A. Sorensen, K. Usami, A. Schliesser, E.S. Polzik, preprint (2013). arXiv:1307.3467
92. R.W. Andrews, R.W. Peterson, T.P. Purdy, K. Cicak, R.W. Simmonds, C.A. Regal, K.W. Lehnert, preprint (2013). arXiv:1310.5276
93. J.M. Taylor, A.S. Sørensen, C.M. Marcus, E.S. Polzik, Phys. Rev. Lett. **107**, 273601 (2011)
94. S.A. McGee, D. Meiser, C.A. Regal, K.W. Lehnert, M.J. Holland, Phys. Rev. A **87**, 053818 (2013)

Index

M. Aspelmeyer et al. (eds.), *Cavity Optomechanics*, Quantum Science and Technology,
DOI: 10.1007/978-3-642-55312-7,

The manufacturer's authorised representative in the EU is Springer
Nature Customer Service Centre GmbH, Europaplatz 3, 69115 Heidelberg,
Germany. If you have any concerns regarding our products, please
contact ProductSafety@springernature.com

Printed and bound by CPI Group (UK) Ltd, Croydon, CR0 4YY
23/09/2026
02221973-0003